Robotics and Control

AF886048

Robotics and Control

R K Mittal

Professor
Department of Mechanical Engineering
Birla Institute of Technology and Science, Pilani

I J Nagrath

Adjunct Professor and Former Deputy Director
Birla Institute of Technology and Science, Pilani

Tata McGraw-Hill Education Private Limited

NEW DELHI

McGraw-Hill Offices

New Delhi New York St Louis San Francisco Auckland Bogotá Caracas
Kuala Lumpur Lisbon London Madrid Mexico City Milan Montreal
San Juan Santiago Singapore Sydney Tokyo Toronto

Information contained in this work has been obtained by Tata McGraw-Hill, from sources believed to be reliable. However, neither Tata McGraw-Hill nor its authors guarantee the accuracy or completeness of any information published herein, and neither Tata McGraw-Hill nor its authors shall be responsible for any errors, omissions, or damages arising out of use of this information. This work is published with the understanding that Tata McGraw-Hill and its authors are supplying information but are not attempting to render engineering or other professional services. If such services are required, the assistance of an appropriate professional should be sought

Tata McGraw-Hill

© 2003, Tata McGraw-Hill Education Private Limited

Thirteenth reprint 2009
RLXLCRBFRALRC

No part of this publication may be reproduced or distributed in any form or by any means, electronic, mechanical, photocopying, recording, or otherwise or stored in a database or retrieval system without the prior written permission of the publishers. The program listings (if any) may be entered, stored and executed in a computer system, but they may not be reproduced for publication

This edition can be exported from India only by the publishers, Tata McGraw-Hill Education Private Limited

ISBN-13: 978-0-07-048293-7
ISBN-10: 0-07-048293-4

Published by Tata McGraw-Hill Education Private Limited, 7 West Patel Nagar, New Delhi 110 008, typeset in Times at Script Makers, 19, A1-B, DDA Market, Paschim Vihar, New Delhi 110 063 and printed at Pushp Print Services, Delhi 110 053

The McGraw·Hill Companies

Preface

Recognizing the impact of the upcoming field of Robotics on industrial automation, space, hazardous tasks, and several other application areas, a Center for Robotics was set up at BITS, Pilani as early as 1990. With increasing developmental activity, the Centre has been renamed as the Centre for Robotics and Intelligent Systems. The philosophy adopted by the Centre has been *design, simulate, fabricate, implement on hardware-software and prove*, which has yielded good results.

In the initial stages, an allied activity of the Centre was to hold *collaborative study* sessions with students and young faculty guided by the authors. The initial study lessons were produced as lecture notes* on Robotics and Control. These notes were periodically updated and fine-tuned based on the feedback and suggestions received from teachers and students. All these efforts and processes have culminated in the form of this book.

The field of Robotics and Control is both interdisciplinary and multidisciplinary as robots are amazingly complex systems comprising mechanical, electrical and electronics hardware and software systems and issues germane to all these. As it is not feasible to give coverage to all these aspects in one text, this book is designed to lay stress on the fundamentals and applications of robotics. Requisite mathematical techniques are treated rigorously.

Plenty of illustrative examples, based on a variety of robot configurations are presented. Physical explanations are provided alongwith practical applications. This would give the reader a clear mental picture of the subject and make him/her capable of carrying out projects involving design, simulation, fabrication and testing of robots. Further, he/she would be able to read and appreciate latest published literature for idea formulations.

The book is primarily intended for students of engineering (also sciences) at senior undergraduate level and could also be used at postgraduate level for those who have not taken this course at undergraduate level. Students can self-study certain portions of the book, which would permit the teacher to proceed at fair speed. This text would also be helpful to the practicing engineers as a systematic source of ready reference.

Chapters and exercises are designed for the readers to test their subject understanding. There are plenty of exercises of varying degree of complexity from

* These notes were produced by Educational Development Division at BITS, Pilani.

which the teacher can select and assign to students. Some of these exercises supplement and complement the topics covered in the chapter. It is assumed that the reader has adequate preparation of standard courses in calculus, differential equations, linear algebra, mechanics, control systems, computers and programming.

A chapter-wise outlay of the text is briefly laid out below.

Chapter 1 introduces the field of robotics and provides a panoramic view of the evolution of robots and robotics, types of robots, progressive advancement in robots, general definitions, robot anatomy and configurations. In addition, the issues of design, manipulation, control and programming of robots are discussed. The future prospects, bio-robotics and humanoid robotics are also outlined. This chapter enables the reader to comprehend what is an industrial robot and what is the current state of art.

Chapter 2 deals with coordinate frames, description of position and orientation in space, mapping from one frame to another, rotation and translation in space, and coordinate transformations, which are essential for mathematical modeling, analysis and synthesis of robot manipulators. Homogeneous transformations and their application in modeling robots are introduced. Fundamental rotation matrices, homogeneous transformation matrices, Euler angle and other representations are also covered to build the base for spatial geometry. The chapter lays the mathematical foundation for understanding what follows.

Chapters 3 and 4 dwell on the symbolic modeling of robots. Kinematic modeling —both forward and inverse kinematics are dealt with. Treatment is in depth with a number of illustrative examples pertaining to various classes of robot configurations. Examples illustrate modeling of some of the well known industrial, educational and generic robot configurations. These configurations have been used throughout the text in examples and exercises.

Chapter 5 treats the differential kinematics, the differential transformations required for controlling the motions, velocities, and accelerations of manipulators and the use of Jacobian for the same. A detailed description of joint space and Cartesian space singularities; their identification, causes and effects are included in this chapter.

Chapter 6 is devoted to dynamic modeling of manipulator to find the forces and torques required to cause motion of the manipulator. Lagrange-Euler and Newton-Euler formulations for obtaining equations of motion are presented. The derivational details of Lagrange-Euler formulation are relegated to Appendix D. This makes for smooth reading of the chapter, which is well illustrated with examples. The two formulations are compared for a simple manipulator to illustrate their relative merits as well as complexities.

Chapter 7 deals with another aspect of motion control to accomplish a specified task. Trajectory planning techniques are discussed for computations of path primitives for motion through specified points/paths. Both joint-space and Cartesian space techniques for generating trajectories for point-to-point and continuous path motion are treated.

Chapter 8 uses the developed robot models and illustrates the methods of controlling a manipulator to track a desired trajectory through space, apply the requisite force, move with desired velocity, etc. Robot control system architectures for independent joint control and centralized control, free motion control and interactive force-torque control are presented. The independent joint control typically used for industrial robot control is also described. Various strategies for control of robotic manipulators including partitioned PD, PID control, computed torque control, force control, impedance control and hybrid control are presented.

Chapter 9 introduces yet another important aspect of robotics, that is, robotic sensors and vision. The meaning of sensing, classification of robotic sensors and various types of robotic sensors are covered in the chapter. Introductory concepts of robotic vision systems, their components, architecture and process of imaging are presented. Image acquisition, representation as well as a number of image processing techniques are introduced. Further, industrial applications of vision controlled robotic systems and the practical aspects of various robotic sensors should help in choosing the right sensor system for a given application.

Chapter 10 on robot applications outlines a wide variety of current and future industrial and non-industrial applications of robots that will affect humans at all levels. Applications in manufacturing industries, in material handling, processing, assembly and inspection are discussed in fair detail. Principles of robot applications, application planning and justification of robots are included. For using robots, safety for all is important and the reader is exposed to the safety issues involved.

Appendix A on linear algebra and trigonometry, and Appendix C on moment of inertia tensor are given to provide a quick review for the essential background material required for the study of robotics. Appendix E gives the detailed list of the different symbols used in the text.

A tutorial on MATLAB and SIMULINK is included as Appendix B. This appendix aims at training the novice user into using MATLAB and SIMULINK for robotic applications. After going through the tutorial, the readers will be able to use MATLAB for solving the exercises in various chapters. Most of the examples in different chapters have been solved using MATLAB version 5.0.

There is a long list of our senior students, who worked with us on projects, for dissertation or as Professional Assistants, and young faculty members working for their Ph.D., all of whom contributed and helped in reading and reviewing the initial drafts of this book and to prepare and solve the exercise problems. A team of undergraduate students has recently reviewed the book with a view to ascertain its suitability as a text. We are grateful to all of them for their invaluable contributions.

The authors are thankful to all their colleagues who also have contributed ideas and suggestions at various stages in the preparation of the text material.

The authors are grateful to the Director and authorities of Birla Institute of Technology and Science, Pilani who have always encouraged this and similar course development efforts, by providing necessary facilities, fund support and proper atmosphere to work in.

The valuable suggestions made by the reviewers, at the request of Tata McGraw-Hill, were of tremendous help in improving the text and are gratefully acknowledged.

It is our bounden duty to put on record the excellent services of our office colleagues and others for their continuous help in mundane tasks. Thanks are due to Sh. Shwetank Srivastava and Sh. Santosh Kumar Saini for wordprocessing numerous versions of the manuscript, for putting together the complete typescript in excellent shape and keeping track of all our floating manuscripts and pieces of writings.

The authors welcome any constructive criticism of the book and will be grateful for any appraisal by the readers.

R K MITTAL

I J NAGRATH

Contents

1

Introduction to Robotics

Mankind has always strived to give life-like qualities to its artifacts in an attempt to find substitutes for himself to carry out his orders and also to work in a hostile environment. The popular concept of a robot is of a machine that looks and works like a human being. This humanoid concept has been inspired by science fiction stories and films in the twentieth century. The industrial robots of today may not look the least bit like a human being although all the research is directed to provide more and more anthropomorphic and human-like features and super-human capabilities in these.

To sum up, machines that can replace human beings as regards to physical work and decision making are categorized as *robots* and their study as *robotics*.

The robot technology is advancing rapidly. The industry is moving from the current state of automation to robotization, to increase productivity and to deliver uniform quality. Robots and robot-like manipulators are now commonly employed in hostile environment, such as at various places in an atomic plant for handling radioactive materials. Robots are being employed to construct and repair space stations and satellites. There are now increasing number of applications of robots such as in nursing and aiding a patient. Microrobots are being designed to do damage control inside human veins. Robot like systems are now employed in heavy earth-moving equipment. It is not possible to put up an exhaustive list of robot applications. One type of robot commonly used in the industry is a robotic manipulator or simply a *manipulator* or a *robotic arm*. It is an open or closed kinematic chain of rigid links interconnected by movable joints. In some configurations, links can be considered to correspond to human anatomy as waist, upper arm, and forearm with joints at shoulder and elbow. At the end of the arm, a wrist joint connects an *end-effector* to the forearm. The end-effector may be a tool and its fixture or a *gripper* or any other device to do the work. The end-effector is similar to the human hand with or without fingers. A robotic arm, as described above, is shown in Fig. 1.1, where various joint movements are also indicated.

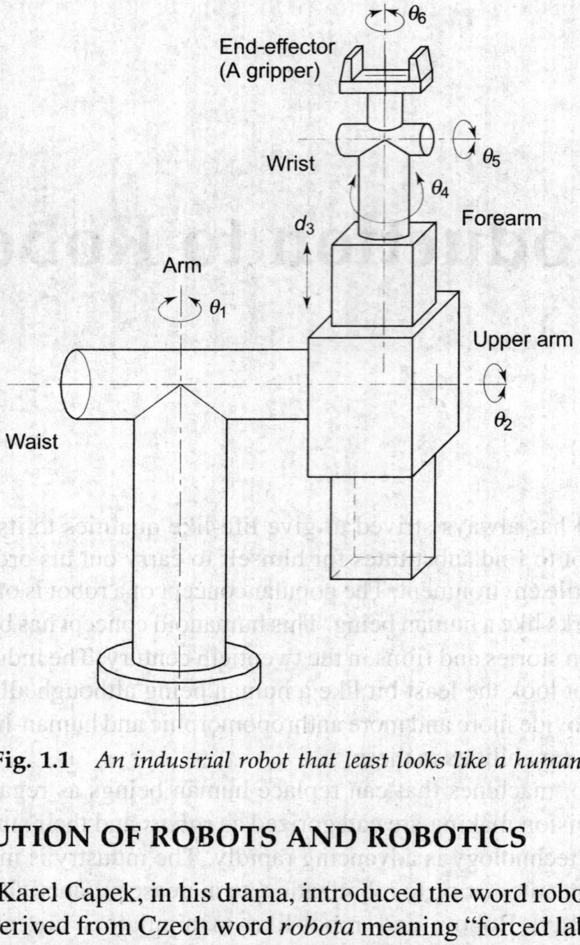

Fig. 1.1 *An industrial robot that least looks like a human*

1.1 EVOLUTION OF ROBOTS AND ROBOTICS

Czech writer, Karel Capek, in his drama, introduced the word robot to the world in 1921. It is derived from Czech word *robota* meaning "forced labourer". Isaac Asimov the well-known Russian science fiction writer, coined the word ***robotics*** in his story "Runaround", published in 1942, to denote the science devoted to study of robots.

The antecedents of the modern reprogrammable automation dates back to the eighteenth century. Perhaps, the best record is of Joseph Jacquard's use of punched cards in mechanical looms, which laid the foundations for NC, CNC, and automats, in addition to robotics.

Numerical control (NC) works on control actions based on stored information that may include start and stop operations, coordinate points, actions, logic for branching, and control sequences. A manufacturing system producing a variety of products in small batches, without requiring major hardware changes, with frequent changes in product models and production schedules, requires flexibility. In the transfer line approach, raw material is automatically transferred from one machine to another till it is converted to the final product. Such a transfer line approach, producing a large quantity of the same product for an extended period of time, may become useless when a major product change is required. It often ends up in abandoning the large capital investment. Contemporary industrial robots are reprogrammable machines that can perform different operations by simply modifying stored data, a feature that evolved from numerical control and is a solution for both of the above situations.

Need of systems to work in hostile environments that human workers cannot easily or safely access (for example radioactive material handling) led to the development of *teleoperated manipulator* in 1940s. The field of "telecherics" deals with the use of remote manipulators controlled by a human being in a "master-slave" configuration. Here, the actual machine (slave) is operated from a distance by a control "joystick" of a geometrically similar machine (master), as shown in Fig. 1.2.

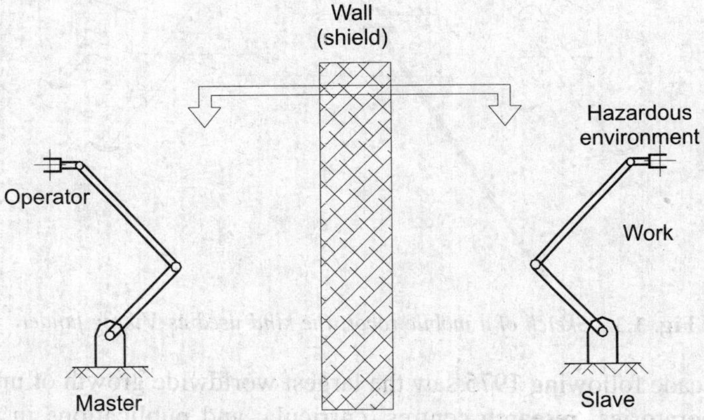

Fig. 1.2 *A master-slave manipulator*

The combination of numerical control and telecherics have evolved the basic concepts of modern industrial robots with human operator and master manipulator of Fig. 1.2 replaced by a programmable controller. This merging created a new field of engineering referred to as ***robotics***, and with it a number of engineering and scientific issues in design, control, and programming have emerged, which are substantially different from those of the existing techniques.

Some of the landmark developments in the field are now enunciated. In 1938–1939, a *jointed mechanical arm* was invented for use in spray painting. A process controller that could be used as a general-purpose playback device for operating machines, was developed in 1946, the year in which first large-scale electronic computer ENIAC was built. The first numerically controlled machine tool was developed in 1952. The patenting of the first manipulator, with the basic concept of teaching/playback, in 1954, set rolling the exponential growth in robotics.

The unmatched quality, reliability, and productivity offered by these robots, although in very limited applications, was recognized by the industry and sparked the formation of several world-wide centres of research in this area by the mid-1960s. The new field of robotics received support from simultaneously developing fields of *artificial intelligence* (AI), artificial vision, and developments in digital microcomputers. In 1967–1968, the first legged and wheeled walking machines using vision and other sensors, were reported. The servomotors were used in place of hydraulic devices in 1970 to power the robots. In 1974, the first servomotor actuated and microcomputer-controlled robots were commercially launched and in 1976, they were used by NASA *Viking* lander to collect samples from the surface of Mars. An elementary sketch of this lander is drawn in Fig. 1.3.

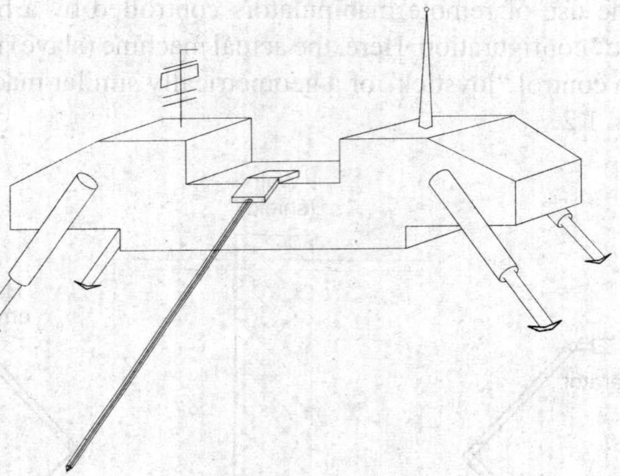

Fig. 1.3 *Sketch of a mobile robot, the kind used as Viking lander*

The decade following 1975 saw the largest worldwide growth of university-based laboratories, research centres, curricula, and publications in robotics. Mobile robotics also grew substantially during this period with designs of legged vehicles based on gait of both human beings and insects. The research activity in robotics started almost 40 years ago. The Robotic Institute of America (RIA), now called Robotic Industries Association (RIA), was formed only in 1975 as an organization of robot manufacturers and users.

The growth, thereafter, in robotics has been closely associated with developments in microcomputers, micro-controllers, sensor technology, vision technology, and artificial intelligence. The year 1997 saw the amalgamation of all these in the success of the Mars mission through "Pathfinder" and "Sojourner".

Recent Japanese exhibition *Robodex 2000* exhibited several varieties of entertainment robots; but there were no robots which could effectively do the onerous chores of house cleaning and dishwashing. Such robots are yet to be designed. None of the varieties exhibited attracted the industry.

Industrial robots are increasingly used in manufacturing plants, medical surgery, and rescue efforts. These require more difficult technology as much higher degree of accuracy, repeatability, flexibility, and reliability is needed for industrial robots.

Robotics today is dealing with research and development in a number of interdisciplinary areas, including kinematics, dynamics, control, motion planning, sensing, programming, and machine intelligence. These topics are introduced in the following sections and constitute the core of material in this book.

1.2 LAWS OF ROBOTICS

Issac Asimov conceived the robots as humanoids, devoid of feelings, and used them in a number of stories. His robots were well-designed, fail-safe machines, whose brains were programmed by human beings. Anticipating the dangers and havoc such a device could cause, he postulated rules for their ethical conduct. Robots were required to perform according to three principles known as "three laws of Robotics", which are as valid for real robots as they were for Asimov's robots, and they are:

1. A robot should not injure a human being or, through inaction, allow a human to be harmed.
2. A robot must obey orders given by humans except when that conflicts with the First Law.
3. A robot must protect its own existence unless that conflicts with the First or Second Law.

These are very general laws and apply even to other machines and appliances. They are always taken care of in any robot design.

1.3 WHAT IS AND WHAT IS NOT A ROBOT

Automation as a technology is concerned with the use of mechanical, electrical, electronic, and computer-based control systems to replace human beings with machines, not only for physical work but also for the intelligent information processing. Industrial automation, which started in the eighteenth century as fixed automation has transformed into flexible and programmable automation in the last 15 or 20 years. Computer Numerically Controlled (CNC) machine tools, transfer, and assembly lines are some examples in this category.

Common people are easily influenced by science fiction and thus, imagine a robot as a humanoid that can walk, see, hear, speak, and do the desired work. But the scientific interpretation of science fiction scenario propounds a robot as an automatic machine that is able to interact with and modify the environment in which it operates. Therefore, it is essential to define what constitutes a robot. Different definitions from diverse sources are available for a robot.

Japan is the world leader in robotics development and robot use. Japan Industrial Robot Association (JIRA) and the Japanese Industrial Standards Committee defines the industrial robot at various levels as:

"*Manipulator:* a machine that has functions similar to human upper limbs, and moves the objects spatially.

Playback robot: a manipulator that is able to perform an operation by reading off the memorized information for an operating sequence, which is learned beforehand.

Intelligent robot: a robot that can determine its own behaviour and conduct through its functions of sense and recognition.

The British Robot Association (BRA) has defined the industrial robot as:

"A reprogrammable device with minimum of four degrees of freedom designed to both manipulate and transport parts, tools, or specialized manufacturing implements through variable programmed motions for performance of specific manufacturing task."

The Robotics Industries Association (RIA) of USA defines the robot as:

"A reprogrammable, multifunctional manipulator designed to move material through variable programmed motions for the performance of a variety of tasks."

The definition adopted by International Standards Organization (ISO) and agreed upon by most of the users and manufacturers is:

"An industrial robot is an automatic, servo-controlled, freely programmable, multipurpose manipulator, with several areas, for the handling of work pieces, tools, or special devices. Variably programmed operations make the execution of a multiplicity of tasks possible."

Despite the fact that a wide spectrum of definitions exist, none covers the features of a robot exhaustively. The RIA definition lays emphasis on programmability, whereas while the BRA qualifies minimum degrees of freedom. The JIRA definition is fragmented. Because of all this, there is still confusion in distinguishing a robot from automation and in describing functions of a robot. To distinguish between a robot and automation, following guidelines can be used.

For a machine to be called a robot, it must be able to respond to stimuli based on the information received from the environment. The robot must interpret the stimuli either passively or through active sensing to bring about the changes required in its environment. The decision-making, performance of tasks and so on, all are done as defined in the programs taught to the robot. The functions of a robot can be classified into three areas:

"Sensing" the environment by external sensors, for example, vision, voice, touch, proximity and so on, "decision-making" based on the information received from the sensors, and "performing" the task decided.

1.4 PROGRESSIVE ADVANCEMENT IN ROBOTS

The growth in the capabilities of robots has been taking rapid strides since the introduction of robots in the industry in early 1960s, but there is still a long way to go to obtain the super-humanoid anthropomorphic robot depicted in fiction. The growth of robots can be grouped into *robot generations*, based on characteristic breakthroughs in robot's capabilities. These generations are overlapping and include futuristic projections.

1.4.1 First Generation

The first generation robots are repeating, nonservo, pick-and-place, or point-to-point kind. The technology for these is fully developed and at present about 80% robots in use in the industry are of this kind. It is predicted that these will continue to be in use for a long time.

1.4.2 Second Generation

The addition of sensing devices and enabling the robot to alter its movements in response to sensuary feedback marked the beginning of second generation. These robots exhibit path-control capabilities. This technological breakthrough came around 1980s and is yet not mature.

1.4.3 Third Generation

The third generation is marked with robots having human-like intelligence. The growth in computers led to high-speed processing of information and, thus, robots also acquired artificial intelligence, self-learning, and conclusion-drawing capabilities by past experiences. On-line computations and control, artificial vision, and active force/torque interaction with the environment are the significant characteristics of these robots. The technology is still in infancy and has to go a long way.

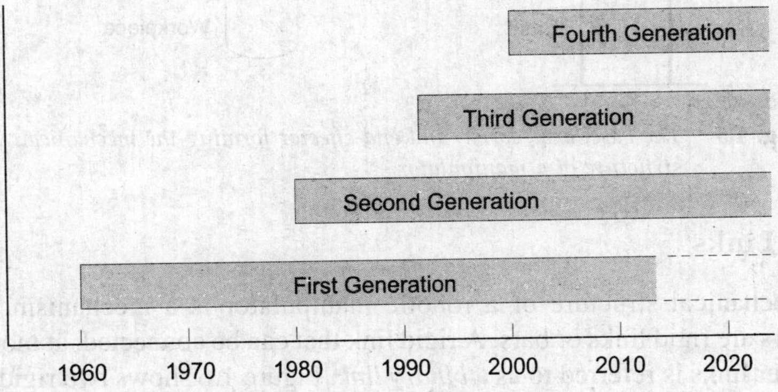

Fig. 1.4 *The four generations of robots*

1.4.4 Fourth Generation

This is futuristic and may be a reality only during this millennium. Prediction about its features is difficult, if not impossible. It may be a true android or an artificial biological robot or a super humanoid capable of producing its own clones. This might provide for fifth and higher generation robots.

A pictorial visualization of these overlapping generations of robots is given in Fig. 1.4.

1.5 ROBOT ANATOMY

As mentioned in the introduction to the chapter, the manipulator or robotic arm has many similarities to the human body. The mechanical structure of a robot is like the skeleton in the human body. The robot anatomy is, therefore, the study of skeleton of robot, that is, the physical construction of the manipulator structure.

The mechanical structure of a manipulator that consists of rigid bodies (links) connected by means of articulations (joints), is segmented into an *arm* that ensures mobility and reachability, a *wrist* that confers orientation, and an end-effector that performs the required task. Most manipulators are mounted on a base fastened to the floor or on the mobile platform of an autonomous guided vehicle (AGV). The arrangement of base, arm, wrist, and end-effector is shown in Fig. 1.5.

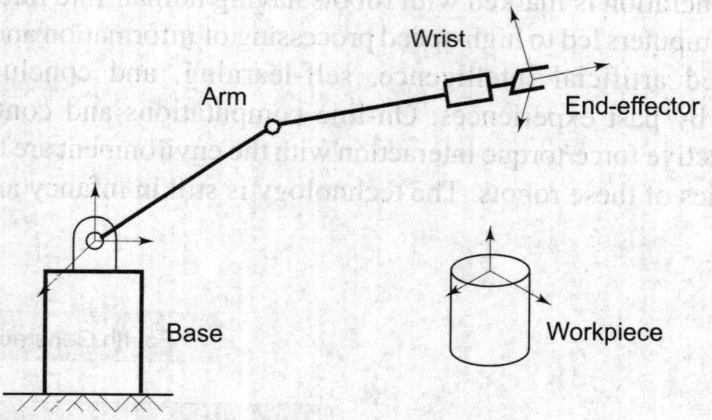

Fig. 1.5 *The base, arm, wrist, and end-effector forming the mechanical structure of a manipulator*

1.5.1 Links

The mechanical structure of a robotic manipulator is a mechanism, whose members are rigid links or bars. A rigid link that can be connected, at most, with two other links is referred to as a *binary link*. Figure 1.6 shows two rigid binary links, 1 and 2, each with two holes at the ends A, B, and C, D, respectively to connect with each other or to other links.

Two links are connected together by a joint. By putting a pin through holes B and C of links 1 and 2, an *open kinematic chain* is formed as shown in Fig. 1.7. The joint formed is called a *pin joint* also known as a *revolute* or *rotary joint*. Relative rotary motion between the links is possible and the two links are said to

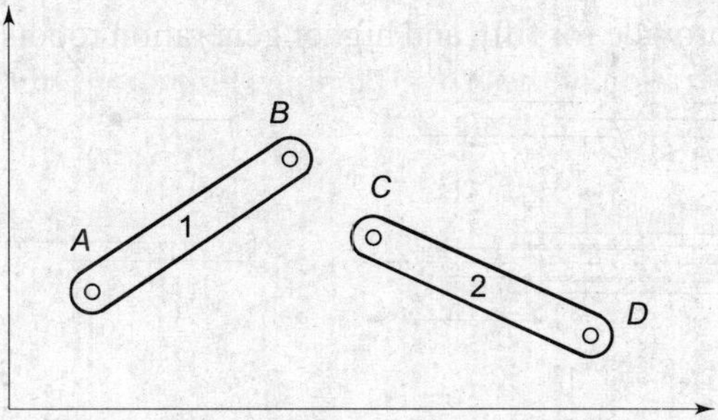

Fig. 1.6 *Two rigid binary links in free space*

be paired. In Fig. 1.7 links are represented by straight lines and rotary joint by a small circle.

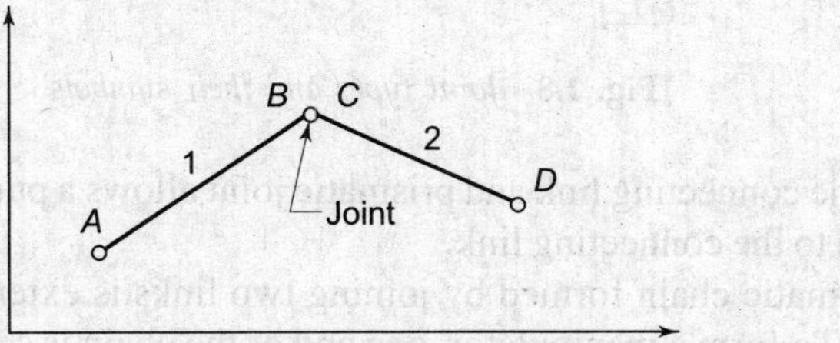

Fig. 1.7 *An open kinematic chain formed by joining two links*

1.5.2 Joints and Joint Notation Scheme

Many types of joints can be made between two links. However, only two basic types are commonly used in industrial robots. These are

- Revolute (R) and
- Prismatic (P).

The relative motion of the adjoining links of a joint is either rotary or linear depending on the type of joint.

Revolute joint: It is sketched in Fig. 1.8(a). The two links are jointed by a pin (pivot) about the axis of which the links can rotate with respect to each other.

Prismatic joint: It is sketched in Fig. 1.8(b). The two links are so jointed that these can slide (linearly move) with respect to each other. Screw and nut (slow linear motion of the nut), rack and pinon are ways to implement prismatic joints.

Other types of possible joints used are: planar (one surface sliding over another surface); cylindrical (one link rotates about the other at 90° angle, Fig. 1.8(c)); and spherical (one link can move with respect to the other in three dimensions). Yet another variant of rotary joint is the 'twist' joint, where two links remain aligned along a straight line but one turns (twists) about the other around the link axis, Fig. 1.8(d).

At a joint, links are connected such that they can be made to move relative to each other by the actuators. A rotary joint allows a pure rotation of one link

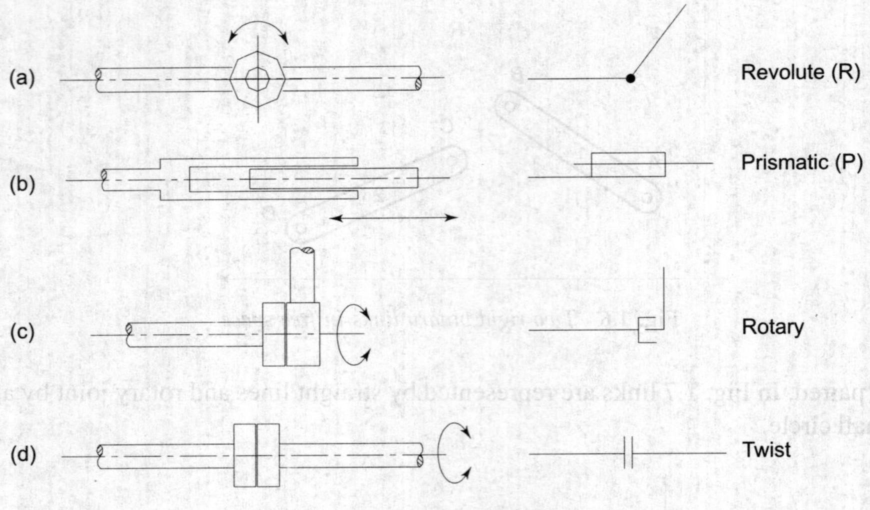

Fig. 1.8 *Joint types and their symbols*

relative to the connecting link and prismatic joint allows a pure translation of one link relative to the connecting link.

The kinematic chain formed by joining two links is extended by connecting more links. To form a manipulator, one end of the chain is connected to the base or ground with a joint. Such a manipulator is an open kinematic chain. The end-effector is connected to the free end of the last link, as illustrated in Fig. 1.5. Closed kinematic chains are used in special purpose manipulators, such as parallel manipulators, to create certain kind of motion of the end-effector.

The kinematic chain of the manipulator is characterized by the degrees of freedom it has, and the space its end-effector can sweep. These parameters are discussed in next sections.

1.5.3 Degrees of Freedom (DOF)

The number of independent movements that an object can perform in a 3-D space is called the number of *degrees of freedom* (DOF). Thus, a rigid body free in space has six degrees of freedom—three for position and three for orientation. These six independent movements pictured in Fig. 1.9 are:

(i) three translations (T_1, T_2, T_3), representing linear motions along three perpendicular axes, specify the position of the body in space.

(ii) three rotations (R_1, R_2, R_3), which represent angular motions about the three axes, specify the orientation of the body in space.

Note from the above that six independent variables are required to specify the location (position and orientation) of an object in 3-D space, that is, $2 \times 3 = 6$. Nevertheless, in a 2-D space (a plane), an object has 3-DOF—two translatory and one rotational. For instance, link 1 and link 2 in Fig. 1.6 have 3-DOF each.

Consider an open kinematic chain of two links with revolute joints at A and B (or C), as shown in Fig. 1.10. Here, the first link is connected to the ground by a

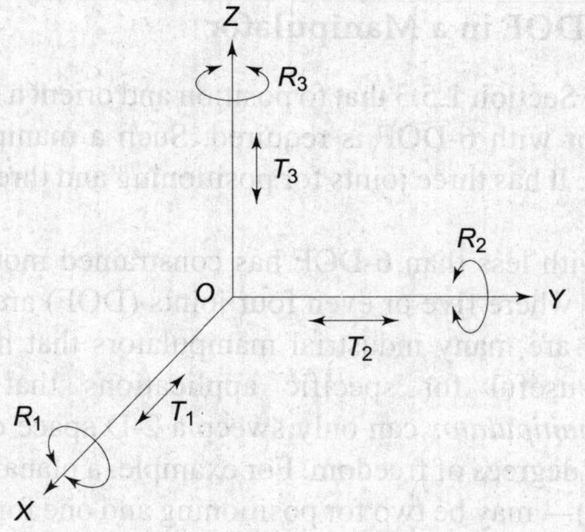

Fig. 1.9 *Representation of six degrees of freedom with respect to a coordinate frame*

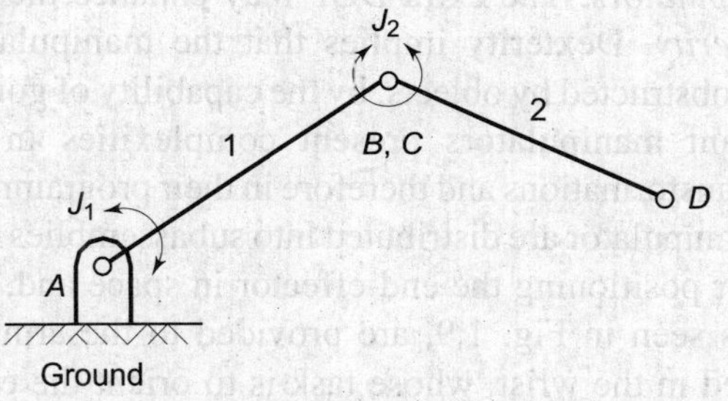

Fig. 1.10 *A two-DOF planar manipulator–two links, two joints*

joint at *A*. Therefore, link 1 can only rotate about joint 1 (J_1) with respect to ground and contributes one independent variable (an angle), or in other words, it contributes one degree of freedom. Link 2 can rotate about joint 2 (J_2) with respect to link 1, contributing another independent variable and so another DOF. Thus, by induction, conclude that an open kinematic chain with one end connected to the ground by a joint and the farther end of the last link free, has as many degrees of freedom as the number of joints in the chain. It is assumed that each joint has only one DOF.

The DOF is also equal to the number of links in the open kinematic chain. For example, in Fig. 1.10, the open kinematic chain manipulator with two DOF has two links and two joints.

The variable defining the motion of a link at a joint is called a *joint-link variable*. Thus, for an ***n***-DOF manipulator ***n*** independent joint-link variables are required to completely specify the location (position and orientation) of each link (and joint), specifying the location of the end-effector in space. Thus, for the two-link, in turn 2-DOF manipulator, in Fig. 1.10, two variables are required to define location of end-point, point *D*.

1.5.4 Required DOF in a Manipulator

It is concluded from Section 1.5.3 that to position and orient a body freely in 3-D space, a manipulator with 6-DOF is required. Such a manipulator is called a *spatial manipulator*. It has three joints for positioning and three for orienting the end-effector.

A manipulator with less than 6-DOF has constrained motion in 3-D space. There are situations where five or even four joints (DOF) are enough to do the required job. There are many industrial manipulators that have five or fewer DOF. These are useful for specific applications that do not require 6-DOF. A *planar manipulator* can only sweep a 2-D space or a plane and can have any number of degrees of freedom. For example, a planar manipulator with three joints (3-DOF)— may be two for positioning and one for orientation —can only sweep a plane.

Spatial manipulators with more than 6-DOF have surplus joints and are known as *redundant manipulators*. The extra DOF may enhance the performance by adding to its *dexterity*. Dexterity implies that the manipulator can reach a subspace, which is obstructed by objects, by the capability of going around these. However, redundant manipulators present complexities in modelling and coordinate frame transformations and therefore in their programming and control.

The DOF of a manipulator are distributed into subassemblies of *arm* and *wrist*. The arm is used for positioning the end-effector in space and, hence, the three positional DOF, as seen in Fig. 1.9, are provided to the arm. The remaining 3-DOF are provided in the wrist, whose task is to orient the end-effector. The type and arrangement of joints in the arm and wrist can vary considerably. These are discussed in the next section.

1.5.5 Arm Configuration

The mechanics of the arm with 3-DOF depends on the type of three joints employed and their arrangement. The purpose of the arm is to position the wrist in the 3-D space and the arm has following characteristic requirements.

- Links are long enough to provide for maximum reach in the space.
- The design is mechanically robust because the arm has to bear not only the load of workpiece but also has to carry the wrist and the end-effector.

According to joint movements and arrangement of links, four well-distinguished basic structural configurations are possible for the arm. These are characterized by the distribution of three arm joints among prismatic and rotary joints, and are named according to the coordinate system employed or the shape of the space they sweep. The four basic configurations are:

(i) Cartesian (rectangular) configuration – all three P joints.
(ii) Cylindrical configuration – one R and two P joints.
(iii) Polar (spherical) configuration – two R and one P joint.
(iv) Articulated (Revolute or Jointed-arm) Configuration – all three R joints.

Each of these arm configurations is now discussed briefly.

(i) Cartesian (Rectangular) Configuration This is the simplest configuration with all three prismatic joints, as shown in Fig. 1.11. It is constructed by three perpendicular slides, giving only linear motions along the three principal axes. There is an upper and lower limit for movement of each link. Consequently, the *endpoint* of the arm is capable of operating in a cuboidal space, called *workspace*.

The workspace represents the portion of space around the base of the manipulator that can be accessed by the arm endpoint. The shape and size of the workspace depends on the arm configuration, structure, degrees of freedom, size of links, and design of joints. The physical space that can be swept by a manipulator (with wrist and end-effector) may be more or less than the arm endpoint workspace. The volume of the space swept is called *work volume*; the surface of the workspace describes the *work envelope*.

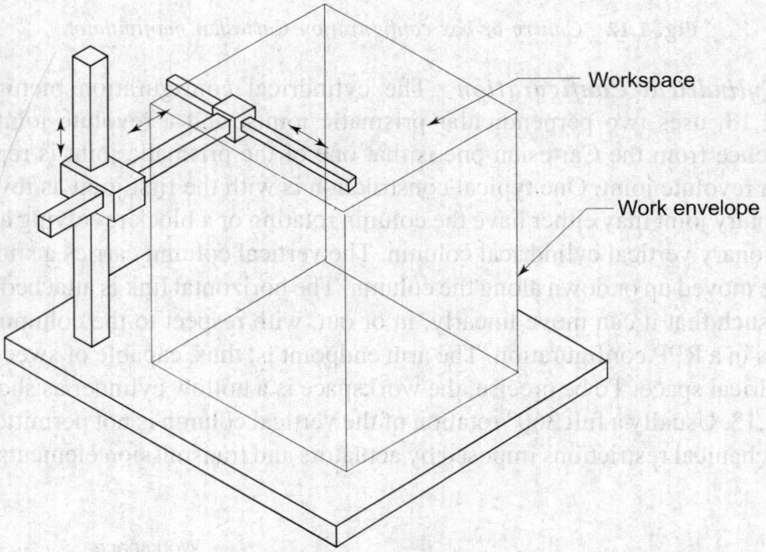

Fig. 1.11 *A 3-DOF Cartesian arm configuration and its workspace*

The workspace of Cartesian configuration is cuboidal and is shown in Fig. 1.11. Two types of constructions are possible for Cartesian arm: a *Cantilevered Cartesian*, as in Fig. 1.11, and a *Gantry* or *box Cartesian*. The latter one has the appearance of a gantry-type crane and is shown in Fig. 1.12. Despite the fact that Cartesian arm gives high precision and is easy to program, it is not preferred for many applications due to limited manipulatability. Gantry configuration is used when heavy loads must be precisely moved. The Cartesian configuration gives large work volume but has a low dexterity.

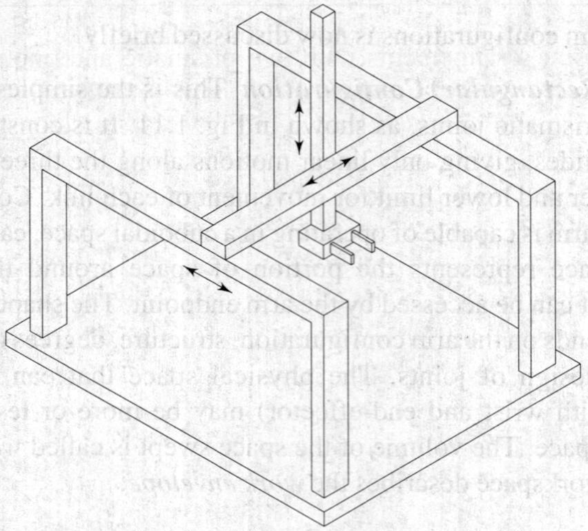

Fig. 1.12 *Gantry or box configuration Cartesian manipulator*

(ii) Cylindrical Configuration The cylindrical configuration pictured in Fig. 1.13, uses two perpendicular prismatic joints, and a revolute joint. The difference from the Cartesian one is that one of the prismatic joint is replaced with a revolute joint. One typical construction is with the first joint as revolute. The rotary joint may either have the column rotating or a block revolving around a stationary vertical cylindrical column. The vertical column carries a slide that can be moved up or down along the column. The horizontal link is attached to the slide such that it can move linearly, in or out, with respect to the column. This results in a RPP configuration. The arm endpoint is, thus, capable of sweeping a cylindrical space. To be precise, the workspace is a hollow cylinder as shown in Fig. 1.13. Usually a full 360° rotation of the vertical column is not permitted due to mechanical restrictions imposed by actuators and transmission elements.

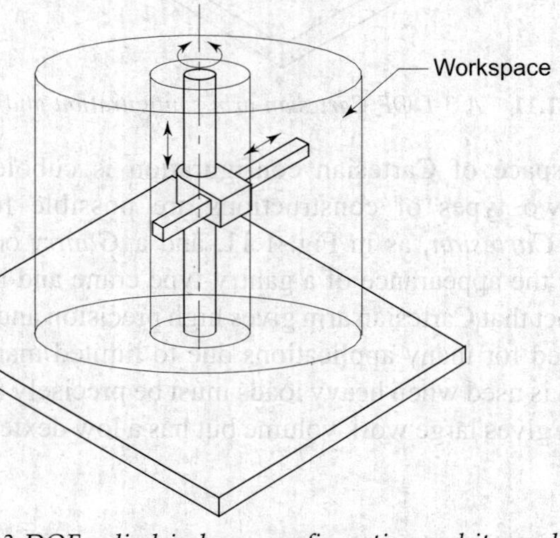

Fig. 1.13 *A 3-DOF cylindrical arm configuration and its workspace*

Many other joint arrangements with two prismatic and one rotary joint are possible for cylindrical configuration, for example, a PRP configuration. Note that all combinations of 1R and 2P are not useful configurations as they may not give suitable workspace and some may only sweep a plane. Such configurations are called *nonrobotic configurations*. It is left for the reader to visualize as to which joint combinations are robotic arm configurations.

The cylindrical configuration offers good mechanical stiffness and the wrist positioning accuracy decreases as the horizontal stroke increases. It is suitable to access narrow horizontal cavities and, hence, is useful for machine-loading operations.

(iii) Polar (Spherical) Configuration The polar configuration is illustrated in Fig. 1.14. It consists of a telescopic link (prismatic joint) that can be raised or lowered about a horizontal revolute joint. These two links are mounted on a rotating base. This arrangement of joints, known as RRP configuration, gives the capability of moving the arm end-point within a partial spherical shell space as work volume, as shown in Fig. 1.14.

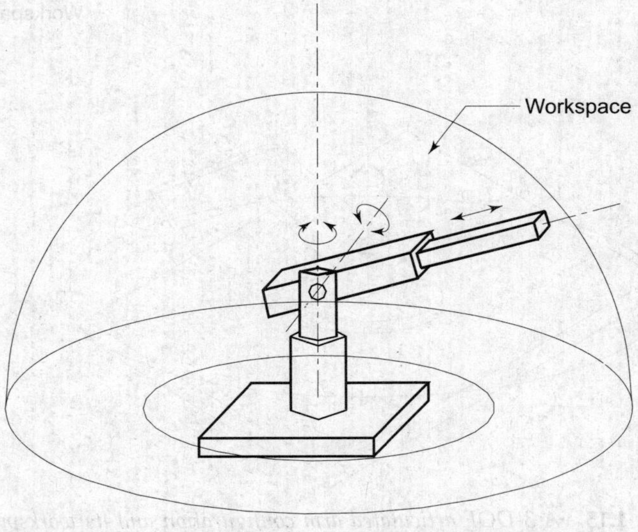

Fig. 1.14 *A 3-DOF polar arm configuration and its workspace*

This configuration allows manipulation of objects on the floor because its shoulder joint allows its end-effector to go below the base. Its mechanical stiffness is lower than Cartesian and cylindrical configurations and the wrist positioning accuracy decreases with the increasing radial stroke. The construction is more complex. Polar arms are mainly employed for industrial applications such as machining, spray painting and so on. Alternate polar configuration can be obtained with other joint arrangements such as RPR, but PRR will not give a spherical work volume.

(iv) Articulated (Revolute or Jointed-arm) Configuration The articulated arm is the type that best simulates a human arm and a manipulator with this type of an arm is often referred as an *anthropomorphic manipulator*. It consists of two straight links, corresponding to the human "forearm" and "upper arm" with two rotary joints corresponding to the "elbow" and "shoulder" joints. These two links are mounted on a vertical rotary table corresponding to the human waist joint. Figure 1.15 illustrates the joint-link arrangement for the *articulated arm*.

This configuration (RRR) is also called revolute because three revolute joints are employed. The work volume of this configuration is spherical shaped, and with proper sizing of links and design of joints, the arm endpoint can sweep a full spherical space. The arm endpoint can reach the base point and below the base, as shown in Fig. 1.15. This anthropomorphic structure is the most dexterous one, because all the joints are revolute, and the positioning accuracy varies with arm endpoint location in the workspace. The range of industrial applications of this arm is wide.

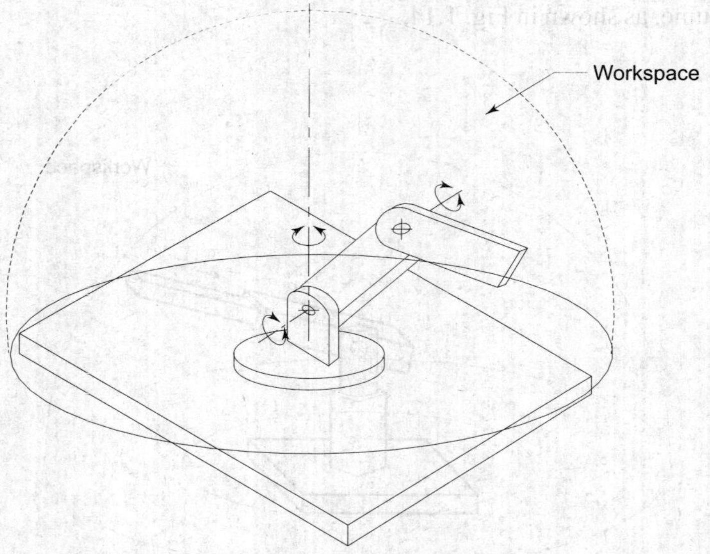

Fig. 1.15 *A 3-DOF articulated arm configuration and its workspace*

(v) Other Configurations New arm configurations can be obtained by assembling the links and joints differently, resulting in properties different from those of basic arm configurations outlined above. For instance, if the characteristics of articulated and cylindrical configurations are combined, the result will be another type of manipulator with revolute motions, confined to the horizontal plane. Such a configuration is called SCARA, which stands for Selective Compliance Assembly Robot Arm.

The SCARA configuration has vertical major axis rotations such that gravitational load, Coriolis, and centrifugal forces do not stress the structure as much as they would if the axes were horizontal. This advantage is very important at high speeds and high precision. This configuration provides high stiffness to the arm in the vertical direction, and high compliance in the horizontal plane, thus making SCARA congenial for many assembly tasks. The SCARA configuration and its workspace are presented pictorially in Fig. 1.16.

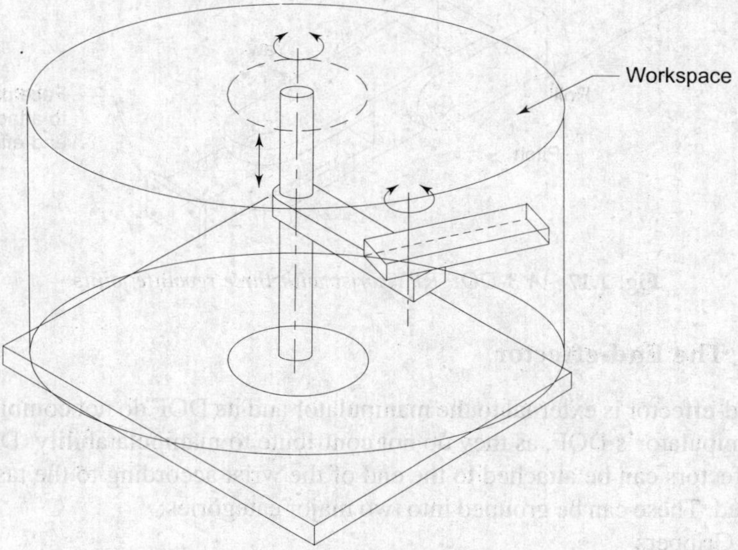

Fig. 1.16 *The SCARA configuration and its workspace*

1.5.6 Wrist Configuration

The arm configurations discussed above carry and position the wrist, which is the second part of a manipulator that is attached to the endpoint of the arm. The wrist subassembly movements enable the manipulator to orient the end-effector to perform the task properly, for example, the gripper (an end-effector) must be oriented at an appropriate angle to pick and grasp a workpiece. For arbitrary orientation in 3-D space, the wrist must possess at least 3-DOF to give three rotations about the three principal axes. Fewer than 3-DOF may be used in a wrist, depending on requirements. The wrist has to be compact and it must not diminish the performance of the arm.

The wrist requires only rotary joints because its sole purpose is to orient the end-effector. A 3-DOF wrist permitting rotation about three perpendicular axes provides for *roll* (motion in a plane perpendicular to the end of the arm), *pitch* (motion in vertical plane passing through the arm), and *yaw* (motion in a horizontal plane that also passes through the arm) motions. This type of wrist is called *roll-pitch-yaw* or *RPY* wrist and is illustrated in Fig. 1.17. A wrist with the highest dexterity is one where three rotary joint axes intersect at a point. This complicates the mechanical design.

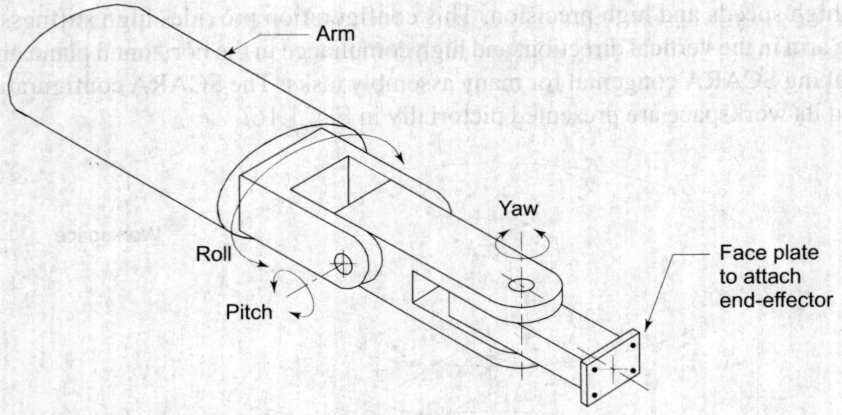

Fig. 1.17 *A 3-DOF RPY wrist with three revolute joints*

1.5.7 The End-effector

The end-effector is external to the manipulator and its DOF do not combine with the manipulator's DOF, as they do not contribute to manipulatability. Different end-effectors can be attached to the end of the wrist according to the task to be executed. These can be grouped into two major categories:

1. Grippers
2. Tools

Grippers are end-effectors to grasp or hold the workpiece during the work cycle. The applications include material handling, machine loading-unloading, palletizing, and other similar operations. Grippers employ mechanical grasping or other alternative ways such as magnetic, vacuum, bellows, or others for holding objects. The proper shape and size of the gripper and the method of holding are determined by the object to be grasped and the task to be performed. Some typical mechanical grippers are shown in Fig. 1.18.

For many tasks to be performed by the manipulator, the end-effector is a *tool* rather than a gripper. For example, a cutting tool, a drill, a welding torch, a spray

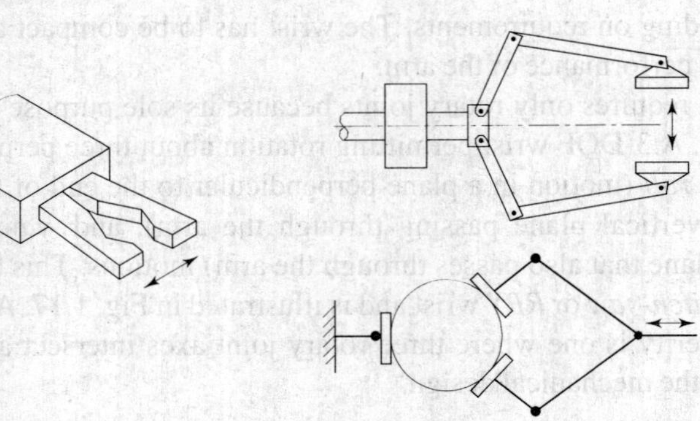

Fig. 1.18 *Some fingered grippers for holding different types of jobs*

gun, or a screwdriver is the end-effector for machining, welding, painting, or assembly task, mounted at the wrist endpoint. The tool is usually directly attached to the end of the wrist. Sometimes, a gripper may be used to hold the tool instead of the workpiece. *Tool changer* devices can also be attached to the wrist end for multi-tool operations in a work cycle.

1.6 HUMAN ARM CHARACTERISTICS

The industrial robot, though not similar to human arm, draws inspiration for its capabilities from the latter. The human arm and its capabilities make the human race class apart from other animals. The design of the human arm structure is a unique marvel and is still a challenge to replicate. Certain characteristics of the human arm are a far cry for today's manipulators. It is, therefore, worth considering briefly, human arm's most important characteristics as these serve as a benchmark for the manipulators.

The human arm's basic performance specifications are defined from the *zero reference position*, which is the stretched right arm and hand straight out and horizontal with the palm in downward direction. The three motions to orient the hand, which is the first part of human arm, are approximately in the following range.

$$-180° \leq \text{Roll} \leq +90°$$
$$-90° \leq \text{Pitch} \leq +50°$$
$$-45° \leq \text{Yaw} \leq +15°$$

Note that to provide the roll motion to the hand, forearm, and the upper-arm, both undergo a twist, while pitch and yaw are provided by the wrist joint. The second part of the human arm consists of upper arm and forearm with shoulder and elbow joints. It has 2-DOF in the shoulder with a ball and socket joint, 1-DOF in the elbow between forearm and upper-arm, with two bones in the forearm and one in upper arm. The 2-DOF shoulder joint provides an approximately hemispherical sweep to the elbow joint. The elbow joint moves the forearm by approximately 170° (from –5° to 165°) in different planes, depending on the orientation of two forearm bones and the elbow joint. For the zero reference position defined above, the forearm and the wrist can only sweep an arc in the horizontal plane.

Another important feature of the human arm is the ratio of the length of the upper arm to that of the forearm, which is around 1.2. Any ratio other than this results in performance impairment. A mechanical structure identical to the human arm, with 2-DOF shoulder joint, three-bones elbow joint, eight-bones wrist joint with complicated geometry of each bone and joint, is yet to be designed and constructed. The technology has to go a long way to replicate human arm's bone shapes, joint mechanisms, mechanism to power and move joints, motion control, safety, and above all, self repair.

The human hand, at the end of arm, with four fingers and a thumb, each with 4-DOF, is another marvel with no parallel. The finger and thumb joints can act independently or get locked, depending on the task involved, offering a very high dexterity to zero dexterity. This, coupled with the joint actuation and control mechanism and tactile sensing provided by the skin makes the human hand a marvel. In contrast, the robot gripper with two or three fingers has almost no dexterity. The human arm's articulation, and to the same extent, the human leg's locomotion are challenges yet to be met.

1.7 DESIGN AND CONTROL ISSUES

Robots are driven to perform more and more variety of highly skilled jobs with minimum human assistance or intervention. This requires them to have much higher mobility, manipulatability, and dexterity than conventional machine tools. The mechanical structure of a robot, which consists of rigid cantilever beams connected by hinged joints forming spatial mechanism, is inherently poor in stiffness, accuracy, and load carrying capacity. The errors accumulate because joints are in a serial sequence. These difficulties are overcome by advanced design and control techniques.

The serial-spatial linkage geometry of a manipulator is described by complex nonlinear transcendental equations. The position and motion of each joint is affected by the position and motion of all other joints. Further, each joint has to be powered independently, rendering modeling, analysis, and design to be quite an involved issue.

The weight and inertial load of each link is carried by the previous link. The links undergo rotary motion about the joints, making centrifugal and Coriolis effects significant. All these make the dynamic behaviour of the robot manipulator complex, highly coupled, and nonlinear. The kinematic and dynamic complexities create unique control problems that make control of a robot a very challenging task and effective control system design a critical issue. The robot control problem has added a new dimension in control research.

The environment in which robots are used poses numerous other complexities as compared to conventional machine tools. The work environment of the latter is well-defined and structured and the machine tools are essentially self-contained to handle workpieces and tools in well-defined locations. The work environment of the robot is often poorly structured, uncertain, and requires effective means to identify locations, workpieces and tools, and obstacles. The robot is also required to interact and coordinate with peripheral devices.

Robots being autonomous systems, require to perform additional tasks of planning and generating their own control commands. The detailed procedure, control strategy, and algorithm must be taught in advance and coded in an appropriate form so that the robot can interpret these and execute these accurately. Effective means to store the data, commands, and manage memory are also needed. Thus, programming and command generation become critical issues in robotics. To monitor it's own motions and to adapt to disturbances and unpredictable environments, robot requires interfacing with internal and external sensors. To utilize the sensory information, effective sensor-based algorithms and advanced control systems are required, in addition to a thorough understanding of the task.

1.8 MANIPULATION AND CONTROL

This section briefly describes the topics, which will be covered in this text, and introduces some terminology in the robotics field.

In the analysis of spatial mechanisms (manipulators), the location of links, joints, and end-effector in 3-D space is continuosly required. Mathematical description of the position and orientation of links in space and manipulation of these is, naturally, one topic of immediate importance.

To describe position and orientation of a body in space, a frame is attached to the body. The position and orientation of this frame with respect to some reference coordinate frame, called *base frame*, mathematically describes the location of the body. Frames are attached to joints, links, end-effector, and workpieces in the environment of the robot to mathematically describe them, as illustrated in Fig. 1.19.

Often, the description of a body in one frame is known, while requirement is the description of the body with respect to another frame. This requires *mapping* or *transforming* or changing the description of its attributes from one frame to another. Conventions and methodologies for description of position and orientation, and the mathematics of transforming these quantities are first discussed in this text.

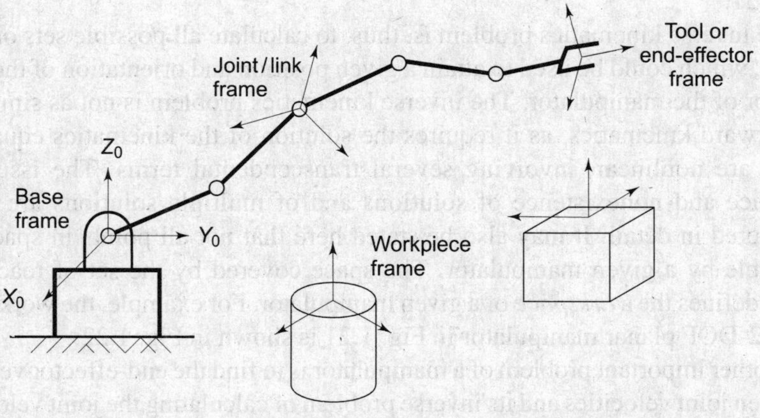

Fig. 1.19 *Attachment of frames for manipulator modelling*

Consider the simplest nontrivial two-link planar manipulator of Fig. 1.20 with link lengths (L_1, L_2) and assume that the joint angles are (θ_1, θ_2) and the coordinates of end-effector point P are (x, y). From simple geometrical analysis for this manipulator, it is possible to compute coordinates (x, y) from the given joint angles (θ_1, θ_2) and for a given location of point P (x, y), joint angles (θ_1, θ_2) can be computed.

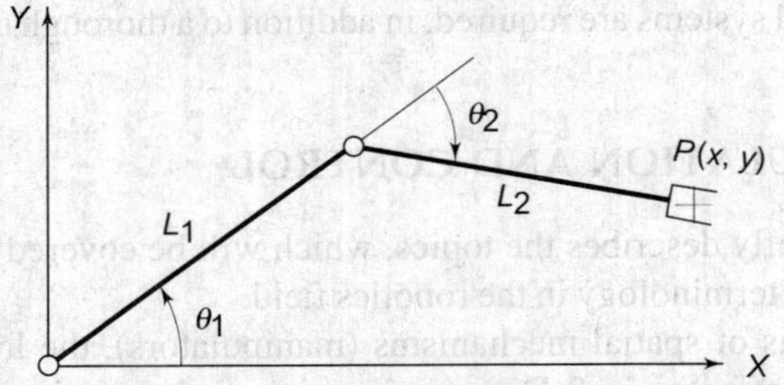

Fig. 1.20 *The 2-DOF two-link planar manipulator*

The basic problem in the study of mechanical manipulation is of computing the position and orientation of end-effector of the manipulator when the joint angles are known. This is referred to as *forward kinematics* problem. The *inverse kinematics* problem is to determine the joint angles, given the position and orientation of the end-effector.

A problem that can be faced in inverse kinematics is that the solution for joint angles may not be unique; there may be multiple solutions. This is illustrated for the simple planar 2-DOF manipulator in Fig. 1.20.

If the 2-DOF manipulator in Fig. 1.20 is used to position some object held in its end-effector to a specified position P_1 (x_1, y_1), the joint angles θ_1 and θ_2 that make the end point coincide with desired location must be found. This is the *inverse kinematics* problem. For the manipulator in Fig. 1.20, there are two sets of joint angles θ_1 and θ_2 that lead to the same endpoint position, as illustrated in Fig. 1.21.

The inverse kinematics problem is, thus, to calculate all possible sets of joint angles, which could be used to attain a given position and orientation of the end-effector of the manipulator. The inverse kinematics problem is not as simple as the forward kinematics, as it requires the solution of the kinematics equations which are nonlinear, involving several transcendental terms. The issues of existence and nonexistence of solutions and of multiple solutions are to be considered in detail. It may also be stated here that not all points in space are reachable by a given manipulator. The space covered by the set of reachable points defines the *workspace* of a given manipulator. For example, the workspace of the 2-DOF planar manipulator in Fig. 1.21 is shown in Fig. 1.22.

Another important problem of a manipulator is to find the end-effector velocity for given joint velocities and its inverse problem of calculating the joint velocities for specified end-effector velocity. These two problems, direct and inverse need the manipulator Jacobian (matrix), which is obtained from the kinematic parameters.

An identical problem of the static force analysis can also be solved through the Jacobian. This problem is stated as: given a desired contact force and moment,

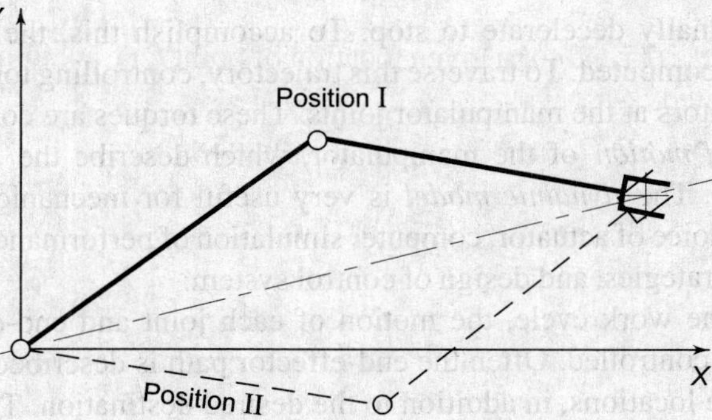

Fig. 1.21 *Two possible joint positions for a given end point position*

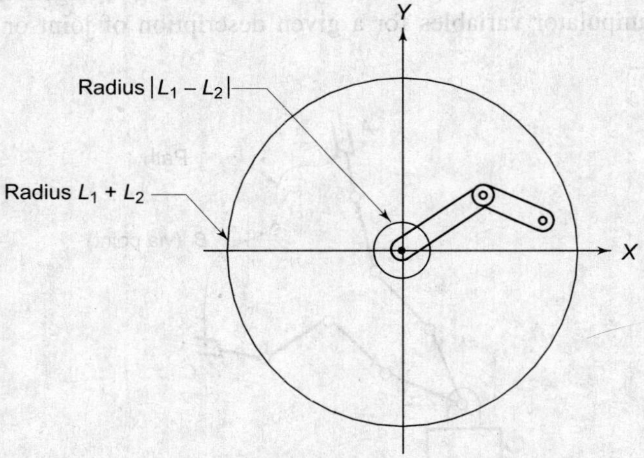

Fig. 1.22 *The workspace of a 2-DOF planar manipulator*

determine the set of joint torques to generate them or vice-versa. Figure 1.23 illustrates the interaction of a manipulator at rest with the environment; the manipulator is exerting a force $\boldsymbol{F}$ on the body.

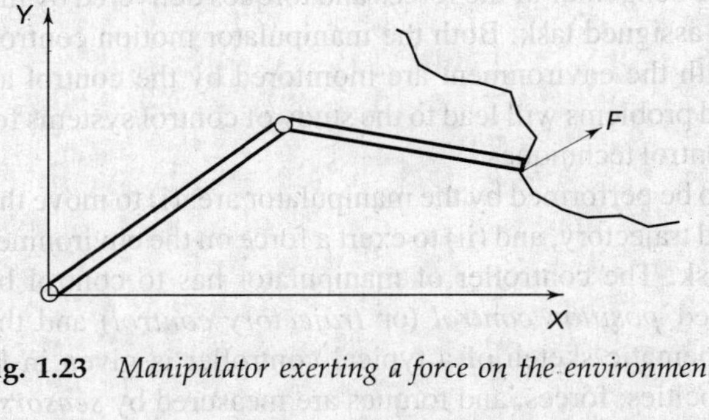

Fig. 1.23 *Manipulator exerting a force on the environment*

To perform an assigned task or to attain a desired position, a manipulator is required to accelerate from rest, travel at specified velocity, traverse a specified path, and finally decelerate to stop. To accomplish this, the trajectory to be followed is computed. To traverse this trajectory, controlling torques are applied by the actuators at the manipulator joints. These torques are computed from the *equations of motion* of the manipulator, which describe the *dynamics* of the manipulator. The *dynamic model* is very useful for mechanical design of the structure, choice of actuator, computer simulation of performance, determination of control strategies, and design of control system.

During the work cycle, the motion of each joint and end-effector must be smooth and controlled. Often the end-effector path is described by a number of intermediate locations, in addition to the desired destination. The term *spline* is used to refer to a smooth function, which passes through a set of specified points. The motion of end-effector through space from point A to point C via point B is illustrated in Fig. 1.24. The goal of *trajectory planning* is to generate time laws for the manipulator variables for a given description of joint or end-effector motion.

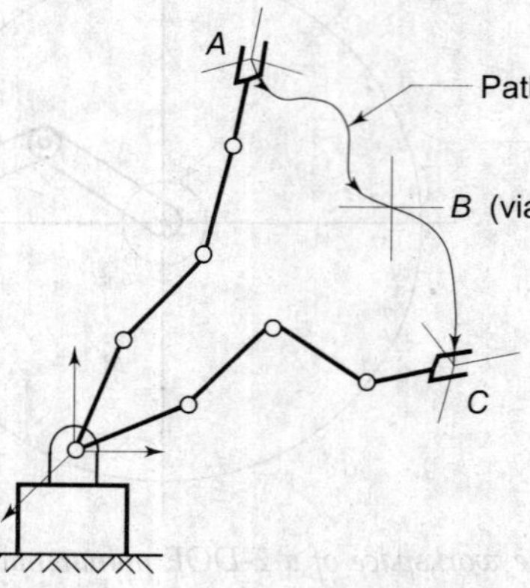

Fig. 1.24 *Trajectory generation for motion from A to C via B*

The dynamic model and the generated trajectory constitute the inputs to the motion-control system of the manipulator. The problem of manipulator control is to find the time behaviour of the forces and torques delivered by the actuators for executing the assigned task. Both the manipulator motion control and its force interaction with the environment are monitored by the control algorithm. The above exposed problems will lead to the study of control systems for manipulator and several control techniques.

The tasks to be performed by the manipulator are: (i) to move the end-effector along a desired trajectory, and (ii) to exert a force on the environment to carry out the desired task. The controller of manipulator has to control both tasks, the former is called *position control* (or *trajectory control*) and the latter *force control*. A schematic sketch of a typical controller is given in Fig. 1.25. The positions, velocities, forces, and torques are measured by *sensors* and based on these measurements and the desired behaviour, the controller determines the inputs to the actuators on the robot so that the end-effector carries out the desired task as closely as possible.

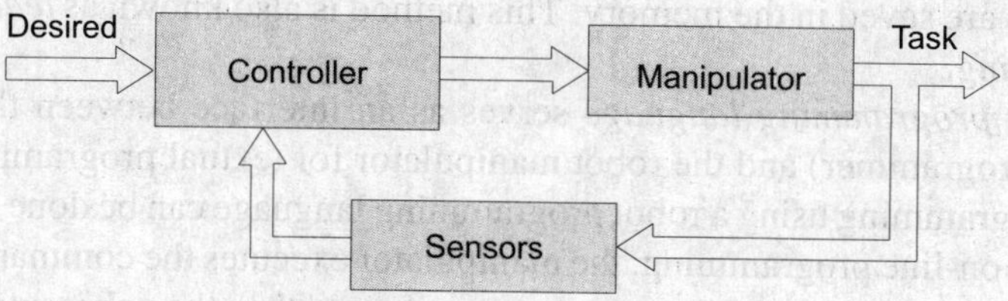

Fig. 1.25 *A schematic sketch of a manipulator control system*

1.9 SENSORS AND VISION

The control of the manipulator demands exact determination of parameters of interest so that the controller can compare them with desired values and accordingly command the actuators of the manipulator. Sensors play the most important role in the determination of actual values of the parameters of interest. For manipulator motion control, joint-link positions, velocities, torques, or forces are required to be sensed and the end-effector position and orientation is required for determining actual trajectory being tracked. The force control requires sensing of joint force/torque and end-effector force/torque.

Sensors used in robotics include simple devices such as a potentiometer as well as sophisticated ones such as a *robotic vision system*. Sensors can be an integral part of the manipulator (*internal sensors*) or they may be placed in the robot's environment or workcell (*external sensors*) to permit the robot to interact with the other activities and objects in the workcell.

The task performance capability of a robot is greatly dependent on the sensors used and their capabilities. Sensors provide intelligence to the manipulator. Sensors used in robotics are *tactile sensor*s or *nontactile sensors*; *proximity or range sensors*; *contact* or *noncontact sensors*, or a vision system.

A robotic vision system imparts enormous capabilities to a robot. The robotic vision or vision sensing provides the capability of viewing the workspace and interpreting what is seen. Vision-equipped robots are used for inspection, part recognition, and identification, sorting, obstacle avoidance, and other similar tasks.

1.10 PROGRAMMING ROBOTS

Robots have no intelligence to learn by themselves. They need to be "taught" what they are expected to do and "how" they should do it. The teaching of the workcycle to a robot is known as *robot programming*. Robots can be programmed in different ways. One is " teach-by-showing" and the other is using textual commands with a suitable interface.

The manipulator is required to execute a specified workcycle and, therefore, must know where to move, how to move, what work to do, where and so on. In teach-by-showing method of programming, the manipulator is made to move through the desired motion path of the entire workcycle and the path and other parameters are saved in the memory. This method is also known as *lead through programming*.

A *robot programming language* serves as an interface between the human user (the programmer) and the robot manipulator for textual programming. The textual programming using a robot programming language can be done on-line or off-line. In on-line programming, the manipulator executes the command as soon as it is entered and the programmer can verify whether the robot executes the desired task. Any discrepancy is, therefore, corrected immediately.

In off-line programming, the robot is not tied-up and can continue doing its task, that is, there is no loss of production. The programmer develops the program and tests it in a simulated graphical environment without the access to the manipulator. After the programmer is satisfied with the correctness of the program, it is uploaded to the manipulator. In off-line and on-line programming, after the program is complete, it is saved and the robot executes it in the 'run' mode relentlessly.

The robot programming languages are built on the lines of conventional computer programming languages and have their own 'vocabulary', 'grammar', and 'syntaxes'. A typical vocabulary includes command verbs for (i) definition of points, paths, frames and so on, (ii) motion of joints-links and end-effector, (iii) control of end-effector, say grippers to open, close and so on; and (iv) interaction with sensors, environment, and other devices.

Each robot-programming language will also require traditional commands and functions for: arithmetic, logical, trigonometric operations; condition testing and looping operations; input-output operations; storage, retrieval, update, and debugging and so on.

The robot programming encompasses all the issues of traditional computer programming or software development and computer programming languages. This is an extensive subject itself and is not included in this book.

1.11 THE FUTURE PROSPECTS

The use of robots in industries has been increasing at the rate of about 25% annually. This growth rate is expected to increase rapidly in the years to come with more capable robots being available to the industry at lesser costs. The favourable factors for this prediction are:

(i) More people in the industry are becoming aware of robot technology and its potential benefits.

(ii) The robotics technology will develop rapidly in the next few years and more user-friendly robots will be available.

(iii) The hardware, software interfacing, and installations will become easier.

(iv) The production of industrial robots will increase and will bring down the unit cost, making deployment of robots justifiable.

(v) The medium and small-scale industries will be able to beneficially utilize the new technology.

All these will increase the customer base and, therefore, demand for the industrial robots and manpower geared with robot technology.

Robot is the technology for the future and with a future. The current research goals and trends indicate that the industrial robots of the future will be more robust, more accurate, more flexible, with more than one arm, more mobile, and will have many more capabilities. The robots will be human friendly and intelligent, capable of responding to voice commands and will be easy to program.

1.11.1 Biorobotics and Humanoid Robotics

A new research field in robotics inspired by biological systems has arisen. The technological developments have made it possible for engineers and robot designers to look for solutions in nature and look forward to achieving one of their most attractive goals to develop a humanoid robot.

Conventional viewpoint in robot design is dominated by its industrial applications, where emphasis is on mechanical properties that go beyond human performance, such as doing stereotype work tirelessly, carrying heavy loads, working in hostile environment, or giving high precision and consistent performance.

The biorobotics is historically connected to service robotics. These robots are conceptualized in a different manner than industrial robots. Their task is usually to help humans in diverse activities from house cleaning to carrying out a surgery, or playing the piano to assisting the disabled and the elderly.

The motion abilities of biological systems, their intelligence, and sensing are far ahead of all the achievements in manmade things till date. Progress in robot technology, rapid technological developments through the remarkable achievements in computer-aided technology in recent years have opened an entirely new research area, where the objective is to analyze and model biological systems behaviour, intelligence, sensing, and motions in order to incorporate properties of biological systems in robots. The ultimate objective is to produce a humanoid robot. The aspirations are not to limit these to service robots but these are to be extended to the industrial robots. It is expected that humanoid robots will be able to communicate with humans and other robots; facilitate robot programming; increase their flexibility and adaptability for executing different tasks; learn from experience; and adapt to different tasks and environments or change of place.

In the implementation of biological behavioural systems, the replication of anthropomorphic characteristics is possibly the answer in every context of development in robotics. The research in anthropomorphic robotics has advanced to development of anthropomorphic components for humanoid robots like anthropomorphic visual and tactile sensors, anthropomorphic actuators and anthropomorphic computing techniques. Replicating the functionality of the human brain is one of the hardest challenges in the biorobotics and in general still one of the most difficult objectives.

1.12 NOTATIONS

In a subject of interdisciplinary nature like robotics encompassing mechanical, electrical and many other disciplines, use of clear and consistent notations is always an issue. In this text we have used the following notations and conventions:

1. Vectors and matrices are written in upper case-bold-italic. Unit vectors are lower case-bold-italic, as an exception. Lower case italic is used for scalars. Vectors are taken as column vectors. Components of a vector or matrix are scalars with single subscript for vector components and double subscripts for matrix components. For example, components of a vector are a_i or b_z and elements of a matrix are a_{ij}.
2. Coordinate frames are enclosed in curved parenthesis { }, for example coordinate frame with axes XYZ is $\{x\ y\ z\}$ or coordinate frame 1 is {1} and square parenthesis [ ] are used for elements of vectors and matrices.
3. The association of a vector to a coordinate frame is indicated by a leading superscript. For example, ${}^0\boldsymbol{P}$ is a position vector $\boldsymbol{P}$ in frame {0}.
4. A trailing subscript on a vector is used, wherever necessary to indicate what the vector represents. For example, $\boldsymbol{P}_{\text{tool}}$, represents the tool position vector and $\boldsymbol{v}_i$ represents velocity vector for link i.
5. Matrices used for transformation from one coordinate frame to another, have a leading superscript and a trailing subscript. For example, ${}^0\boldsymbol{T}_1$ denotes the coordinate transformation matrix, which transforms coordinates from frame {1} to frame {0}.
6. Trailing superscripts on matrices are used for inverse or transpose of a matrix, for example, $\boldsymbol{R}^{-1}$ or $\boldsymbol{R}^T$ and on vectors for transpose of a vector, for example, if $\boldsymbol{P}$ is a column vector $\boldsymbol{P}^T$ is a row vector.
7. Many trigonometric functions are required in mathematical models. The sines and cosines of an angle θ_i can take any of the forms: $\cos\theta_i = C\theta_i = C_i$ and $\sin\theta_i = S\theta_i = S_i$. Some more shortened forms are $V\theta_i$ for $(1 - \cos\theta_i)$ and S_{ij} for $\sin(\theta_i + \theta_j)$.

A complete list of symbols used in the text is available in Appendix E.

1.13 BIBLIOGRAPHICAL REFERENCE TEXTS

Literature production in the nascent field of robotics has been conspicuous in the last twenty years, both in terms of research monographs and textbooks. The number of scientific and technical journals dedicated to robotics are also few, though the robotics field has simulated an ever-increasing number of scholars and has established a truly respectable international research community.

This chapter, therefore, includes a selection of journals, reference texts and monographs related to the field. The bibliography references cited here are representative of publications dealing with topics of interest in robotics and related fields.

General and Specialized Texts

The following texts include general and specialized books on robotics and allied subjects. The texts on robotics share an affinity of contents with this text and may provide the complimentary reading for the material in this text. Other books and monographs render supplementary material for those readers who wish to make a thorough study in the robotics.

1. Issac Asimov, *The Complete Robot*, Doubleday & Company, Garden City, New York, 1982.
2. Amitabh Bhattacharya, *Robotics and their Applications in India: A State of the Art Report*, Department of Science and Technology, New Delhi, 1987.
3. J.J. Craig, *Introduction to Robotics, Mechanics and Control*, 2nd edition, Addison-Wesley, 1989.
4. R.C. Dorf and S. Nof, Editors, *The International Encyclopedia of Robotics*, John C. Wiley and Sons, 1988.
5. D.M. Etter, *Engineering Problem Solving with MATLAB*, Prentice-Hall Englewood Cliffs, 1993.
6. Daniel T. Finkbeiner, II, *Introduction of Matrices and Linear Transformation*, D.B. Taraporevala Sons & Co. Pvt. Ltd., Mumbai, 1968.
7. K.S. Fu, R.C. Gonzalez and C.S.G. Lee, *Robotics: Control, Sensing, Vision, and Intelligence*, McGraw-Hill, 1987.
8. M.P. Groover, M.Weiss, R.N.Nagel and N.G.Odrey, *Industrial Robotics —Technology, Programming and Applications*, McGraw-Hill, 1986.
9. R.S. Hartenberg and J. Denavit, *Kinematic Synthesis of Linkages*, McGraw-Hill, 1964.
10. P.A. Janakiraman, *Robotics and Image Processing*, Tata McGraw-Hill Co. Ltd., New Delhi, 1995.
11. R.D. Klafter, Thomas A. Chmielewski and Michael Negin, *Robotic Engineering: An Integrated Approach*, Prentice-Hall, 1994.
12. J.C. Latombe, *Robot Motion Planning*, Kluwer Academic Publishers, 1991.
13. N.E. Leonard and W.S. Levone, *Using MATLAB to Analysis and Design Control Systems*, Addison-Wesley, 1995.
14. Fairhurst C. Micheal, *Computer Vision for Robotic Systems*, Prentice-Hall International (UK), 1988.
15. I.J. Nagrath and M. Gopal, *Control Systems Engineering*, New Age International Ltd., 3rd ed., 1999.
16. R.P. Paul, *Robot Manipulators: Mathematics, Programming, and Control*, MIT Press, Cambridge, Mass, 1981.
17. E. I. Rivin, *Mechanical Design of Robots*, McGraw-Hill, 1988.

18. R.J. Schilling, *Fundamentals of Robotics: Analysis and Control*, Prentice-Hall of India, New Delhi, 1996.
19. L. Sciavicco and B. Siciliano, *Modeling and Control of Robot Manipulators*, The McGraw-Hill Companies Inc., New York, 1996.
20. J.E. Shigley and Jr.J.J. Uicker, *Theory of Machines and Mechanisms*, 2nd edition, McGraw-Hill, 1995.
21. M.W. Spong, F.L. Lewis and C. Abdallah, *Robot Control: Dynamics, Motion Planning, and Analysis*, IEEE Press, New York, 1993.
22. W. Stadler, *Analytical Robotics and Mechatronics*, McGraw-Hill Inc., New York, 1995.
23. T. Yoshikawa, *Foundations of Robotics: Analysis and Control*, Prentice-Hall of India, 1998. (MIT Press, Cambridge, Mass., 1990).

Dedicated and Related Journals

Some of the following journals and magazines are dedicated to robotics while other prestigious journals give substantial space to robotics and occasionally or routinely publish papers in the robotics field, on allied topics or contain articles on various aspects of robots and robotics.

1. *Advanced Robotics*, Published (8 issues) by Robotics Society of Japan.
2. American Society of Mechanical Engineers (ASME), New York Publications:
   - *Journal of Mechanical Design*, Quarterly.
   - *Mechanical Engineering Magazine*, Monthly.
   - *Journal of Applied Mechanics*, Bimonthly.
   - *Journal of Dynamics Systems, Measurement and Control*, Quarterly.
   - *Journal of Mechanisms, Transmission and Automaton in Design*.
3. *Artificial Intelligence*, Published monthly by Elsevier Science.
4. *Computers Graphics, Vision & Image Processing*, Published monthly by Academic Press.
5. *Institute of Electrical and Electronic Engineering* (IEEE) Inc., New York Publications:
   - *IEEE Computer Magazine,* Monthly.
   - *IEEE Control Systems Magazine*, Bimonthly.
   - *IEEE/ASME Journal of Microelectromechanical Systems*, Bimonthly.
   - *IEEE Robotics and Automation Magazine*, Quarterly.
   - *IEEE Sensors Journal,* Bimonthly.
   - *IEEE Transactions on Biomedical Engineering*, Monthly.
   - *IEEE Transactions on Control Systems Technology*, Quarterly.
   - *IEEE Transactions on Pattern Analysis and Machine Intelligence*, Monthly.
   - *IEEE Transactions on Robotics and Automation*, Bimonthly.
   - *IEEE Transactions on System, Man and Cybernetics,* Bimonthly.
   - *Proceedings of IEEE.*

6. *The Industrial Robots*, Published monthly by Society of Manufacturing Engineers.
7. *International Journal of Robotics Research*, Published Monthly by Sage Science Press, USA.
8. *International Journal of Robotics & Automation*, Published quarterly by ACTA Press (IASTED).
9. *Journal of Manufacturing Technology*, Published by Indian Institute of Production Engineers.
10. *International Journals of Robotics,* Published bimonthly by Robotics Society of Japan.
11. *Journal of Robotic Systems*, Published monthly by Wiley InterScience, John Wiley & Sons Inc., New York.
12. *Journal of Robotics and Computer Integrated Manufacturing*, Published by monthly by Elsevier Science Ltd.
13. *Journal of Intelligent and Robotic Systems—Theory and Applications* (*incorporating Mechatronics Systems Engineering*), Published monthly by Kluwer Press.
14. *Robotica*, (International Journal of Information, Educational Research in Robotics and Artificial Intelligence), Published monthly by Cambridge University Press, UK.
15. *Robotics and Autonomous Systems,* Published quarterly by Elsevier Science, North Holland.

Selected References

The technical articles cited below are on topics introduced in this chapter and other topics of interest in robotics and related fields. The references at the end of later chapters are specific to the topics discussed in the chapter.

1. C. Buhler, "Robotics for rehabilitation—a European (?) Perspective," *Robotica*, **16**, 487–489, 1998.
2. R.A. Brooks, "Solving the Path Finding Problem, by Good Representation of Free Space," *IEEE Tr on Systems, Man, and Cybernetics*, **13**, 190–197, 1983.
3. Susan Bonner and Kang G. Shin, "A Comparative Study of Robot Languages," *IEEE Computer Magazine*, **15**(12), 82–96, Dec. 1982.
4. K.D. Chaney and J.K. Davidson, "A Synthesis Method for Placing Workpieces in RPR Planar Robotic Workcells", *ASME Journal of Mechanical Design*, **120**, 262–268, 1998.
5. P. Dario, E. Guglielmelli and C. Laschi, "Humanoids and Personal Robots: Design and Experiments," *J. of Robotic Systems*, **18**(12), 673–690, 2001.
6. Y. Eiichi, "Local Communication of Multiple Mobile Robots," *Journal of Robotic Systems*, **15**(7), 407–409, 1998.
7. C. Gosselin, "Determination of the Workspace of 6-DoF Parallel Manipulators," *J of Mechanical Design*, **114**, 331–335, Sep 1992.

8. W. Hsu, C.S.G. Lee and A. Lim, "A Computer-aided Product Redesign System for Robotic Assembly," *Robotica*, ***16***, 239–249, 1998.
9. A. Kumar and K.J. Waldron, "The Workspaces of Mechanical Manipulators," *J of Mechanical Design*, ***103***, 665–672, Jul 1981.
10. K.B. Kim, "Introduction to the Special Issue on Design and Applications in Robotics," *Robotica*, ***17***(1), 1–2, 1999.
11. M.H. Lee, "Tactile sensing: New Direction, New Challenges," *Int. J. of Robotic Research*, ***19***(7), 636–343, 2001.
12. J.Y.S. Luh, "Conventional Controller Design for Industrial Robots—A Tutorial," *IEEE Tr on Systems, Man, and Cybernetics*, ***13***(3), 298–316, 1983.
13. F.C. Park and R.W. Brockett, "Kinematic Dexterity of Robotic Mechanisms," *The International Journal of Robotics Research*, ***13***(1), 1–15, Feb 1994.
14. R.P. Paul, B.E. Shimano and G. Mayer, "Kinematic Control Equations for Simple Manipulators," *IEEE Tr on Systems, Man, and Cybernetics*, ***11***(6), 449–455, 1981.
15. B.K. Rout and R.K. Mittal, A Review on Minimizing variability in product performance with the help of computational tools by controlling active and noise factor for Robust design, *Proc. of National Seminar on Current Applications of Computers in Design Engineering*, Jodhpur, India, 81–90, March 3–4, 2001.
16. B. Siciliano and J. Lenarcic, "Biorobotics and Human Robotics," editorial for special issue of *J. of Robotic Systems*, ***18***(12), 671–672, 2001.
17. C. Schafer and R Dillmann, "Kinematic Design of a Humanoid Robot Wrist," *J. of Robotic Systems*, ***18***(12), 747–754, 2001.
18. N.N. Sharma and R.K. Mittal, "Examination of Design Model's Influences on Control Performance of Robotic Manipulator", *Proc. of 9th IASTED International Conference: Applied Informatics (AI2001)*, Innsbruck, Austria, 131–136, Feb. 19–23, 2001.
19. Ken Taylor, Barney Dalton and James Trevelyan, "Web-based Telerobotics," *Robotica*, ***17***(1), 49–57, 1999.
20. T. Yoshikawa, "Manipulability of Robotic Mechanisms," *The International Journal of Robotics Research*, ***4***(2), 3–9, 1985.

EXERCISES

1.1 Name the four basic components of a robot system.

1.2 Describe the functions of four basic components of a robot.

1.3 Define the degree of freedom.

1.4 Name the four basic arm configurations that are used in robotic manipulators.

1.5 Where is the end-effector connected to the manipulator?

1.6 Give all possible classifications of robots.

1.7 Describe the role of arm and wrist of a robotic manipulator.
1.8 Define the term work envelope.
1.9 Draw the side view of the workspace of a typical
 (a) Cylindrical configuration arm.
 (b) Polar configuration arm.
 (c) Articulated configuration arm.
1.10 Briefly describe the four basic configurations of arm in robotic manipulators.
1.11 Define the following
 (a) Load carrying capacity
 (b) Work volume
 (c) End-effector.
1.12 What is the range of number of axes that may be found in industrial manipulators?
1.13 How many degrees of freedom are normally provided in the arm of a manipulator?
1.14 How many degrees of freedom can a wrist have? What is the purpose of these degrees of freedom?
1.15 Discuss the differences between polar arm and articulated arm configurations.
1.16 What are the advantages and disadvantages of cylindrical arm configuration over a polar arm configuration?
1.17 For each of the following tasks, state whether a gripper or an end-of-arm tooling is appropriate:
 (a) Welding.
 (b) Scraping paint from a glass pane.
 (c) Assembling two parts.
 (d) Drilling a hole.
 (e) Tightening a nut of automobile engine.
1.18 An end-effector attached to a robot makes the robot "specialized" for a particular task. Explain the statement.
1.19 Make a chart showing the major industrial applications of robots.
1.20 Make a chronological chart showing the major developments in the field of robotics.
1.21 Prepare a state of art report on robotics in India.
1.22 Who are the users of robots in India? Prepare a status report on industrial applications of robots in India and project the demand for the future.
1.23 Find out the applications of robots in space exploration.
1.24 Discuss reasons for using a robot instead of human being to perform a specific task.
1.25 Discuss the possible applications of robots other than industrial applications. Prepare a report and indicate the weakest areas.
1.26 What are the socioeconomic issues in using robots to replace human workers from the workplace? Explain.

1.27 What are the control issues in robotic control? Explain briefly.

1.28 Describe the methods of teaching robots.

1.29 How are robots different from conventional machine tools? Discuss the design and control issues involved in the two cases and compare.

1.30 Explore the anatomy of the human wrist joint and analyze it for type of motions provided, number of degrees of freedom, number of joints, type of joints, etc.

1.31 A robot is required to perform an assembly of a shaft into a bearing placed in an arbitrary position. How many degrees of freedom are required for a manipulator to perform this task? If the bearing is placed in a fixed plane, say a horizontal plane, what will be the required number of degrees of freedom? Explain.

1.32 Study the human arm anatomy and describe the features a humanoid robot should have.

2

Coordinate Frames, Mapping, and Transforms

The robot (manipulator or arm) consists of several rigid links, connected together by joints, to achieve the required motion in space and perform the desired task. The modeling of robot comprises of establishing a special relationship between the manipulator and the manipulated object. The position of links in space and their motion are described by spatial geometry.

A systematic and generalized approach for mathematical modeling of position and orientation of links in space with respect to a reference frame is carried out with the help of vector and matrix algebra. Because the motion of each link can be described with respect to a reference coordinate frame, it is convenient to have a coordinate frame attached to the body of each link.

2.1 COORDINATE FRAMES

In a 3-D space, a coordinate frame is a set of three orthogonal right-handed axes *X, Y, Z,* called *principal axes*. Such a frame is shown in Fig. 2.1 with the origin of the principal axes at '*O*' along with three unit vectors $\hat{x}$, $\hat{y}$, $\hat{z}$ along these axes. This frame is labelled as $\{x\ y\ z\}$ or by a number as $\{1\}$ using a numbering scheme. Other frames in the space are similarly labelled.

Any point *P* in a 3-D space can be defined with respect to this coordinate frame by a vector $\overrightarrow{OP}$ (a directed line from origin *O* to point *P* pointing towards *P*). In vector notation

$$\vec{P} = \overrightarrow{OP} = p_x\hat{x} + p_y\hat{y} + p_z\hat{z} \tag{2.1}$$

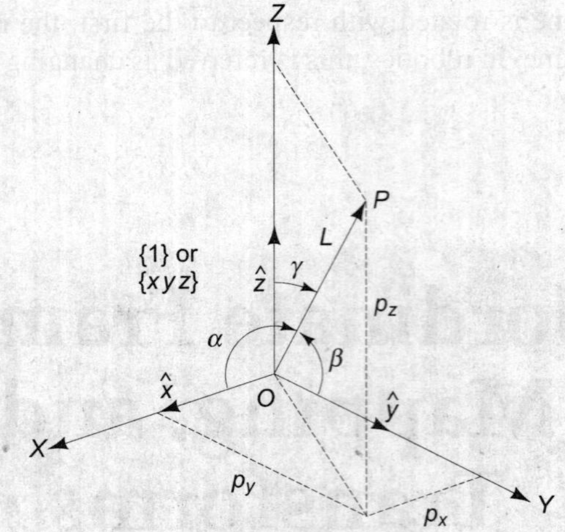

Fig. 2.1 *Position and orientation of a point P in a coordinate frame*

where p_x, p_y, p_z are the components of the vector $\overrightarrow{OP}$ along the three coordinate axes or the projections of the vector $\overrightarrow{OP}$ on the axes *X, Y, Z*, respectively. A *frame-space* notation is introduced as ${}^1\boldsymbol{P}$ to refer to the point *P* (or vector $\overrightarrow{OP}$) with respect to frame {1} with its components in the frame as ${}^1p_x$, ${}^1p_y$, and ${}^1p_z$, that is,

$$ {}^1\boldsymbol{P} = {}^1p_x\hat{\boldsymbol{x}} + {}^1p_y\hat{\boldsymbol{y}} + {}^1p_z\hat{\boldsymbol{z}} \tag{2.2} $$

In vector-matrix notation, this equation can be written in terms of the vector components only as:

$$ {}^1\boldsymbol{P} = \begin{bmatrix} {}^1p_x \\ {}^1p_y \\ {}^1p_z \end{bmatrix} = \begin{bmatrix} {}^1p_x & {}^1p_y & {}^1p_z \end{bmatrix}^T \tag{2.3} $$

Observe that the leading superscript refers to the coordinate frame number (frame {1} in this case) and $[A]^T$ indicates the transpose of matrix $\boldsymbol{A}$. In addition, the direction of the position vector $\overrightarrow{OP}$ can be expressed by the direction cosines:

$$ \cos\alpha = \frac{{}^1p_x}{L}, \ \cos\beta = \frac{{}^1p_y}{L}, \ \cos\gamma = \frac{{}^1p_z}{L} $$

with $$ L = |\vec{\boldsymbol{P}}| = |\overrightarrow{\boldsymbol{OP}}| = \sqrt{\left({}^1p_x\right)^2 + \left({}^1p_y\right)^2 + \left({}^1p_z\right)^2} \tag{2.4} $$

where α, β, and γ are, respectively, the right handed angles measured from the coordinate axes to the vector $\overrightarrow{OP}$, which has a length *L*.

2.1.1 Mapping

Mappings refer to changing the description of a point (or vector) in space from one frame to another frame. The second frame has three possibilities in relation to the first frame:

(a) Second frame is rotated with respect to the first; the origin of both the frames is same. In robotics, this is referred as changing the orientation.
(b) Second frame is moved away from the first, the axes of both frames remain parallel, respectively. This is a translation of the origin of the second frame from the first frame in space.
(c) Second frame is rotated with respect to the first and moved away from it, that is, the second frame is translated and its orientation is also changed.

These situations are modelled in the following sections. It is important to note that mapping changes the description of the point and not the point itself.

2.1.2 Mapping between Rotated Frames

Consider two frames, frame {1} with axes *X, Y, Z*, and frame {2} with axes *U, V, W* with a common origin, as shown in Fig. 2.2. A point *P* in space can be described by the two frames and can be expressed as vectors ${}^1\boldsymbol{P}$ and ${}^2\boldsymbol{P}$,

$$ {}^1\boldsymbol{P} = {}^1p_x\hat{\boldsymbol{x}} + {}^1p_y\hat{\boldsymbol{y}} + {}^1p_z\hat{\boldsymbol{z}} \tag{2.5}$$

$$ {}^2\boldsymbol{P} = {}^2p_u\hat{\boldsymbol{u}} + {}^2p_v\hat{\boldsymbol{v}} + {}^2p_w\hat{\boldsymbol{w}} \tag{2.6}$$

where ${}^2p_u$, ${}^2p_v$, ${}^2p_w$ are projections of point *P* on frame {2} or {*u v w*} (the *U, V, W* coordinates). Because the point *P* is same, its two descriptions given by Eqs. (2.5) and (2.6) are related.

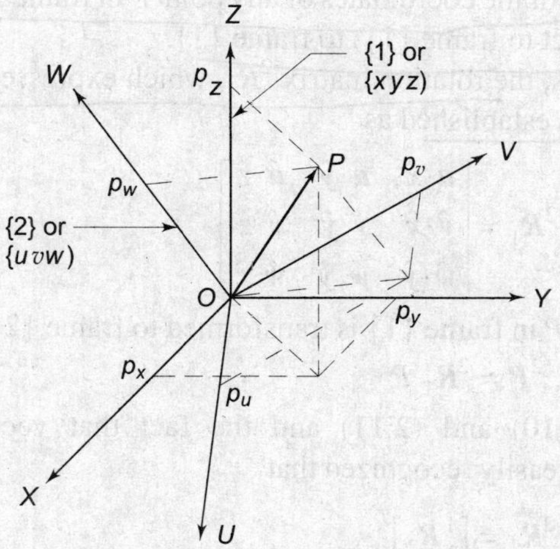

Fig. 2.2 *Representation of a point P in two frames {x y z} and {u v w} rotated with respect to each other*

Now, let the problem be posed as, "The description of point *P* in frame {2} is known and its description in frame {1} is to be found (or vice-versa)." This is accomplished by projecting the vector ${}^2\boldsymbol{P}$ on to the coordinates of frame {1}. Projections of ${}^2\boldsymbol{P}$ on frame {1} are obtained by taking the dot product of ${}^2\boldsymbol{P}$ with the unit vectors of frame {1}. Thus, substituting for ${}^2\boldsymbol{P}$ from Eq. (2.6) gives

$$
\begin{aligned}
{}^1p_x &= \hat{\boldsymbol{x}}\cdot{}^2\boldsymbol{P} = \hat{\boldsymbol{x}}\cdot{}^2p_u\hat{\boldsymbol{u}} + \hat{\boldsymbol{x}}\cdot{}^2p_v\hat{\boldsymbol{v}} + \hat{\boldsymbol{x}}\cdot{}^2p_w\hat{\boldsymbol{w}} \\
{}^1p_y &= \hat{\boldsymbol{y}}\cdot{}^2\boldsymbol{P} = \hat{\boldsymbol{y}}\cdot{}^2p_u\hat{\boldsymbol{u}} + \hat{\boldsymbol{y}}\cdot{}^2p_v\hat{\boldsymbol{v}} + \hat{\boldsymbol{y}}\cdot{}^2p_w\hat{\boldsymbol{w}} \\
{}^1p_z &= \hat{\boldsymbol{z}}\cdot{}^2\boldsymbol{P} = \hat{\boldsymbol{z}}\cdot{}^2p_u\hat{\boldsymbol{u}} + \hat{\boldsymbol{z}}\cdot{}^2p_v\hat{\boldsymbol{v}} + \hat{\boldsymbol{z}}\cdot{}^2p_w\hat{\boldsymbol{w}}
\end{aligned}
\tag{2.7}
$$

This can be written in matrix form as

$$
\begin{bmatrix} {}^1p_x \\ {}^1p_y \\ {}^1p_z \end{bmatrix} = \begin{bmatrix} \hat{\boldsymbol{x}}\cdot\hat{\boldsymbol{u}} & \hat{\boldsymbol{x}}\cdot\hat{\boldsymbol{v}} & \hat{\boldsymbol{x}}\cdot\hat{\boldsymbol{w}} \\ \hat{\boldsymbol{y}}\cdot\hat{\boldsymbol{u}} & \hat{\boldsymbol{y}}\cdot\hat{\boldsymbol{v}} & \hat{\boldsymbol{y}}\cdot\hat{\boldsymbol{w}} \\ \hat{\boldsymbol{z}}\cdot\hat{\boldsymbol{u}} & \hat{\boldsymbol{z}}\cdot\hat{\boldsymbol{v}} & \hat{\boldsymbol{z}}\cdot\hat{\boldsymbol{w}} \end{bmatrix} \begin{bmatrix} {}^2p_x \\ {}^2p_y \\ {}^2p_z \end{bmatrix} \tag{2.8}
$$

In compressed vector-matrix notation Eq. (2.8) is written as

$$
{}^1\boldsymbol{P} = {}^1\boldsymbol{R}_2\,{}^2\boldsymbol{P} \tag{2.9}
$$

where

$$
{}^1\boldsymbol{R}_2 = \begin{bmatrix} \hat{\boldsymbol{x}}\cdot\hat{\boldsymbol{u}} & \hat{\boldsymbol{x}}\cdot\hat{\boldsymbol{v}} & \hat{\boldsymbol{x}}\cdot\hat{\boldsymbol{w}} \\ \hat{\boldsymbol{y}}\cdot\hat{\boldsymbol{u}} & \hat{\boldsymbol{y}}\cdot\hat{\boldsymbol{v}} & \hat{\boldsymbol{y}}\cdot\hat{\boldsymbol{w}} \\ \hat{\boldsymbol{z}}\cdot\hat{\boldsymbol{u}} & \hat{\boldsymbol{z}}\cdot\hat{\boldsymbol{v}} & \hat{\boldsymbol{z}}\cdot\hat{\boldsymbol{w}} \end{bmatrix} \tag{2.10}
$$

Because frames {1} and {2} have the same origin, they can only be rotated with respect to each other, therefore, $\boldsymbol{R}$ is called a *rotation matrix* or *rotational transformation matrix*. It contains only the dot products of unit vectors of the two frames and is independent of the point P. Thus, rotation matrix ${}^1\boldsymbol{R}_2$ can be used for transformation of the coordinates of any point P in frame {2} (which has been rotated with respect to frame {1}) to frame {1}.

On similar lines, the rotation matrix ${}^2\boldsymbol{R}_1$, which expresses frame {1} as seen from frame {2}, is established as

$$
{}^2\boldsymbol{R}_1 = \begin{bmatrix} \hat{\boldsymbol{u}}\cdot\hat{\boldsymbol{x}} & \hat{\boldsymbol{u}}\cdot\hat{\boldsymbol{y}} & \hat{\boldsymbol{u}}\cdot\hat{\boldsymbol{z}} \\ \hat{\boldsymbol{v}}\cdot\hat{\boldsymbol{x}} & \hat{\boldsymbol{v}}\cdot\hat{\boldsymbol{y}} & \hat{\boldsymbol{v}}\cdot\hat{\boldsymbol{z}} \\ \hat{\boldsymbol{w}}\cdot\hat{\boldsymbol{y}} & \hat{\boldsymbol{w}}\cdot\hat{\boldsymbol{y}} & \hat{\boldsymbol{w}}\cdot\hat{\boldsymbol{z}} \end{bmatrix} \tag{2.11}
$$

Hence, a point P in frame {1} is transformed to frame {2} using

$$
{}^2\boldsymbol{P} = {}^2\boldsymbol{R}_1\,{}^1\boldsymbol{P} \tag{2.12}
$$

From Eqs. (2.10) and (2.11) and the fact that vector dot product is commutative, it is easily recognized that

$$
{}^2\boldsymbol{R}_1 = [{}^1\boldsymbol{R}_2]^T \tag{2.13}
$$

From Eqs. (2.9), (2.12), and (2.13), ${}^2\boldsymbol{P}$ is expressed as

$$
{}^2\boldsymbol{P} = [{}^1\boldsymbol{R}_2]^{-1}\,{}^1\boldsymbol{P} = {}^2\boldsymbol{R}_1{}^1\boldsymbol{P} = [{}^1\boldsymbol{R}_2]^T\,{}^1\boldsymbol{P} \tag{2.14}
$$

Therefore, it is concluded that

$$
{}^2\boldsymbol{R}_1 = [{}^1\boldsymbol{R}_2]^{-1} = [{}^1\boldsymbol{R}_2]^T
$$

or, in general, for any rotational transformation matrix $\boldsymbol{R}$

$$\boldsymbol{R}^{-1} = \boldsymbol{R}^T \quad \text{and} \quad \boldsymbol{R}\boldsymbol{R}^T = \boldsymbol{I} \tag{2.15}$$

where $\boldsymbol{I}$ is the 3×3 identity matrix.

2.1.3 Mapping between Translated Frames

Consider two frames, frame {1} and frame {2}, with origins O_1 and O_2 such that the axes of frame {1} are parallel to axes of frame {2}, as shown in Fig. 2.3. A point P in space can be expressed as vectors $\overrightarrow{\boldsymbol{O_1P}}$ and $\overrightarrow{\boldsymbol{O_2P}}$ with respect to the frames {1} and {2}, respectively.

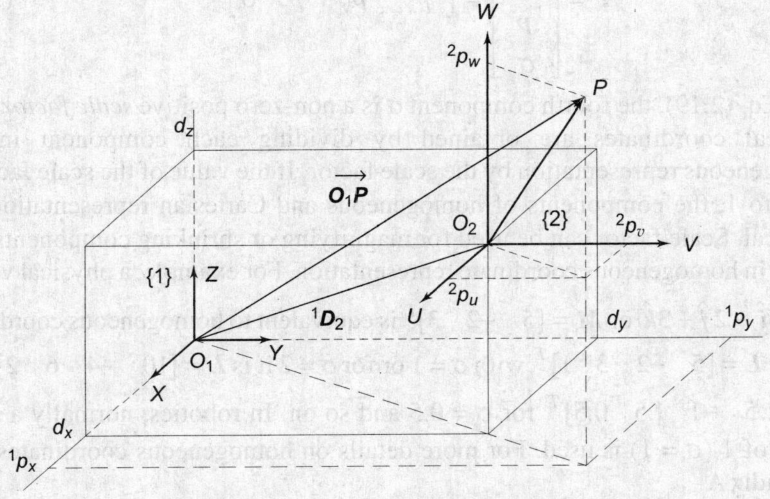

Fig. 2.3 *Translation of frames: frame {2} is translated with respect to frame {1} by distance* $^1\boldsymbol{D}_2$

The two vectors are related as

$$\overrightarrow{\boldsymbol{O_1P}} = \overrightarrow{\boldsymbol{O_2P}} + \overrightarrow{\boldsymbol{O_1O_2}} \tag{2.16}$$

or in the notation introduced earlier Eq. (2.16) becomes

$$^1\boldsymbol{P} = {}^2\boldsymbol{P} + {}^1\boldsymbol{D}_2 \tag{2.17}$$

where $^1\boldsymbol{D}_2 = \overrightarrow{\boldsymbol{O_1O_2}}$ is the translation of origin of frame {2} with respect to frame {1}. Because $^2\boldsymbol{P} = [^2p_u \ {}^2p_v \ {}^2p_w]^T$, substituting $^2\boldsymbol{P}$ and $^1\boldsymbol{D}_2$ in Eq. (2.17) gives

$$^1\boldsymbol{P} = (^2p_u + d_x)\hat{\boldsymbol{x}} + (^2p_v + d_y)\hat{\boldsymbol{y}} + (^2p_w + d_z)\hat{\boldsymbol{z}} \tag{2.18}$$

As $\quad {}^1\boldsymbol{P} = {}^1p_x\hat{\boldsymbol{x}} + {}^1p_y\hat{\boldsymbol{y}} + {}^1p_z\hat{\boldsymbol{z}}$, this gives

$$^1p_x = {}^2p_u + d_x; \quad {}^1p_y = {}^2p_v + d_y; \quad {}^1p_z = {}^2p_z + d_z$$

which is verified from Fig. 2.3.

Translation is qualitatively different from rotation in one important respect. In rotation, the origin of two coordinate frames is same. This invariance of the origin characteristic allows the representation of rotations in 3-D space as a 3×3 rotation matrix $\boldsymbol{R}$. However, in translation, the origins of translated frame and original frame are not coincident and translation is represented by a 3×1 vector, ${}^1\boldsymbol{D}_2$.

A powerful representation of translation is in a 4-D space of *homogeneous coordinates*. In these coordinates, point P in space with respect to frame {1} is denoted as [refer to Fig. 2.1 and Eq. (2.3)]:

$$ {}^1\boldsymbol{P} = \begin{bmatrix} {}^1p_x \\ {}^1p_y \\ {}^1p_z \\ \sigma \end{bmatrix} = \begin{bmatrix} {}^1p_x & {}^1p_y & {}^1p_z & \sigma \end{bmatrix}^T \tag{2.19} $$

In Eq. (2.19), the fourth component σ is a non-zero positive *scale factor*. The physical coordinates are obtained by dividing each component in the homogeneous representation by the scale factor. If the value of the scale factor σ is set to 1, the components of homogeneous and Cartesian representation are identical. Scale factor can be used for magnifying or shrinking components of a vector in homogeneous coordinate representation. For example, a physical vector $\overrightarrow{\boldsymbol{M}} = 5\hat{\boldsymbol{i}} - 2\hat{\boldsymbol{j}} + 3\hat{\boldsymbol{k}}$ or $\boldsymbol{M} = [5 \quad -2 \quad 3]^T$ is equivalent to homogeneous coordinate vector $\boldsymbol{L} = [5 \quad -2 \quad 3 \quad 1]^T$ with $\sigma = 1$ or for $\sigma = 2$ it is $\boldsymbol{L} = [10 \quad -4 \quad 6 \quad 2]^T$ or $\boldsymbol{L} = [2.5 \quad -1 \quad 1.5 \quad 0.5]^T$ for $\sigma = 0.5$ and so on. In robotics, normally a scale factor of 1 ($\sigma = 1$) is used. For more details on homogeneous coordinates, see Appendix **A**.

Using the homogeneous coordinates, Eq. (2.17) is written in the vector-matrix form as:

$$ {}^1\boldsymbol{P} = \begin{bmatrix} 1 & 0 & 0 & d_x \\ 0 & 1 & 0 & d_y \\ 0 & 0 & 1 & d_z \\ 0 & 0 & 0 & 1 \end{bmatrix} \begin{bmatrix} {}^2p_u \\ {}^2p_v \\ {}^2p_w \\ 1 \end{bmatrix} $$

or

$$ {}^1\boldsymbol{P} = {}^1\boldsymbol{T}_2 \, {}^2\boldsymbol{P} \tag{2.20} $$

Here, ${}^1\boldsymbol{T}_2$ is a 4×4 homogeneous *transformation matrix* for translation of origin by ${}^1\boldsymbol{D}_2 = \overrightarrow{\boldsymbol{O}_1\boldsymbol{O}_2} = [d_x \quad d_y \quad d_z \quad 1]^T$. It is easily seen that Eq. (2.20) is same as Eq. (2.18). The 4×4 transformation matrix in Eq. (2.20) is called the *basic homogeneous translation matrix*.

2.1.4 Mapping between Rotated and Translated Frames

Consider now, the general case of two frames, frame {1} and frame {2}. Frame {2} is rotated and translated with respect to frame {1} as shown in Fig. 2.4. The distance between the two origins is vector $\overrightarrow{\boldsymbol{O}_1\boldsymbol{O}_2}$ or ${}^1\boldsymbol{D}_2$. Assume a point P described with respect to frame {2} as ${}^2\boldsymbol{P}$, it is required to refer it to frame {1}, that is, to find ${}^1\boldsymbol{P}$.

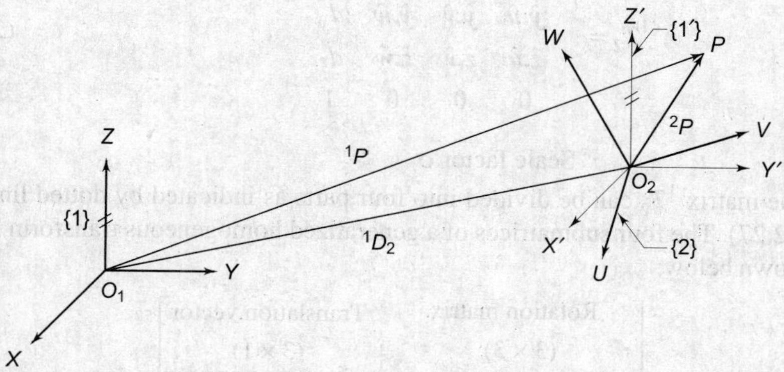

Fig. 2.4 *Mapping between two frames–translated and rotated with respect to each other*

In terms of vectors in Fig. 2.4,

$$\overrightarrow{O_1P} = \overrightarrow{O_2P} + \overrightarrow{O_1O_2} \tag{2.21}$$

Vector $\overrightarrow{O_2P}$ in frame {2} is ${}^2\boldsymbol{P}$; therefore, it must be transformed to frame {1}. First, consider an intermediate frame {1′} with its origin coincident with O_2. The frame {1′} is rotated with respect to frame {2} such that its axes are parallel to axes of frame {1}. Thus, frame {1′} is related to frame {2} by pure rotation. Hence, using Eq. (2.9), point P is expressed in frame {1′} as

$${}^{1'}\boldsymbol{P} = {}^{1'}\boldsymbol{R}_2\,{}^2\boldsymbol{P} \tag{2.22}$$

Because frame {1′} is aligned with frame {1}, ${}^{1'}\boldsymbol{R}_2 = {}^1\boldsymbol{R}_2$. Hence

$$\overrightarrow{O_2P} = {}^{1'}\boldsymbol{P} = {}^1\boldsymbol{R}_2\,{}^2\boldsymbol{P} \tag{2.23}$$

Substituting this in Eq. (2.21) and converting to vector-matrix notation,

$${}^1\boldsymbol{P} = {}^1\boldsymbol{R}_2\,{}^2\boldsymbol{P} + {}^1\boldsymbol{D}_2 \tag{2.24}$$

The vector $\overrightarrow{O_1O_2}$ or ${}^1\boldsymbol{D}_2$ has components $(d_x\, d_y\, d_z)$ in frame {1} as

$$\overrightarrow{O_1O_2} = {}^1\boldsymbol{D}_2 = [d_x \quad d_y \quad d_z]^T \tag{2.25}$$

Using the homogeneous coordinates, from Eqs. (2.10) and (2.20), the two terms on the right-hand side of Eq. (2.24) can be combined into a single 4×4 matrix, which is then written as

$${}^1\boldsymbol{P} = {}^1\boldsymbol{T}_2\,{}^2\boldsymbol{P} \tag{2.26}$$

Here, ${}^1\boldsymbol{P}$ and ${}^2\boldsymbol{P}$ are 4×1 vectors as in Eq. (2.19) with a scale factor of 1 and $\boldsymbol{T}$ is 4×4 matrix referred to as the *homogeneous transformation matrix* (or *homogeneous transform*). It describes both the position and orientation of frame {2} with respect to frame {1} or any frame with respect to any other frame. The components of ${}^1\boldsymbol{T}_2$ matrix are as under

$$
{}^1\boldsymbol{T}_2 = \left[\begin{array}{ccc:c} \hat{x}.\hat{u} & \hat{x}.\hat{v} & \hat{x}.\hat{w} & d_x \\ \hat{y}.\hat{u} & \hat{y}.\hat{v} & \hat{y}.\hat{w} & d_y \\ \hat{z}.\hat{u} & \hat{z}.\hat{v} & \hat{z}.\hat{w} & d_z \\ \hdashline 0 & 0 & 0 & 1 \end{array}\right] \tag{2.27}
$$

(Labels on Eq. (2.27): ${}^1\boldsymbol{R}_2$ marks the 3 × 3 block, ${}^1\boldsymbol{D}_2$ the last column, and "Scale factor σ" the bottom-right element.)

The matrix ${}^1\boldsymbol{T}_2$ can be divided into four parts as indicated by dotted lines in Eq. (2.27). The four submatrices of a generalized homogeneous transform $\boldsymbol{T}$ are as shown below:

$$
\boldsymbol{T} = \left[\begin{array}{c:c} \begin{array}{c}\text{Rotation matrix}\\ (3\times 3)\end{array} & \begin{array}{c}\text{Translation vector}\\ (3\times 1)\end{array} \\ \hdashline \begin{array}{c}\text{Perspective}\\ \text{transformation matrix}\\ (1\times 3)\end{array} & \begin{array}{c}\text{Scale factor}\\ (1\times 1)\end{array} \end{array}\right] \tag{2.28}
$$

Perspective transformation matrix is useful in vision systems and is set to zero vector wherever no perspective views are involved. The scale factor σ has non-zero positive ($\sigma > 0$) values and is called *global scaling* parameter. $\sigma > 1$ is useful for reducing and $0 < \sigma < 1$ is useful for enlarging. For robotic study presented here $\sigma = 1$ is used. For describing the position and orientation of frame {2} with respect to frame {1}, $\boldsymbol{T}$ takes the form

$$
{}^1\boldsymbol{T}_2 = \left[\begin{array}{ccc:c} & {}^1\boldsymbol{R}_2 & & {}^1\boldsymbol{D}_2 \\ \hdashline 0 & 0 & 0 & 1 \end{array}\right] \tag{2.29}
$$

In the reverse problem when ${}^1\boldsymbol{P}$ is known and ${}^2\boldsymbol{P}$ is to be found, Eq. (2.26), takes the form

$$
{}^2\boldsymbol{P} = {}^2\boldsymbol{T}_1\, {}^1\boldsymbol{P} \tag{2.30}
$$

where ${}^2\boldsymbol{T}_1 = \left[{}^1\boldsymbol{T}_2\right]^{-1}$.

2.2 DESCRIPTION OF OBJECTS IN SPACE

The location of an object is completely specified in 3-D space by describing both its position and its orientation. Consider a body B in space whose location is to be specified with respect to a known reference frame {0}. Let a frame with origin O_1, frame {1}, be attached to the body B, as shown in Fig. 2.5. The homogeneous transform ${}^0\boldsymbol{T}_1$ completely describes the location (position and orientation) of the body B, that is, the position vector component ${}^0\boldsymbol{D}_1$ of ${}^0\boldsymbol{T}_1$ describes its position, while the rotation matrix component ${}^0\boldsymbol{R}_1$ describes the orientation of the body with respect to frame {0}.

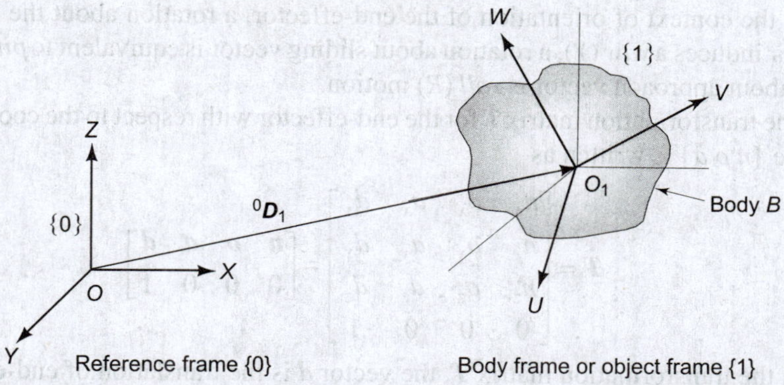

Fig. 2.5 *Description of a body (object) in space using homogeneous transform*

In a robotic arm, the location of links is specified by assigning frames to each link, starting from the base to the tool or end-effector. While the convention for assigning frames to links will be discussed in the next chapter, here, the convention for assigning frames to the end-effector using *normal*, *sliding,* and *approach* vectors, which are yaw, pitch, and roll vectors, respectively, is explained.

The end-effector coordinate frame is shown in Fig. 2.6. The axes of the frame are defined as: (i) z-axis is the approach vector $\hat{\boldsymbol{a}}$, that is, the direction in which the end-effector approaches towards the target, (ii) y-axis is the direction of the sliding vector, that is, the direction of opening and closing of the end-effector as it manipulates objects. It is also called *orientation* vector $\hat{\boldsymbol{o}}$, and (iii) x-axis is the normal vector $\hat{\boldsymbol{n}}$, which is orthogonal to the approach and sliding vectors in right-handed manner, that is, $\hat{\boldsymbol{o}} \times \hat{\boldsymbol{a}}$.

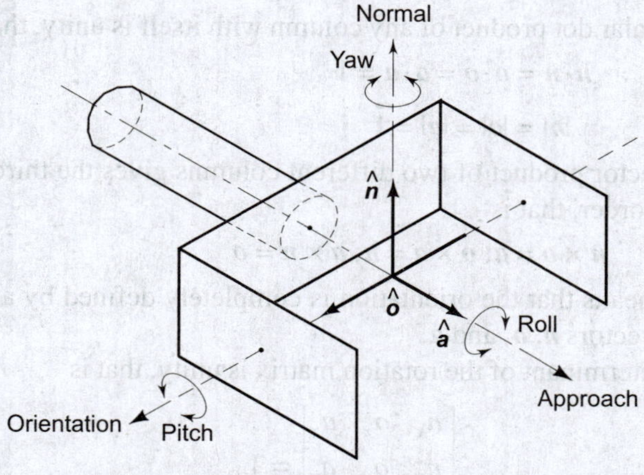

Fig. 2.6 *Assigning a frame to end-effector—approach, orientation and normal directions; and roll, pitch, and yaw motions*

In the context of orientation of the end-effector, a rotation about the normal vector induces a *yaw* (*Y*), a rotation about sliding vector is equivalent to *pitch* (*P*), and about approach vector is *roll* (*R*) motion.

The transformation matrix $\boldsymbol{T}$ for the end-effector with respect to the coordinate frame $\{n\ o\ a\}$ is written as

$$\boldsymbol{T} = \begin{bmatrix} n_x & o_x & a_x & d_x \\ n_y & o_y & a_y & d_y \\ n_z & o_z & a_z & d_z \\ 0 & 0 & 0 & 1 \end{bmatrix} = \begin{bmatrix} \boldsymbol{n} & \boldsymbol{o} & \boldsymbol{a} & \boldsymbol{d} \\ 0 & 0 & 0 & 1 \end{bmatrix} \tag{2.31}$$

In the transformation matrix $\boldsymbol{T}$, the vector $\boldsymbol{d}$ is the translation of end-effector frame from the reference frame and vectors ($\boldsymbol{n}$; $\boldsymbol{o}$, $\boldsymbol{a}$) describe the orientation of end-effector. The vectors $\boldsymbol{n}$, $\boldsymbol{o}$, and $\boldsymbol{a}$ represent the *X*, *Y*, *Z* axes of the end-effector frame. The matrix $\boldsymbol{T}$ in Eq. (2.31) is same as in Eq. (2.27) and would apply for any coordinate frame and, hence, to any joint of the manipulator.

The orientation of the end-effector is specified by the 3×3 rotation submatrix $\boldsymbol{R}$. From Eqs. (2.27) and (2.31), the end-effector rotation matrix is

$$\boldsymbol{R} = \begin{bmatrix} \hat{\boldsymbol{x}}\cdot\hat{\boldsymbol{u}} & \hat{\boldsymbol{x}}\cdot\hat{\boldsymbol{v}} & \hat{\boldsymbol{x}}\cdot\hat{\boldsymbol{w}} \\ \hat{\boldsymbol{y}}\cdot\hat{\boldsymbol{u}} & \hat{\boldsymbol{y}}\cdot\hat{\boldsymbol{v}} & \hat{\boldsymbol{y}}\cdot\hat{\boldsymbol{w}} \\ \hat{\boldsymbol{z}}\cdot\hat{\boldsymbol{u}} & \hat{\boldsymbol{z}}\cdot\hat{\boldsymbol{v}} & \hat{\boldsymbol{z}}\cdot\hat{\boldsymbol{w}} \end{bmatrix} = \begin{bmatrix} n_x & o_x & a_x \\ n_y & o_y & a_y \\ n_z & o_z & a_z \end{bmatrix} \tag{2.32}$$

This is the general rotation matrix. Its properties are enumerated below:

- The vectors $\boldsymbol{n}$, $\boldsymbol{o}$, and $\boldsymbol{a}$ are in three mutually perpendicular directions and hence the rotation matrix $\boldsymbol{R}$ is an *orthogonal transformation*. Because the vectors in the dot products are all unit vectors, it is also called *orthonormal transformation*.
- The scalar dot product of two different columns is zero, that is,

$$\boldsymbol{n}\cdot\boldsymbol{o} = \boldsymbol{o}\cdot\boldsymbol{a} = \boldsymbol{a}\cdot\boldsymbol{n} = 0 \tag{2.33}$$

- The scalar dot product of any column with itself is unity, that is,

$$\boldsymbol{n}\cdot\boldsymbol{n} = \boldsymbol{o}\cdot\boldsymbol{o} = \boldsymbol{a}\cdot\boldsymbol{a} = 1$$

or

$$|\boldsymbol{n}| = |\boldsymbol{o}| = |\boldsymbol{a}| = 1 \tag{2.34}$$

- The vector product of two different columns gives the third column in a cyclic order, that is

$$\boldsymbol{n}\times\boldsymbol{o} = \boldsymbol{a};\ \boldsymbol{o}\times\boldsymbol{a} = \boldsymbol{n};\ \boldsymbol{a}\times\boldsymbol{n} = \boldsymbol{o} \tag{2.35}$$

  This means that the orientation is completely defined by any two of the three vectors $\boldsymbol{n}$, $\boldsymbol{o}$, and $\boldsymbol{a}$.
- The determinant of the rotation matrix is unity, that is

$$\begin{vmatrix} n_x & o_x & a_x \\ n_y & o_y & a_y \\ n_z & o_z & a_z \end{vmatrix} = 1 \tag{2.36}$$

- The inverse and transpose relationships are as in Eq. (2.15).

The rotation matrix $\boldsymbol{R}$ has nine elements in total, which are subjected to the orthogonality constraints [Eqs. (2.33)–(2.35)]. Thus, only three of the nine elements are independent or, the rotation matrix representation has redundancy.

2.3 TRANSFORMATION OF VECTORS

In the previous section, concepts of mapping and the spatial transformation of frames were developed. These concepts will now be applied to vector transformations. Transformation of vectors, rotation and/or translation is distinctly different from mapping, where the description of a point from one frame to another is changed. Different situations of transformation of vectors are discussed now.

2.3.1 Rotation of Vectors

Let us consider a vector ${}^1\boldsymbol{P}$, which is rotated by an angle θ to give new vector ${}^1\boldsymbol{Q}$. If $\boldsymbol{R}(\theta)$ is the rotation that describes the rotation θ about k-axis (which can be x-y-or z-axis), then

$$ {}^1\boldsymbol{Q} = \boldsymbol{R}(\theta)\ {}^1\boldsymbol{P} \tag{2.37} $$

For the rotation matrix $\boldsymbol{R}(\theta)$ no super- or subscripts are used because both ${}^1\boldsymbol{P}$ and ${}^1\boldsymbol{Q}$ are in the same frame {1}.

Equation (2.37) is similar to Eq. (2.12) in mathematical form but both have different interpretation. The distinction is that when vector ${}^1\boldsymbol{P}$ is rotated with reference to frame {1}, it may be considered either as the vector rotation, as shown in Fig. 2.7(a), to give ${}^1\boldsymbol{Q}$ or as the rotation of the frame in "opposite" direction to give rotated frame {$u\ v\ w$}, as shown in Fig. 2.7(b), for rotation about x-axis.

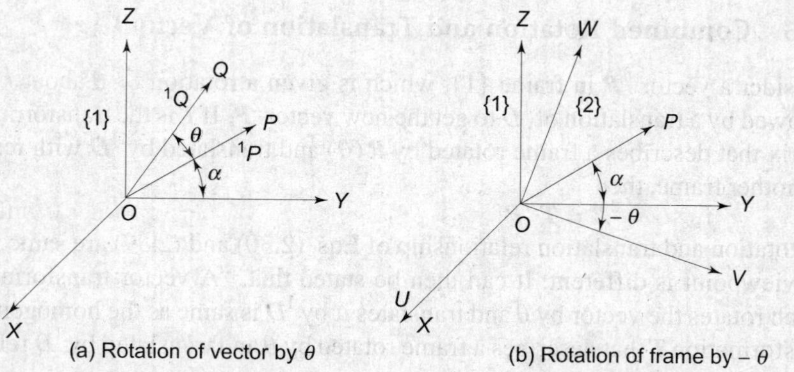

Fig. 2.7 *The equivalence of rotation of a vector and a frame*

The operations involved in two cases are identical, only the viewpoint is different. This allows using the rotational transformation matrices for vector rotations. It is also noted that, "The rotation matrix $\boldsymbol{R}(\theta)$ which rotates a vector through some angle θ about k-axis, is the same as the rotational transformation matrix, which describes a frame rotated by θ relative to the reference frame."

2.3.2 Translation of Vectors

Suppose, a vector ${}^1\boldsymbol{P}$ is translated by a vector ${}^1\boldsymbol{D}$ to get ${}^1\boldsymbol{Q}$, as shown in Fig. 2.8(a), then the vector ${}^1\boldsymbol{Q}$ is given by

$$ {}^1\boldsymbol{Q} = {}^1\boldsymbol{P} + {}^1\boldsymbol{D} \tag{2.38} $$

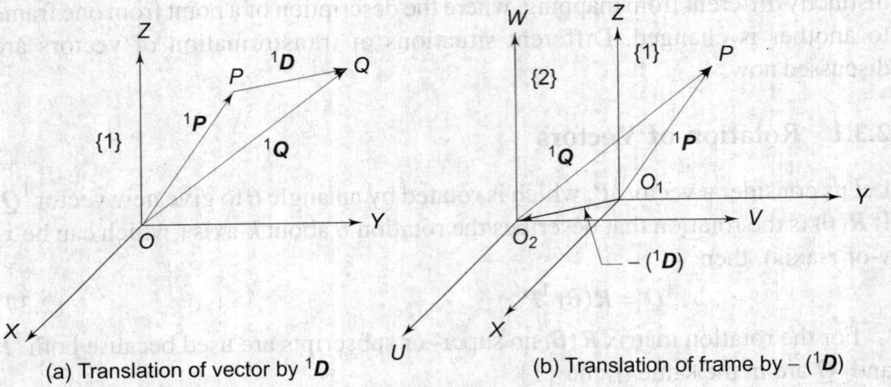

Fig. 2.8 *Translation of vector ${}^1\boldsymbol{P}$ by distance ${}^1\boldsymbol{D}$*

Here, as in case of rotations, instead of moving the vector "forward" by ${}^1\boldsymbol{D}$, the frame can be moved in the opposite sense, as shown in Fig. 2.8(b), which is equivalent to the problem of mapping. This explains why Eq. (2.38) is similar to Eq. (2.17) obtained by mapping between translated frames.

2.3.3 Combined Rotation and Translation of Vectors

Consider a vector ${}^1\boldsymbol{P}$ in frame {1}, which is given a rotation of θ about k-axis followed by a translation of ${}^1\boldsymbol{D}$ to get the new vector ${}^2\boldsymbol{P}$. If $\boldsymbol{T}$ is the transformation matrix that describes a frame rotated by $\boldsymbol{R}(\theta)$ and translated by ${}^1\boldsymbol{D}$ with respect to another frame, then

$$ {}^2\boldsymbol{P} = \boldsymbol{T}\,{}^1\boldsymbol{P} \tag{2.39} $$

Rotation and translation relationship of Eqs. (2.30) and (2.39) are same, only the viewpoint is different. It can then be stated that, "A vector transformation which rotates the vector by θ and translates it by ${}^1\boldsymbol{D}$ is same as the homogeneous transformation $\boldsymbol{T}$ that describes a frame rotated by θ and translated by ${}^1\boldsymbol{D}$ relative to the reference frame."

2.3.4 Composite Transformation

Three frames, with each frame rotated and translated from its preceding frame, are depicted in Fig. 2.9. It is proposed to find the transform, which relates ${}^3\boldsymbol{P}$ in frame {3} to ${}^1\boldsymbol{P}$, as it is seen from frame {1}.

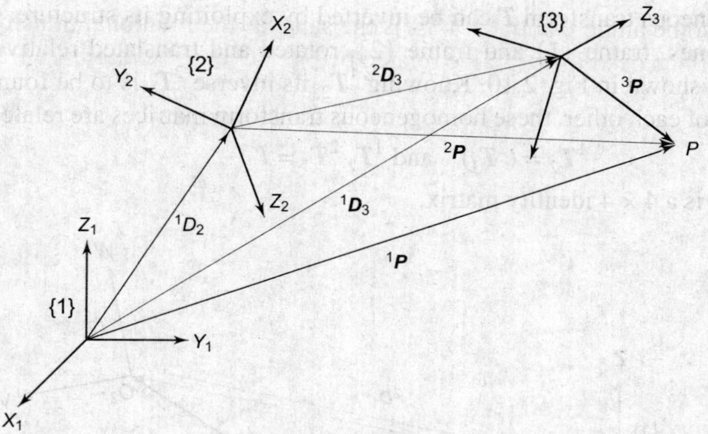

Fig. 2.9 *Composite transformation of three frames*

The transformation matrix $\boldsymbol{T}$ can be used to progressively map $^3\boldsymbol{P}$, the point P in frame {3}, to frame {2}, and then to frame {1} as

$$^2\boldsymbol{P} = {}^2\boldsymbol{T}_3\, {}^3\boldsymbol{P} \tag{2.40}$$

and

$$^1\boldsymbol{P} = {}^1\boldsymbol{T}_2\, {}^2\boldsymbol{P} \tag{2.41}$$

These lead to the overall transformation as

$$^1\boldsymbol{P} = {}^1\boldsymbol{T}_2\, {}^2\boldsymbol{T}_3\, {}^3\boldsymbol{P} \tag{2.42}$$

or

$$^1\boldsymbol{P} = {}^1\boldsymbol{T}_3\, {}^3\boldsymbol{P} \tag{2.43}$$

The overall transformation between frame {3} and frame {1} is obtained from Eqs. (2.42) and (2.43) as

$$^1\boldsymbol{T}_3 = {}^1\boldsymbol{T}_2\, {}^2\boldsymbol{T}_3 \tag{2.44}$$

It easily follows that the transformation from frame $\{i\}$ to frame {1} is

$$^1\boldsymbol{T}_i = {}^1\boldsymbol{T}_2\, {}^2\boldsymbol{T}_3 \ldots {}^j\boldsymbol{T}_{j+1} \ldots {}^{i-1}\boldsymbol{T}_i \tag{2.45}$$

or in general from frame $\{i\}$ to frame $\{j\}$, $(i > j)$

$$^j\boldsymbol{T}_i = {}^j\boldsymbol{T}_{j+1}\, {}^{j+1}\boldsymbol{T}_{j+2} \ldots {}^{i-1}\boldsymbol{T}_i$$

or

$$^j\boldsymbol{T}_i = \prod_{k=j}^{i-1} {}^k\boldsymbol{T}_{k+1} \tag{2.46}$$

That is, the individual homogeneous transformation matrices can be multiplied together to obtain composite homogeneous transformation matrix. Matrix multiplication being not commutative, the order of multiplication, in above equations is fixed and cannot be altered.

2.4 INVERTING A HOMOGENEOUS TRANSFORM

In robotic analysis, often $^i\boldsymbol{T}_j$ is required, while $^j\boldsymbol{T}_i$ is known. This is found by computing the inverse of $^j\boldsymbol{T}_i$. The inverse of the 4 × 4 transformation matrix can be computed using the conventional methods of matrix inversion. However, the homogeneous transform $\boldsymbol{T}$ can be inverted by exploiting its structure. Consider two frames, frame {1} and frame {2}, rotated and translated relative to each other as shown in Fig. 2.10. Knowing ${}^1\boldsymbol{T}_2$, its inverse ${}^2\boldsymbol{T}_1$ is to be found. Being inverse of each other, these homogeneous transform matrices are related as

$$ {}^1\boldsymbol{T}_2 = ({}^2\boldsymbol{T}_1)^{-1} \text{ and } {}^1\boldsymbol{T}_2\, {}^2\boldsymbol{T}_1 = \boldsymbol{I} $$

where $\boldsymbol{I}$ is a 4 × 4 identity matrix.

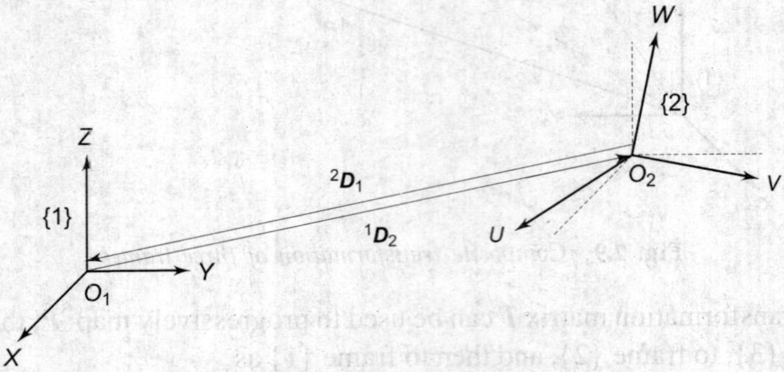

Fig. 2.10 *Inverting a homogeneous transform*

Homogenous transforms ${}^1\boldsymbol{T}_2$ and ${}^2\boldsymbol{T}_1$ can be written in partitioned form from Eq. (2.29) as

$$ {}^1\boldsymbol{T}_2 = \left[\begin{array}{ccc|c} & {}^1\boldsymbol{R}_2 & & {}^1\boldsymbol{D}_2 \\ \hline 0 & 0 & 0 & 1 \end{array}\right] \tag{2.47} $$

and

$$ {}^2\boldsymbol{T}_1 = \left[\begin{array}{ccc|c} & {}^2\boldsymbol{R}_1 & & {}^2\boldsymbol{D}_1 \\ \hline 0 & 0 & 0 & 1 \end{array}\right] \tag{2.48} $$

The rotation sub-matrix $\boldsymbol{R}$ has the property ${}^2\boldsymbol{R}_1 = {}^1\boldsymbol{R}_2^T$ (Eq. 2.13). Therefore, the mapping of a point $\boldsymbol{P}$ from frame {2} to frame {1}, is

$$ {}^1\boldsymbol{P} = {}^1\boldsymbol{D}_2 + {}^1\boldsymbol{R}_2\, {}^2\boldsymbol{P} \tag{2.49} $$

Premultiplying both sides by ${}^2\boldsymbol{R}_1$ gives

$$ {}^2\boldsymbol{R}_1\, {}^1\boldsymbol{P} = {}^2\boldsymbol{R}_1\, {}^1\boldsymbol{D}_2 + {}^2\boldsymbol{R}_1\, {}^1\boldsymbol{R}_2\, {}^2\boldsymbol{P} $$

As ${}^2\boldsymbol{R}_1\, {}^1\boldsymbol{R}_2 = \boldsymbol{I}$, it gives

$$ {}^2\boldsymbol{R}_1\, {}^1\boldsymbol{P} = {}^2\boldsymbol{R}_1\, {}^1\boldsymbol{D}_2 + {}^2\boldsymbol{P} $$

or

$$ {}^2\boldsymbol{P} = {}^2\boldsymbol{R}_1\, {}^1\boldsymbol{P} - {}^2\boldsymbol{R}_1\, {}^1\boldsymbol{D}_2 \tag{2.50} $$

The mapping of a point P from frame {1} to frame {2} is

$$ {}^2\boldsymbol{P} = {}^2\boldsymbol{R}_1\, {}^1\boldsymbol{P} + {}^2\boldsymbol{D}_1 \tag{2.51} $$

Comparing Eqs. (2.50) and (2.51), gives

$$ {}^2\boldsymbol{D}_1 = -\, {}^2\boldsymbol{R}_1\, {}^1\boldsymbol{D}_2 $$

or

$$ {}^2\boldsymbol{D}_1 = -\, {}^1\boldsymbol{R}_2^T\, {}^1\boldsymbol{D}_2 \tag{2.52} $$

Substituting Eq. (2.52) in Eq. (2.48), gives

$$ {}^2\boldsymbol{T}_1 = \left[{}^1\boldsymbol{T}_2\right]^{-1} = \left[\begin{array}{ccc|c} & {}^1\boldsymbol{R}_2^T & & -{}^1\boldsymbol{R}_2^{T}\,{}^1\boldsymbol{D}_2 \\ \hline 0 & 0 & 0 & 1 \end{array}\right] \tag{2.53} $$

This gives an easy way of computing inverse of a homogeneous transform taking full advantage of the structure inherent in the transform.

2.5 FUNDAMENTAL ROTATION MATRICES

In the previous section, the background to describe the orientation of frame {2} with respect to frame {1} has been developed. These are now applied to rotation matrices in different situations. A frame {2} may be rotated about one or more of the principal axes, an arbitrary axis, or by some fixed angles relative to frame {1}. Each of these situations is discussed in this section.

2.5.1 Principal Axes Rotation

To determine the orientation of frame {2}, which is rotated about one of the three principal axes of frame {1}, consider, for example, the rotation of frame {2} with respect to frame {1} by angle θ about the *z*-axis of frame {1}, as shown in a 3-D view in Fig. 2.11 (a) and on *xy*-plane in Fig. 2.11 (b). The corresponding rotation matrix ${}^1\boldsymbol{R}_2$, known as the *fundamental rotation matrix*, is denoted by the symbol $\boldsymbol{R}_z(\theta)$ or $\boldsymbol{R}(z, \theta)$ or $\boldsymbol{R}_{z,\,\theta}$.

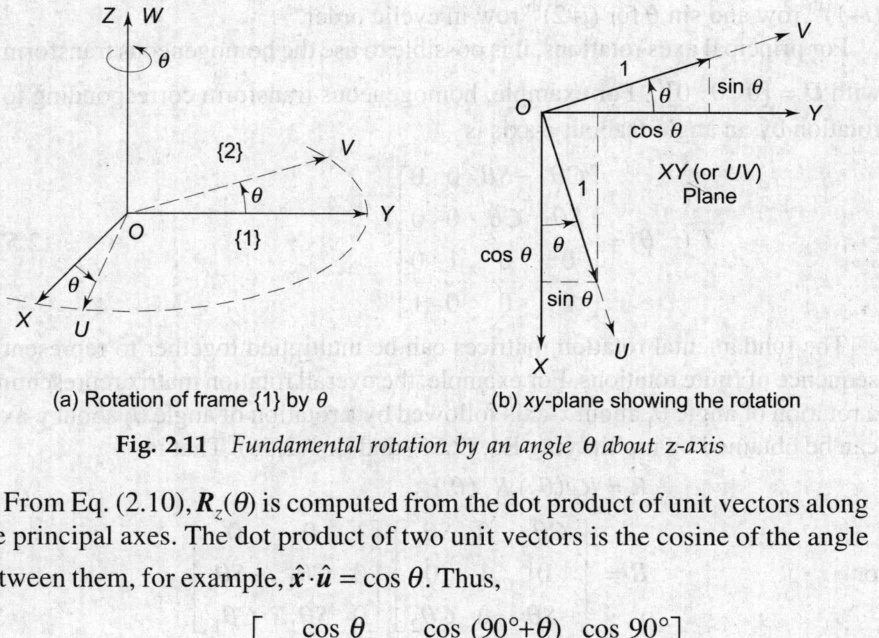

Fig. 2.11 *Fundamental rotation by an angle θ about z-axis*

From Eq. (2.10), $\boldsymbol{R}_z(\theta)$ is computed from the dot product of unit vectors along the principal axes. The dot product of two unit vectors is the cosine of the angle between them, for example, $\hat{\boldsymbol{x}} \cdot \hat{\boldsymbol{u}} = \cos\theta$. Thus,

$$ \boldsymbol{R}_z(\theta) = \begin{bmatrix} \cos\theta & \cos(90°+\theta) & \cos 90° \\ \cos(90°-\theta) & \cos\theta & \cos 90° \\ \cos 90° & \cos 90° & \cos 0° \end{bmatrix} $$

or
$$\boldsymbol{R}_z(\theta) = \begin{bmatrix} C\theta & -S\theta & 0 \\ S\theta & C\theta & 0 \\ 0 & 0 & 1 \end{bmatrix} \tag{2.54}$$
where $S\theta = \sin\theta$ and $C\theta = \cos\theta$.

Equation (2.54) is the fundamental rotation matrix for a rotation of angle θ about z-axis of the frame. Similarly, fundamental rotation matrices for rotation about x-axis and y-axis can be obtained and these are:

$$\boldsymbol{R}_x(\theta) = \begin{bmatrix} 1 & 0 & 0 \\ 0 & C\theta & -S\theta \\ 0 & S\theta & C\theta \end{bmatrix} \tag{2.55}$$

and
$$\boldsymbol{R}_y(\theta) = \begin{bmatrix} C\theta & 0 & S\theta \\ 0 & 1 & 0 \\ -S\theta & 0 & C\theta \end{bmatrix} \tag{2.56}$$

The rotation matrices $\boldsymbol{R}_x$, $\boldsymbol{R}_y$, and $\boldsymbol{R}_z$ exhibit a pattern and using this pattern these matrices can be easily written. The rotation matrix for rotation about k^{th} principal axis $\boldsymbol{R}_k(\theta)$ can be obtained as follows: The elements of ith row and i^{th} column for $i = 1, 2$, or 3 for $k = x, y$, or z respectively, of 3×3 matrix $\boldsymbol{R}_k(\theta)$ are zero except the element (i, i), which is 1. The other two diagonal elements are $\cos\theta$. The remaining two off-diagonal elements are $\pm\sin\theta$, with $-\sin\theta$ for $(i+1)^{\text{th}}$ row and $\sin\theta$ for $(i+2)^{\text{th}}$ row in cyclic order.

For principal axes rotations, it is possible to use the homogeneous transform $\boldsymbol{T}$ with $\boldsymbol{D} = [0 \ \ 0 \ \ 0]^T$. For example, homogeneous transform corresponding to a rotation by an angle θ about z-axis is

$$\boldsymbol{T}(z, \theta) = \left[\begin{array}{ccc:c} C\theta & -S\theta & 0 & 0 \\ S\theta & C\theta & 0 & 0 \\ 0 & 0 & 1 & 0 \\ \hdashline 0 & 0 & 0 & 1 \end{array}\right] \tag{2.57}$$

The fundamental rotation matrices can be multiplied together to represent a sequence of finite rotations. For example, the overall rotation matrix representing a rotation of angle θ_1 about x-axis followed by a rotation of angle θ_2 about y-axis can be obtained by multiplying Eq. (2.55) and Eq. (2.56). That is,

$$\boldsymbol{R} = R_y(\theta_2)\, R_x(\theta_1)$$

or
$$\boldsymbol{R} = \begin{bmatrix} C\theta_2 & 0 & S\theta_2 \\ 0 & 1 & 0 \\ -S\theta_2 & 0 & C\theta_2 \end{bmatrix} \begin{bmatrix} 1 & 0 & 0 \\ 0 & C\theta_1 & -S\theta_1 \\ 0 & S\theta_1 & C\theta_1 \end{bmatrix}$$

or
$$\boldsymbol{R} = \begin{bmatrix} C_2 & S_1S_2 & C_1S_2 \\ 0 & C_1 & -S_1 \\ -S_2 & S_1C_2 & C_1C_2 \end{bmatrix} \tag{2.58}$$

where $C_i = C\theta_i = \cos\theta_i$ and $S_i = S\theta_i = \sin\theta_i$.

It is important to note the sequence of multiplication of **R** matrices. A different sequence may not give the same result and obviously will not correspond to same orientation of the rotated frame. This is because the matrix product is not commutative. In view of this, it can be concluded that two rotations in general do not result in same orientation and the resultant rotation matrix depends on the order of rotations.

Another significant variable is how the rotations are performed. There are two alternatives:

(i) to perform successive rotations about the principal axes of the fixed frame.

(ii) to perform successive rotations about the current principal axes of a moving frame.

The successive rotations in either case, in general, do not produce identical results.

Figure 2.12 shows the effect of two successive rotations of 90° to an object about the principal axes of the fixed frame. It is observed that the final orientation of the object is different when same two rotations are made but the order of rotations is changed.

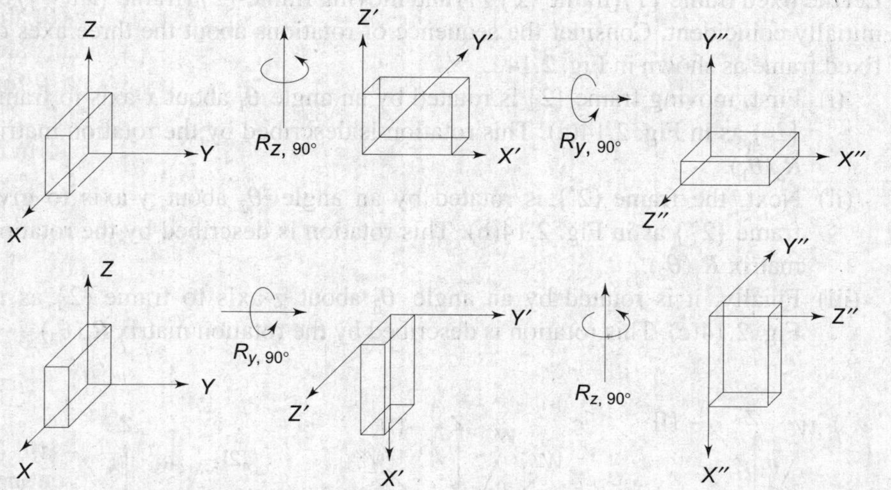

Fig. 2.12 *Effect of order of rotations of a cuboid about principal axes of a fixed frame*

Similarly, the order of rotations about the principal axes of the moving frame also produces different final orientation of the object. This is illustrated in Fig. 2.13.

The representation of orientation of rotated frames for different types of rotations is discussed next.

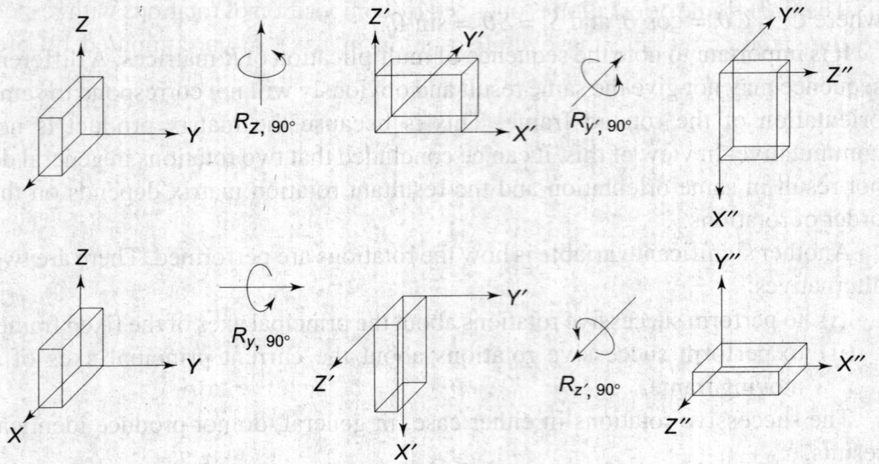

Fig. 2.13 *Effect of order of rotations of a cuboid about axes of the moving frame*

2.5.2 Fixed Angle Representation

Let the fixed frame {1}(frame $\{x\,y\,z\}$) and moving frame {2} (frame $\{u\,v\,w\}$) be initially coincident. Consider the sequence of rotations about the three axes of fixed frame as shown in Fig. 2.14.

(i) First, moving frame {2} is rotated by an angle θ_1 about x-axis to frame {2′} as in Fig. 2.14(a). This rotation is described by the rotation matrix $\boldsymbol{R}_x(\theta_1)$.

(ii) Next, the frame {2′} is rotated by an angle θ_2 about y-axis to give frame {2″} as in Fig. 2.14(b). This rotation is described by the rotation matrix $\boldsymbol{R}_y(\theta_2)$.

(iii) Finally, it is rotated by an angle θ_3 about z-axis to frame {2} as in Fig. 2.14(c). This rotation is described by the rotation matrix $\boldsymbol{R}_z(\theta_3)$.

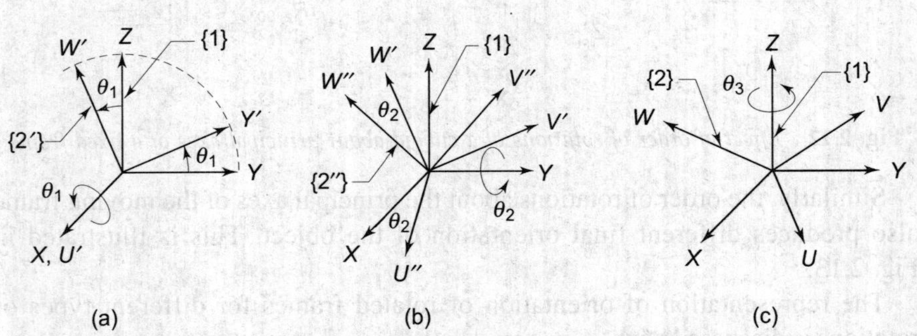

Fig. 2.14 *Three rotations of θ_1, θ_2, and θ_3 about fixed axes*

This convention for specifying orientation is known as *fixed angle representation* because each rotation is specified about an axis of fixed reference frame. The above three rotations are referred as *XYZ-fixed angle* rotations.

The final frame orientation is obtained by composition of rotations with respect to the fixed frame and the overall rotation matrix ${}^1\boldsymbol{R}_2$ is computed by pre-multiplication of the matrices of elementary rotations, that is,

$$\boldsymbol{R}_{xyz}(\theta_3\,\theta_2\,\theta_1) = {}^1\boldsymbol{R}_2 = \boldsymbol{R}_z(\theta_3)\boldsymbol{R}_y(\theta_2)\boldsymbol{R}_x(\theta_1) \quad \text{(rotation ordering right to left)} \tag{2.59}$$

Substituting the results of Eqs. (2.54)–(2.56) in Eq. (2.59) for fixed angle rotations, the final rotation matrix is

$$\boldsymbol{R}_{xyz}(\theta_3\,\theta_2\,\theta_1) = \begin{bmatrix} C_3 & -S_3 & 0 \\ S_3 & C_3 & 0 \\ 0 & 0 & 1 \end{bmatrix}\begin{bmatrix} C_2 & 0 & S_2 \\ 0 & 1 & 0 \\ -S_2 & 0 & C_2 \end{bmatrix}\begin{bmatrix} 1 & 0 & 0 \\ 0 & C_1 & -S_1 \\ 0 & S_1 & C_1 \end{bmatrix}$$

or
$$\boldsymbol{R}_{xyz}(\theta_3\,\theta_2\,\theta_1) = \begin{bmatrix} C_2C_3 & S_1S_2C_3 - C_1S_3 & C_1S_2C_3 + S_1S_3 \\ C_2S_3 & S_1S_2S_3 + C_1C_3 & C_1S_2S_3 - S_1C_3 \\ -S_2 & S_1C_2 & C_1C_2 \end{bmatrix} \tag{2.60}$$

The final frame orientation for any set of rotations performed about the axes of the fixed frame (e.g. *ZYX*, *ZXZ* etc.) can be obtained by multiplying the rotation matrices in a consistent order as indicated in Eq. (2.59). In fixed angle representation, order of rotations *XYZ* or *ZYX* are equivalent, that is, $\boldsymbol{R}_{xyz}(\theta_1\,\theta_2\,\theta_3) = \boldsymbol{R}_{zyx}(\theta_1\,\theta_2\,\theta_3)$.

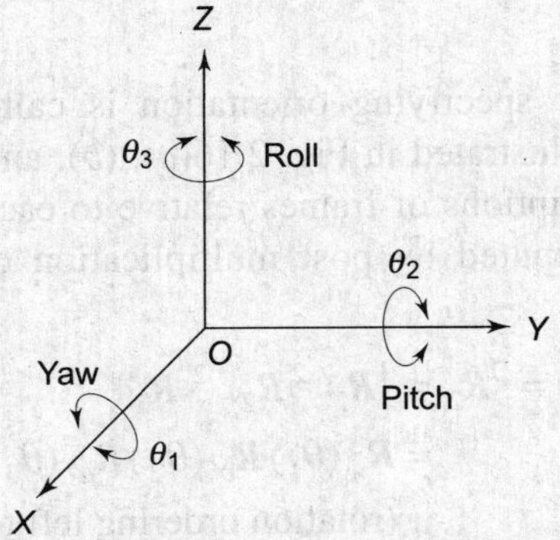

Fig. 2.15 *Representation of roll, pitch, and yaw (RPY) rotations*

The three rotations about the three fixed principal axes in fixed angle rotation produce the motions, which are also known as *roll*, *pitch,* and *yaw* motions, as shown in Fig. 2.15. The *XYZ*-fixed angle transformation in Eq. (2.60) is equivalent to *roll-pitch-yaw* (RPY) transformation.

2.5.3 Euler Angle Representations

The moving frame, instead of rotating about the principal axes of the fixed frame, can rotate about its own principal axes. Consider alternate rotations of frame {2} with respect to frame {1}, as shown in Fig. 2.16, starting from the position when the two frames are coincident.

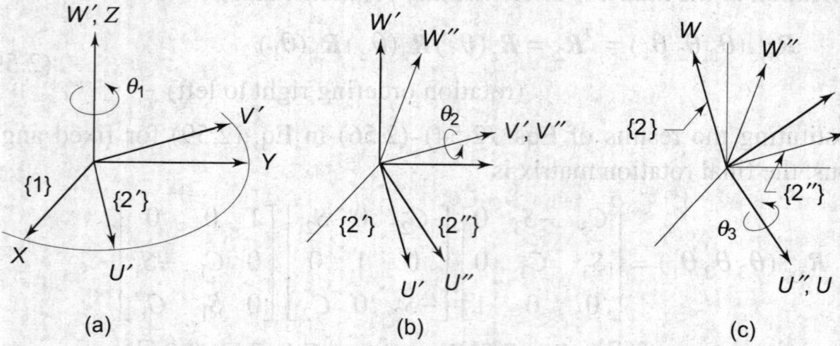

Fig. 2.16 *Euler angle representation for three rotations of θ_1, θ_2, and θ_3*

(i) To begin with, frame {2} is rotated by an angle θ_1 about its w-axis coincident with z-axis of frame {1}. The rotated frame is now {2′} and the rotation is described by the rotation matrix $\boldsymbol{R}_w(\theta_1)$.

(ii) Next, moving frame {2′} is rotated by an angle θ_2 about v'-axis, the rotated v-axis to frame {2″}. This rotation is described by the rotation matrix $\boldsymbol{R}_{v'}(\theta_2)$.

(iii) Finally, frame {2″} is rotated by an angle θ_3 about its u''-axis, the rotated u-axis to give frame {2}. This rotation is described by the rotation $\boldsymbol{R}_{u''}(\theta_3)$.

This convention for specifying orientation is called *WVU-Euler angle representation* and is illustrated in Fig. 2.16(*a*), (*b*), and (*c*). Viewing each of these rotations as descriptions of frames relative to each other, the equivalent rotation matrix is computed by post multiplication of the matrices of the elementary rotations as

$$\begin{aligned}\boldsymbol{R}_{wvu}(\theta_1\theta_2\theta_3) &= {}^1\boldsymbol{R}_2 = {}^1\boldsymbol{R}_{2'}\;{}^{2'}\boldsymbol{R}_{2''}\;{}^{2''}\boldsymbol{R}_2 \\ &= \boldsymbol{R}_w(\theta_1)\boldsymbol{R}_{v'}(\theta_2)\boldsymbol{R}_{u''}(\theta_3) \qquad (2.61)\end{aligned}$$

(rotation ordering left to right)

The rotations are performed about the current axes of the moving frame {*uvw*}. Using the results of Eqs. (2.54)–(2.56), the resulting frame orientation or the rotation matrix is

$$\boldsymbol{R}_{wvu}(\theta_1\theta_2\theta_3) = \begin{bmatrix} C_1 & -S_1 & 0 \\ S_1 & C_1 & 0 \\ 0 & 0 & 1 \end{bmatrix}\begin{bmatrix} C_2 & 0 & S_2 \\ 0 & 1 & 0 \\ -S_2 & 0 & C_2 \end{bmatrix}\begin{bmatrix} 1 & 0 & 0 \\ 0 & C_3 & -S_3 \\ 0 & S_3 & C_3 \end{bmatrix}$$

or $$\boldsymbol{R}_{wvu}(\theta_1\theta_2\theta_3) = \begin{bmatrix} C_2C_3 & S_1S_2C_3 - C_1S_3 & C_1S_2C_3 + S_1S_3 \\ C_2S_3 & S_1S_2S_3 + C_1C_3 & C_1S_2S_3 - S_1C_3 \\ -S_2 & S_1C_2 & C_1C_2 \end{bmatrix} \qquad (2.62)$$

It is observed that this result is exactly same as that obtained for fixed angle representation, Eq. (2.60), but the rotations about the fixed axes were performed in opposite order. In general, three rotations performed about fixed axes give the same final orientation as obtained by the same three rotations performed in the opposite order about the moving axes. Hence,

$$\boldsymbol{R}_{xyz}(\theta_3\,\theta_2\,\theta_1) = \boldsymbol{R}_{wvu}(\theta_1\theta_2\theta_3) = \boldsymbol{R}_{zyx}(\theta_1\,\theta_2\,\theta_3) \tag{2.63}$$

Another most widely used Euler angle representation consists of the so called *ZYZ* representation for rotations about the axes of the current frame. The sequences of elementary rotations corresponding to this representation are:

(i) A rotation by angle θ_1 about the *w*-axis (or *z*-axis of the fixed frame), that is, $\boldsymbol{R}_w(\theta_1)$.

(ii) The second rotation by angle θ_2 about the rotated *v*-axis, that is $\boldsymbol{R}_{v'}(\theta_2)$. These two rotations are same as the previous case in Fig. 2.13.

(iii) Finally, a rotation of angle θ_3 about the rotated *w*-axis, that is $\boldsymbol{R}_{w''}(\theta_3)$.

The resulting rotation matrix is

$$\boldsymbol{R}_{wvw}(\theta_1\theta_2\theta_3) = {}^1\boldsymbol{R}_2 = \boldsymbol{R}_w(\theta_1)\boldsymbol{R}_{v'}(\theta_2)\boldsymbol{R}_{w''}(\theta_3)$$

$$= \begin{bmatrix} C_1 & -S_1 & 0 \\ S_1 & C_1 & 0 \\ 0 & 0 & 1 \end{bmatrix} \begin{bmatrix} C_2 & 0 & S_2 \\ 0 & 1 & 0 \\ -S_2 & 0 & C_2 \end{bmatrix} \begin{bmatrix} C_3 & -S_3 & 0 \\ S_3 & C_3 & 0 \\ 0 & 0 & 1 \end{bmatrix} \tag{2.64}$$

$$= \begin{bmatrix} C_1C_2C_3 - S_1S_3 & -C_1C_2S_3 - S_1C_3 & C_1S_2 \\ S_1C_2C_3 + C_1S_3 & -S_1C_2S_3 + C_1C_3 & S_1S_2 \\ -S_2C_3 & S_2S_3 & C_2 \end{bmatrix}$$

The above Euler angle rotation matrix can also be obtained by rotations about the fixed frame as: a rotation by angle θ_3 about *z*-axis followed by a rotation by angle θ_2 about *y*-axis and finally a rotation angle θ_1 again about *z*-axis. The reader should verify this.

In all, twelve distinct sets of Euler angles and twelve sets of fixed angles are possible, with regard to sequence of elementary rotations. Other alternative Euler angle representations are also in vogue. For each of these, the rotation matrix can be found on similar lines.

2.5.4 Equivalent Angle Axis Representation

A third representation of orientation is by a single rotation about an arbitrary axis. A coordinate frame can be rotated about an arbitrary axis *k* passing through the origin of fixed reference frame {1}. The rotation matrix for this case is obtained by viewing the rotation as a sequence of rotations of frame {2} (along with *k*-axis) about the principal axes of frame {1}.

Consider frame {2}, initially coincident with frame {1}. Frame {2} is rotated by an angle θ about *k*-axis, in frame {1}, as shown in Fig. 2.17. The rotation of frame {2} is decomposed into rotations about the principal axes of frame {1}.

First, suitable rotations are made about the principal axes of frame {1} so as to align the axis k with x-axis. Next, the rotation of angle θ is made about the k-axis (which is coincident with x-axis). Then, by reverse rotations about the axes of frame {1}, k-axis is returned to its original location.

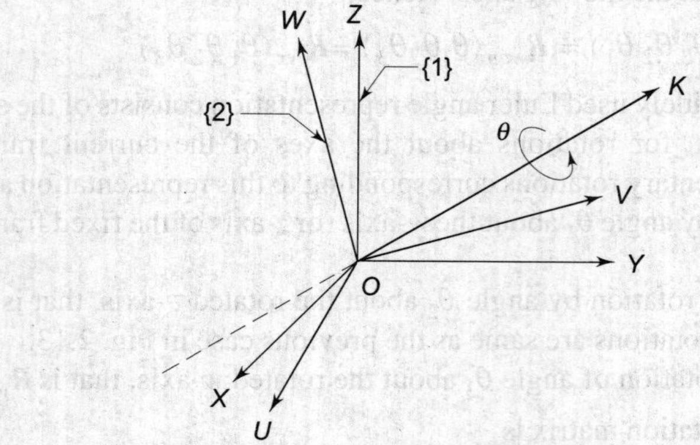

Fig. 2.17 *Equivalent angle axis representation*

These rotations are illustrated with the help of a vector $\boldsymbol{P}$, initially in the direction of k-axis, in Fig. 2.18.

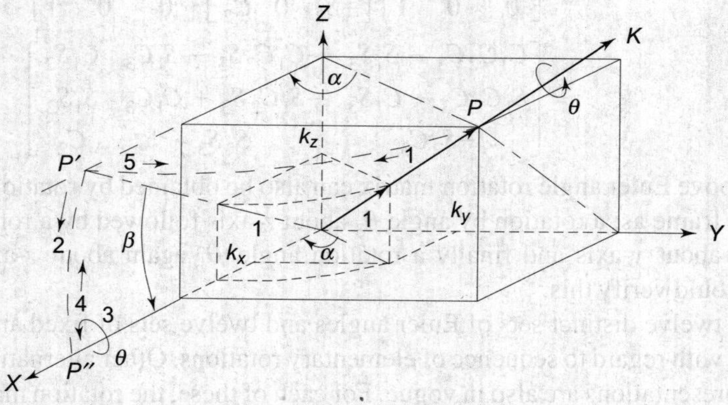

Fig. 2.18 *Rotations of frame about k-axis*

First, rotate the vector $\boldsymbol{P}$ (along with axis k and frame {2} of Fig. 2.17) by an angle $-\alpha$ about z-axis such that this rotation causes the axis k to lie in xz-plane of frame {1}. This rotation, marked as "1" in Fig. 2.18 and is written as

$$^1\boldsymbol{R}_2 = \boldsymbol{R}_z(-\alpha) \tag{2.65}$$

Next, vector $\boldsymbol{P}$ (along with rotated axis k) is rotated about y-axis by an angle β so that axis k aligns with x-axis, rotation "2". At the end of this rotation,

$$^1\boldsymbol{R}_2 = \boldsymbol{R}_y(\beta)\,\boldsymbol{R}_z(-\alpha) \tag{2.66}$$

Now a rotation "3" of angle θ about the rotated axis k, which is rotation about x-axis, is made. The resulting rotation matrix is then

$$ {}^1\boldsymbol{R}_2 = \boldsymbol{R}_x(\theta)\, \boldsymbol{R}_y(\beta)\, \boldsymbol{R}_x(-\alpha) \tag{2.67} $$

Finally, the rotations "4" and "5" of $-\beta$ and α are made about y- and z-axes, respectively, in the opposite sense and reverse order so as to restore the k-axis to its original position leaving frame {2} in the rotated position. This gives

$$ {}^1\boldsymbol{R}_2 = \boldsymbol{R}_k(\theta) = \boldsymbol{R}_z(\alpha)\, \boldsymbol{R}_y(-\beta)\, \boldsymbol{R}_x(\theta)\, \boldsymbol{R}_y(\beta)\, \boldsymbol{R}_z(-\alpha) \tag{2.68} $$

Substituting the values for the fundamental rotation matrices from Eqs. (2.54)–(2.56) gives,

$$ {}^1\boldsymbol{R}_2 = \begin{bmatrix} C\alpha & -S\alpha & 0 \\ S\alpha & C\alpha & 0 \\ 0 & 0 & 1 \end{bmatrix} \begin{bmatrix} C\beta & 0 & -S\beta \\ 0 & 1 & 0 \\ S\beta & 0 & C\beta \end{bmatrix} \begin{bmatrix} 1 & 0 & 0 \\ 0 & C\theta & -S\theta \\ 0 & S\theta & C\theta \end{bmatrix} \begin{bmatrix} C\beta & 0 & S\beta \\ 0 & 1 & 0 \\ -S\beta & 0 & C\beta \end{bmatrix} \begin{bmatrix} C\alpha & S\alpha & 0 \\ -S\alpha & C\alpha & 0 \\ 0 & 0 & 1 \end{bmatrix} \tag{2.69} $$

The angles α and β can be eliminated from Eq. (2.69) using the geometry. From Fig. 2.18, following are easily observed for the unit vector $\hat{\boldsymbol{k}} = \begin{bmatrix} k_x & k_y & k_z \end{bmatrix}^T$

$$ \sin\alpha = \frac{k_y}{\sqrt{k_x{}^2 + k_y{}^2}},\ \cos\alpha = \frac{k_x}{\sqrt{k_x{}^2 + k_y{}^2}} $$

$$ \sin\beta = k_z,\ \cos\beta = \sqrt{k_x{}^2 + k_y{}^2} \tag{2.70} $$

Substituting these in Eq. (2.69) and simplifying gives

$$ {}^1\boldsymbol{R}_2 = \boldsymbol{R}_k(\theta) = \begin{bmatrix} k_x^2 V\theta + C\theta & k_x k_y V\theta - k_z S\theta & k_x k_z V\theta + k_y S\theta \\ k_x k_y V\theta + k_z S\theta & k_y^2 V\theta + C\theta & k_y k_z V\theta - k_x S\theta \\ k_x k_z V\theta - k_y S\theta & k_y k_z V\theta + k_x S\theta & k_z^2 V\theta + C\theta \end{bmatrix} \tag{2.71} $$

where k_x, k_y, k_z are the projections of a unit vector $\hat{\boldsymbol{k}}$ on frame $\{xyz\}$, and $V\theta = 1 - \cos\theta$.

This is an important rotation matrix and must be thoroughly understood. The principal axes fundamental rotations can be obtained from Eq. (2.71). For example, if k-axis is aligned with z-axis, $\boldsymbol{R}_k(\theta)$ becomes $\boldsymbol{R}_z(\theta)$ with $k_x = k_y = 0$ and $k_z = 1$. Substituting these k_x, k_y, k_z in Eq. (2.71) gives

$$ \boldsymbol{R}_k(\theta) = \boldsymbol{R}_z(\theta) = \begin{bmatrix} C\theta & -S\theta & 0 \\ S\theta & C\theta & 0 \\ 0 & 0 & 1 \end{bmatrix} \tag{2.72} $$

which is same as Eq. (2.54) derived before.

Note that any combination of rotations about the principal axes of a coordinate frame is always equivalent to a single rotation by some angle θ about some arbitrary axis k. To find the direction $\boldsymbol{K}$, consider the general rotational transformation matrix $\boldsymbol{R}$ of Eq. (2.32). It is required to determine θ and $\hat{\boldsymbol{k}}$. Equating Eqs. (2.32) and (2.71), one gets nine equations in four unknowns k_x, k_y, k_z, and θ, which can be easily computed (see Review Question 2.21).

The concepts of transformation developed in this chapter will be required for analysis of manipulators for various aspects of robotics covered in rest of the chapters. To enhance the understanding, several examples are worked out involving different concepts of transformations.

SOLVED EXAMPLES

Example 2.1 *Use of Transformations*

The coordinates of point P in frame {1} are $[3.0 \;\; 2.0 \;\; 1.0]^T$. The position vector P is rotated about the z-axis by 45°. Find the coordinates of point Q, the new position of point P.

Solution The 45° rotation of P about the z-axis of frame {1} from Eq. (2.54) is

$$\boldsymbol{R}_z(45°) = \begin{bmatrix} \cos 45° & -\sin 45° & 0 \\ \sin 45° & \cos 45° & 0 \\ 0 & 0 & 1 \end{bmatrix} = \begin{bmatrix} 0.707 & -0.707 & 0 \\ 0.707 & 0.707 & 0 \\ 0 & 0 & 1 \end{bmatrix} \tag{2.73}$$

For the rotation of vectors, Eq. (2.37) gives

$${}^1\boldsymbol{Q} = \boldsymbol{R}_z(45°)\,{}^1\boldsymbol{P}$$

Substituting values of $\boldsymbol{R}_z$ and ${}^1\boldsymbol{P}$,

$${}^1\boldsymbol{Q} = \begin{bmatrix} 0.707 & -0.707 & 0 \\ 0.707 & 0.707 & 0 \\ 0 & 0 & 1 \end{bmatrix} \begin{bmatrix} 3.0 \\ 2.0 \\ 1.0 \end{bmatrix} = \begin{bmatrix} 0.707 \\ 3.535 \\ 1 \end{bmatrix} \tag{2.74}$$

Thus, the coordinates of the new point Q relative to frame {1} are $[0.707 \;\; 3.535 \;\; 1.0]^T$ or the new position vector is

$$\boldsymbol{Q} = [0.707 \;\; 3.535 \;\; 1.0]^T \tag{2.75}$$

Example 2.2 *Homogeneous Transformation*

Frame {2} is rotated with respect to frame {1} about the x-axis by an angle of 60°. The position of the origin of frame {2} as seen from frame {1} is ${}^1\boldsymbol{D}_2 = [7.0 \;\; 5.0 \;\; 7.0]^T$. Obtain the transformation matrix ${}^1\boldsymbol{T}_2$, which describes frame {2} relative to frame {1}. Also, find the description of point P in frame {1} if ${}^2\boldsymbol{P} = [2.0 \;\; 4.0 \;\; 6.0]^T$.

Solution The homogeneous transform matrix describing frame {2} with respect to frame {1}, Eq. (2.29), is

$$ {}^1\boldsymbol{T}_2 = \left[\begin{array}{ccc|c} \multicolumn{3}{c|}{{}^1\boldsymbol{R}_2} & {}^1\boldsymbol{D}_2 \\ \hline 0 & 0 & 0 & 1 \end{array}\right] $$

Because frame {2} is rotated relative to frame {1} about x-axis by 60°, Eq. (2.55) gives

$$ {}^1\boldsymbol{R}_2 = \begin{bmatrix} 1 & 0 & 0 \\ 0 & \cos 60^\circ & -\sin 60^\circ \\ 0 & \sin 60^\circ & \cos 60^\circ \end{bmatrix} = \begin{bmatrix} 1 & 0 & 0 \\ 0 & 0.500 & -0.866 \\ 0 & 0.866 & 0.500 \end{bmatrix} \tag{2.76} $$

Substituting ${}^1\boldsymbol{R}_2$ and ${}^1\boldsymbol{D}_2$ in the above equation

$$ {}^1\boldsymbol{T}_2 = \left[\begin{array}{ccc|c} 1 & 0 & 0 & 7.000 \\ 0 & 0.500 & -0.866 & 5.000 \\ 0 & 0.866 & 0.500 & 7.000 \\ \hline 0 & 0 & 0 & 1 \end{array}\right] \tag{2.77} $$

Given: ${}^2\boldsymbol{P} = [2.0 \quad 2.0 \quad 6.0]^T$, point P in frame {1} is given by

$$ {}^1\boldsymbol{P} = {}^1\boldsymbol{T}_2\, {}^2\boldsymbol{P} $$

Substituting values

$$ {}^1\boldsymbol{P} = \begin{bmatrix} 1 & 0 & 0 & 7.000 \\ 0 & 0.500 & -0.866 & 5.000 \\ 0 & 0.866 & 0.500 & 7.000 \\ 0 & 0 & 0 & 1 \end{bmatrix} \begin{bmatrix} 2.0 \\ 4.0 \\ 6.0 \\ 1 \end{bmatrix} = \begin{bmatrix} 9.000 \\ 1.804 \\ 13.464 \\ 1 \end{bmatrix} \tag{2.78} $$

or

$$ {}^1\boldsymbol{P} = [9.000 \quad 1.804 \quad 13.464 \quad 1]^T \tag{2.79} $$

The 3 × 1 position vector of point P in frame {1} in physical coordinates is then

$$ {}^1\boldsymbol{P} = [9.000 \quad 1.804 \quad 13.464]^T \tag{2.80} $$

Example 2.3 *Transformation of Vector and Frames*

Consider a point P in space. Determine the new location of this point after rotating it by an angle of 45° about z-axis and then translating it by –1 unit along x-axis and –2 units along z-axis. Pictorially show the transformation of the vector. What will be the equivalent frame transformation for this vector transformation? Show the transformation of frames.

Solution Figure 2.19 shows a point P and a vector from origin as ${}^1\boldsymbol{P}$ in frame {1} and its new location after the rotational and transnational transformation as ${}^1\boldsymbol{Q}$. The relation between ${}^1\boldsymbol{Q}$ and ${}^1\boldsymbol{P}$ is described by Eq. (2.26) as

$$ {}^1\boldsymbol{Q} = \boldsymbol{T}\, {}^1\boldsymbol{P} $$

where

$$ \boldsymbol{T} = \left[\begin{array}{ccc|c} \multicolumn{3}{c|}{\boldsymbol{R}(\theta)} & \boldsymbol{D} \\ \hline 0 & 0 & 0 & 1 \end{array}\right] $$

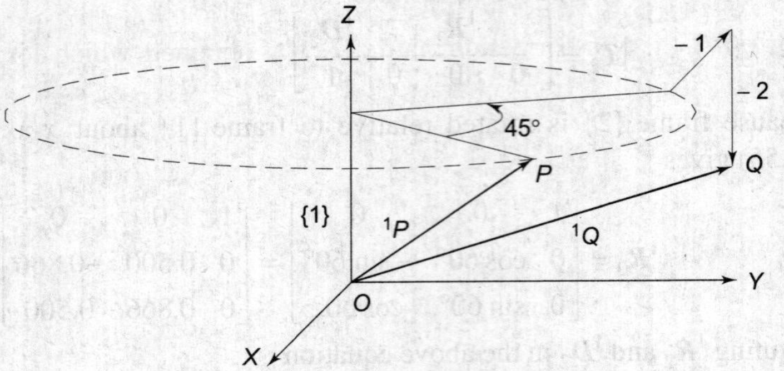

Fig. 2.19 *Transformation of point P in space*

Substituting values gives

$$^{1}\boldsymbol{Q} = \begin{bmatrix} \cos 45^{\circ} & -\sin 45^{\circ} & 0 & -1 \\ \sin 45^{\circ} & \cos 45^{\circ} & 0 & 0 \\ 0 & 0 & 1 & -2 \\ 0 & 0 & 0 & 1 \end{bmatrix} {}^{1}\boldsymbol{P} = \begin{bmatrix} 0.707 & -0.707 & 0 & -1 \\ 0.707 & 0.707 & 0 & 0 \\ 0 & 0 & 1 & -2 \\ 0 & 0 & 0 & 1 \end{bmatrix} {}^{1}\boldsymbol{P} \quad (2.81)$$

The transformation in Eq. (2.81) can be regarded as transformation of two frames {1} and {2}. Assuming frame {1} and frame {2} to be initially coincident, the final position of frame {2} is obtained by translating it by +2 units along z_1-axis (motion 1), and +1 unit along x_1-axis (motion 2) and then rotating it by an angle of –45° about z_1-axis (motion 3). The two frames are shown in Fig. 2.20.

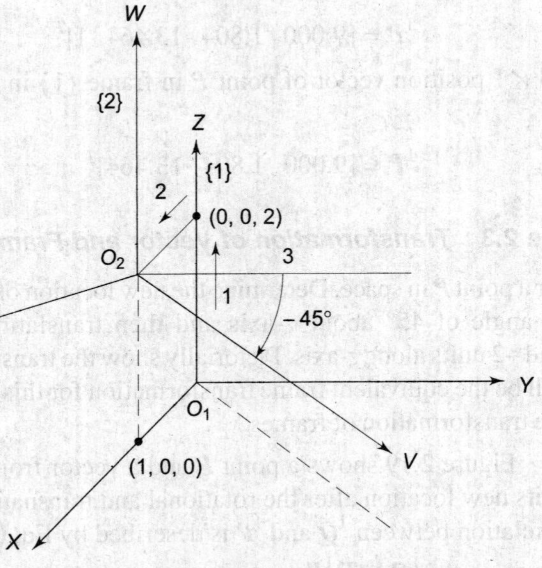

Fig. 2.20 *Transformation of frames corresponding to transformation of vectors*

Example 2.4 *Description of Frames*

In Example 2.2, the transformation matrix ${}^1\boldsymbol{T}_2$ was obtained, which describes the position and orientation of frame {2} relative to frame {1}. Using this matrix, determine the description of frame {1} relative to frame {2}.

Solution The homogeneous transformation for describing frame {1} relative to frame {2}, ${}^2\boldsymbol{T}_1$ is given by

$$ {}^2\boldsymbol{T}_1 = \left[\begin{array}{ccc|c} & {}^2\boldsymbol{R}_1 & & {}^2\boldsymbol{D}_1 \\ \hline 0 & 0 & 0 & 1 \end{array}\right] = \left[{}^1\boldsymbol{T}_2\right]^{-1} \tag{2.82} $$

The inverse of ${}^1\boldsymbol{T}_2$ is given by Eq. (2.53), that is,

$$ {}^2\boldsymbol{T}_1 = \left[\begin{array}{ccc|c} & {}^1\boldsymbol{R}_2^{\ T} & & -{}^1\boldsymbol{R}_2^{\ T}\,{}^1\boldsymbol{D}_2 \\ \hline 0 & 0 & 0 & 1 \end{array}\right] $$

From ${}^1\boldsymbol{T}_2$ in Example 2.2, Eq. (2.77),

$$ {}^1\boldsymbol{R}_2 = \begin{bmatrix} 1 & 0 & 0 \\ 0 & 0.500 & -0.866 \\ 0 & 0.866 & 0.500 \end{bmatrix} $$

and $$ {}^1\boldsymbol{D}_2 = [7.0 \quad 5.0 \quad 7.0]^T $$

Hence, $$ {}^2\boldsymbol{R}_1 = {}^1\boldsymbol{R}_2^T = \begin{bmatrix} 1 & 0 & 0 \\ 0 & 0.500 & 0.866 \\ 0 & -0.866 & 0.500 \end{bmatrix} \tag{2.83} $$

and the position of the origin of frame {1} with respect to frame {2} is given by

$$ {}^2\boldsymbol{D}_1 = -{}^1\boldsymbol{R}_2^{T}\,{}^1\boldsymbol{D}_2 $$

Substituting values

$$ {}^2\boldsymbol{D}_1 = -\begin{bmatrix} 1 & 0 & 0 \\ 0 & 0.500 & 0.866 \\ 0 & -0.866 & 0.500 \end{bmatrix} \begin{bmatrix} 7.0 \\ 5.0 \\ 7.0 \end{bmatrix} = \begin{bmatrix} -7.000 \\ -8.562 \\ 0.830 \end{bmatrix} \tag{2.84} $$

Therefore,

$$ {}^2\boldsymbol{T}_1 = \begin{bmatrix} 1 & 0 & 0 & -7.000 \\ 0 & 0.500 & 0.866 & -8.562 \\ 0 & -0.866 & 0.500 & 0.830 \\ 0 & 0 & 0 & 1 \end{bmatrix} \tag{2.85} $$

Example 2.5 *Euler Angle Rotations*

In Euler angle representation, the equivalent rotation matrix relating the two frames is specified by a set of *ZYX*-Euler angle rotations. Consider now the inverse of this problem: Given the rotation matrix ${}^1\boldsymbol{R}_2$, relating the orientation of frame {2} with respect to frame {1}. Determine the corresponding set of *ZYX*-Euler angle rotations.

Solution Let the given rotation matrix which specifies the orientation of frame {2} with respect to frame {1} be

$$ {}^1\boldsymbol{R}_2 = \begin{bmatrix} r_{11} & r_{12} & r_{13} \\ r_{21} & r_{22} & r_{23} \\ r_{31} & r_{32} & r_{33} \end{bmatrix} \tag{2.86}$$

The equivalent rotation matrix for a set of *ZYX*-Euler angle rotation ($\theta_1\ \theta_2\ \theta_3$) is given by Eq. (2.62),

$$ {}^1\boldsymbol{R}_2 = \begin{bmatrix} C_2C_3 & S_1S_2C_3 - C_1S_3 & S_1S_3 + C_1S_2C_3 \\ C_2S_3 & S_1S_2S_3 + C_1C_3 & -S_1C_3 + C_1S_2S_3 \\ -S_2 & S_1C_2 & C_1C_2 \end{bmatrix} \tag{2.87}$$

Equating the corresponding elements of these matrices gives nine equations in three independent variables, θ_1, θ_2, θ_3. Apart from the redundancy in equations, additional complication is that these are transcendental in nature.

Equating elements (1,1) and (2,1) in Eq (2.86) with corresponding elements in Eq. (2.87) gives,

$$C_2C_3 = r_{11} \text{ and } C_2S_3 = r_{21}$$

Squaring and adding gives

$$C_2 = \cos\theta_2 = \pm\sqrt{r_{11}^2 + r_{21}^2} \tag{2.88}$$

Combining with the element (3,1), ($-S_2 = r_{31}$), the angle θ_2 is computed as

$$\tan\theta_2 = \frac{S_2}{C_2}$$

which gives

$$\theta_2 = A\tan2(-r_{31}, \pm\sqrt{r_{11}^2 + r_{21}^2}) \tag{2.89}$$

where $Atan2(a, b)$ is a two-argument *arc tangent* function (see Appendix **A**).

The solution for θ_1 and θ_3 depends on value of θ_2. Here, two cases arise which are worked out as follows:

Case 1 $\theta_2 \neq 90°$

From the elements (1,1) and (2,1) in Eqs. (2.86) and (2.87), θ_3 is obtained as

$$\theta_3 = A\tan2\left(\frac{r_{21}}{C_2}, \frac{r_{11}}{C_2}\right) \tag{2.90}$$

and from elements (3,2) and (3,3), θ_1 is

$$\theta_1 = A\tan2\left(\frac{r_{32}}{C_2}, \frac{r_{33}}{C_2}\right) \tag{2.91}$$

Note that there is one set of solution corresponding to each value of θ_2.

Case 2 $\theta_2 = \pm 90°$

For $\theta_2 = \pm 90°$, the solution obtained in Case 1 degenerates. However, it is possible to find only the sum or difference of θ_3 and θ_1. Comparing elements (1, 2) and (2, 2)

$$r_{12} = S_1S_2C_3 - C_1S_3$$

and $$r_{22} = S_1S_2S_3 + C_1C_3 \tag{2.92}$$

If $\theta_2 = +90°$, these equations reduce to

$$r_{12} = \sin(\theta_1 - \theta_3)$$

$$r_{22} = \cos(\theta_1 - \theta_3)$$

and $$\theta_1 - \theta_3 = \text{Atan2}(r_{12}, r_{22}) \tag{2.93}$$

Choosing $\theta_3 = 0°$ gives the particular solution

$$\theta_2 = 90°;\ \theta_3 = 0° \text{ and } \theta_1 = \text{Atan2}(r_{12}, r_{22}) \tag{2.94}$$

With $\theta_2 = -90°$, the solution is

$$r_{12} = -\sin(\theta_1 + \theta_2)$$

$$r_{22} = \cos(\theta_1 + \theta_2)$$

and $$\theta_1 + \theta_2 = \text{Atan2}(-r_{12}, r_{22}) \tag{2.95}$$

Choosing $\theta_2 = 0°$ gives the particular solution

$$\theta_2 = -90°;\ \theta_3 = 0° \text{ and } \theta_1 = \text{Atan2}(-r_{12}, r_{22}) \tag{2.96}$$

Example 2.6 *Multiple Rotations of a Frame*

Frame {1} and {2} have coincident origins and differ only in orientation. Frame {2} is initially coincident with frame {1}. Certain rotations are carried out about the axis of the fixed frame {1}: first rotation about *x*-axis by 45° then about *y*-axis by 30° and finally about *x*-axis by 60°. Obtain the equivalent rotation matrix ${}^1\boldsymbol{R}_2$.

Solution Rotations are in order *X*-*Y*-*X* about the fixed axes; hence, it is a case of fixed angle representation. Therefore,

$${}^1\boldsymbol{R}_2 = \boldsymbol{R}_x(60°)\,\boldsymbol{R}_y(30°)\,\boldsymbol{R}_x(45°) \tag{2.97}$$

or

$${}^1\boldsymbol{R}_2 = \begin{bmatrix} 1 & 0 & 0 \\ 0 & \cos 60° & -\sin 60° \\ 0 & \sin 60° & \cos 60° \end{bmatrix} \begin{bmatrix} \cos 30° & 0 & \sin 30° \\ 0 & 1 & 0 \\ -\sin 30° & 0 & \cos 30° \end{bmatrix} \begin{bmatrix} 1 & 0 & 0 \\ 0 & \cos 45° & -\sin 45° \\ 0 & \sin 45° & \cos 45° \end{bmatrix}$$

On multiplication,

$${}^1\boldsymbol{R}_2 = \begin{bmatrix} 0.866 & 0.354 & 0.354 \\ 0.433 & -0.177 & -0.884 \\ -0.25 & 0.919 & 0.306 \end{bmatrix} \tag{2.98}$$

The reader must verify that the same orientation could have been obtained by performing the same rotations about the moved *xyx*-axes of the moving frame but in the opposite order. This convention is also referred as *XYX*-Euler angle representation.

Example 2.7 *Equivalent Axis Representation*

Two coordinate frames {1} and {2} are initially coincident. Frame {2} is rotated by 45° about a vector $\hat{\boldsymbol{k}} = [0.5\ 0.866\ 0.707]^T$ passing through the origin. Determine the new description of frame {2}.

Solution Substituting $\hat{\boldsymbol{k}}$ and $\theta = 45°$ in Eq. (2.71) yields the rotation matrix ${}^1\boldsymbol{R}_2$ for rotation about *k*-axis as

$$\boldsymbol{R}_k(45°) = {}^1\boldsymbol{R}_2 = \begin{bmatrix} 0.780 & -0.373 & 0.716 \\ 0.627 & 0.927 & -0.174 \\ -0.509 & 0.533 & 0.854 \end{bmatrix} \tag{2.99}$$

Since there is no translation of frame {2}, the position vector is $\boldsymbol{D} = [0\,0\,0\,1]^T$ and the description of frame {2} with respect to frame {1} is:

$${}^1\boldsymbol{T}_2 = \begin{bmatrix} 0.780 & -0.373 & 0.716 & 0 \\ 0.627 & 0.927 & -0.174 & 0 \\ -0.509 & 0.533 & 0.854 & 0 \\ 0 & 0 & 0 & 1 \end{bmatrix} \tag{2.100}$$

Example 2.8 *Screw Transformation*

The moving coordinate frame {*u v w*} undergoes a "screw transformation", that is, it is translated by 4 units along *z*-axis and rotated by an angle of 180° about same axis of stationary reference coordinate frame {*x y z*}. Coordinates *XYZ* and *UVW* are initially coincident.

(a) Obtain the homogeneous transformation matrix for the screw transformation.
(b) Show the coordinate frames before and after the transformation.
(c) If the order of transformations is reversed, will the homogeneous screw transformation matrix change?

Solution (a) In screw transformation the moving frame is translated and rotated about same axis. The overall transformation matrix for the given situation is

$$\boldsymbol{T} = \boldsymbol{T}(z, \pi)\, \boldsymbol{T}\,(0, 0, 4)$$

or

$$\boldsymbol{T} = \begin{bmatrix} -1 & 0 & 0 & 0 \\ 0 & -1 & 0 & 0 \\ 0 & 0 & 1 & 0 \\ 0 & 0 & 0 & 1 \end{bmatrix} \begin{bmatrix} 1 & 0 & 0 & 0 \\ 0 & 1 & 0 & 0 \\ 0 & 0 & 1 & 4 \\ 0 & 0 & 0 & 1 \end{bmatrix} = \begin{bmatrix} -1 & 0 & 0 & 0 \\ 0 & -1 & 0 & 0 \\ 0 & 0 & 1 & 4 \\ 0 & 0 & 0 & 1 \end{bmatrix} \tag{2.101}$$

Assume a unit vector along *y*-axis. This is given by $\hat{\boldsymbol{p}} = [0\ 1\ 0]^T$. This vector moves with the moving frame and undergoes the two transformations specified. Its position after given translation and rotation will be

$$P' = TP = \begin{bmatrix} -1 & 0 & 0 & 0 \\ 0 & -1 & 0 & 0 \\ 0 & 0 & 1 & 4 \\ 0 & 0 & 0 & 1 \end{bmatrix} \begin{bmatrix} 0 \\ 1 \\ 0 \\ 1 \end{bmatrix} = \begin{bmatrix} 0 \\ -1 \\ 4 \\ 1 \end{bmatrix} \tag{2.102}$$

(b) The initial and final positions of two frames and point P are is shown in Fig. 2.21.

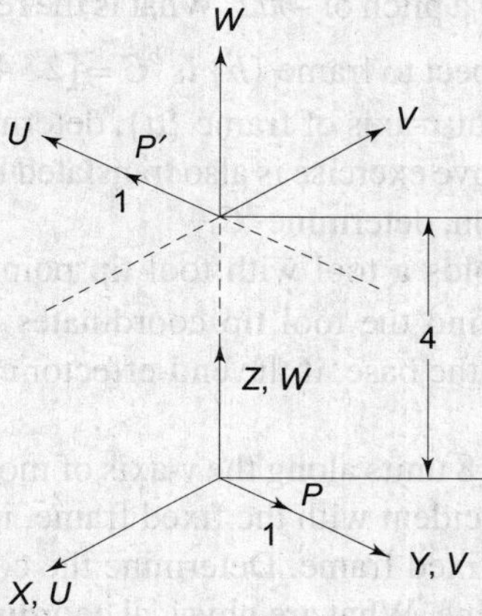

Fig. 2.21 *Screw transformation of point P.*

(c) If the order of transformations is reversed, that is, rotation followed by translation, the overall transformation matrix will not change. This can be easily verified by the reader and is the property of screw transformation.

EXERCISES

2.1 The coordinates of point P with respect to a moving coordinate frame are given as $\boldsymbol{P} = [0.5 \quad 0.8 \quad 1.3 \quad 1]^T$. What are the coordinates of P with respect to fixed coordinate frame, if the moving frame is rotated by 90° about z-axis of the fixed frame?

2.2 Determine the rotation matrix for a rotation of 45° about y-axis, followed by a rotation of 120° about z-axis, and a final rotation of 90° about x-axis.

2.3 A vector $\boldsymbol{P} = 3\hat{\boldsymbol{i}} - 2\hat{\boldsymbol{j}} + 5\hat{\boldsymbol{k}}$ is first rotated by 90° about x-axis, then by 90° about z-axis. Finally, it is translated by $-3\hat{\boldsymbol{i}} + 2\hat{\boldsymbol{j}} - 5\hat{\boldsymbol{k}}$. Determine the new position of vector $\boldsymbol{P}$.

2.4 Find the new location of point G, initially at $\boldsymbol{G} = [3 \quad 0 \quad -1 \quad 1]^T$, if (i) it is rotated by π about z-axis and then translated by 3 units along y-axis, and (ii) it is first translated by 3 units along y-axis and then rotated by π about z-axis. Are the two locations same? Explain why the final position in two cases is same or different.

2.5 A moving frame is rotated about x-axis of the fixed coordinate frame by $\pi/6$ radians. The coordinates of a point Q in fixed coordinate frame are given by $\boldsymbol{Q} = [2 \;\; 0 \;\; 3]^T$. What will be the coordinates of a point Q with respect to the moving frame?

2.6 Show that determinant of the rotation matrix $\boldsymbol{R}$, is +1 for a right-hand coordinate system and –1 for a left-hand coordinate system.

2.7 The end-effector of a robot is rotated about fixed axes starting with a yaw of $-\pi/2$, followed by a pitch of $-\pi/2$. What is the resulting rotation matrix?

2.8 A vector $\boldsymbol{C}$ with respect to frame $\{b\}$ is ${}^b\boldsymbol{C} = [2 \;\; 4 \;\; -5]^T$. If frame $\{b\}$ is rotated by $-\pi/4$ about x-axis of frame $\{a\}$, determine ${}^a\boldsymbol{C}$.

2.9 If vector $\boldsymbol{C}$ in the above exercise is also translated by 4 units in $-y$ direction in addition to rotation, determine ${}^a\boldsymbol{C}$.

2.10 The end-effector holds a tool with tool tip point P having coordinates $\boldsymbol{P} = [0 \;\; 0 \;\; 1.2]^T$. Find the tool tip coordinates with respect to a fixed coordinate frame at the base, if the end-effector coordinates are given by Eq. (2.31).

2.11 A point Q is located 8 units along the y-axis of moving frame. The mobile frame, initially coincident with the fixed frame, is rotated by $\pi/3$ radians about the z-axis of fixed frame. Determine the coordinates of point Q in fixed coordinate frame. What are physical coordinates of point Q in fixed coordinate frame?

2.12 The end-effector of a manipulator is a gripper. The gripper is relocated from initial point $[2 \;\; 0 \;\; 4 \;\; 1]^T$ to $[4 \;\; 0 \;\; 0 \;\; 1]^T$. Determine the direction of axis k and the angle of rotation about this k-axis.

2.13 Show that a rotation by θ about axis k (Eq. (2.71)) can be used to get the fundamental rotation by choosing axis k to be axis x- or y- or z-axis, respectively.

2.14 An end-effector is rotated by 60° about an axis whose unit vector is $\hat{k} = \left[1/\sqrt{2} \;\; 1/\sqrt{2} \;\; 1 \;\; 1\right]^T$. Find the homogeneous transformation matrix representing this rotation.

2.15 The end-point of a link of a manipulator is at $\boldsymbol{P} = [2 \;\; 2 \;\; 6 \;\; 1]^T$. The link is rotated by 90° about x-axis, then by –180° about its own w-axis, and finally by –90° about its own v-axis. Find the resulting homogeneous transformation matrix and the final location of end-point.

2.16 For a rotation of 90° about z-axis followed by a rotation of 90° about y-axis followed by a rotation of 90° about x-axis, determine an equivalent k-axis of rotation and rotation angle θ about this axis.

2.17 Determine the transformation matrix $\boldsymbol{T}$ that represents a translation of a unit along x-axis, followed by a rotation of angle α about x-axis followed by a rotation of θ about the rotated z-axis.

2.18 Two frames, $\{A\}$ and $\{B\}$, are initially coincident. Frame $\{B\}$ undergoes the following four motions in sequence with respect to axes of frame $\{A\}$:

(i) A rotation of θ about z-axis,
(ii) A translation of d along z-axis,
(iii) A translation of a along x-axis, and finally
(iv) A rotation of α about x-axis.

Determine the final homogeneous transformation matrix to describe frame $\{B\}$, after the transformations, with respect to the frame $\{A\}$.

2.19 The homogeneous transformation matrices between frames $\{i\}$–$\{j\}$ and $\{i\}$–$\{k\}$ are:

$$ {}^{j}\boldsymbol{T}_i = \begin{bmatrix} 0.866 & -0.500 & 0 & 11 \\ 0.500 & 0.866 & 0 & -1 \\ 0 & 0 & 1 & 8 \\ 0 & 0 & 0 & 1 \end{bmatrix}; \ {}^{k}\boldsymbol{T}_i = \begin{bmatrix} 1 & 0 & 0 & 0 \\ 0 & 0.866 & -0.500 & 10 \\ 0 & 0.500 & 0.866 & -20 \\ 0 & 0 & 0 & 1 \end{bmatrix} \quad (2.103)$$

Determine ${}^{j}\boldsymbol{T}_k$.

2.20 Show that the inverse of the homogeneous transformation matrix with no perspective transformation, that is, if $\boldsymbol{T}$ is

$$\boldsymbol{T} = \begin{bmatrix} n_x & o_x & a_x & d_x \\ n_y & o_y & a_y & d_y \\ n_z & o_z & a_z & d_z \\ 0 & 0 & 0 & 1 \end{bmatrix} = \begin{bmatrix} \boldsymbol{n} & \boldsymbol{o} & \boldsymbol{a} & \boldsymbol{d} \\ 0 & 0 & 0 & 1 \end{bmatrix} \quad (2.104)$$

$\boldsymbol{T}^{-1}$ is given by

$$\boldsymbol{T}^{-1} = \begin{bmatrix} n_x & n_y & n_z & -\boldsymbol{d}\cdot\boldsymbol{n} \\ o_x & o_y & o_z & -\boldsymbol{d}\cdot\boldsymbol{o} \\ a_x & a_y & a_z & -\boldsymbol{d}\cdot\boldsymbol{a} \\ 0 & 0 & 0 & 1 \end{bmatrix} \quad (2.105)$$

2.21 For a given equivalent rotation matrix $\boldsymbol{R}$, show that the equivalent angle of rotation θ about k-axis and the direction of axis k are given by

$$\theta = \cos^{-1}\left[\frac{(r_{11} + r_{22} + r_{33}) - 1}{2}\right]$$

$$\begin{bmatrix} k_x \\ k_y \\ k_z \end{bmatrix} = \frac{1}{2\sin\theta}\begin{bmatrix} r_{32} - r_{23} \\ r_{13} - r_{31} \\ r_{21} - r_{12} \end{bmatrix} \quad (2.106)$$

where r_{ij} are the elements of the known 3×3 orientation matrix $\boldsymbol{R}$,

$$\boldsymbol{R} = \begin{bmatrix} r_{11} & r_{12} & r_{13} \\ r_{21} & r_{22} & r_{23} \\ r_{31} & r_{32} & r_{33} \end{bmatrix} \quad (2.107)$$

Assume that $\sin\theta \neq 0$

2.22 The rotation matrix ${}^1\boldsymbol{R}_2$, relating the orientation of frame {2} with respect to frame {1} is given by

$$ {}^1\boldsymbol{R}_2 = \begin{bmatrix} 0.87 & -0.43 & 0.25 \\ 0.50 & 0.75 & -0.43 \\ 0 & 0.50 & 0.87 \end{bmatrix} \tag{2.108} $$

Determine the corresponding set of *ZYX*-Euler angles

2.23 If the rotation matrix ${}^1\boldsymbol{R}_2$ in Exercise 2.22 corresponds to the fixed angle rotations, determine the corresponding set of roll, pitch, and yaw angles.

2.24 In a roll-pitch-roll convention, roll stands for rotation (δ) about *z*-axis, pitch for rotation (λ) about new *y*-axis, and roll again (α) about new *z*-axis. The roll-pitch-roll geometry can be represented by Euler angles. Show that the overall rotation matrix $\boldsymbol{R}_{RPR}(\delta, \lambda, \alpha)$ is given by

$$ \boldsymbol{R}_{RPR}(\delta, \lambda, \alpha) = \begin{bmatrix} C\delta\, C\lambda\, C\alpha - S\delta\, S\alpha & -C\delta\, C\lambda\, S\alpha - S\delta\, C\alpha & C\delta\, S\lambda \\ S\delta\, C\lambda\, C\alpha + C\delta\, S\alpha & -S\delta\, C\lambda\, S\alpha + C\delta\, C\alpha & S\delta\, S\lambda \\ -S\lambda\, C\alpha & S\lambda\, S\alpha & C\lambda \end{bmatrix} \tag{2.109} $$

where $C\delta = \cos\delta$, $C\lambda = \cos\lambda$, $C\alpha = \cos\alpha$, $S\delta = \sin\delta$, $S\lambda = \sin\lambda$, and $S\alpha = \sin\alpha$.

2.25 A frame is given two rotations, one about *x*-axis by 60° and one about *y*-axis by 45°. Show that $\boldsymbol{R}_x \boldsymbol{R}_y \neq \boldsymbol{R}_y \boldsymbol{R}_x$. Explain why.

2.26 Determine the orientation matrix for
(a) *ZXZ* fixed angle rotations.
(b) *ZXZ* Euler angle rotations.

2.27 For the rotations about an arbitrary axis *k*, show that

$$ \boldsymbol{R}_{-k}(-\theta) = \boldsymbol{R}_k(\theta) \tag{2.110} $$

that is, the rotation by angle $-\theta$ about $-k$-axis produces the same effect as those of a rotation by angle θ about *k*-axis.

2.28 Show that the Euler angles θ_1, θ_2, and θ_3 in Eq. (2.64) can be computed for a known rotation matrix

$$ \boldsymbol{R} = \begin{bmatrix} r_{11} & r_{12} & r_{13} \\ r_{21} & r_{22} & r_{23} \\ r_{31} & r_{32} & r_{33} \end{bmatrix} \tag{2.111} $$

as

$$ \begin{aligned} \theta_1 &= A\tan 2(r_{23}, r_{13}) \\ \theta_2 &= A\tan 2\left(\sqrt{r_{13}^2 + r_{23}^2}, r_{33}\right) \\ \theta_3 &= A\tan 2(r_{32}, -r_{31}) \end{aligned} \tag{2.112} $$

in the range $0 \le \theta_2 \le \pi$.

2.29 A point P is moving with a uniform velocity ${}^2\boldsymbol{v} = [12 \quad 5 \quad 25]^T$ relative to frame{2}. If the transformation of frame {2} to frame {1} is given by

$$
{}^1\boldsymbol{T}_2 = \begin{bmatrix} 1 & 0 & 0 & 6 \\ 0 & 0.866 & -0.500 & 10 \\ 0 & 0.500 & 0.866 & -20 \\ 0 & 0 & 0 & 1 \end{bmatrix} \tag{2.113}
$$

compute ${}^1\boldsymbol{v}$.

2.30 Explain why homogeneous coordinates are required in modeling of robotic manipulators.

2.31 Explain why homogeneous transformations are required in modeling of robotic manipulators.

2.32 What are global and local scale factors? When these are useful? Give one situation each where global scale factor is less than one and more than one.

2.33 What do you understand by screw transformations? Where these transformations can be useful?

2.34 What are fundamental rotation matrices? Obtain the three fundamental rotations matrices for rotations about axes x, y and z from the rotation matrix for rotation about an arbitrary axis k.

BIBLIOGRAPHY

1. J. Denavit and R.S. Hartenberg, "A Kinematic Notation for Lower-Pair Mechanisms Based on Matrices," *ASME J. of Applied Mechanics*, 215–221, Jun 1955.
2. A.J. Laub and G.R. Shiflett, "A linear algebra approach to the analysis of rigid body displacement from initial and final position data," *J. Appl. Mech.*, ***49***, 213–216, 1982.
3. R.J. Schilling and H. Lee, *Engineering Analysis: A Vector Space Approach*, Wiley, New York, 1988.
4. D. VanArsdale, "Homogeneous Transformation Matrices for Computer Graphics," *Computers & Graphics*, ***18***(2), 177–191, 1994.
5. D.E. Whitney, "The Mathematics of Coordinated Control of Prosthetic Arms and Manipulators, "*ASME J. of Dynamic Systems, Measurement and Control*, ***94***, 303–309, 1972.

3

Symbolic Modeling of Robots—Direct Kinematic Model

A robotic manipulator is designed to perform a task in the 3-D space. The tool or end-effector is required to follow a planned trajectory to manipulate objects or carry out the task in the workspace. This requires control of position of each link and joint of the manipulator to control both the position and orientation of the tool. To program the tool motion and joint-link motions, a mathematical model of the manipulator is required to refer to all geometrical and/or time-based properties of the motion. Kinematic model describes the spatial position of joints and links, and position and orientation of the end-effector. The derivatives of kinematics deal with the mechanics of motion without considering the forces that cause it. The relationships between the motions and the forces and/or torques that cause them is the dynamics problem.

In designing a robot manipulator, kinematics and dynamics play a vital role. The mathematical tools of spatial descriptions developed in the previous chapter are used in the modeling of robotic manipulators. The *kinematic model* gives relations between the position and orientation of the end-effector and spatial positions of joint-links. The *differential kinematics* of manipulators refers to differential motion, that is, velocity, acceleration, and all higher order derivatives of position variables. The problem of completely describing the position and orientation of a manipulator, the kinematic model, is considered in this and the next chapter. The velocities and accelerations associated with motion would be discussed in Chapter 5 and the forces/torques which cause the motion in Chapter 6.

3.1 MECHANICAL STRUCTURE AND NOTATIONS

The anatomy of the manipulator was discussed in Chapter 1. A manipulator consists of a chain of rigid bodies, called *links*, connected to each other by joints, which allow linear or revolute motion between connected links each of which exhibits just one degree of freedom (DOF). Joints with more than one DOF are not common. A joint with m degrees of freedom can be modeled as m joints with one degree of freedom each connected with (m–1) links of zero length. Most industrial robotic manipulators are open serial kinematic chains, that is, each link is connected to two other links, at the most, without the formation of closed loops. In open chain robots, all joints are motorized (active). Some robots may have closed kinematic chains such as parallelogram linkages and require different considerations to model them.

The number of degrees of freedom a manipulator possesses is the number of independent parameters required to completely specify its position and orientation in space. Because each joint has only one degree of freedom, the degrees of freedom of a manipulator are equal to number of joints.

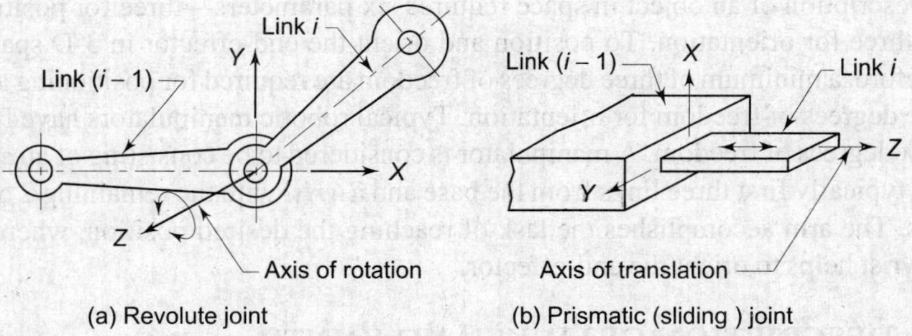

Fig. 3.1 *Two common types of joints and axis of motion (joint axis)*

Single DOF joints between links of a manipulator can be classified as revolute or prismatic. A *revolute joint*, denoted as R-joint, allows rotational motion between connected links. A *prismatic joint*, denoted as P-joint, also known as sliding or rectilinear joint, permits translational motion between the connected links. Each joint has a *joint axis* with respect to which, the motion of joint is described, as shown in Fig. 3.1. In the case of revolute joints, the axis of relative rotation is the joint axis. For the prismatic joint, the axis of relative translational motion is the joint axis. By convention, the z-axis of a coordinate frame is aligned with the joint axis.

The links of a manipulator are numbered outwardly starting from the immobile base as link 0, first moving body as link 1, to the last link out to the free end as link n. Link n is the "*tool*" or "*end-effector*". The joints are numbered, similarly, with joint 1 between link 0 and link 1 and so on, out to the joint n between link (n–1) and link n. The numbering scheme for labelling links and joints is shown in Fig. 3.2 for a 3-DOF manipulator arm which is an open serial kinematic chain of rigid bodies having three revolute joints. Thus, an n-DOF manipulator arm consists of $(n+1)$ links (including link 0) connected by n joints.

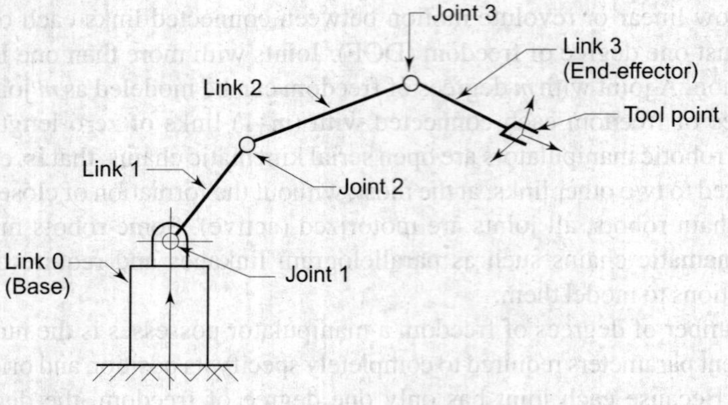

Fig. 3.2 *A 3-DOF manipulator arm—numbering of links and joints*

Description of an object in space requires six parameters—three for position and three for orientation. To position and orient the end-effector in 3-D space, therefore, a minimum of three degrees of freedom are required for positioning and three degrees of freedom for orientation. Typical robotic manipulators have five or six degrees of freedom. A manipulator is considered to be consisting of an *arm* with typically first three links from the base and a *wrist* with the remaining 2 or 3 links. The arm accomplishes the task of reaching the desired position, whereas the wrist helps to orient the end-effector.

3.2 DESCRIPTION OF LINKS AND JOINTS

The n-DOF robotic manipulator is modelled as a chain of rigid links interconnected by revolute and/or prismatic joints. To describe the position and orientation of a link in space, a coordinate frame is attached to each link, namely, frame $\{i\}$ to link i. The position and orientation of frame $\{i\}$, relative to previous frame $\{i–1\}$, can be described by a homogeneous transformation matrix as discussed in the previous chapter.

In this section, the parameters required to completely specify the position and orientation of links and joints of a manipulator are discussed. Every link of the manipulator is connected to two other links with joints at either end, with the exception of the base and the end-effector, the first and the last link (recall that immobile base is link 0), which have only one joint. Figure 3.3 shows link i of a manipulator with associated joint axes $(i–1)$ and i.

From a geometric viewpoint, the link defines the relative position and orientation of joint axes at its two ends. For the two axes $(i–1)$ and i, there exist a mutual perpendicular, which gives the shortest distance between the two axes. This shortest distance along the common normal is defined as the *link length* and is denoted as a_i. The angle between the projection of axis (i–1) and axis i, on a plane perpendicular to the common normal AB, is known as the *link twist* and is denoted by α_i. The link twist α_i is measured from axis (i–1) to axis i in the right-hand sense about AB, as shown in Fig. 3.3.

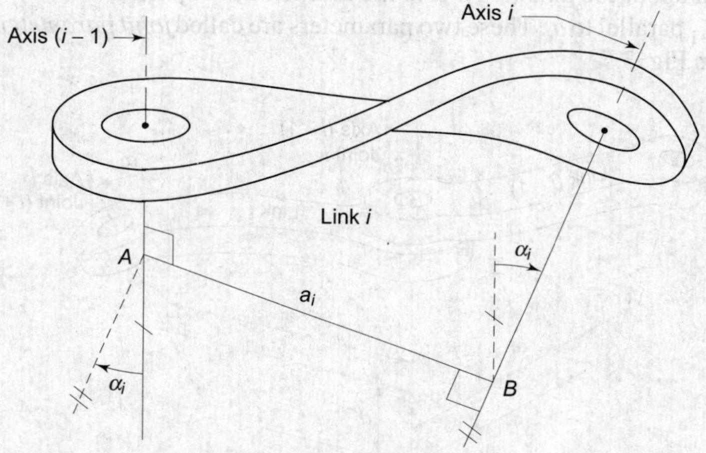

Fig. 3.3 *Description of link parameters a_i and α_i*

These two parameters, a_i and α_i are known as *link parameters* and are constant for a given link. For industrial robots, the links are usually straight, that is, the two joint axes are parallel, giving link length equal to physical link dimension and link twist equal to zero. Another common link geometry is straight link with link twist angle as multiple of $\pi/2$ radians. Sometimes, the link may have a bend such that the axis of joint (i–1) and joint i intersect and in this case the link length of link i is zero although physical link dimension is not zero. Figure 3.4 shows a straight link with link twist of $\pi/2$ radians.

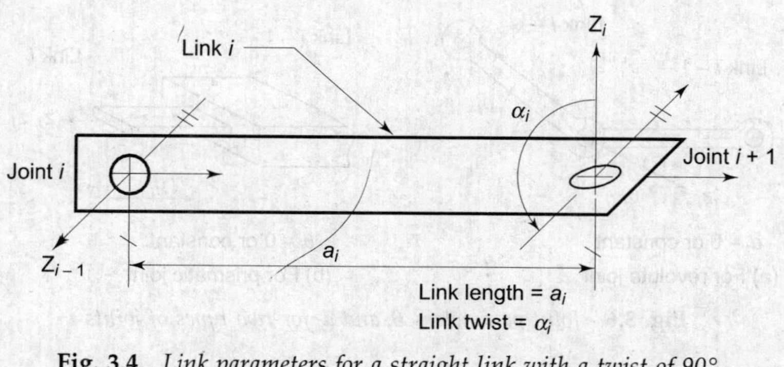

Fig. 3.4 *Link parameters for a straight link with a twist of 90°*

For two links connected by either a revolute or a prismatic joint, the relative position of these links is measured by the displacement at the joint, which is either *joint distance* or *joint angle*, depending on the type of joint. Joint distance (d_i) is the perpendicular distance between the two adjacent common normals a_{i-1} and a_i measured along axis (i–1). In other words, joint distance is the translation needed along joint axis (i–1) to make a_{i-1} intersect with a_i. Joint angle (θ_i) is the angle between the two adjacent common normals a_{i-1} and a_i, measured in right-handed direction about the axis (i–1). It is the rotation about joint axis (i–1) needed to make a_{i-1} parallel to a_i. These two parameters are called *joint parameters* and are shown in Fig. 3.5.

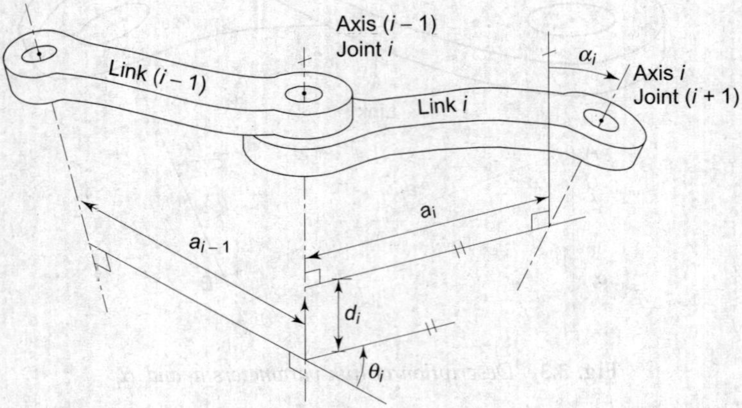

Fig. 3.5 *Description of joint-link parameters for joint i and link i*

A 1-DOF joint requires only one variable to describe its position. Thus, for every 1-DOF joint, it will always be the case that one of the two joint parameters (θ_i and d_i) is fixed and the other is a variable. The displacement of a joint is measured by either angle θ_i or distance d_i depending on the type of joint. The joint displacements for a revolute and prismatic joints are shown in Figs. 3.6(a) and (b), respectively.

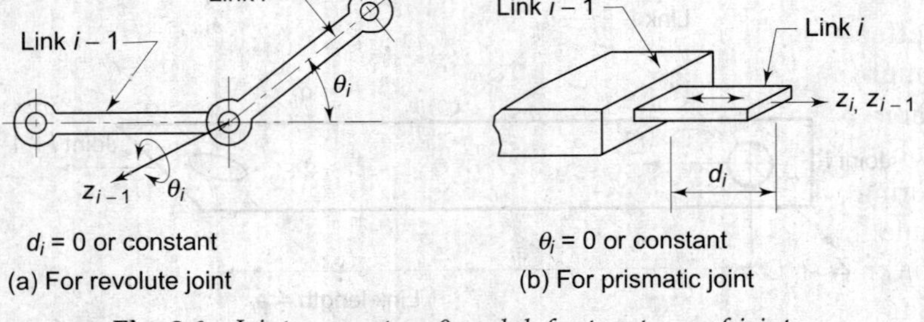

Fig. 3.6 *Joint parameters θ_i and d_i for two types of joints*

For a revolute joint, d_i is zero or constant and θ_i varies, while for a prismatic joint θ_i is zero or constant and d_i varies, describing the relative position of links. The varying parameter is known as *joint variable* and a generalized parameter q is used to denote the joint displacement (variable) of either type of joint. The generalized joint displacement variable is defined as

$$q_i = \begin{cases} \theta_i, & \text{if joint } i \text{ is revolute} \\ d_i, & \text{if joint } i \text{ is prismatic.} \end{cases} \tag{3.1}$$

3.3 KINEMATIC MODELING OF THE MANIPULATOR

With the definition of fixed and variable kinematic parameters for each link, kinematic models can be defined. This model is the analytical description of the spatial geometry of motion of the manipulator with respect to a fixed (inertial) reference frame, as a function of time. In particular, the relation between the joint-variables and the position and orientation of the end-effector is the kinematic model. It is required to control position and orientation of the end-effector, in 3-D space, so that it can follow a defined trajectory or manipulate objects in the workspace. The kinematic modeling problem is split into two problems as:

1. Given the set of joint-link parameters, the problem of finding the position and orientation of the end-effector with respect to a known (immobile or inertial) reference frame for an n-DOF manipulator is the first problem. This is referred to as *direct (or forward) kinematic model* or *direct kinematics*. This model gives the position and orientation of the end-effector as a function of the joint variables and other joint-link constant parameters.
2. For a given position and orientation of the end-effector (of the n-DOF manipulator), with respect to an immobile or inertial reference frame, it is required to find a set of joint variables that would bring the end-effector in the specified position and orientation.
   This is the second problem and is referred to as the *inverse kinematic model* or *inverse kinematics*.

The problem of manipulator control requires both the direct and inverse kinematic models of the manipulator. The block diagram for both the models is illustrated in Fig. 3.7, wherein the commonality is the joint-link fixed and variable parameters. The task to be performed by a manipulator is stated in terms of the end-effector location in space. The values of joint variables required to accomplish the task are computed using the inverse kinematic model. To find the location of end-effector in space, at any instant of time, the joint variable values are substituted in the direct kinematic model. This chapter addresses the problem of formulation of direct kinematic model. The inverse kinematic model formulation will be discussed in the next chapter.

For kinematic modeling, frames are assigned to each link of the manipulator starting from the base to the end-effector. The homogeneous transformation matrices relating the frames attached to successive links describe the spatial relationship between adjacent links. The composition of these individual transform matrices determines the overall transform matrix, describing tool frame with respect to base frame.

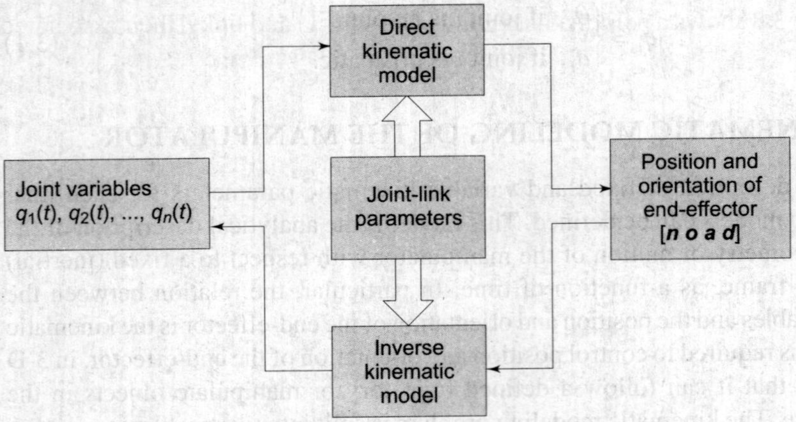

Fig. 3.7 *The direct and inverse kinematic models*

3.4 DENAVIT-HARTENBERG NOTATION

The definition of a manipulator with four joint-link parameters for each link and a systematic procedure for assigning right-handed orthonormal coordinate frames, one to each link in an open kinematic chain, was proposed by Denavit and Hartenberg (1955) and is known as *Denavit-Hartenberg (DH) notation*. This notation is presented in this section and followed throughout the text.

A frame $\{i\}$ is rigidly attached to distal end of link i and it moves with link i. An n-DOF manipulator will have $(n + 1)$ frames with the frame $\{0\}$ or base frame acting as the reference inertial frame and frame $\{n\}$ being the "*tool frame*".

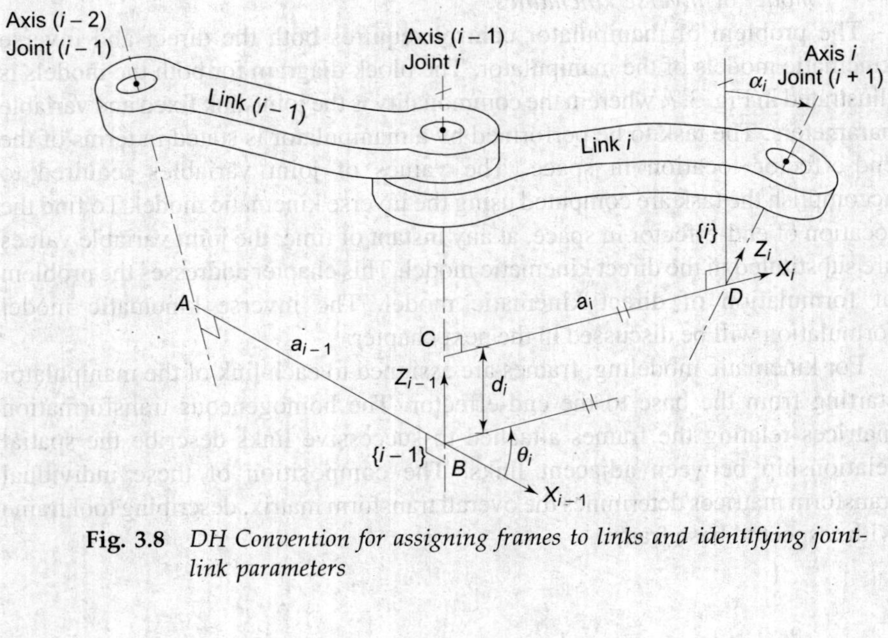

Fig. 3.8 *DH Convention for assigning frames to links and identifying joint-link parameters*

Figure 3.8 shows a pair of adjacent links, link (i–1) and link i, their associated joints, joints (i–1), i and (i+1), and axes (i–2), (i–1), and i, respectively. Line AB, in the figure, is the common normal to (i–2)- and (i–1)-axes and line CD is the common normal to (i–1)- and i-axes. A frame $\{i\}$ is assigned to link i as follows:

(i) The z_i-axis is aligned with axis i, its direction being arbitrary. The choice of direction defines the positive sense of joint variable θ_i.

(ii) The x_i-axis is perpendicular to axis z_{i-1} and z_i and points away from axis z_{i-1}, that is, x_i-axis is directed along the common normal CD.

(iii) The origin of the ith coordinate frame, frame $\{i\}$, is located at the intersection of axis of joint (i+1), that is, axis i, and the common normal between axes (i–1) and i (common normal is CD), as shown in the figure.

(iv) Finally, y_i-axis completes the right-hand orthonormal coordinate frame $\{i\}$.

Note that the frame $\{i\}$ for link i is at the distal end of link i and moves with the link.

With respect to frame $\{i–1\}$ and frame $\{i\}$, the four DH-parameters — two link parameters (a_i, α_i) and two joint parameters (d_i, θ_i) — are defined as:

(a) Link Length (a_i) — distance measured along x_i-axis from the point of intersection of x_i-axis with z_{i-1}-axis (point C) to the origin of frame $\{i\}$, that is, distance CD.

(b) Link twist (α_i) — angle between z_{i-1}- and z_i-axes measured about x_i-axis in the right-hand sense.

(c) Joint distance (d_i) — distance measured along z_{i-1}-axis from the origin of frame $\{i–1\}$ (point B) to the intersection of x_i- axis with z_{i-1}-axis (point C), that is, distance BC.

(d) Joint angle (θ_i) — angle between x_{i-1}- and x_i-axes measured about the z_{i-1}-axis in the right-hand sense.

The convention outlined above does not result in a unique attachment of frames to links because alternative choices are available. For example, joint axis i has two choices of direction to point z_i-axis, one pointing upward (as in Fig. 3.8) and other pointing downward. To minimize such options and get a consistent set of frames, an algorithm is presented below to assign frames to all links of a manipulator.

Algorithm 3.1 ➢ Link Frame Assignment

This algorithm assigns frames and determines the DH-parameters for each link of an n-DOF manipulator. Both, the first link 0 and the last link n, are connected to only one other link and, thus, have more arbitrariness in frame assignment. For this reason, the first (frame $\{0\}$) and the last (frame $\{n\}$) frames are assigned after assigning frames to intermediate links, link 1 to link (n–1).

The displacement of each joint-link is measured with respect to a frame, therefore the *zero position* of each link needs to be clearly defined. The zero position for a revolute joint is when the joint angle θ is zero, while for a prismatic joint it is when the joint displacement is minimal; it may or may not be zero. When all the joints are in zero position, the manipulator is said to be in *home position*. Thus, the home position of an n-DOF manipulator is the position where the $n \times 1$ vector of joint variables is equal to the zero vector, that is, $q_i = 0$ for $i = 1, 2, \ldots, n$. Before assigning frames, the zero position of each joint, that is, the home position of the manipulator must be decided. The frames are then assigned imagining the manipulator in home position.

Because of mechanical constraints, the range of joint motion possible is restricted and, in some cases, this may result in a home position that is unreachable. In such cases, the home position is redefined by changing the initial manipulator joint positions and/or frame assignments. The new home position can be obtained by adding a constant value to the joint angle in case of revolute joint and to the joint displacement in case of prismatic joint. This shifting of the home position is illustrated in Example 3.3.

The algorithm is divided into four parts. The first segment gives steps for labelling scheme and the second one describes the steps for frame assignment to intermediate links 1 to (n–1). The third and fourth segments give steps for frame {0} and frame {n} assignment, respectively.

Step 0 Identify and number the joints starting with base and ending with end-effector. Number the links from 0 to n starting with immobile base as 0 and ending with last link as n.

Step 1 Align axis z_i with axis of joint (i+1) for $i = 0, 1, \ldots, n-1$.

Assigning frames to intermediate links – link 1 to link (n–1) For each link i repeat steps 2 and 3.

Step 2 The x_i-axis is fixed perpendicular to both z_{i-1}- and z_i-axes and points away from z_{i-1}. The origin of frame {i} is located at the intersection of z_i- and x_i-axes. Three situations are possible:

Case (i) If z_{i-1}- and z_i-axes intersect, choose the origin at the point of their intersection. The x_i-axis will be perpendicular to the plane containing z_{i-1}- and z_i-axes. This will give a_i to be zero.

Case (ii) If z_{i-1}- and z_i-axes are parallel or lie in parallel planes then their common normal is not uniquely defined. If joint i is revolute then x_i-axis is chosen along that common normal, which passes through origin of frame {i–1}. This will fix the origin and make d_i zero. If joint i is prismatic, x_i-axis is arbitrarily chosen as any convenient common normal and the origin is located at the distal end of the link i.

Case (iii) If z_{i-1}- and z_i-axes coincide, the origin lies on the common axis. If joint i is revolute, origin is located to coincide with origin of frame {i–1} and x_i-axis coincides with x_{i-1}-axis to cause d_i to be zero. If joint i is prismatic, x_i-axis is chosen parallel to x_{i-1}-axis to make a_i to be zero. The origin is located at distal end of link i.

Step 3 The y_i-axis has no choice and is fixed to complete the right-handed orthonormal coordinate frame {i}.

Assigning frame to link 0, the immobile base – frame {0}

Step 4 The frame {0} location is arbitrary. Its choice is made based on simplification of the model and some convenient reference in workspace. The x_0-axis, which is perpendicular to z_0-axis, is chosen to be parallel to x_1-axis in the home position to make $\theta_1 = 0$. The origin of frame {0} is located based on type of joint 1. If joint 1 is revolute, the origin of frame {0} can be chosen at a convenient reference such as, floor, work table, and so on, giving a constant value for parameter d_1 or at a suitable location along axis of joint 1 so as to make d_1 zero. If joint 1 is prismatic, parallel x_0- and x_1-axes will make θ_1 to be zero and origin of frame {0} is placed arbitrarily.

Step 5 The y_0-axis completes the right-handed orthonormal coordinate frame {0}.

Link n, the end-effector, frame assignment – frame {n}

Step 6 The origin of frame $\{n\}$ is chosen at the tip of the manipulator, that is, a convenient point on the last link (the end-effector). This point is called the "tool point" and the frame $\{n\}$ is the tool frame.

Step 7 The z_n-axis is fixed along the direction of z_{n-1}-axis and pointing away from the link n. It is the direction of "approach."

Step 8 If joint n is prismatic, take x_n parallel to x_{n-1}-axis. If joint n is revolute, the choice of x_n is similar to step 4, that is, x_n is perpendicular to both z_{n-1}- and z_n-axes. x_n direction is the "normal" direction. The y_n-axis is chosen to complete the right-handed orthonormal frame $\{n\}$. The y_n-axis is the "orientation" or "sliding" direction.

Once the frames are assigned to each link, the joint-link parameters (θ_i, d_i, α_i, a_i) can be easily identified for each link, using which the direct kinematic model is developed in the next section.

In fixing the frames, it is desirable to make as many of the joint-link parameters zero as possible because the amount of computations necessary in later analysis is dependent on these. Hence, whenever there is a choice in frame assignment, emphasis is on making a choice, which results in as many zero parameters as possible.

3.5 KINEMATIC RELATIONSHIP BETWEEN ADJACENT LINKS

To find the transformation matrix relating two frames attached to the adjacent links, consider frame $\{i–1\}$ and frame $\{i\}$ as shown in Fig. 3.9. These two frame are associated with link (i–1) and i but for clarity the links are not shown in the figure. The kinematic joint-link parameters involved (θ_i, d_i, α_i, a_i) are shown therein. Points B, C, D and frame $\{i–1\}$ and $\{i\}$ are the same as in Fig. 3.8.

The transformation of frame $\{i–1\}$ to frame $\{i\}$ consists of four basic transformations as shown in Fig. 3.9.

(a) A rotation about z_{i-1}-axis by an angle θ_i;
(b) Translation along z_{i-1}-axis by distance d_i;
(c) Translation by distance α_i along x_i-axis, and
(d) Rotation by an angle α_i about x_i-axis.

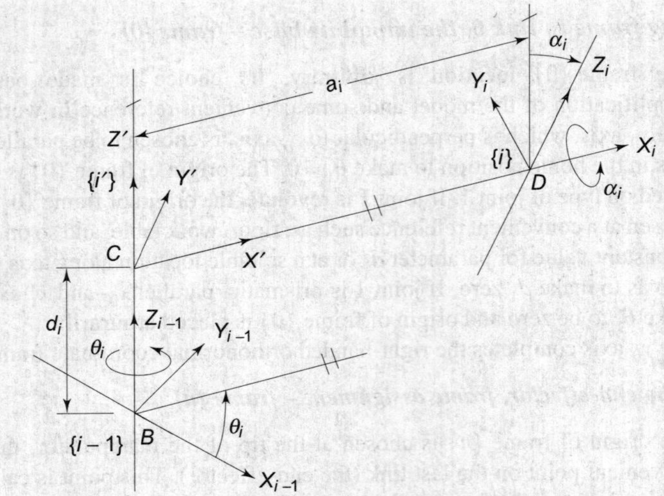

Fig. 3.9 *Geometric relationship between adjacent links*

Using the spatial coordinate transformations discussed in Chapter 2, the composite transformation matrix, which describes frame $\{i\}$ with respect to frame $\{i–1\}$, is obtained using Eq. (2.46) as

$$^{i-1}\boldsymbol{T}_i = \boldsymbol{T}_z(\theta_i)\boldsymbol{T}_z(d_i)\boldsymbol{T}_x(a_i)\boldsymbol{T}_x(\alpha_i) \tag{3.2}$$

From Eqs. (2.20), (2.54), and (2.55),

$$^{i-1}\boldsymbol{T}_i = \begin{bmatrix} C\theta_i & -S\theta_i & 0 & 0 \\ S\theta_i & C\theta_i & 0 & 0 \\ 0 & 0 & 1 & 0 \\ 0 & 0 & 0 & 1 \end{bmatrix} \begin{bmatrix} 1 & 0 & 0 & 0 \\ 0 & 1 & 0 & 0 \\ 0 & 0 & 1 & d_i \\ 0 & 0 & 0 & 1 \end{bmatrix} \begin{bmatrix} 1 & 0 & 0 & a_i \\ 0 & 1 & 0 & 0 \\ 0 & 0 & 1 & 0 \\ 0 & 0 & 0 & 1 \end{bmatrix} \begin{bmatrix} 1 & 0 & 0 & 0 \\ 0 & C\alpha_i & -S\alpha_i & 0 \\ 0 & S\alpha_i & C\alpha_i & 0 \\ 0 & 0 & 0 & 1 \end{bmatrix}$$

or

$$^{i-1}\boldsymbol{T}_i = \begin{bmatrix} C\theta_i & -S\theta_i C\alpha_i & S\theta_i S\alpha_i & a_i C\theta_i \\ S\theta_i & C\theta_i C\alpha_i & -C\theta_i S\alpha_i & a_i S\theta_i \\ 0 & S\alpha_i & C\alpha_i & d_i \\ 0 & 0 & 0 & 1 \end{bmatrix} \tag{3.3}$$

where $C\theta_i = \cos\theta_i$, $S\theta_i = \sin\theta_i$, $C\alpha_i = \cos\alpha_i$, and $S\alpha_i = \sin\alpha_i$.

The transformation from frame $\{i–1\}$ to frame $\{i\}$ can also be obtained by considering an intermediate coordinate frame $\{i'\}$ located at point C, as shown in Fig. 3.9. From the figure, the transformation from frame $\{i\}$ to frame $\{i'\}$ consists of a rotation and a translation about x_i-axis and the transformation from frame $\{i'\}$ to frame $\{i–1\}$ consists of a rotation and a translation about $z_{i–1}$-axis. The two homogeneous transformations are:

$$^{i'}\boldsymbol{T}_i = \begin{bmatrix} 1 & 0 & 0 & a_i \\ 0 & C\alpha_i & -S\alpha_i & 0 \\ 0 & S\alpha_i & C\alpha_i & 0 \\ 0 & 0 & 0 & 1 \end{bmatrix}$$

$$ {}^{i-1}\boldsymbol{T}_{i'} = \begin{bmatrix} C\theta_i & -S\theta_i & 0 & 0 \\ S\theta_i & C\theta_i & 0 & 0 \\ 0 & 0 & 1 & d_i \\ 0 & 0 & 0 & 1 \end{bmatrix} \tag{3.4} $$

The composite transformation from frame $\{i\}$ to frame $\{i–1\}$ is, thus, obtained as

$$ {}^{i-1}\boldsymbol{T}_i = {}^{i-1}\boldsymbol{T}_{i'}\, {}^{i'}\boldsymbol{T}_i $$

Substituting from Eq. (3.4) gives the basic link transformation matrix as:

$$ {}^{i-1}\boldsymbol{T}_i = \left[\begin{array}{ccc:c} C\theta_i & -S\theta_i C\alpha_i & S\theta_i S\alpha_i & a_i C\theta_i \\ S\theta_i & C\theta_i C\alpha_i & -C\theta_i S\alpha_i & a_i S\theta_i \\ 0 & S\alpha_i & C\alpha_i & d_i \\ \hdashline 0 & 0 & 0 & 1 \end{array}\right] \tag{3.5} $$

This is identical to Eq. (3.3) as it should be. This is an important result for modeling manipulators.

The homogeneous transformation matrix ${}^{i-1}\boldsymbol{T}_i$ describes the position and orientation of frame $\{i\}$ relative to frame $\{i–1\}$ and completely specifies the geometric relationship between these links in terms of four DH-parameters $(\theta_i, d_i\ \alpha_i\ a_i)$. Of these four parameters, only one is a variable for link i, the displacement variable q_i (θ_i or d_i) and other three are constant. The matrix ${}^{i-1}\boldsymbol{T}_i(q_i)$ is known as link i transformation matrix. As shown before, the 3×3 upper left corner submatrix of Eq. (3.5) gives the orientation of coordinate axes of frame $\{i\}$, while the 3×1 upper right corner sub-matrix represents the position of the origin of frame $\{i\}$.

3.6 MANIPULATOR TRANSFORMATION MATRIX

In this section, the last step in formulating the forward kinematic model of a manipulator is discussed. This model describes position and orientation of the last link (tool frame) with reference to the base frame as a function of joint displacements q_1 through q_n. An n-DOF manipulator consists of $(n+1)$ links from base to tool point and a frame is assigned to each link. Figure 3.10 shows the $(n+1)$ frames, frame $\{0\}$ to frame $\{n\}$, attached to the links of the manipulator.

The position and orientation of the tool frame relative to the base frame can be found by considering the n consecutive link transformation matrices relating frames fixed to adjacent links. Thus,

$$ {}^{0}\boldsymbol{T}_n = {}^{0}\boldsymbol{T}_1(q_1)\ {}^{1}\boldsymbol{T}_2(q_2) \ldots\ {}^{n-1}\boldsymbol{T}_n(q_n) \tag{3.6} $$

where ${}^{i-1}\boldsymbol{T}_i(q_i)$ for $i = 1, 2, \ldots, n$ is the homogeneous link transformations matrix between frames $\{i–1\}$ and $\{i\}$ and is given by Eq. (3.5).

The tool frame, frame $\{n\}$, can also be considered as a translated and rotated frame with respect to base frame $\{0\}$. The transformation between these two frames is denoted by end-effector transformation matrix $\boldsymbol{T}$, Eq. (2.31), in terms of

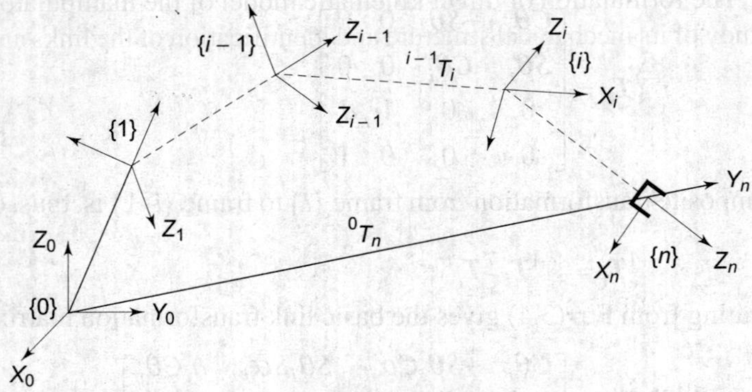

Fig 3.10 *Location of end-effector frame relative to base frame*

tool frame orientation ($\boldsymbol{n}, \boldsymbol{o}, \boldsymbol{a}$) and its displacement ($\boldsymbol{d}$) from the base frame {0}. In Fig. 3.10, frame {n} is the tool frame, thus, $\boldsymbol{T}$ is equal to $^0\boldsymbol{T}_n$, or

$$\boldsymbol{T} = {}^0\boldsymbol{T}_n = {}^0\boldsymbol{T}_1\,{}^1\boldsymbol{T}_2 \ldots {}^{n-1}\boldsymbol{T}_n \tag{3.7}$$

Equation (3.7) is known as the kinematic model of the n-DOF manipulator. It provides the functional relationship between the tool frame (or end-effector) position and orientation and displacement of each link q_i, which may be angular or linear, depending on joint being revolute or prismatic. That is,

$$\boldsymbol{T} = f(q_i), \quad i = 1, 2, \ldots, n \tag{3.8}$$

or

$$\begin{bmatrix} n_x & o_x & a_x & d_x \\ n_y & o_y & a_y & d_y \\ n_z & o_z & a_z & d_z \\ 0 & 0 & 0 & 1 \end{bmatrix} = \begin{bmatrix} r_{11} & r_{12} & r_{13} & r_{14} \\ r_{21} & r_{22} & r_{23} & r_{24} \\ r_{31} & r_{32} & r_{33} & r_{34} \\ 0 & 0 & 0 & 1 \end{bmatrix} \tag{3.9}$$

where coefficient r_{ij} are functions of joint displacements q_i. For the known joint displacements q_i for $i = 1, 2, \ldots, n$, the end-effector orientation ($\boldsymbol{n}\ \boldsymbol{o}\ \boldsymbol{a}$) and position $\boldsymbol{d}$ can be computed from Eq. (3.8).

Several examples are now worked out to clarify the concepts of the direct kinematic modeling. The first example is a simple one, a 2-DOF manipulator, the others are of some common configurations of manipulator arm and wrist, and the last example illustrates the kinematic modeling of a 6-DOF industrial manipulator.

SOLVED EXAMPLES

Example 3.1 *A 2-DOF planar manipulator arm*

Obtain the position and orientation of the tool point P with respect to the base for the 2-DOF, RP planar manipulator shown in Fig. 3.11.

Solution The formulation of direct kinematic model of the manipulator begins with the study of its mechanical structure and identification of the links and joints. The frames are then assigned using Algorithm 3.1. This example is a simple one and illustrates the basic steps involved in formulation of kinematic model.

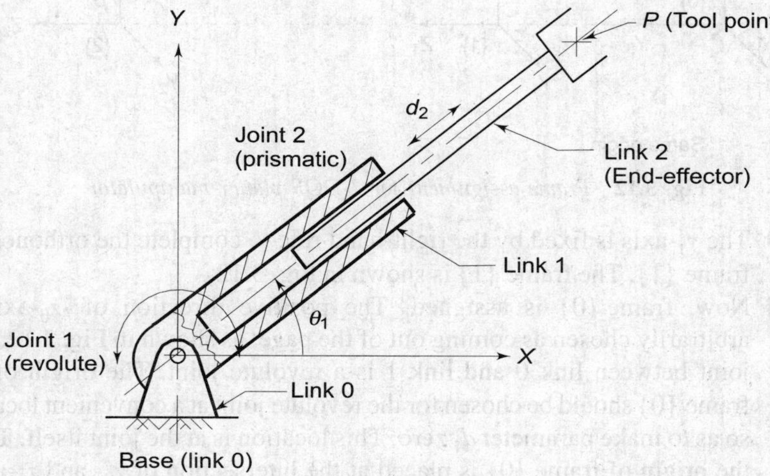

Fig. 3.11 *A 2-DOF planar manipulator arm with one rotary and one prismatic joint*

The planar configuration of this manipulator can be employed to manipulate objects within a plane, the *xy*-plane. The first joint is a revolute joint and the second one is prismatic. It is easy to see that it has a circular area as workspace. The size of the two links determines the radius of inner and outer circles of the workspace area. Point *P* may or may not traverse a full circle, depending on the mechanical design of joint and joint range available at joint 1.

The axis of joint 1 is perpendicular to the plane of workspace, while axis of joint 2 lies in the plane. The two joint axes intersect each other. The home position is considered as the horizontal position ($\theta_1 = 0$) and prismatic link completely retracted in, corresponding to radius of inner boundary of workspace. The step-by-step frame assignment is carried out, according to Algorithm 3.1, as explained below.

Step 0 The two joints are numbered as 1 and 2 and links as 0, 1, and 2 starting with the immobile base as 0.

Step 1 Joint axes z_0 and z_1 are aligned with the axes of joint 1 and 2, respectively.

The joint-link labelling and joint axes are shown in Figs. 3.11 and 3.12. Frames are assigned to intermediate links first, and then to the first and last links. In this example, there is only one intermediate link, link 1.

Step 2 For frame {1} of link 1, the z_1-axis is fixed in step 1 above. Because z_0- and z_1-axes intersect, the origin of frame {1} is fixed at the point of their intersection, according to Step 2 case (i) of the Algorithm 3.1. The x_1-axis is set in the direction of perpendicular to plane containing z_0- and

z_1-axes. Note that α_1, d_1, and a_1 will be defined after frame {0} is fixed. The variable parameter for this is θ_1.

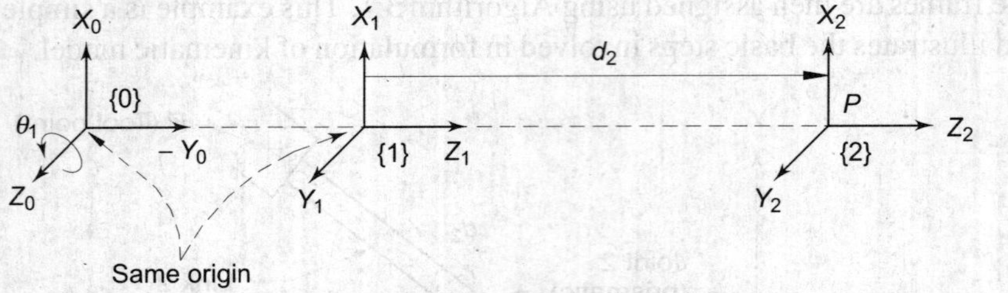

Fig. 3.12 *Frame assignment for 2-DOF planar manipulator*

Step 3 The y_1-axis is fixed by the right-hand rule to complete the orthonormal frame {1}. The frame {1} is shown in Fig. 3.12.

Step 4 Now, frame {0} is assigned. The positive direction of z_0-axis is arbitrarily chosen as coming out of the page, as shown in Fig. 3.12. The joint between link 0 and link 1 is a revolute joint. The origin of the frame {0} should be chosen for the revolute joint at a convenient location so as to make parameter d_1 zero. This location is at the joint itself. Thus, the origin of frame {0} is placed at the intersection of z_0- and z_1-axes. This is also situated the origin of frame {1} or two origins coincide giving $a_1 = 0$, and $d_1 = 0$. The x_0-axis is chosen parallel to x_1-axis, and it coincides with x_1-axis. The rotation of z_0-axis to z_1-axes about x_1-axis defines the twist angle α_1 as 90°. Thus, the choice of frame {0} and {1} defines parameters as $\alpha_1 = 90°$, $a_1 = 0$ and $d_1 = 0$.

Step 5 The y-axis is fixed to complete the orthonormal frame {0}.

Step 6 The origin frame {2}, the last frame, is fixed to the tool point P of the last link (the end-effector). The choice of this origin defines the joint variable d_2 as distance measured from origin of frame {1}.

Step 7 The direction of z_2-axis is chosen to be same as z_1-axis pointing away from link 2.

Step 8 Joint 2 is prismatic and, hence, x_2-axis is chosen to be parallel to x_1-axis. The y_2-axis is fixed to complete the frame {2}. Once the frame {2} is defined, the parameters get the values as: $a_2 = 0$, $\alpha_2 = 0$, and $\theta_2 = 0$.

The complete frame assignment is shown in Fig. 3.12. The coinciding frames, frame {0} and frame {1} are drawn away from each other for clarity but marked as "same origin" and there is zero distance between their origins.

The assigned frames define the four DH-parameters for each link so as to completely specify the geometric structure of the given manipulator. The joint-link parameters are tabulated in Table 3.1. For each link, the displacement variable q_i is identified and placed in the displacement variable column. It is important to note that each row of the joint-link parameter table has exactly one variable and there is no row without a variable. Any deviation from these conditions indicates an error in frame assignment and/or joint-link parameter identification. Note that out of six constant joint-link parameters, five are zero and the sixth is 90°. The two displacement variables are θ_1 and d_2.

Table 3.1 *Joint-link parameters for the RP manipulator arm*

Link i	a_i	α_i	d_i	θ_i	*Displacement Variable* q_i	$C\theta_i$	$S\theta_i$	$C\alpha_i$	$S\alpha_i$
1	0	90°	0	θ_1	θ_1	C_1	S_1	0	1
2	0	0	d_2	0	d_2	1	0	1	0

The next step is to obtain the individual transformation matrices ${}^0\boldsymbol{T}_1$ and ${}^1\boldsymbol{T}_2$ for relating successive links. These are obtained by substituting the values of the joint-link parameters in Eq. (3.5). To facilitate writing of transformation matrices, four columns defining cos θ_i, sin θ_i, cos α_i, and sin α_i are appended to the joint-link parameter table and values are filled in for each row. The two transformation matrices are, therefore,

$$ {}^0\boldsymbol{T}_1(\theta_1) = \begin{bmatrix} C_1 & 0 & S_1 & 0 \\ S_1 & 0 & -C_1 & 0 \\ 0 & 1 & 0 & 0 \\ 0 & 0 & 0 & 1 \end{bmatrix} \tag{3.10} $$

$$ {}^1\boldsymbol{T}_2(d_2) = \begin{bmatrix} 1 & 0 & 0 & 0 \\ 0 & 1 & 0 & 0 \\ 0 & 0 & 1 & d_2 \\ 0 & 0 & 0 & 1 \end{bmatrix} \tag{3.11} $$

Each of the above transformation matrices is a function of only one variable, the displacement variable for the link. Finally, The forward kinematic model is obtained by combining the individual transform matrices. Thus, ${}^0\boldsymbol{T}_2$ the transformation of tool frame, frame {2}, with respect to base frame, frame {0} is obtained by substituting individual matrices, Eqs. (3.10) and (3.11) in Eq. (3.6). The final result after simplifying is:

$$ {}^0\boldsymbol{T}_2 = {}^0\boldsymbol{T}_1\,{}^1\boldsymbol{T}_2 = \begin{bmatrix} C_1 & 0 & S_1 & d_2S_1 \\ S_1 & 0 & -C_1 & -d_2C_1 \\ 0 & 1 & 0 & 0 \\ 0 & 0 & 0 & 1 \end{bmatrix} \tag{3.12} $$

This overall transformation, Eq. (3.12), is equal to the end-effector transformation matrix, Eq. (2.31), and the direct kinematic model in matrix form is:

$$ \begin{bmatrix} n_x & o_x & a_x & d_x \\ n_y & o_y & a_y & d_y \\ n_z & o_z & a_z & d_z \\ 0 & 0 & 0 & 1 \end{bmatrix} = \begin{bmatrix} C_1 & 0 & S_1 & d_2S_1 \\ S_1 & 0 & -C_1 & -d_2C_1 \\ 0 & 1 & 0 & 0 \\ 0 & 0 & 0 & 1 \end{bmatrix} \tag{3.13} $$

This kinematic model can also be expressed by 12 equations as:

$$
\begin{aligned}
n_x &= C_1 \\
n_y &= S_1 \\
n_z &= 0 \\
o_x &= 0 \\
o_y &= 0 \\
o_z &= 1 \\
a_x &= S_1 \\
a_y &= -C_1 \\
a_z &= 0 \\
d_x &= d_2 S_1 \\
d_y &= -d_2 C_1 \\
d_z &= 0
\end{aligned} \tag{3.14}
$$

From Eq. (3.13) or Eq. (3.14), the orientation and position of the tool point P can be computed for given values of displacement variables θ_1 and d_2 at any instant of time. For example, for $\theta_1 = 120°$ and $d_2 = 200$ mm the end-effector transformation matrix will be

$$
\boldsymbol{T}_E = \begin{bmatrix} -0.5 & 0 & 0.866 & 173.2 \\ 0.866 & 0 & 0.5 & 100.0 \\ 0 & 1 & 0 & 0 \\ 0 & 0 & 0 & 1 \end{bmatrix}
$$

It is assumed that θ_1 and d_2 chosen above are within the available range of joint motions.

Example 3.2 *Kinematic model of a cylindrical arm*

Formulate the forward kinematic model of the three-degree of freedom (RPP) manipulator arm shown in Fig. 3.13.

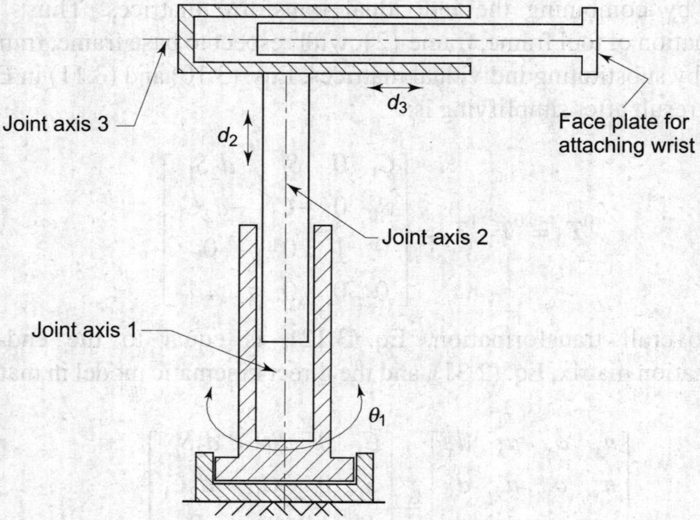

Fig. 3.13 *Mechanical structure of a 3-DOF cylindrical (RPP) manipulator arm*

Solution The cylindrical configuration manipulator arm has three joints—the first joint is revolute, while the next two are prismatic. The axes of the first two joints coincide. This configuration has a cylindrical workspace as discussed in Chapter 1.

As in the previous example, begin with fixing home position, labelling links, joints and assigning frames using Algorithm 3.1. The details of step-by-step frame assignment are left for the reader. The final frame assignment is shown in Fig. 3.14.

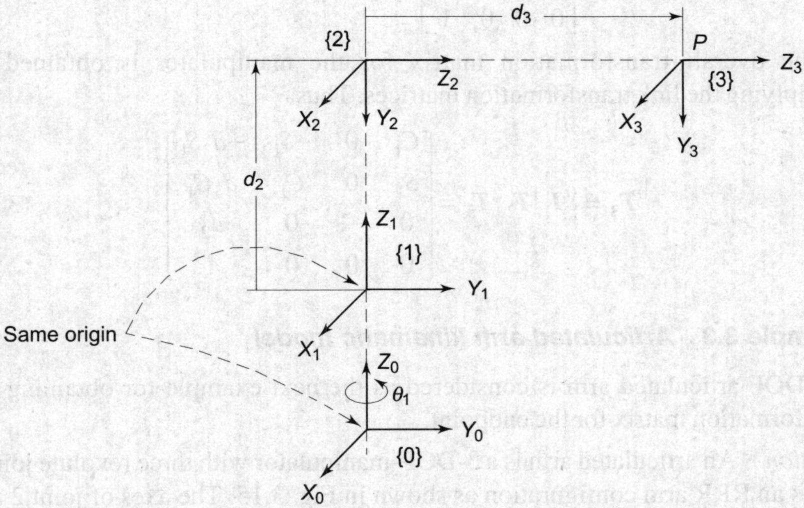

Fig. 3.14 *Frame assignment for the cylindrical manipulator arm*

Next, the joint-link parameters are identified and these are tabulated in Table 3.2.

Table 3.2 *Joint-link parameters for the RPP manipulator arm*

Link i	a_i	α_i	d_i	θ_i	q_i	$C\theta_i$	$S\theta_i$	$C\alpha_i$	$S\alpha_i$
1	0	0	0	θ_1	θ_1	C_1	S_1	1	0
2	0	–90°	d_2	0	d_2	1	0	0	–1
3	0	0	d_3	0	d_3	1	0	1	0

The transformation matrices for transformation of each link (frame) with respect to the previous one is obtained as:

$$ {}^0\boldsymbol{T}_1(\theta_1) = \begin{bmatrix} C_1 & -S_1 & 0 & 0 \\ S_1 & C_1 & 0 & 0 \\ 0 & 0 & 1 & 0 \\ 0 & 0 & 0 & 1 \end{bmatrix} \qquad (3.15)$$

$$
{}^{1}\boldsymbol{T}_2(d_2) = \begin{bmatrix} 1 & 0 & 0 & 0 \\ 0 & 0 & 1 & 0 \\ 0 & -1 & 0 & d_2 \\ 0 & 0 & 0 & 1 \end{bmatrix} \tag{3.16}
$$

$$
{}^{2}\boldsymbol{T}_3(d_3) = \begin{bmatrix} 1 & 0 & 0 & 0 \\ 0 & 1 & 0 & 0 \\ 0 & 0 & 1 & d_3 \\ 0 & 0 & 0 & 1 \end{bmatrix} \tag{3.17}
$$

The overall transformation matrix for the manipulator is obtained by multiplying the link transformation matrices. Thus,

$$
{}^{0}\boldsymbol{T}_3 = {}^{0}\boldsymbol{T}_1\,{}^{1}\boldsymbol{T}_2\,{}^{2}\boldsymbol{T}_3 = \begin{bmatrix} C_1 & 0 & -S_1 & -d_3S_1 \\ S_1 & 0 & C_1 & d_3C_1 \\ 0 & -1 & 0 & d_2 \\ 0 & 0 & 0 & 1 \end{bmatrix} \tag{3.18}
$$

Example 3.3 *Articulated arm kinematic model*

A 3-DOF articulated arm is considered as the next example for obtaining the transformation matrix for the endpoint.

Solution An articulated arm is a 3-DOF-manipulator with three revolute joints, that is an RRR arm configuration as shown in Fig. 3.15. The axes of joint 2 and joint 3 are parallel and axis of joint 1 is perpendicular to these two. At the end of the arm, a faceplate is provided to attach the wrist.

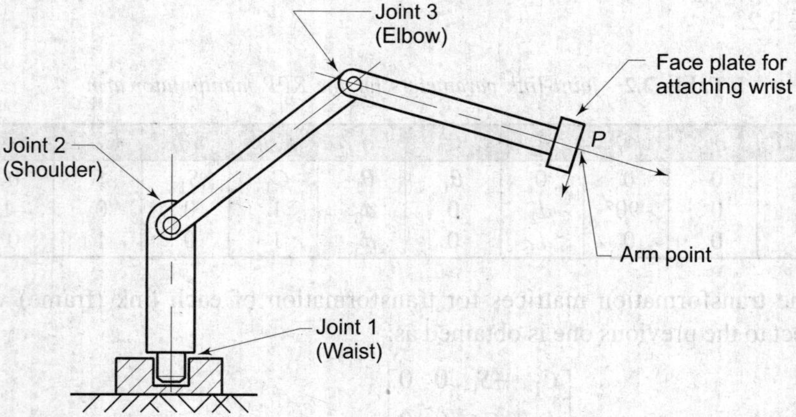

Fig. 3.15 *A 3-DOF articulated arm with three revolute joints*

To determine the "arm point" transformation matrix, the frames are assigned first as shown in Fig. 3.16. The resulting joint-link parameters are tabulated in Table 3.3. For all the three joints, joint-offsets are assumed to be zero.

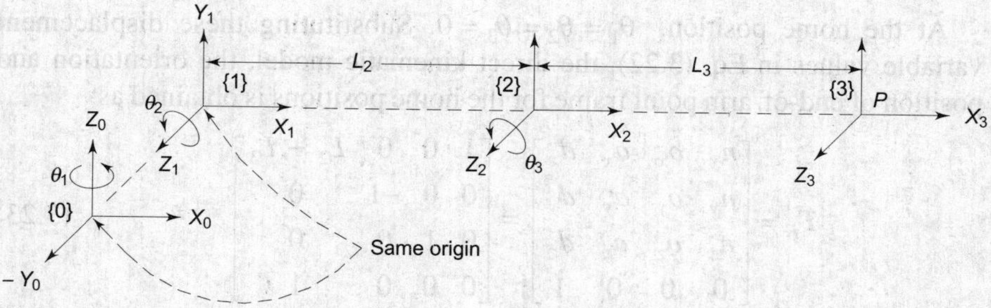

Fig. 3.16 *Frame assignment for articulated arm*

Table 3.3 *Joint-link parameters for articulated arm*

Link i	a_i	α_i	d_i	θ_i	q_i	$C\alpha_i$	$S\alpha_i$
1	0	90°	0	θ_1	θ_1	0	1
2	L_2	0	0	θ_2	θ_2	1	0
3	L_3	0	0	θ_3	θ_3	1	0

The link transformation matrices are

$$ {}^0\boldsymbol{T}_1(\theta_1) = \begin{bmatrix} C_1 & 0 & S_1 & 0 \\ S_1 & 0 & -C_1 & 0 \\ 0 & 1 & 0 & 0 \\ 0 & 0 & 0 & 1 \end{bmatrix} \tag{3.19} $$

$$ {}^1\boldsymbol{T}_2(\theta_2) = \begin{bmatrix} C_2 & -S_2 & 0 & L_2C_2 \\ S_2 & C_2 & 0 & L_2S_2 \\ 0 & 0 & 1 & 0 \\ 0 & 0 & 0 & 1 \end{bmatrix} \tag{3.20} $$

$$ {}^2\boldsymbol{T}_3(\theta_3) = \begin{bmatrix} C_3 & -S_3 & 0 & L_3C_3 \\ S_3 & C_3 & 0 & L_3S_3 \\ 0 & 0 & 1 & 0 \\ 0 & 0 & 0 & 1 \end{bmatrix} \tag{3.21} $$

The overall transformation matrix for the endpoint of the arm is, therefore,

$$ {}^0\boldsymbol{T}_1 = {}^0\boldsymbol{T}_1\,{}^1\boldsymbol{T}_2\,{}^2\boldsymbol{T}_3 = \begin{bmatrix} C_1C_{23} & -C_1S_{23} & S_1 & C_1(L_3C_{23} + L_2C_2) \\ S_1C_{23} & -S_1S_{23} & -C_1 & S_1(L_3C_{23} + L_2C_2) \\ S_{23} & C_{23} & 0 & L_3S_{23} + L_2S_2 \\ 0 & 0 & 0 & 1 \end{bmatrix} \tag{3.22} $$

where C_{23} and S_{23} refer to $\cos(\theta_2+\theta_3)$ and $\sin(\theta_2+\theta_3)$, respectively.

At the home position, $\theta_1 = \theta_2 = \theta_3 = 0$. Substituting these displacement variable values in Eq. (3.22), the direct kinematic model, the orientation and position of end-of-arm point frame for the home positions is obtained as:

$$\boldsymbol{T}_E = \begin{bmatrix} n_x & o_x & a_x & d_x \\ n_y & o_y & a_y & d_y \\ n_z & o_z & a_z & d_z \\ 0 & 0 & 0 & 1 \end{bmatrix} = \begin{bmatrix} 1 & 0 & 0 & L_2 + L_3 \\ 0 & 0 & -1 & 0 \\ 0 & 1 & 0 & 0 \\ 0 & 0 & 0 & 1 \end{bmatrix} \tag{3.23}$$

From Eq. (3.23), it is observed that in the home position the arm point frame, frame {3}, has its x-axis (x_3-axis) in the same direction as x_0-axis, y_3-axis in the z_0-axis direction, and z_3-axis in the negative y_0-axis direction. The origin of frame {3} is translated by a distance of (L_2+L_3) in the x_0-axis direction. This means that if, initially frame {3} is coincident with frame {0}, its home position and orientation is obtained by translating the origin by (L_2+L_3) along x_0-axis and rotating it by +90° about x_0-axis. The position and orientation of frame {3} obtained from Eq. (3.23) matches with the coordinate system established in Fig. 3.16, verifying the correctness of the model obtained.

Home position of the articulated arm corresponding to the frame assigned in Fig. 3.16, that is, $\theta_1 = \theta_2 = \theta_3 = 0$, is drawn in Fig. 3.17(a). An alternate home position can be obtained by adding constant angles to θ_2 and θ_3. For example, if we added +90° to joint angle θ_2 and –90° to joint angle θ_3, the new home position is drawn in Fig. 3.17(b).

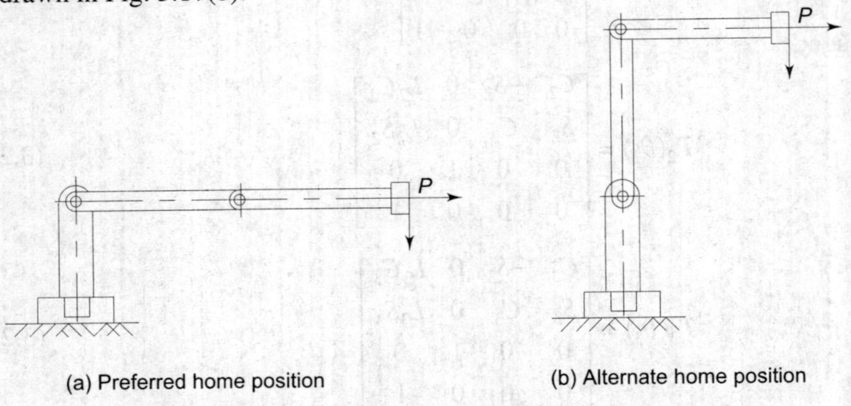

Fig. 3.17 *Two possible home positions for the articulated arm*

For this alternate home position of the manipulator the new joint displacements θ'_2 and θ'_3 are defined by adding +90° to joint angle θ_2 and –90° to joint angle θ_3, respectively.

Frame assignment and the kinematic model formulation for this new home position with displacement variables θ'_1, θ'_2, and θ'_3 is left as an exercise for the reader. The joint-link parameters for this home position are tabulated in Table 3.4.

Table 3.4 *Joint-link parameters for the articulated arm with new home position*

Link i	a_i	α_i	d_i	θ_i	q_i	$C\alpha_i$	$S\alpha_i$
1	0	90°	0	θ_1	$\theta'_1 = \theta_1$	0	1
2	L_2	0	0	θ'_2	$\theta'_2 = \theta_2 + 90°$	1	0
3	L_3	0	0	θ'_3	$\theta'_3 = \theta_3 - 90°$	1	0

Example 3.4 *RPY wrist kinematics*

For the 3-DOF roll-pitch-yaw (RPY) wrist shown in Fig. 3.18 obtain the direct kinematic model.

Solution The 3-DOF RPY wrist has three revolute (RRR) joints, which provide any arbitrary orientation to the end-effector in 3-D space.

To get the direct kinematic model, it is assumed that the arm end-point is stationary and can be considered as the stationary base frame, frame {0}, for the wrist.

The joints are labelled and joint axes are identified as shown in Fig. 3.18. Observe that for the "home position" shown in figure the axes of joint 1 and joint 2 are perpendicular to each other and intersect at joint 2. The axes of joint 2 and joint 3 are also mutually perpendicular but are in parallel planes. The three joint displacements θ_1, θ_2, and θ_3 are along three mutually perpendicular directions: roll, pitch, and yaw.

The frame assignment for the four frames, frames {0} to frame {3} is carried out next and is explained frame by frame in the paragraphs below.

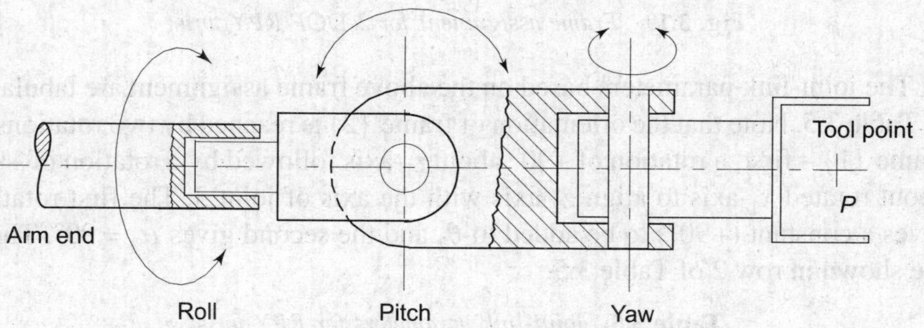

Fig. 3.18 *A 3-DOF freedom roll, pitch and yaw (RPY) wrist*

The frame {1} for link 1 is fixed with x_1-axis perpendicular to both z_0- and z_1-axes and its origin is fixed at joint 2, the point where axes z_0 and z_1 intersect, as per step 2(i) of Algorithm 3.1. For frame {3}, the axes z_1 and z_2 are perpendicular to each other but do not intersect as they lie in parallel planes. Because the joint is revolute, the common normal, which passes through origin of frame {1}, gives the direction of x_2-axis, as per step 2(ii) of Algorithm 3.1. The origin of frame {2} is fixed at the intersection of x_2- and z_2-axes and is located at origin of frame {1}, giving $a_2 = 0$ and $d_2 = 0$.

The base frame, frame {0} is fixed with its origin coinciding with origin of frame {1} and choosing x_0-axis parallel to x_1-axis to give $a_1 = 0$ and $d_1 = 0$. The physical distance between joint 1 and 2 can be accounted by increasing the size of last link of arm appropriately.

The origin of the last frame, frame {3}, is normally fixed to the tool point for convenience. If this is done, the distance between origins of frame {2} and frame {3}, corresponding to the size of the end-effector will be nonzero in the kinematic model. To simplify the kinematic model the origin of frame {3} can also be chosen to coincide with the origin of frame {2}, giving $a_3 = d_3 = 0$. The constant dimension of the end-effector can be accounted later by applying a constant translational transformation.

Now, the x_3- and z_3-axes of frame {3} are fixed to coincide with x_2- and z_3-axes. The complete frame assignment is shown in Fig. 3.19. Note that the origins of all four frames are coincident and that the orientation of frame{3}, the tool point frame is different from the conventional orientation where z-axis is taken in the approach direction.

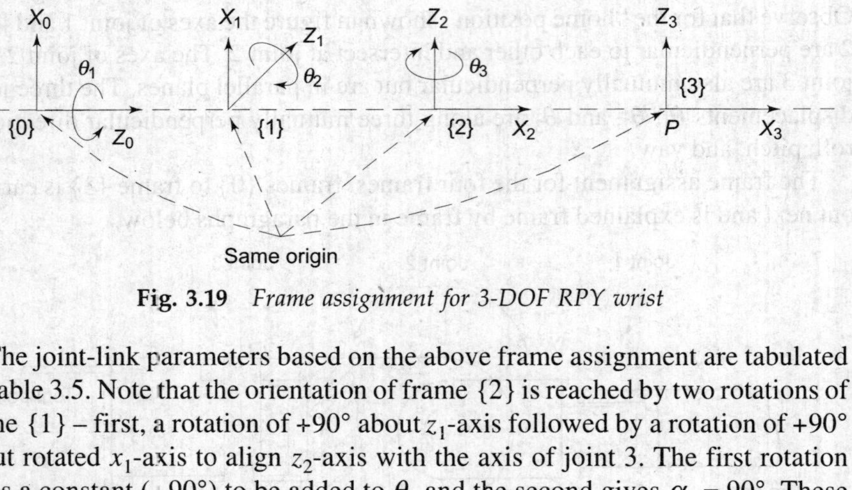

Fig. 3.19 *Frame assignment for 3-DOF RPY wrist*

The joint-link parameters based on the above frame assignment are tabulated in Table 3.5. Note that the orientation of frame {2} is reached by two rotations of frame {1} – first, a rotation of +90° about z_1-axis followed by a rotation of +90° about rotated x_1-axis to align z_2-axis with the axis of joint 3. The first rotation gives a constant (–90°) to be added to θ_2 and the second gives $\alpha_2 = 90°$. These are shown in row 2 of Table 3.5.

Table 3.5 *Joint-link parameters for RPY wrist*

Link i	a_i	α_i	d_i	θ_i	q_i	$C\theta_i$	$S\theta_i$	$C\alpha_i$	$S\alpha_i$
1	0	90°	0	θ_1	θ_1	C_1	S_1	0	1
2	0	90°	0	θ_2+ 90°	θ_2	$-S_2$	C_2	0	1
3	0	0	0	θ_3	θ_3	C_3	S_3	1	0

The transformation matrices are now obtained from Table 3.5 as

$$ {}^0\boldsymbol{T}_1 = \begin{bmatrix} C_1 & 0 & S_1 & 0 \\ S_1 & 0 & -C_1 & 0 \\ 0 & 1 & 0 & 0 \\ 0 & 0 & 0 & 1 \end{bmatrix} \qquad (3.24) $$

$$^{1}T_2 = \begin{bmatrix} -S_2 & 0 & C_2 & 0 \\ C_2 & 0 & S_2 & 0 \\ 0 & 1 & 0 & 0 \\ 0 & 0 & 0 & 1 \end{bmatrix} \tag{3.25}$$

$$^{2}T_3 = \begin{bmatrix} C_3 & -S_3 & 0 & 0 \\ S_3 & C_3 & 0 & 0 \\ 0 & 0 & 1 & 0 \\ 0 & 0 & 0 & 1 \end{bmatrix} \tag{3.26}$$

and the overall transformation matrix for the RPY wrist is

$$^{0}T_3 = {}^{0}T_1\,{}^{1}T_2\,{}^{2}T_3 = \begin{bmatrix} -C_1S_2C_3 + S_1S_3 & C_1S_2S_3 + S_1C_3 & C_1C_2 & 0 \\ -S_1S_2C_3 - C_1S_3 & S_1S_2S_3 - C_1C_3 & S_1C_2 & 0 \\ C_2C_3 & -C_2S_3 & S_2 & 0 \\ 0 & 0 & 0 & 1 \end{bmatrix} \tag{3.27}$$

Checking the correctness of this model for the home position is left to the reader.

Example 3.5 *Kinematics of a 3-DOF Polar Arm*

For the 3-DOF (RRP) manipulator arm shown in Fig. 3.20, obtain the orientation and position of tool point *P* of the joint variable vector is $\boldsymbol{q} = [90°\ -45°\ 100\ \text{mm}]^T$ with $x_1 = 50$ mm and $x_2 = 40$ mm.

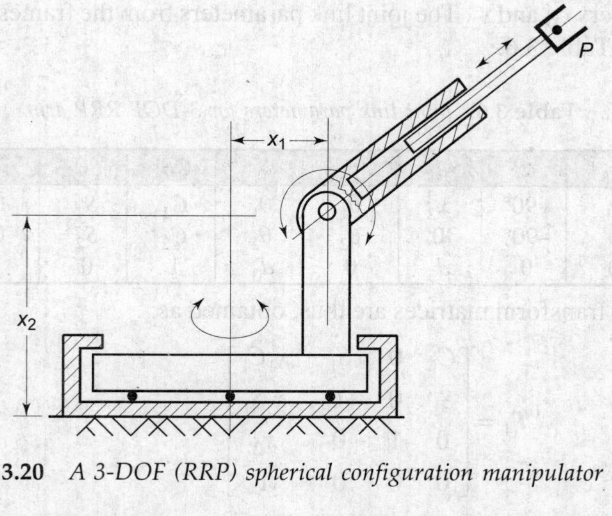

Fig. 3.20 *A 3-DOF (RRP) spherical configuration manipulator arm*

Solution The 3-DOF RRP manipulator is constructed by fixing the 2-DOF arm of Example 3.1, Fig. 3.11 on a rotary table.

Assume in home position of the arm, the last two links are vertical. For this home position, the frame assignment using Algorithm 3.1 will be as shown in Fig. 3.21.

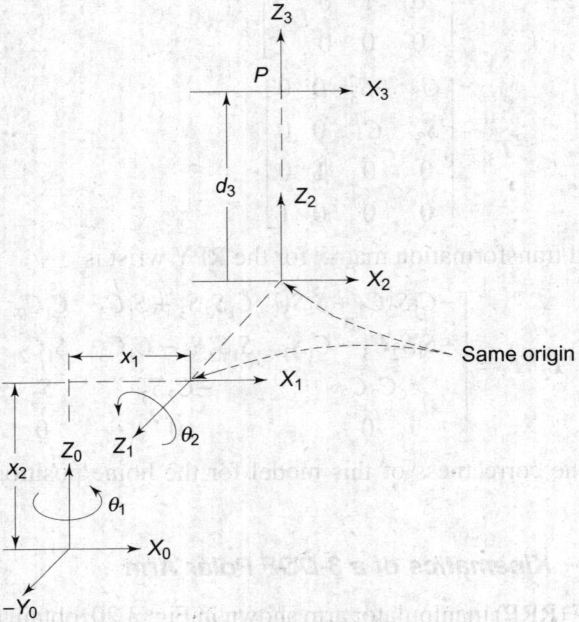

Fig. 3.21 *Frame assignment for home position of arm in Fig. 3.20*

Note the location of frame {1} with respect to frame {0} and the inclusion to two parameters x_1, and x_2. The joint link parameters from the frames assigned are tabulated in Table 3.6.

Table 3.6 *Joint link parameters for 3-DOF RRP arm*

i	a_i	α_i	d_i	θ_i	q_i	$C\theta_i$	$S\theta_i$	$C\alpha_i$	$S\alpha_i$
1	x_1	+90°	x_2	θ_1	θ_1	C_1	S_1	0	1
2	0	–90°	0	θ_2	θ_2	C_2	S_2	0	1
3	0	0	d_3	0	d_3	1	0	1	0

The three transform matrices are thus, obtained as:

$$
{}^0\boldsymbol{T}_1 = \begin{bmatrix} C_1 & 0 & S_1 & x_1C_1 \\ S_1 & 0 & -C_1 & x_1S_1 \\ 0 & 1 & 0 & x_2 \\ 0 & 0 & 0 & 1 \end{bmatrix} \tag{3.28}
$$

$$
{}^1\boldsymbol{T}_2 = \begin{bmatrix} C_2 & 0 & -S_2 & 0 \\ S_2 & 0 & C_2 & 0 \\ 0 & -1 & 0 & 0 \\ 0 & 0 & 0 & 1 \end{bmatrix} \tag{3.29}
$$

$$^2\boldsymbol{T}_3 = \begin{bmatrix} 1 & 0 & 0 & 0 \\ 0 & 1 & 0 & 0 \\ 0 & 0 & 1 & d_3 \\ 0 & 0 & 0 & 1 \end{bmatrix} \tag{3.30}$$

and the overall transformation for the arm is

$$\text{or} \quad \boldsymbol{T} = {}^0\boldsymbol{T}_1\, {}^1\boldsymbol{T}_2\, {}^2\boldsymbol{T}_3 = \begin{bmatrix} C_1C_2 & -S_1 & -C_1S_2 & -C_1S_2/d_3 + x_1C_1 \\ S_1C_2 & C_1 & -S_1S_2 & -S_1S_2/d_3 + x_1S_1 \\ S_2 & 0 & C_2 & C_2/d_3 - x_2 \\ 0 & 0 & 0 & 1 \end{bmatrix} \tag{3.31}$$

The orientation of the end-effector (point *P*) is described in terms of joint variables by the rotation submatrix of $\boldsymbol{T}$, that is,

$$\boldsymbol{R} = \begin{bmatrix} C_1C_2 & -S_1 & -C_1S_2 \\ S_1C_2 & C_1 & -S_1S_2 \\ S_2 & 0 & C_2 \end{bmatrix} \tag{3.32}$$

For $\theta_1 = 90°$ and $\theta_2 = -45°$, the orientation of *P* is

$$\begin{bmatrix} n_x & o_x & a_x \\ n_y & o_y & a_y \\ n_z & o_z & a_z \end{bmatrix} = \begin{bmatrix} 0 & -1 & 0 \\ 0.707 & 0 & 0.707 \\ -0.707 & 0 & 0.707 \end{bmatrix} \tag{3.33}$$

and the position of end-effector is described by the translation vector $\boldsymbol{D}$ as

$$\boldsymbol{D} = \begin{bmatrix} -C_1S_2d_3 + x_1C_1 \\ -S_1S_2d_3 + x_1S_1 \\ d_3C_2 - x_2 \end{bmatrix} \tag{3.34}$$

for given joint-link parameters and $\boldsymbol{q}$, the position of *P* is

$$\boldsymbol{D} = [0 \quad 20.711 \quad 110.711]^T \tag{3.35}$$

Example 3.6 ***SCARA manipulator kinematics***

Obtain the direct kinematics equation of the 4-DOF Selective Compliance Assembly Robot Arm (SCARA) robots.

Solution The SCARA manipulator is a widely used industrial robot for assembly operations. Figure 3.22 shows the configuration of the SCARA manipulator, which is a four-axis horizontal-jointed articulated arm configuration as discussed in Chapter 1. The first two joints are revolute to establish the horizontal position of the tool. The third joint is prismatic which determines the vertical position of tool. Finally, the last joint is revolute which controls the tool orientation. Thus, SCARA has RRPR configuration and gives an upright cylindrical workspace. It may be useful to recall that the RPP manipulator discussed in Example 3.2 also has a cylindrical workspace, but the two are different (i) in the number of possible ways to reach a point within the work volume, and (ii) the orientation of tool point. In Example 3.2, the tool point orientation is radial, where as the tool point of the SCARA robot is always oriented in the vertical direction. The frame assignment for the SCARA manipulator is carried out as follows:

All the joints and links are identified and labeled from 1 to 4 and 0 to 4, respectively. The axis of each joint is vertical and frames are fixed next.

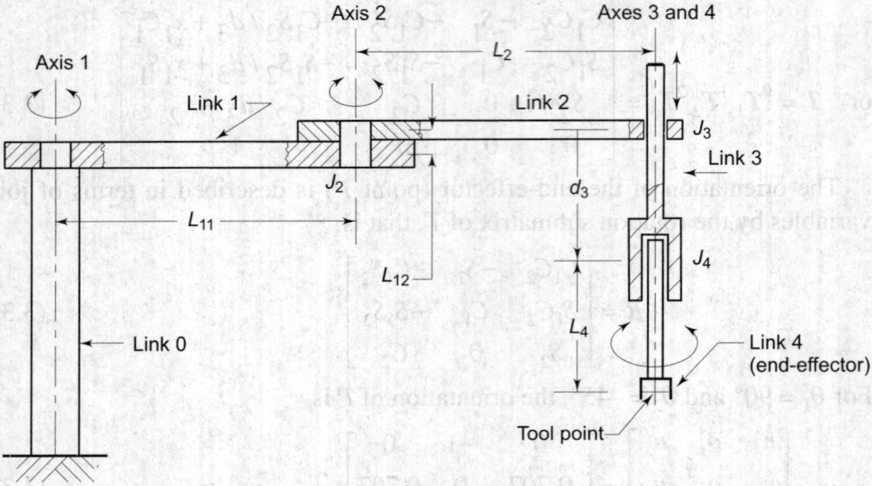

Fig 3.22 *A 4-DOF SCARA manipulator in home position*

Frame {1} Joints 1 and 2 axes are aligned with z_0- and z_1-axes, respectively. As z_0- and z_1-axes are parallel, choose x_1-axis as the common normal to z_0 and z_1 directed along the link 2. The origin of frame {1} is placed on z_1-axis at the joint as shown in Fig. 3.23. The y_1-axis completes right-handed orthonormal frame {1}. This gives $\alpha_1 = 0$.

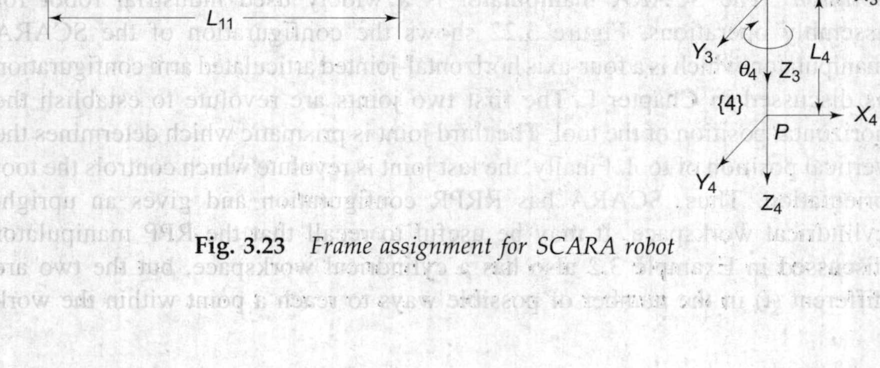

Fig. 3.23 *Frame assignment for SCARA robot*

Frame {2} Joint 3 axis is aligned with z_2-axis. Since z_1- and z_2-axes are parallel, set x_2-axis as the common normal which makes d_1 zero. This fixes the origin of frame {2} and results in $\alpha_2 = 0$, $d_2 = 0$ and $a_2 = L_2$.

Frame {3} Joint 3 is prismatic and z_3-axis coincides with z_2-axis, thus $\alpha_2 = 0$. Origin of frame {3} is placed at distal end of the prismatic link 3. Since the link displacement is downwards, it is convenient to point z_3-axis downward. This gives $\alpha_3 = 180°$. The x_3-axis is chosen parallel to x_2-axis making $\theta_3 = 0$.

Frame {0} z_0-axis has already been aligned along joint 1 axis. The x_0-axis is fixed along the link 1 axis such that x_0-axis remains parallel to x_1-axis. Since joint 1 is revolute, origin of frame {0} is placed at joint 1 at the intersection of axes z_0 and x_0 as shown in Fig. 3.23. The right-handed orthonormal coordinate frame {0} is computed with y_0-axis. The position of origin of frame {0} relative to frame {1} given by a translation by L_{11} in x_0 direction and L_{12} in z_0 direction, this frame assignment gives $a_1 = L_{11}$ and $d_1 = L_{12}$. Here L_{11} corresponds to the length of link 1 and the distance L_{12} is known as joint offset. Alternate locations of the origin of frame {0} are possible.

Frame {4} z_4-axis is chosen parallel to z_3-axis and origin of frame {4} is located at the tool tip. Joint 4 being revolute, x_4-axis is chosen to be parallel to x_3-axis giving $\alpha_4 = 0$, $a_4 = 0$ and $d_4 = L_4$.

The complete frame assignment is shown in Fig. 3.23 from which all the joint-link parameters are determined and tabulated in Table 3.7.

Table 3.7 *Joint-link parameters for SCARA robot*

Link i	a_i	α_i	d_i	θ_i	q_i	$C\theta_i$	$S\theta_i$	$C\alpha_i$	$S\alpha_i$
1	L_{11}	0	L_{12}	θ_1	θ_1	C_1	S_1	1	0
2	L_2	0	0	θ_2	θ_2	C_2	S_2	1	0
3	0	180°	d_3	0	d_3	1	0	–1	0
4	0	0	L_4	θ_4	θ_4	C_4	S_4	1	0

We now obtain the individual transformation matrices that specify the relationship between adjacent links as below:

$$ {}^0T_1(\theta_1) = \begin{bmatrix} C_1 & -S_1 & 0 & L_{11}C_1 \\ S_1 & C_1 & 0 & L_{11}S_1 \\ 0 & 0 & 1 & L_{12} \\ 0 & 0 & 0 & 1 \end{bmatrix} \tag{3.36} $$

$$ {}^1T_2(\theta_2) = \begin{bmatrix} C_2 & -S_2 & 0 & L_2C_2 \\ S_2 & C_2 & 0 & L_2S_2 \\ 0 & 0 & 1 & 0 \\ 0 & 0 & 0 & 1 \end{bmatrix} \tag{3.37} $$

$$^2T_3(d_3) = \begin{bmatrix} 1 & 0 & 0 & 0 \\ 0 & -1 & 0 & 0 \\ 0 & 0 & -1 & d_3 \\ 0 & 0 & 0 & 1 \end{bmatrix} \tag{3.38}$$

$$^3T_4(\theta_4) = \begin{bmatrix} C_4 & -S_4 & 0 & 0 \\ S_4 & C_4 & 0 & 0 \\ 0 & 0 & 1 & L_4 \\ 0 & 0 & 0 & 1 \end{bmatrix} \tag{3.39}$$

The manipulator transformation matrix obtained after multiplication of the above four matrices and simplification using trigonometric identities, is

$$^0T_4 = \begin{bmatrix} C_{124} & S_{124} & 0 & L_2C_{12} + L_{11}C_1 \\ S_{124} & -C_{124} & 0 & L_2S_{12} + L_{11}S_1 \\ 0 & 0 & -1 & L_{12} + d_3 - L_4 \\ 0 & 0 & 0 & 1 \end{bmatrix} \tag{3.40}$$

Here, C_{124} denotes cos $(\theta_1 + \theta_2 - \theta_4)$, S_{124} denotes sin $(\theta_1 + \theta_2 - \theta_4)$, C_{12} denotes cos $(\theta_1 + \theta_2)$, and S_{12} denotes sin $(\theta_1 + \theta_2)$.

The fact that the third column of the matrix $^0T_4$ is always $[0 \;\; 0 \;\; -1 \;\; 0]^T$, means that at any instant the *z*-axis of the tool frame (approach vector) is in the direction of negative *z*-axis of the base frame. This is a characteristic of SCARA robots that are designed to manipulate objects from directly above. In addition, the SCARA wrist possesses only 1-DOF, the roll motion, to orient the tool.

Example 3.7 *Kinematics of a 5-DOF Industrial Manipulator*

In many industrial applications of robots five degrees of freedom are sufficient to carry out the industrial tasks effectively. Many common industrial manipulators are, therefore, constructed with a 3-DOF arm and a 2-DOF wrist with roll and pitch motions. One such manipulator with an articulated arm is shown in Fig. 3.24. All the five axes of the manipulator are revolute.

Obtain the kinematic model of the manipulator and test it for the home position. Determine the position and orientation of the end-effector (tool point *P*) if the joint variable vector is $\boldsymbol{q} = [\pi/4 \;\; -3\pi/4 \;\; \pi/2 \;\; \pi/4 \;\; \pi]^T$ and the link parameters are $L_1 = 50$, $L_2 = L_3 = 140$ and $L_5 = 20$.

Solution To determine the kinematic model of the manipulator, the home position of the manipulator is chosen such that link 2 of the arm is vertical and link 3 and the wrist are horizontal. For this home position the joint variable vector will be $\boldsymbol{q}_{\text{home}} = [0 \;\; 0 \;\; 0 \;\; 0 \;\; 0]^T$. For the home position, the frame assignment is carried out. Applying steps 0 to 8 of the Algorithm 3.1, the frames to all joints-links are assigned and are shown in Fig. 3.25.

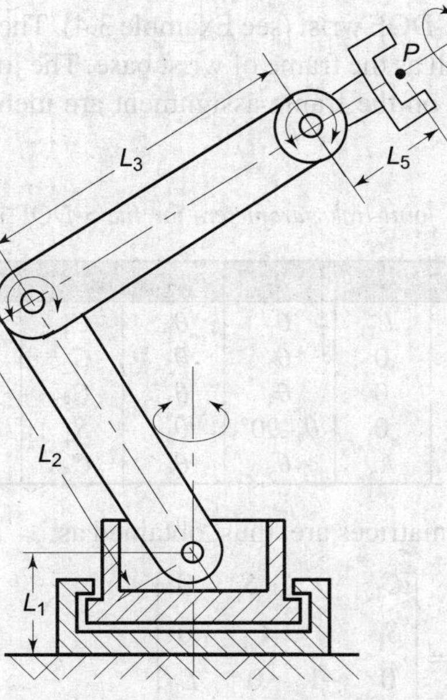

Fig. 3.24 *A five degree of freedom industrial manipulator*

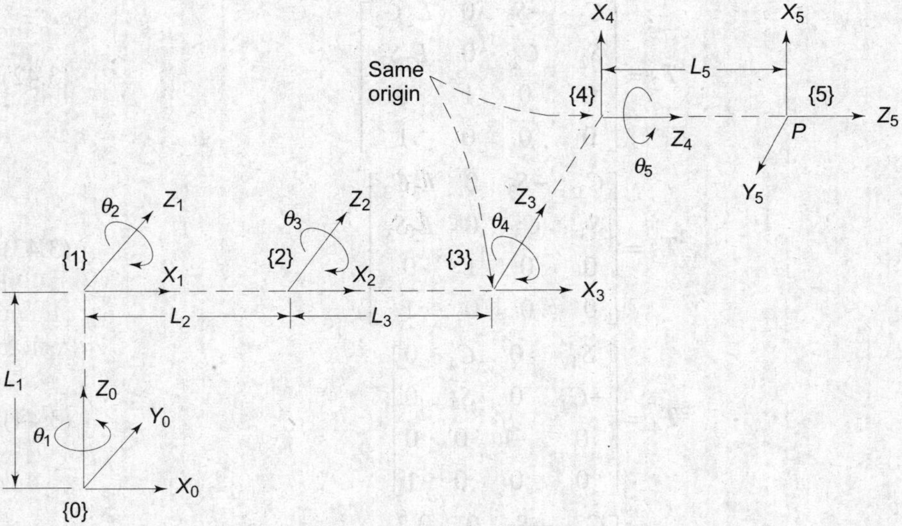

Fig. 3.25 *Frame assignment for 5-DOF industrial manipulator*

Note that the origins of frame {3} and frame {4} coincide but are shown separated to make the diagram clear. The orientation of frame {4} is obtained by two rotations of frame {3}, first by –90° about z_3-axis and, second by –90° about rotated x_3-axis. Recall that the frame assignment by Algorithm 3.1 is not unique, for example, a different frame assignment will be obtained by choosing the z-axis of any or all of the joins in the opposite direction. In Fig. 3.25 frames {0} to {3} correspond to the 3-DOF of articulated arm (see Example 3.3) and frame {3} to

{5} correspond to the 2-DOF wrist (see Example 3.4). The frame {3} is the frame of arm end-point as well as the frame of wrist base. The joint-link parameters for the manipulator based on the frame assignment are identified and tabulated in Table 3.8.

Table 3.8 *Joint-link parameters for the 5-DOF manipulator*

i	a_i	α_i	d_i	θ_i	q_i	$C\theta_i$	$S\theta_i$	$C\alpha_i$	$S\alpha_i$
1	0	–90°	L_1	θ_1	θ_1	C_1	S_1	0	–1
2	L_2	0	0	θ_2	θ_2	C_2	S_2	1	0
3	L_3	0	0	θ_3	θ_3	C_3	S_3	1	0
4	0	–90°	0	θ_4–90°	θ_4	S_4	$-C_4$	0	–1
5	0	0	L_5	θ_5	θ_5	C_5	S_5	1	0

The transformation matrices are, thus, obtained as

$$^0\boldsymbol{T}_1 = \begin{bmatrix} C_1 & 0 & -S_1 & 0 \\ S_1 & 0 & C_1 & 0 \\ 0 & -1 & 0 & L_1 \\ 0 & 0 & 0 & 1 \end{bmatrix} \tag{3.41}$$

$$^1\boldsymbol{T}_2 = \begin{bmatrix} C_2 & -S_2 & 0 & L_2C_2 \\ S_2 & C_2 & 0 & L_2S_2 \\ 0 & 0 & 1 & 0 \\ 0 & 0 & 0 & 1 \end{bmatrix} \tag{3.42}$$

$$^2\boldsymbol{T}_3 = \begin{bmatrix} C_3 & -S_3 & 0 & L_3C_3 \\ S_3 & C_3 & 0 & L_3S_3 \\ 0 & 0 & 1 & 0 \\ 0 & 0 & 0 & 1 \end{bmatrix} \tag{3.43}$$

$$^3\boldsymbol{T}_4 = \begin{bmatrix} S_4 & 0 & C_4 & 0 \\ -C_4 & 0 & S_4 & 0 \\ 0 & -1 & 0 & 0 \\ 0 & 0 & 0 & 1 \end{bmatrix} \tag{3.44}$$

$$^4\boldsymbol{T}_5 = \begin{bmatrix} C_5 & -S_5 & 0 & 0 \\ S_5 & C_5 & 0 & 0 \\ 0 & 0 & 1 & L_5 \\ 0 & 0 & 0 & 1 \end{bmatrix} \tag{3.45}$$

The transformation matrix for the arm is

$$
{}^0T_3 = {}^0T_1\,{}^1T_2\,{}^2T_3 = \begin{bmatrix} C_1C_{23} & -C_1S_{23} & -S_1 & C_1(L_2C_2 + L_3C_{23}) \\ S_1C_{23} & -S_1S_{23} & C_1 & S_1(L_2C_2 + L_3C_{23}) \\ -S_{23} & -C_{23} & 0 & L_1 - L_2S_2 - L_3S_{23} \\ 0 & 0 & 0 & 1 \end{bmatrix} \quad (3.46)
$$

and the transformation matrix for the wrist is

$$
{}^3T_5 = {}^3T_4\,{}^4T_5 = \begin{bmatrix} S_4C_5 & -S_4S_5 & C_4 & L_5C_4 \\ -C_4C_5 & C_4S_5 & S_4 & L_5S_4 \\ -S_5 & -C_5 & 0 & 0 \\ 0 & 0 & 0 & 1 \end{bmatrix} \quad (3.47)
$$

The overall transformation matrix for the manipulator is, therefore,

$$
{}^0T_5 = {}^0T_3\,{}^3T_5
$$

$$
= \begin{bmatrix} C_1S_{234}C_5 + S_1S_5 & -C_1S_{234}S_5 + S_1C_5 & C_1C_{234} & C_1(L_2C_2 + L_3C_{23} + L_5C_{234}) \\ S_1C_{234}C_5 - C_1S_5 & -S_1S_{234}S_5 - C_1C_5 & S_1C_{234} & S_1(L_2C_2 + L_3C_{23} + L_5C_{234}) \\ -C_{234}C_5 & C_{234}S_5 & -S_{234} & L_1 - L_2S_2 - L_3S_{23} - L_5S_{234} \\ 0 & 0 & 0 & 1 \end{bmatrix}
$$

(3.48)

Equation 3.48 gives the kinematic model of the manipulator. Note that the approach vector $\boldsymbol{a}$ and position vector $\boldsymbol{p}$ are independent of the wrist roll (θ_5). The transformation matrix for the home position of the manipulator is obtained by substituting $\boldsymbol{q}_{\text{home}}$ in Eq. (3.48). The resulting home position matrix for the end-effector is

$$
T_{\text{home}} = \begin{bmatrix} 0 & 0 & 1 & L_2 + L_3 + L_5 \\ 0 & -1 & 0 & 0 \\ 1 & 0 & 0 & L_1 \\ 0 & 0 & 0 & 1 \end{bmatrix} \quad (3.49)
$$

The home position transformation matrix gives the orientation and position of the end-effector frame, frame {5} or tool frame at point P. The orientation of frame {5} is given by the 3×3 rotation submatrix of Eq. (3.49) and is described as follows: the frame {5} is rotated relative to frame {0} such that x_5-axis is parallel and in same direction to z-axis of base frame, frame {0} or z_0-axis; y_5-axis is parallel to y_0-axis but in opposite direction; and z_5-axis is in x_0-axis direction. The position of point P or the origin of end-effector frame, frame {5}, is given by 3×1 displacement matrix of Eq. (3.49) and it is displaced by $[L_2 + L_3 + L_5 \quad 0 \quad L_1]^T$ from the base frame, frame {0}.

Finally, the position and orientation of the end-effector, point P, for the given joint variable vector $\boldsymbol{q} = [\pi/4 \quad -3\pi/4 \quad \pi/2 \quad \pi/4 \quad \pi]^T$ and link parameters $L_1 = 50$, $L_2 = L_3 = 140$ and $L_5 = 20$ is obtained by substituting these values of joint-variables in Eq. (3.48). This gives

$$T_E = \begin{bmatrix} 0 & -0.707 & 0.707 & 14.14 \\ 0 & 0.707 & 0.707 & 14.14 \\ -1 & 0 & 0 & 248 \\ 0 & 0 & 0 & 1 \end{bmatrix} \tag{3.50}$$

Example 3.8 *Stanford manipulator kinematics*

The last example considered is of a 6-DOF industrial manipulator. Formulate the direct kinematic model of the six degrees of freedom Stanford manipulator shown in Fig. 3.26.

Solution The Stanford arm, shown in Fig. 3.26, is characterized by a three degree of freedom arm and three degree of freedom wrist. The first three joints, two revolute and one prismatic, constitute the arm of the spherical (RRP) configuration (see Section 1.6.5). The last three revolute joints constitute a wrist of RRR configuration. The first three links are larger in size and are used to position the wrist and the last three links for the wrist are small in size and are used to orient the end-effector. Joint 1 is a revolute joint, which rotates the whole body about the vertical axis (joint axis 1). Joint 2 is also revolute and moves about horizontal axis (joint axis 2) in a plane perpendicular to axis of joint 1.

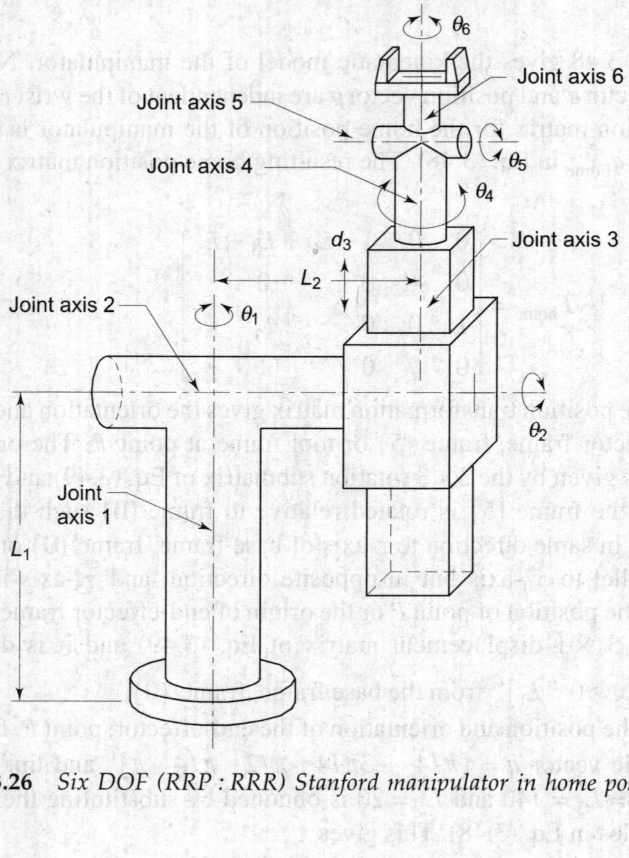

Fig. 3.26 *Six DOF (RRP : RRR) Stanford manipulator in home position*

Joint 3 is a prismatic joint that causes translational motion along joint axis 3 in a plane perpendicular to joint axis 2. The last three revolute joint motions are roll (joint 4), pitch (joint 5) and roll (joint 6) motions about joint axes 4, 5 and 6, respectively, which orient the end-effector. This wrist configuration is known as "Euler wrist", see Section 2.5.3. Observe that joint axes 4, 5 and 6 intersect at a point. Reader must note the differences between the RPY wrist discussed in Example 3.4 and the Euler wrist.

According to step 0, Algorithm 3.1, the six joints are numbered from 1 to 6 starting with joint 1 between link 0 (the immobile base) and link 1. The orientation of each joint axis and joint variables are identified and labeling corresponds to the home position as shown in Fig. 3.26. The reader must note that this is the home position of the manipulator in which all the joint variables are zero or at minimum.

The step-by-step frame assignment for each joint-link of the manipulator according to Algorithm 3.1 is explained below. Frames are first assigned to the intermediate links, links 1 to 5 and then to the link 0 and link 6. Refer Fig. 3.27 for the frame assignment.

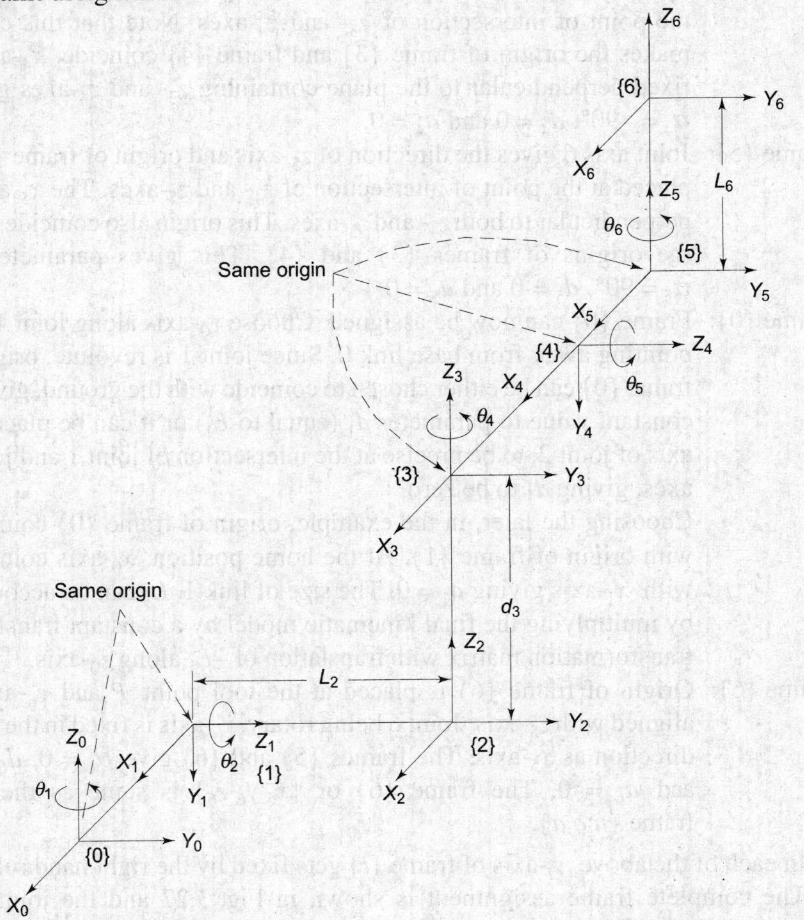

Fig. 3.27 *Frame assignment for chosen home position for Example 3.8*

Frame {1}: Align z_0-axis with joint axis 1 and z_1-axis with joint axis 2. Because z_0- and z_1-axes intersect (step 2(i) Algorithm 3.1), fix origin of frame {1} at the point of their intersection and set x_1-axis perpendicular to the plane containing z_0- and z_1-axes.
The y_1-axis completes the right-hand orthonormal coordinate frame {1}. This frame assignment gives $\alpha_1 = -90°$.

Frame {2}: Align z_2-axis with joint axis 3. Fix origin of frame {2} at the intersection of z_1- and z_2-axes and set x_2-axis perpendicular to the plane containing z_1- and z_2-axes. Frames {1} and {2} give $\alpha_2 = 90°$, $d_2 = L_2$ and $a_2 = 0$.

Frame {3}: Align z_3-axis with axis 4. Since z_2- and z_3-axes coincide (step 2(iii)) and joint 3 being prismatic, choose x_3-axis in the same direction as x_2-axis making constant parameter α_3 to be zero. The origin is placed at the distal end of the link 3 to coincide with axis 4. From frames {2} and {3}, $\theta_3 = 0$, $\alpha_3 = 0$ and $a_3 = 0$. Note that for prismatic joint d_3 is the joint variable.

Frame {4}: z_4-axis is aligned with joint axis 5 and origin of frame {4} is fixed at the point of intersection of z_3- and z_4-axes. Note that this choice makes the origin of frame {3} and frame {4} coincide. x_4-axis is fixed perpendicular to the plane containing z_3- and z_4-axes giving $\alpha_4 = -90°$, $d_4 = 0$ and $a_4 = 0$.

Frame {5}: Joint axis 6 gives the direction of z_5-axis and origin of frame {5} is placed at the point of intersection of z_4- and z_5-axes. The x_5-axis is perpendicular to both z_4- and z_5-axes. This origin also coincides with the origins of frames {3} and {4}. This gives parameters as $\alpha_5 = 90°$, $d_5 = 0$ and $a_5 = 0$.

Frame {0}: Frame {0} can now be assigned. Choose z_0-axis along joint 1 axis pointing away from base link 0. Since joint 1 is revolute, origin of frame {0} can be either chosen to coincide with the ground, giving a constant value to parameter d_1 (equal to L_1) or it can be placed on axis of joint 2, to be precise at the intersection of joint 1 and joint 2 axes, giving d_1 to be zero.
Choosing the later, in the example, origin of frame {0} coincides with origin of frame {1}. At the home position, x_0-axis coincides with x_1-axis giving $a_1 = 0$. The size of link 1, L_1 can be accounted by multiplying the final kinematic model by a constant translation transformation matrix with translation of $-L_1$ along z_0-axis.

Frame {6}: Origin of frame {6} is placed at the tool point P and z_6-axis is aligned with z_5-axis. Joint 6 being rotary, x_6-axis is fixed in the same direction as x_5-axis. The frames {5} and {6} give $\alpha_6 = 0$, $d_6 = L_6$ and $a_6 = 0$. The frame {6} or $\{x_6\, y_6\, z_6\}$ is same as the tool frame $\{n\, o\, a\}$.

In each of the above, y_i-axis of frame $\{i\}$ gets fixed by the right hand rule.

The complete frame assignment is shown in Fig. 3.27 and the joint-link parameters are tabulated in Table 3.9. In the frame assignments, wherever alternatives exist, the choice has been made so that as many fixed parameters as possible are zeros. This results in the simplest possible direct kinematic model that has a small number of non-zero parameters. As in earlier example the displacement variable for each link is identified and tabulated in Table 3.9. It is noted that 12 out of 18 constants parameters are zero.

Table 3.9 *Joint-link parameters for Stanford manipulator*

Link i	a_i	α_i	d_i	θ_i	q_i	$C\theta_i$	$S\theta_i$	$C\alpha_i$	$S\alpha_i$
1	0	–90°	0	θ_1	θ_1	C_1	S_1	0	–1
2	0	90°	L_2	θ_2	θ_2	C_2	S_2	0	1
3	0	0	d_3	0	d_3	1	0	1	0
4	0	–90°	0	θ_4	θ_4	C_4	S_4	0	–1
5	0	90°	0	θ_5	θ_5	C_5	S_5	0	1
6	0	0	L_6	θ_6	θ_6	C_6	S_6	1	0

The next step is to find the individual transformation matrices ${}^{i-1}T_i(q_i)$ between successive links by substituting the joint-link parameter values in Eq. (3.5). As usual to facilitate writing these matrices, four columns defining $\cos\theta_i$, $\sin\theta_i$, $\cos\alpha_i$ and $\sin\alpha_i$ are appended to the joint-link parameters table, Table 3.9, for each link.

The six link transformation matrices are, thus,

$$ {}^0\boldsymbol{T}_1(\theta_1) = \begin{bmatrix} C_1 & 0 & -S_1 & 0 \\ S_1 & 0 & C_1 & 0 \\ 0 & -1 & 0 & 0 \\ 0 & 0 & 0 & 1 \end{bmatrix} \tag{3.51} $$

$$ {}^1\boldsymbol{T}_2(\theta_2) = \begin{bmatrix} C_2 & 0 & S_2 & 0 \\ S_2 & 0 & -C_2 & 0 \\ 0 & 1 & 0 & L_2 \\ 0 & 0 & 0 & 1 \end{bmatrix} \tag{3.52} $$

$$ {}^2\boldsymbol{T}_3(d_3) = \begin{bmatrix} 1 & 0 & 0 & 0 \\ 0 & 1 & 0 & 0 \\ 0 & 0 & 1 & d_3 \\ 0 & 0 & 0 & 1 \end{bmatrix} \tag{3.53} $$

$$ {}^3\boldsymbol{T}_4(\theta_4) = \begin{bmatrix} C_4 & 0 & -S_4 & 0 \\ S_4 & 0 & C_4 & 0 \\ 0 & -1 & 0 & 0 \\ 0 & 0 & 0 & 1 \end{bmatrix} \tag{3.54} $$

$$
{}^4T_5(\theta_5) = \begin{bmatrix} C_5 & 0 & S_5 & 0 \\ S_5 & 0 & -C_5 & 0 \\ 0 & 1 & 0 & 0 \\ 0 & 0 & 0 & 1 \end{bmatrix} \tag{3.55}
$$

$$
{}^5T_6(\theta_6) = \begin{bmatrix} C_6 & -S_6 & 0 & 0 \\ S_6 & C_6 & 0 & 0 \\ 0 & 0 & 1 & L_6 \\ 0 & 0 & 0 & 1 \end{bmatrix} \tag{3.56}
$$

Finally, obtain the transformation of the tool frame with respect to the base frame by substituting the individual transform matrices in Eq. (3.6) as

$$
{}^0T_6 = {}^0T_1\,{}^1T_2\,{}^2T_3\,{}^3T_4\,{}^4T_5\,{}^5T_6 \tag{3.57}
$$

Substituting Eq. (3.51) to Eq. (3.56) in Eq. (3.57) and multiplying gives the final result for ${}^0T_6$ as in Eq. (3.58).

The manipulator (over all) transformation matrix ${}^0T_6$ represents the position and orientation of end-effector as a function of the six displacement variables θ_1, θ_2, d_3, θ_4, θ_5 and θ_6, other parameters are constant joint-link parameters.

Out of the 16 elements of the 4×4 homogeneous transformation matrix only 12 elements of upper 3×4 matrix are significant. Hence, only a maximum of 12 equations can be non-trivial. These equations give the forward kinematic model of the manipulator.

$$
{}^0T_6 = \begin{bmatrix}
\begin{matrix} C_1C_2C_4C_5C_6 \\ -S_1S_4C_5C_6 \\ -C_1S_2S_5C_6 \\ -C_1C_2S_4S_6 \\ -S_1C_4S_6 \end{matrix} &
\begin{matrix} C_1C_2C_4C_5S_6 \\ +S_1S_4C_5S_6 \\ +C_1S_2S_5S_6 \\ -C_1C_2S_4C_6 \\ -S_1C_4C_6 \end{matrix} &
\begin{matrix} C_1C_2C_4S_5 \\ -S_1S_4S_5 \\ +C_1S_2C_5 \end{matrix} &
\begin{matrix} C_1C_2C_4S_5L_6 \\ -S_1S_4S_5L_6 \\ +C_1S_2C_5L_6 \\ +C_1S_2d_3 - S_1L_2 \end{matrix} \\
\begin{matrix} S_1C_2C_4C_5C_6 \\ +C_1S_4C_5C_6 \\ -S_1S_2S_5C_6 \\ -S_1C_2S_4S_6 \\ +C_1C_4S_6 \end{matrix} &
\begin{matrix} -S_1C_2C_4C_5S_6 \\ -C_1S_4C_5S_6 \\ +S_1S_2S_5S_6 \\ -S_1C_2S_4C_6 \\ +C_1C_4C_6 \end{matrix} &
\begin{matrix} S_1C_2C_4S_5 \\ +C_1S_4S_5 \\ +S_1S_2C_5 \end{matrix} &
\begin{matrix} S_1C_2C_4S_5L_6 \\ +C_1S_4S_5L_6 \\ +S_1S_2C_5L_6 \\ +S_1S_2d_3 + C_1L_2 \end{matrix} \\
\begin{matrix} S_2C_4C_5C_6 \\ -C_2S_5C_6 \\ +S_2S_4S_6 \end{matrix} &
\begin{matrix} S_2C_4C_5S_6 \\ +C_2S_5S_6 \\ +S_2S_4C_6 \end{matrix} &
\begin{matrix} -S_2C_4S_5 \\ +C_2C_5 \end{matrix} &
\begin{matrix} -S_2C_4S_5L_6 \\ +C_2C_5L_6 + C_2d_3 \end{matrix} \\
0 & 0 & 0 & 1
\end{bmatrix} \tag{3.58}
$$

For given values of the joint displacements q_i and other constant joint-link parameters the end-effector position and orientation is obtained by equating ${}^0T_6$ to the end-effector orientation and position matrix T, Eq. (2.31), that is,

$$T = \begin{bmatrix} n_x & o_x & a_x & d_x \\ n_y & o_y & a_y & d_y \\ n_z & o_z & a_z & d_z \\ 0 & 0 & 0 & 1 \end{bmatrix} \tag{3.59}$$

Equating the elements of matrices in Eqs. (3.58) and (3.59), gives 12 equations as:

$$\begin{aligned}
n_x &= C_1C_2C_4C_5C_6 - S_1S_4C_5C_6 - C_1S_2S_5C_6 - C_1C_2S_4S_6 - S_1C_4S_6 \\
n_y &= S_1C_2C_4C_5C_6 + C_1S_4C_5C_6 - S_1S_2S_5C_6 - S_1C_2S_4S_6 + C_1C_4S_6 \\
n_z &= S_2C_4C_5C_6 - C_2S_5C_6 + S_2S_4S_6 \\
o_x &= C_1C_2C_4C_5S_6 + S_1S_4C_5S_6 + C_1S_2S_5S_6 - C_1C_2S_4C_6 - S_1C_4C_6 \\
o_y &= -S_1C_2C_4C_5S_6 - C_1S_4C_5S_6 + S_1S_2S_5S_6 - S_1C_2S_4C_6 + C_1C_4C_6 \\
o_z &= S_2C_4C_5S_6 + C_2S_5S_6 + S_2S_4C_6 \\
a_x &= C_1C_2C_4S_5 - S_1S_4S_5 + C_1S_2C_5 \\
a_y &= S_1C_2C_4S_5 + C_1S_4S_5 + S_1S_2C_5 \\
a_z &= -S_2C_4S_5 + C_2C_5 \\
d_x &= C_1C_2C_4S_5L_6 - S_1S_4S_5L_6 + C_1S_2C_5L_6 + C_1S_2d_3 - S_1L_2 \\
d_y &= S_1C_2C_4S_5L_6 + C_1S_4S_5L_6 + S_1S_2C_5L_6 + S_1S_2d_3 + C_1L_2 \\
d_z &= -S_2C_4S_5L_6 + C_2C_5L_6 + C_2d_3
\end{aligned} \tag{3.60}$$

Using this direct kinematic model for the home position ($\theta_1 = \theta_2 = \theta_4 = \theta_5 = \theta_6 = 0°$ and $d_3 = L_3$), assuming L_3 is minimum size of prismatic link, the end-effector position and orientation compute as

$$T = \left[\begin{array}{c:ccc} n_x & o_x & a_x & d_x \\ n_y & o_y & a_y & d_y \\ n_z & o_z & a_z & d_z \\ \hdashline 0 & 0 & 0 & 1 \end{array}\right] = \begin{bmatrix} 1 & 0 & 0 & 0 \\ 0 & 1 & 0 & L_2 \\ 0 & 0 & 1 & L_3 + L_6 \\ 0 & 0 & 0 & 1 \end{bmatrix} \tag{3.61}$$

This agrees with the coordinate system established in Fig. 3.27 and serves as a good check for the correctness of the model. It is important to note that the above 12 equations (Eq. (3.60)) require 10 transcendental function calls, 36 multiplications and 35 additions. Considerable computational saving can be obtained by calculating only 9 elements of the upper right 3×3 submatrix demarked by dotted lines in $\boldsymbol{T}$ above, Eq. (3.61). The first column of $\boldsymbol{T}$ can be obtained as a vector cross product of second and third columns. Further, if L_6 is made zero by shifting the end-effector frame origin (as was done in Example 3.4), this will reduce the computations for d_x, d_y and d_z significantly. This is left for the reader to verify.

EXERCISES

3.1 What are the parameters for a link for kinematic modeling? Which of these parameters are variable and which are constant for (a) a revolute joint, and (b) a prismatic joint?

3.2 Compute the manipulator transformation matrix for the 3-DOF manipulator arm with Cartesian (PPP) configuration. Three prismatic joints are perpendicular to each other and a possible frame assignment is given in Fig. E3.2.

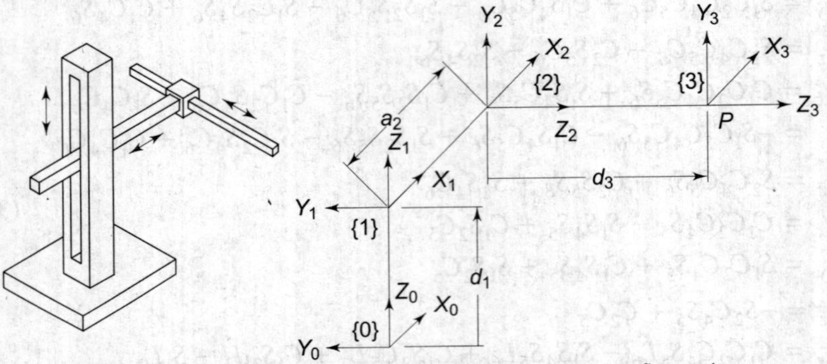

Fig. E3.2 *Frame assignment for a 3-DOF Cartesian configuration arm*

3.3 A 3-DOF cylindrical configuration arm of a manipulator has two prismatic joints and one revolute joint. A typical cylindrical arm with joint offset is shown in Fig. E3.3. Using the Algorithm 3.1 carry out frame assignments and tabulate the joint-link parameters.

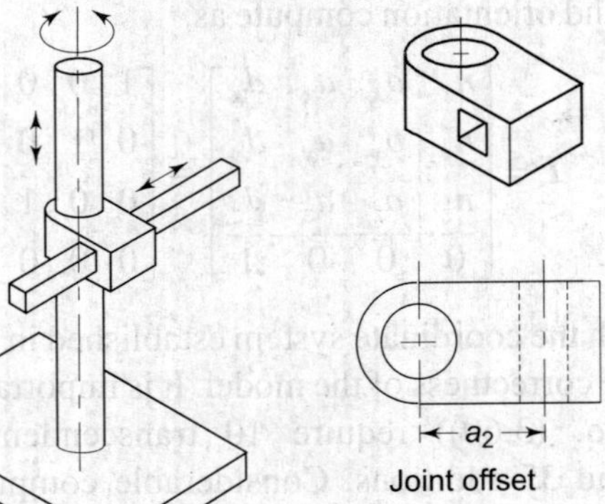

Fig. E3.3 *A 3-DOF cylindrical configuration arm*

3.4 For the cylindrical manipulator arm in Exercise 3.3, obtain the individual link transformation matrices and the overall arm transformation matrix. Verify your answer for the home position.

3.5 The 3R wrist in Example 3.8 is an Euler wrist. Obtain the wrist transformation matrix for the Euler wrist, see Example 3.4.

3.6 For the 3-DOF-manipulator arm shown in Fig. E3.6, assign frames and obtain the joint-link parameters. Also, determine the position of the tool tip with respect to the base frame {0}. Compare this kinematic model with Exercise 3.3.

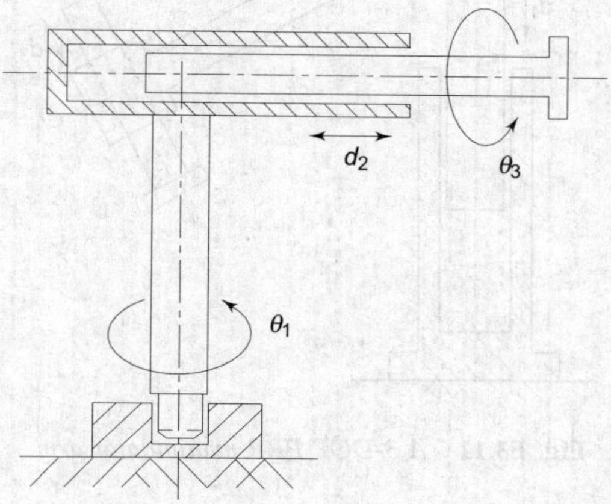

Fig. E3.6 *A 3-DOF RPR-configuration manipulator*

3.7 For the manipulator discussed in Exercise 3.6, obtain the rotation matrix that describes the orientation of the tool relative to the base. Also, compute the value of this rotation matrix for $\boldsymbol{q} = [60° \ 20 \ \ 30°]^T$.

3.8 For the manipulator discussed in Exercise 3.6, determine the transformation matrices relating successive links. Also, determine its forward kinematic model.

3.9 Obtain the tool transformation matrix for the 3-DOF articulated arm, shown in Fig. 3.16, if its home position is shifted to that shown in Fig. 3.17(b). Also, determine the tool position for $\boldsymbol{q} = [0 \ -90° \ 90°]^T$ and show that it is the same as the original home position.

3.10 Obtain the forward kinematic model for the SCARA manipulator of Example 3.6 taking the base frame at the table, that is, base of the column.

3.11 For the 3-DOF robotic manipulator arm shown in Fig. E3.11, assign frames to each of the links and determine the joint-link parameters and, therefrom, obtain the direct kinematic model.

3.12 Given that the position of an object relative to the base frame of manipulator in Fig. E3.11 as ${}^0\boldsymbol{P} = [7.0 \ 10.0 \ 5.0]^T$, determine the position of this object relative to the tool frame.

3.13 For the SCARA robot discussed in Example 3.6, compute the rotation matrix describing the tool relative to the base when the 4×1 joint-space vector is $\boldsymbol{q} = [45° \ 45° \ 5 \ \ 30°]^T$. Also, find the orientation of the tool in *ZYX*-Euler angle representation for the given joint space vector.

3.14 For a 5-DOF, RRR-RR articulated configuration manipulator shown in Fig. E3.14 obtain the forward kinematic model.

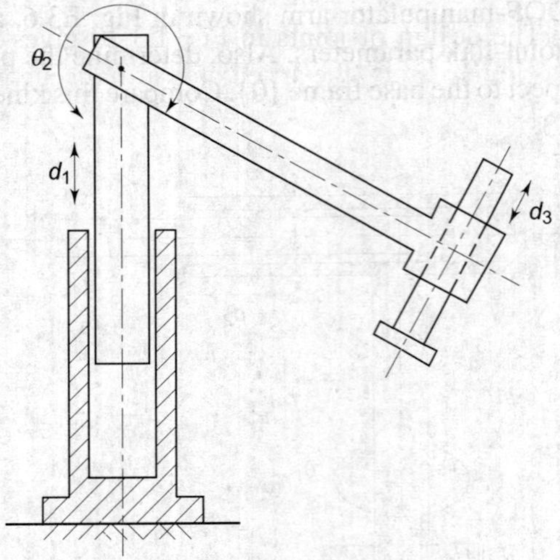

Fig. E3.11 *A 3-DOF PRP manipulator arm*

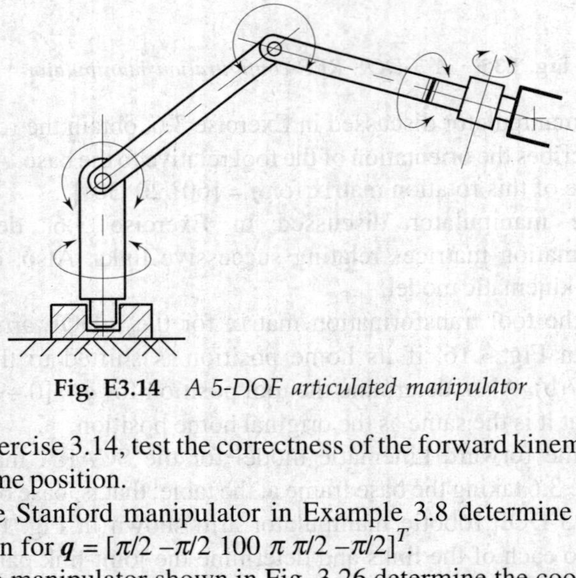

Fig. E3.14 *A 5-DOF articulated manipulator*

3.15 For Exercise 3.14, test the correctness of the forward kinematic model for the home position.

3.16 For the Stanford manipulator in Example 3.8 determine the tool point position for $\boldsymbol{q} = [\pi/2\ -\pi/2\ 100\ \pi\ \pi/2\ -\pi/2]^T$.

3.17 For the manipulator shown in Fig. 3.26 determine the coordinates of the tool point for the joint displacements $\boldsymbol{q} = [30°\ -20°\ 50\ 0°\ -108°\ 34°]^T$.

3.18 Consider the 6-DOF manipulator constructed by attaching an Euler wrist (see Example 3.8) to the faceplate of the 3-DOF articulated arm shown in Fig. 3.15. Attach frames to each link of this manipulator as per Algorithm 3.1 and obtain its forward kinematic model.

3.19 Repeat Exercise 3.18 for the manipulator by fixing a roll-pitch-yaw wrist to the faceplate of manipulator shown in Fig. 3.15.

3.20 A 3-DOF articulated configuration arm of a manipulator has all three revolute joints. In a typical articulated arm, the joint design determines the joint range. The design of joints in Fig. E3.20 gives almost 360° joint range but has joint offsets (the articulated arm discussed in Example 3.3 has no joint offsets). Using the Algorithm 3.1 carry out frame assignments, tabulate the joint-link parameters and obtain the forward kinematic model of the arm.

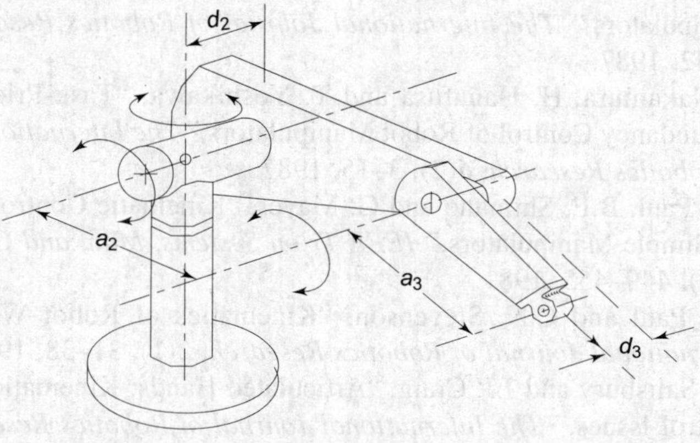

Fig. E3.20 *A 3-DOF articulated arm with joint offsets*

3.21 Why DH convention (Algorithm 3.1) does not give unique frame assignment for a given manipulator? Explain.

3.22 "The forward kinematic model of a manipulator depends on the choice of home position of the manipulator." Comment on this statement.

3.23 Using the DH notation for frame assignment, is it possible to have the a link with zero link length whereas the physical link on the manipulator will have a finite link length?

3.24 What problems will be encountered if the frames are arbitrarily assigned to develop the forward kinematic model of a manipulator?

3.25 Why is it important to choose a frame assignment for an n-DOF manipulator that gives a maximum number of zero joint-link parameters?

3.26 The frame assignment of a manipulator is so carried out that it is found that one of the link's rotation is about y-axis of the frame instead of z-axis. Can the transformation matrix of Eq. (3.3) be used for determining the transformation matrix for the link?

SELECTED BIBLIOGRAPHY

1. J. Denavit and R.S. Hartenberg, "A Kinematic Notation for Lower-Pair Mechanisms Based on Matrices," *ASME. J. of Applied Mechanics*, 215–221, Jun 1955.
2. Colin G. Johnson and Duncan Marsh, "Modeling Robot Manipulators with Multivariate B-Splines," *Robotica*, ***17***(3), 239–247, 1999.
3. Wisama Khalil and Denis Creusot, "SYMORO+: A System for the Symbolic Modeling of Robots," *Robotica*, ***15***, 153–161, 1997.

4. C.S.G. Lee, "Robot Arm Kinematics, Dynamics and Control," *IEEE Computer*, ***15***(12), 62–80, 1982.
5. J.Y.S. Luh, "An Anatomy of Industrial Robots and their Controls," *IEEE Tr on Automatic Control*, ***28***(2), 133–153, Feb 1983.
6. Y. Nakamura and H. Hanafusa, "Optimal Redundancy Control of Robot Manipulators," *The International Journal of Robotics Research*, ***6***(1), 32–42, 1987.
7. Y. Nakamura, H. Hanafusa and T. Yoshikawa, "Task-Priority Based Redundancy Control of Robot Manipulators," *The International Journal of Robotics Research*, ***6***(2), 3–15, 1987.
8. R.P. Paul, B.E. Shimano and G. Mayer, "Kinematic Control Equations for Simple Manipulators," *IEEE Tr on Systems, Man, and Cybernetics*, ***11***(6), 449–455, 1981.
9. R.P. Paul and C.N. Stevenson, "Kinematics of Robot Wrists," *The International Journal of Robotics Research*, ***2***(1), 31–38, 1983.
10. J.K. Salisbury and J.J. Craig, "Articulated Hands: Kinematic and Force Control Issues," *The International Journal of Robotics Research*, ***1***(1), 4–17, Sep 1982.
11. O.R. Williams, J. Angeles and F. Bulca, "Design Philosophy of an Isotropic Six-Axis Serial Manipulator," *Robotics and Computer-Integrated Manufacturing*, ***10***(4), 275–286, 1993.

4

The Inverse Kinematics

The direct kinematic model discussed in Chapter 3 determines the position and orientation of the end-effector (tool) for given values of joint-link displacements. In other words, it answers the question "Where is the origin of the end-effector?" The direct kinematic model, thus, specifies the end-effector frame, frame $\{n\}$ relative to the base frame $\{0\}$, for the n-DOF manipulator, which is expressed as

$$ {}^0\boldsymbol{T}_n(q_1, q_2, \ldots, q_n) = \prod_{i=1}^{n} {}^{i-1}\boldsymbol{T}_i = \begin{bmatrix} n_x & o_x & a_x & d_x \\ n_y & o_y & a_y & d_y \\ n_z & o_z & a_z & d_z \\ 0 & 0 & 0 & 1 \end{bmatrix} = \boldsymbol{T} \tag{4.1} $$

The joint displacements $(q_1, q_2, \ldots, q_n)$ that lead the end-effector to a certain position and orientation $\boldsymbol{T}$ can be found by solving the kinematic model equations for unknown joint displacements. Moving each joint by the respective joint displacement, the location (position and orientation) of the end-effector is achieved. This is the inverse kinematic problem already defined in the previous chapter, Section 3.3. It is possible that for the desired end-effector location, multiple or no solutions may exist. A rigorous definition for the inverse kinematic problem is:

"The determination of all possible and feasible sets of joint variables, which would achieve the specified position and orientation of the manipulator's end-effector with respect to the base frame."

In practice, a robot manipulator control requires knowledge of the end-effector position and orientation for the instantaneous location of each joint as well as knowledge of the joint displacements required to place the end-effector in a new location. Therefore, direct and inverse kinematics are the fundamental problems of utmost importance in the robot manipulator's position control. Many industrial applications such as welding and certain types of assembly operations require that a specific path should be negotiated by the end-effector. To achieve this, it is necessary to find corresponding motions of each joint, which will produce the desired tool-tip motion. This is a typical case of inverse kinematic application.

The manipulator transformation matrix $\boldsymbol{T}$ represents the orientation $\boldsymbol{R}$ and position $\boldsymbol{D}$ of the end-effector with respect to the base frame as:

$$\boldsymbol{T} = \left[\begin{array}{ccc|c} & \boldsymbol{R} & & \boldsymbol{D} \\ \hline 0 & 0 & 0 & 1 \end{array}\right] \tag{4.2}$$

The position and orientation of the end-effector is collectively referred as *configuration* of the end-effector. The configuration of the end-effector is represented by three position components as displacements along three orthogonal axes of base frame and three rotations about the base frame axes. These six components can be represented by a six dimensional space called *configuration space* or *Cartesian space*. The kinematic description of orientation of the end-effector with respect to the base frame can be according to any of the conventions outlined in Chapter 2. The configuration, or position and orientation, of the end-effector is a function of joint displacement variables q_1, q_2, ... q_n as shown in Fig. 4.1.

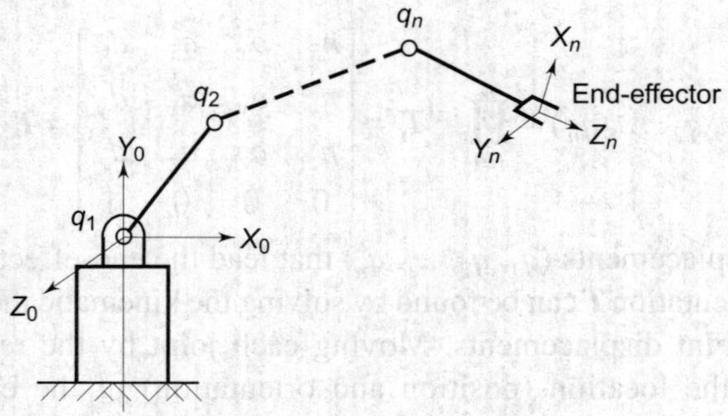

Fig. 4.1 *Configuration of end-effector as a function of joint displacements*

For an n-DOF manipulator the set of n joint displacement variables is represented by a $n \times 1$ vector. The set of all $n \times 1$ joint displacement vectors generates the *joint vector space* or *joint space*. The Cartesian space and joint space representations of a manipulator's end-effector position and orientation are related to each other by mappings shown in Fig. 4.2. The direct kinematic model is the mapping of joint space to Cartesian space, and the mapping from the Cartesian space to the joint space is the inverse problem.

In other words, the inverse kinematics is the determination of the set of positions and orientations in Cartesian space that are reachable by the origin of the end-effector frame as the joint displacements vector $\boldsymbol{q}$ ranges over the joint-space.

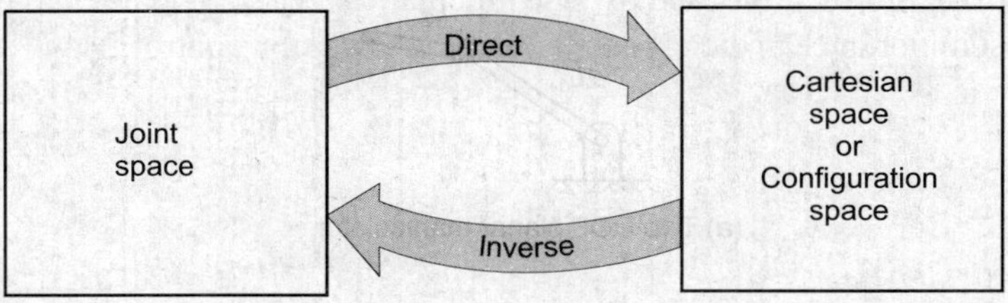

Fig. 4.2 *Mappings between kinematic descriptions*

The inverse kinematic problem is more difficult than the direct problem because no systematic procedures exist for its solution. Inverse problem of every manipulator has to be worked out separately. In this chapter, techniques to solve the inverse problem are presented. First, the concepts of manipulator workspace, solvability of kinematic equations, and existence of multiple solutions are discussed.

4.1 MANIPULATOR WORKSPACE

The workspace of a manipulator is defined as the volume of space in which the manipulator is able to locate its end-effector. The workspace gets specified by the existence or nonexistence of solutions to the inverse problem. The region that can be reached by the origin of the end-effector frame with at least one orientation is called the *reachable workspace* (RWS). If a point in workspace can be reached only in one orientation, the manipulatability of the end-effector is very poor and it is not possible to do any practical work satisfactorily with just one fixed orientation. It is, therefore, necessary to look for the points in workspace, which can be reached in more than one orientation. Some points within the RWS can be reached in more than one orientation. The space where the end-effector can reach every point from all orientations is called *dexterous workspace* (DWS). It is obvious that the dexterous workspace is either smaller (subset) or same as the reachable workspace.

As an example, consider a two-link nontrivial (2-DOF)-planar manipulator having link lengths L_1 and L_2, as shown in Fig. 4.3(a). The RWS for this manipulator is plane annular space with radii $r_1 = L_1 + L_2$ and $r_2 = |L_1 - L_2|$, as shown in Fig. 4.3(b). The DWS for this case is null. Inside the RWS there are two possible orientations of the end-effector for a given position, while on the boundaries of RWS, end-effector has only one possible orientation. For the special case of $L_1 = L_2$, the RWS is a circular area and DWS is a point at the center, as shown in Fig. 4.3(c). It can be shown that for a 3-DOF redundant planar manipulator having link lengths L_1, L_2, and L_3 with $(L_1 + L_2) > L_3$, the RWS is a circle of radius $(L_1 + L_2 + L_3)$, while the DWS is a circle of radius $(L_1 + L_2 - L_3)$.

The reachable workspace of an n-DOF manipulator is the geometric locus of the points that can be achieved by the origin of the end-effector frame as determined by the position vector of direct kinematic model. To locate the tool point or end-effector at an arbitrary position with an arbitrary orientation in 3-D space, a minimum of 6-DOF are required. Thus, for a 6-DOF-manipulator arm,

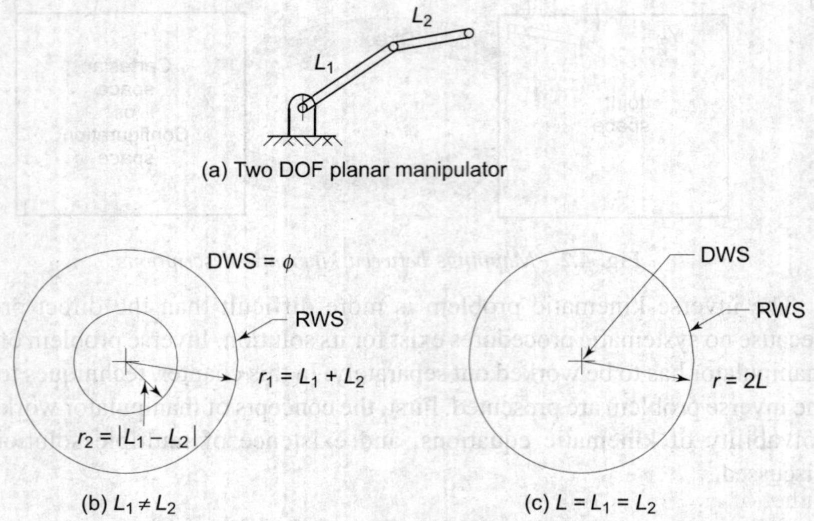

Fig. 4.3 *Workspace a of a two-link planar manipulator*

the dexterous workspace may almost approach the reachable workspace. The reachable work envelops of standard manipulator configurations have been discussed in Chapter 1.

The manipulator workspace is characterized by the mechanical joint limits in addition to the configuration and the number of degrees of freedom of the manipulator. It is important to note that in Fig. 4.3, the workspaces are specified assuming a full 360° rotational joint range for each revolute joint. In practice, the joint range of revolute motion is much less than 360° for the revolute joints and is severely limited for prismatic joints, due to mechanical constraints. This limitation greatly reduces the workspace of the manipulator and the shape of workspace may not be similar to the ideal case.

To understand the effect of mechanical joint limits on the workspace, consider the 2-DOF planar manipulator with $L_1 > L_2$ and joint limits on θ_1 and θ_2 as:

$$-60° \leq \theta_1 \leq 60° \tag{4.3}$$
$$-100° \leq \theta_2 \leq 100°$$

For these joint limits, considering $\theta_1 = \theta_2 = 0$ as home position, the annular workspace in Fig. 4.3(b) gets severely limited. The workspace, obtained geometrically, is not annular any more, rather it has a complex shape and is defined by contour ABCDEFA in Fig. 4.4.

Thus, the factors that decide the workspace of a manipulator apart from the number of degrees of freedom are the manipulator's configuration, link lengths, and the allowed range of joint motions.

4.2 SOLVABILITY OF INVERSE KINEMATIC MODEL

Inverse kinematics is complex because the solution is to be found for nonlinear simultaneous equations, involving transcendental (harmonic sine and cosine)

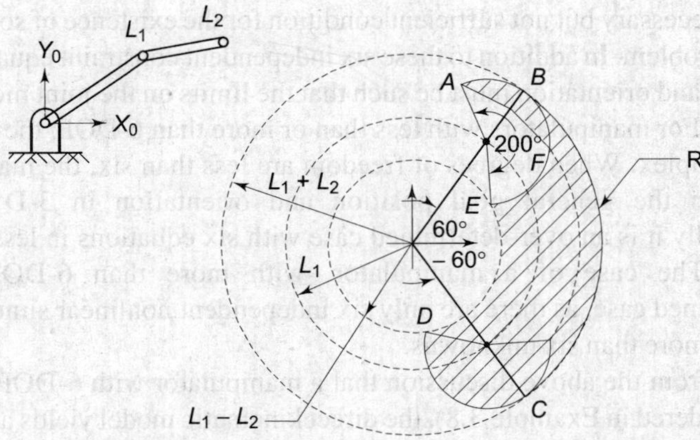

Fig 4.4 *Reachable workspace of a two-link planar manipulator with joint limits*

functions. The number of simultaneous equations is also generally more than the number of unknowns, making some of the equations mutually dependent. These conditions lead to the possibility of multiple solutions or nonexistence of any solution for the given end-effector position and orientation. The existence of solutions, multiple solutions, and methods of solutions are discussed in the following sections.

4.2.1 Existence of Solutions

The conditions for existence of solutions to the inverse kinematic problem are examined first. It is obvious that if the desired point P lies outside the reachable workspace then no solution exists. Even when P is within reachable workspace, not all orientations are realizable, unless P lies within dexterous workspace. If the wrist has fewer than 3-DOF to orient the end-effector, then certain classes of orientations are not realizable. To examine the reasons for this consider Eq. (4.1), the direct kinematic model.

As the last row of the matrix ${}^0\boldsymbol{T}_n$ in Eq. (4.1) is always constant, it can yield a maximum of 12 simultaneous equations, which are nonlinear algebraic equations involving transcendental functions in n unknowns (the joint variables). Out of these, nine equations arise from the rotation matrix (3 × 3) and three from the displacement vector. The nine equations of the rotation matrix involve only three unknowns corresponding to the orientation of the end-effector expressed by Euler angles or roll-pitch-yaw angles. This means that there are only six (three from orientation and three from displacement) independent constraints in n unknowns. This leads to a very important conclusion. For a manipulator to have all position and orientation solutions, the number of DOF n (equal to the number of unknowns) must at least match the number of independent constraints. That is, for general dexterous manipulation

$$n \geq 6 \tag{4.4}$$

This is a necessary but not sufficient condition for the existence of solutions to the inverse problem. In addition to these six independent constraint equations, the tool position and orientation must be such that the limits on the joint motions are not violated. For manipulators with less than or more than 6-DOF, the solutions are more complex. When degrees of freedom are less than six, the manipulator cannot attain the general goal position and orientation in 3-D space—mathematically it is an over-determined case with six equations in less than six unknowns. The case of a manipulator, with more than 6-DOF, is an underdetermined case, as there are only six independent nonlinear simultaneous equations in more than six unknowns.

It is seen from the above discussion that a manipulator with 6-DOF (such as the one considered in Example 3.8), the direct kinematic model yields a set of six independent equations in six unknowns. These six equations form a determinate set of simultaneous equations, which can be quite difficult to solve. It may be recalled that in direct kinematic model it was emphasized to choose the frames, which make as many of the joint-link parameters as possible, to zero. This leads to less complex kinematic equations and, hence, relatively simpler inverse solutions. For the case of a general mechanism with all nonzero-link parameters, the direct kinematic equations are much more complex and so will be the inverse solutions.

4.2.2 Multiple Solutions

The existence of multiple solutions is a common situation encountered in solving inverse kinematic problem. Multiple solutions pose further problem because the robot system has to have a capability to choose one, probably the best one. Multiple solutions can arise because of different factors. Some common situations, which lead to multiple solutions, are discussed as follows.

Consider the 2-DOF planar arm of Fig. 4.3(a) with a wrist having just one DOF. There are two sets of values of joint displacements (θ_1, θ_2) and (θ'_1, θ'_2), as illustrated in Fig. 4.5, which lead to the same end-effector position and orientation

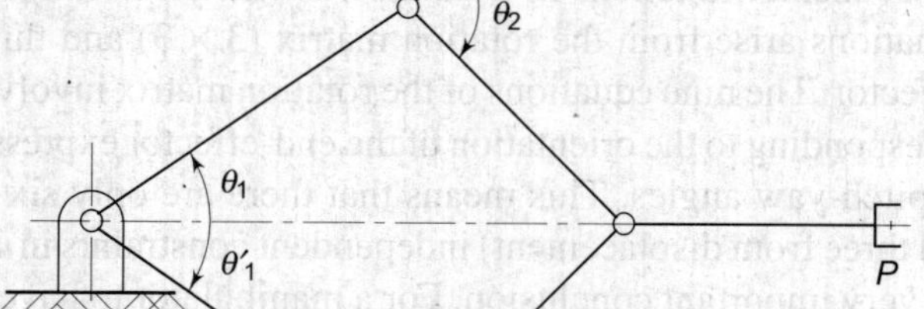

Fig. 4.5 *Multiple solutions due to parallel axes of revolute joints*

for point P. In the configuration space, both solutions are identical as they produce same configuration (position and orientation) of end-effector but are clearly distinct in joint space.

The solution (θ_1, θ_2) is 'elbow-up' position, while solution (θ'_1, θ'_2) is the 'elbow-down' position. Out of these, elbow-up solution may be preferred as in elbow-down solution joint-link may collide with objects lying on the work surface or the work table itself. The two solutions are obtained because the axes of two consecutive revolute joints of the arm are parallel. If more than two joint axes are parallel, the numbers of solutions multiply.

Another cause for multiple solutions is the existence of trigonometric functions in the equations. The harmonic nature of sine and cosine functions gives same magnitude for angles in multiples of π radians. For example, yaw, pitch, and roll motions of the RPY wrist for two sets of joint displacements $(\theta_1, \theta_2, \theta_3)$ and $(\theta'_1, \theta'_2, \theta'_3)$ with $\theta'_1 = 180° + \theta_1$, $\theta'_2 = 180° - \theta_2$ and $\theta'_3 = 180 + \theta_3$ will lead to the same orientation of the wrist. This can be easily verified.

Similarly, if the three motions of the wrist are roll, pitch, and roll; two sets of joint displacements $(\theta_1, \theta_2, \theta_3)$ and $(\theta'_1, \theta'_2, \theta'_3)$ with $\theta'_1 = 180° + \theta_1$, $\theta'_2 = -\theta_2$ and $\theta'_3 = 180° + \theta_3$ will lead to the same orientation of the wrist. With multiple solutions for positioning and orienting the end-effector, the number of solutions may multiply factorially.

The number of solutions also depends on the number of nonzero joint-link parameters and the range of joint motions allowed. In general, the number of ways to reach a certain goal is directly related to the number of nonzero link parameters. For example, for a completely general rotary-jointed, 6-DOF manipulator with all six $a_i \neq 0$, up to sixteen solutions are possible. A manipulator is said to be solvable, if it is possible to find all the solutions to its inverse kinematics problem for a given position and orientation.

Multiple solutions also arise from number of degree of freedom. For example, a manipulator with more than 6-DOF may have infinitely many solutions to the inverse kinematic problem. A manipulator with more degrees of freedom than are necessary is called *kinematically redundant* manipulator. The SCARA configuration is an example of redundant manipulator. It has one redundant degree of freedom in horizontal plane because only two joints (2-DOF) are needed to establish any horizontal position. Redundant manipulators have added flexibility, which can be useful in avoiding obstacles or reaching inaccessible locations, as illustrated in Fig. 4.6.

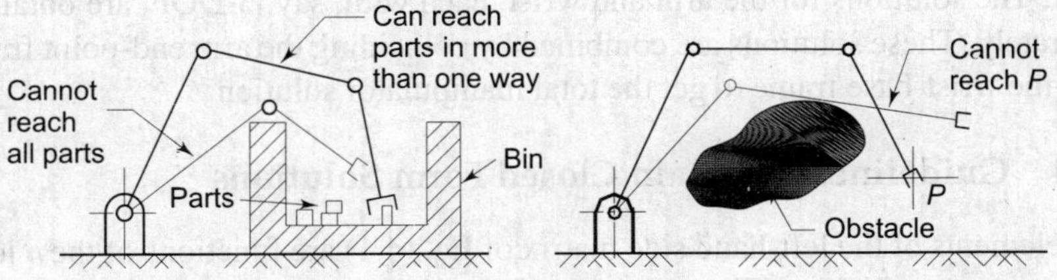

Fig. 4.6 *Use of redundant manipulator to avoid obstacles or reach around them*

4.3 SOLUTION TECHNIQUES

There are two approaches to the solutions to the inverse problem: *closed form solutions* and *numerical solutions*. In the closed form solution, joint displacements are determined as explicit functions of the position and orientation of the end-effector. In numerical methods, iterative algorithms such as the Newton-Raphson method are used. The numerical methods are computationally intensive and by nature slower compared to closed-form methods. Iterative solutions do not guarantee convergence to the correct solution in singular and degenerate cases. Iterative numerical techniques are not discussed in this text.

The "*closed form*" in the present context means a solution method based on analytical algebraic or kinematic approach, giving expressions for solving unknown joint displacements. The closed form solutions may not be possible for all kinds of structures. *A sufficient (but not necessary) condition for a 6-DOF manipulator to possess closed form solutions is that either its three consecutive joint axes intersect or its three consecutive joint axes are parallel.* The kinematic equations under either of these conditions can be reduced to algebraic equations of degree less than or equal to four for which closed form solutions exist. Almost every industrial manipulator manufactured today satisfies one of these conditions so that closed form solutions may be obtained. Manipulator arms with other kinematic structures may be solvable by analytical methods.

4.4 CLOSED FORM SOLUTIONS

Twelve equations, out of which only six are independent, are obtained by equating the elements of the manipulator transformation matrix with end-effector configuration matrix $\boldsymbol{T}$. At the same time, only six of the twelve elements of $\boldsymbol{T}$ specified by the end-effector position and orientation are independent. For a manipulator with less than 6-DOF, the number of independent equations may also be fewer than six. Several approaches such as, inverse transform, screw algebra, and kinematic approach and so on, can be used for solving these equations but none of them is general so as to solve the equations for every manipulator. A composite approach based on direct inspection, algebra, and inverse transform is presented here, which can be used to solve the inverse equations for a class of simple manipulators.

Another useful technique to reduce the complexity is dividing the problem into two smaller parts — the inverse kinematics of arm and the inverse kinematics of wrist. The solutions for the arm and wrist, each with, say, 3-DOF, are obtained separately. These solutions are combined by coinciding the arm end-point frame with the wrist-base frame to get the total manipulator solution.

4.4.1 Guidelines to Obtain Closed Form Solutions

The elements of the left-hand side matrix of Eq. (4.1) are functions of the n joint displacement variables. The elements of the right-hand side matrix $\boldsymbol{T}$ are the desired position and orientation of the end-effector and are either zero or constant. As the matrix equality implies element-by-element equality, 12 equations are obtained. To find the solution for n joint displacement variables from these 12 equations, the following guidelines are helpful.

(a) Look for equations involving only one joint variable. Solve these equations first to get the corresponding joint variable solutions.

(b) Next, look for pairs or set of equations, which could be reduced to one equation in one joint variable by application of algebraic and trigonometric identities.

(c) Use *arc tangent* (Atan2) function instead of *arc cosine* or *arc sine* functions. The two argument Atan2(y, x) function returns the accurate angle in the range of $-\pi \leq \theta \leq \pi$ by examining the sign of both y and x and detecting whenever either x or y is zero.

(d) Solutions in terms of the elements of the position vector components of ${}^0\boldsymbol{T}_n$ are more efficient than those in terms of elements of the rotation matrix, as latter may involve solving more complex equations.

(e) In the inverse kinematic model, the right-hand side of Eq. (4.1) is known, while the left-hand side has n unknowns ($q_1, q_2, \ldots, q_n$). The left-hand side consists of product of n link transformation matrices, that is

$$ {}^0\boldsymbol{T}_n = {}^0\boldsymbol{T}_1\ {}^1\boldsymbol{T}_2\ {}^2\boldsymbol{T}_3 \ldots\ {}^{n-1}\boldsymbol{T}_n = \boldsymbol{T} \tag{4.5} $$

Recall that each ${}^{i-1}\boldsymbol{T}_i$ is a function of only one unknown q_i. Premultiplying both sides by the inverse of ${}^0\boldsymbol{T}_1$ yields

$$ {}^1\boldsymbol{T}_n = {}^1\boldsymbol{T}_2\ {}^2\boldsymbol{T}_3 \ldots\ {}^{n-1}\boldsymbol{T}_n = [{}^0\boldsymbol{T}_1]^{-1}\boldsymbol{T} \tag{4.6} $$

The left-hand side of Eq. (4.6) has now (n–1) unknowns ($q_2, q_3, \ldots, q_n$) and the right-hand side matrix has only one unknown, the q_1. The matrix elements on the right-hand side are zero, constant, or function of the joint variable q_1. A new set of 12 equations is obtained and it may now be possible to determine q_1 from the elements of resulting equations using guideline (a) or (b) above. Similarly, by postmultiplying both sides of Eq. (4.5) by inverse of ${}^{n-1}\boldsymbol{T}_n$, unknown q_n can be determined. This process can be repeated by solving for one unknown at a time, sequentially from q_1 to q_n or q_n to q_1, until all unknowns are found. This is known as *inverse transform* approach.

The closed form solutions for the inverse kinematic model has been using the above guidelines are now illustrated with examples. For some of these examples the direct kinematic models have been obtained in Chapter 3. The multiple solutions and conditions for existence of solutions are also discussed. The first example considered is of the 3-DOF articulated arms solved in Example 3.3.

SOLVED EXAMPLES

Example 4.1 *Articulated arm inverse kinematics*

For the 3-DOF articulated arm, whose kinematic model has been obtained in Example 3.3, determine the joint displacements for known position and orientation of the end of the arm point.

Solution Let the known position and orientation of the endpoint of arm be given by

$$\boldsymbol{T} = \begin{bmatrix} r_{11} & r_{12} & r_{13} & r_{14} \\ r_{21} & r_{22} & r_{23} & r_{24} \\ r_{31} & r_{32} & r_{33} & r_{34} \\ 0 & 0 & 0 & 1 \end{bmatrix} \tag{4.7}$$

where each r_{ij} has a numeric value.

To obtain the solutions for joint variables (θ_1, θ_2, θ_3), in Eq. (4.7) $\boldsymbol{T}$ is equated to overall transformation matrix for the 3-DOF articulated arm ${}^0\boldsymbol{T}_3$ derived in Example 3.3, that is

$$\begin{bmatrix} C_1C_{23} & -C_1S_{23} & -S_1 & C_1(L_3C_{23} + L_2C_2) \\ S_1C_{23} & -S_1S_{23} & C_1 & S_1(L_3C_{23} + L_2C_2) \\ S_{23} & C_{23} & 0 & L_3S_{23} + L_2S_2 \\ 0 & 0 & 0 & 1 \end{bmatrix} = \begin{bmatrix} r_{11} & r_{12} & r_{13} & r_{14} \\ r_{21} & r_{22} & r_{23} & r_{24} \\ r_{31} & r_{32} & r_{33} & r_{34} \\ 0 & 0 & 0 & 1 \end{bmatrix} \tag{4.8}$$

Equation (4.8) gives 11 nontrivial equations for the three unknown joint variables, θ_1, θ_2, and θ_3 appearing on the left-hand side. The determination of solution for these three joint variables for known r_{ij} is the inverse kinematic problem and is worked out as follows.

Step 1 Applying guideline (a), an inspection of elements of the matrices on both the sides of Eq. (4.8) gives that θ_1 can be obtained from element 3 of row 1. The element (1,3) of left-hand side matrix has a term ($-S_1$) in only one variable θ_1 and a constant r_{13} on right-hand side and, hence, it can give angle θ_1 from $-\sin\theta_1 = r_{13}$. However, according to guideline (c) this is not preferred as correct quadrant of the angle can not be found. Alternatively, applying guideline (b), θ_1 can be isolated by dividing element (2, 1) by (1, 1) or (2, 2) by (1, 2) or (1, 3) by (2, 3) or (2, 4) by (1, 4). Out of these, the last one is preferred as per guideline (d). Thus, equating element (1, 4) and (2, 4), on both sides of the matrix two equations are obtained as

$$C_1(L_1C_{23} + L_2C_2) = r_{14} \tag{4.9}$$

$$S_1(L_3C_{23} + L_2C_2) = r_{24} \tag{4.10}$$

Dividing Eq. (4.10) by Eq. (4.9) gives

$$\frac{S_1}{C_1} = \frac{r_{24}}{r_{14}} \tag{4.11}$$

Therefore, according to guidline (c),

$$\theta_1 = \text{Atan2}(r_{24}, r_{14}) \tag{4.12}$$

Step 2 The other two unknowns, θ_2 and θ_3 cannot be obtained directly. To get a solution for θ_2 and θ_3, inverse transform approach, guideline (e), is used. To isolate θ_3, both sides of Eq. (4.8) are postmultiplied by $({}^2\boldsymbol{T}_3)^{-1}$. This will give

$${}^0\boldsymbol{T}_1\,{}^1\boldsymbol{T}_2 = \boldsymbol{T}[{}^2\boldsymbol{T}_3]^{-1} \tag{4.13}$$

From Eq. (3.21), ${}^2\boldsymbol{T}_3$ is

$$
{}^2\boldsymbol{T}_3 = \begin{bmatrix} C_3 & -S_3 & 0 & L_3C_3 \\ S_3 & C_3 & 0 & L_3S_3 \\ 0 & 0 & 1 & 0 \\ 0 & 0 & 0 & 1 \end{bmatrix} \tag{4.14}
$$

The inverse of ${}^2\boldsymbol{T}_3$ is obtained using Eq. (2.53) as

$$
[{}^2\boldsymbol{T}_3]^{-1} = \left[\begin{array}{ccc:c} & {}^2\boldsymbol{R}_3^T & & -{}^2\boldsymbol{R}_3^T\,{}^2\boldsymbol{D}_3 \\ \hdashline 0 & 0 & 0 & 1 \end{array}\right] = \begin{bmatrix} C_3 & S_3 & 0 & -L_3 \\ -S_3 & C_3 & 0 & 0 \\ 0 & 0 & 1 & 0 \\ 0 & 0 & 0 & 1 \end{bmatrix} \tag{4.15}
$$

Substituting ${}^0\boldsymbol{T}_1$ and ${}^1\boldsymbol{T}_2$ from Eqs. (3.19) and (3.20), Example 3.3, $\boldsymbol{T}$ from Eq. (4.7) and $[{}^2\boldsymbol{T}_3]^{-1}$ from Eq. (4.15), in Eq. (4.13) gives

$$
\begin{bmatrix} C_1C_2 & -C_1S_2 & S_1 & L_2C_1C_2 \\ S_1C_2 & -S_1S_2 & -C_1 & L_2S_1C_2 \\ S_2 & C_2 & 0 & L_2S_2 \\ 0 & 0 & 0 & 1 \end{bmatrix} = \begin{bmatrix} C_3r_{11} - S_3r_{12} & S_3r_{11} + C_3r_{12} & r_{13} & -L_3r_{11} + r_{14} \\ C_3r_{21} - S_3r_{22} & S_3r_{21} - S_3r_{22} & r_{23} & -L_3r_{21} + r_{24} \\ C_3r_{31} - S_3r_{32} & S_3r_{31} - S_3r_{32} & r_{32} & -L_3r_{31} + r_{34} \\ 0 & 0 & 0 & 1 \end{bmatrix} \tag{4.16}
$$

Note that the left-hand side of Eq. (4.16) has only θ_1 and θ_2 terms and the right-hand side has only θ_3 terms. A close examination of both sides reveals that equations obtained from elements (1, 4), (2, 4), and (3, 4) are only function of θ_1 and θ_2. Thus, with use of some algebra and trigonometric identities, θ_1 can be eliminated and solution for θ_2 is obtained. Equating the elements (1, 4), (2, 4), and (3, 4) of the two matrices, three equations obtained are

$$
L_2C_1C_2 = -L_3r_{11} + r_{14} \tag{4.17}
$$

$$
L_2S_1C_2 = -L_3r_{21} + r_{24} \tag{4.18}
$$

$$
L_2S_2 = -L_3r_{31} + r_{34} \tag{4.19}
$$

By squaring Eqs. (4.17) and (4.18), and adding gives,

$$
L_2^2C_2^2(C_1^2 + S_1^2) = (-L_3r_{11} + r_{14})^2 + (-L_3r_{21} + r_{24})^2
$$

From this θ_1 is eliminated because $C_1^2 + S_1^2 = 1$, thus

$$
L_2C_2 = \pm\sqrt{(-L_3r_{11} + r_{14})^2 + (-L_3r_{21} + r_{24})^2} \tag{4.20}
$$

Dividing Eq. (4.19) by Eq. (4.20), gives

$$
\frac{S_2}{C_2} = \frac{-L_3r_{31} + r_{34}}{\pm\sqrt{(-L_3r_{11} + r_{14})^2 + (-L_3r_{21} + r_{24})^2}} \tag{4.21}
$$

Hence,

$$
\theta_2 = \text{Atan2}\left((-L_3r_{31} + r_{34}), \pm\sqrt{(-L_3r_{11} + r_{14})^2 + (-L_3r_{21} + r_{24})^2}\right) \tag{4.22}
$$

Step 3 The solution for θ_3 is obtained by first solving for $(\theta_2 + \theta_3)$. Dividing element (3,1) of Eq. (4.8) by element (3,2) gives

$$\frac{S_{23}}{C_{23}} = \frac{r_{31}}{r_{32}} \tag{4.23}$$

or $$\theta_2 + \theta_3 = \text{Atan2}(r_{31}, r_{32}) \tag{4.24}$$

Thus,

$$\theta_3 = \text{Atan2}\,(r_{31}, r_{32}) - \theta_2 \tag{4.25}$$

Equations (4.12), (4.22), and (4.25) give the complete solution for the 3-DOF articulated arm as expressions for the joint displacements θ_1, θ_2, and θ_3 in terms of known arm end-point position and orientation. Note that the above solution is one of the possible sets of expressions. Alternate expressions for θ_1, θ_2, and θ_3 would be obtained if instead of equating the chosen elements of the matrices, other elements are used, or instead of isolating θ_3, θ_1 is isolated by premultiplying both sides with $({}^0\boldsymbol{T}_1)^{-1}$. It is also possible to find solution without use of the inverse matrix approach and instead of using algebra and trigonometry. For instance, after solving for θ_1, $(\theta_2 + \theta_3)$ can be obtained from elements (3, 1) and (3, 2) and through trigonometric manipulation, θ_2 and θ_3 are obtained.

Example 4.2 *Inverse kinematics of RPY wrist*

For the 3-DOF RPY wrist kinematic model was obtained in Example 3.4, Eq. (3.27), as

$${}^0\boldsymbol{T}_3 = \begin{bmatrix} -C_1S_2C_3 + S_1S_3 & C_1S_2S_3 + S_1C_3 & C_1C_2 & 0 \\ -S_1S_2C_3 - C_1S_3 & S_1S_2S_3 - C_1C_3 & S_1C_2 & 0 \\ C_2C_3 & -C_2S_3 & S_2 & 0 \\ 0 & 0 & 0 & 1 \end{bmatrix} \tag{4.26}$$

Determine the solution for the three joint variables for a given end-effector orientation matrix $\boldsymbol{T}_E$.

$$\boldsymbol{T}_E = \begin{bmatrix} n_x & o_x & a_x & 0 \\ n_y & o_y & a_y & 0 \\ n_z & o_z & a_z & 0 \\ 0 & 0 & 0 & 1 \end{bmatrix} \tag{4.27}$$

Solution The overall transformation matrix ${}^0\boldsymbol{T}_3$ and end-effector matrix $\boldsymbol{T}_E$ represent the same transformations. Thus, equating Eqs. (4.26) and (4.27) gives

$$\begin{bmatrix} n_x & o_x & a_x & 0 \\ n_y & o_y & a_y & 0 \\ n_z & o_z & a_z & 0 \\ 0 & 0 & 0 & 1 \end{bmatrix} = \begin{bmatrix} -C_1S_2C_3 + S_1S_3 & C_1S_2S_3 + S_1C_3 & C_1C_2 & 0 \\ -S_1S_2C_3 - C_1S_3 & S_1S_2S_3 - C_1C_3 & S_1C_2 & 0 \\ C_2C_3 & -C_2S_3 & S_2 & 0 \\ 0 & 0 & 0 & 1 \end{bmatrix} \tag{4.28}$$

The elements of the matrix on left-hand side of matrix equation are known (given), while, the elements of matrix on right-hand side have three unknown joint variables θ_1, θ_2 and θ_3. To get the solution for these joint variables, the more consistent analytical approach (guideline (e)) is used here.

Guideline (e) suggests premultiplying the matrix equation, Eq. (4.28) by inverse of transformation matrix ${}^0\boldsymbol{T}_1$ involving the unknown θ_1 and from the elements of the resultant matrix equation determine the unknown. Recall that the right-hand side of Eq. (4.28) is the product of three transformation matrices ${}^0\boldsymbol{T}_1$, ${}^1\boldsymbol{T}_2$ and ${}^2\boldsymbol{T}_3$ each involving one unknown θ_1, θ_2 and θ_3, respectively.

This process is continued successively, that is, moving one unknown (by its inverse transform) from right-hand side of the matrix equation to the left-hand side of the matrix equation and solving it, then moving the next unknown to the left-hand side, until all unknown are solved.

To solve for θ_1, both sides of Eq. (4.28) are premultiplied by ${}^0\boldsymbol{T}_1^{-1}$. From Eqs. (3.24) – (3.26)

$$\begin{bmatrix} C_1 & S_1 & 0 & 0 \\ 0 & 0 & 1 & 0 \\ S_1 & -C_1 & 0 & 0 \\ 0 & 0 & 0 & 1 \end{bmatrix}\begin{bmatrix} n_x & o_x & a_x & 0 \\ n_y & o_y & a_y & 0 \\ n_z & o_z & a_z & 0 \\ 0 & 0 & 0 & 1 \end{bmatrix} = \begin{bmatrix} -S_2 & 0 & C_2 & 0 \\ C_2 & 0 & S_2 & 0 \\ 0 & 1 & 0 & 0 \\ 0 & 0 & 0 & 1 \end{bmatrix}\begin{bmatrix} C_3 & -S_3 & 0 & 0 \\ S_3 & C_3 & 0 & 0 \\ 0 & 0 & 1 & 0 \\ 0 & 0 & 0 & 1 \end{bmatrix}$$

or

$$\begin{bmatrix} C_1 n_x + S_1 n_y & C_1 o_x + S_1 o_y & C_1 a_x + S_1 a_y & 0 \\ n_z & o_z & a_z & 0 \\ S_1 n_x - C_1 n_y & S_1 o_x - C_1 o_y & S_1 a_x - C_1 a_y & 0 \\ 0 & 0 & 0 & 1 \end{bmatrix} = \begin{bmatrix} -S_2 C_3 & S_2 S_3 & C_2 & 0 \\ C_2 C_3 & -C_2 S_3 & S_2 & 0 \\ S_3 & C_3 & 0 & 0 \\ 0 & 0 & 0 & 1 \end{bmatrix} \tag{4.29}$$

The left-hand side of Eq. (4.29) has one unknown (θ_1) and the right-hand side has two unknown (θ_2 and θ_3). Scanning the elements of both the matrices in Eq. (4.29), the equation in one unknown (θ_1) is obtained by equating elements (3, 3). That is,

$$S_1 a_x - C_1 a_y = 0 \tag{4.30}$$

or

$$\frac{S_1}{C_1} = \tan\theta_1 = \frac{a_y}{a_x}$$

which gives

$$\theta_1 = \text{Atan2}(a_y, a_x) \tag{4.31}$$

The process of further premultiplication is not necessary because the solutions for the remaining two unknowns (θ_2 and θ_3) can be obtained from Eq. (4.29). Equating (1, 3) and (2, 3) elements on both sides in Eq. (4.29) gives

$$C_2 = C_1 a_x + S_1 a_y$$
$$S_2 = a_z \tag{4.32}$$

From these two equations the solution for θ_2 is obtained as

$$\theta_2 = \text{Atan2}(a_z,\ C_1 a_x + S_1 a_y) \tag{4.33}$$

Equating elements (3,1) and (3, 2) of Eq. (4.29) gives

$$S_3 = S_1 n_x - C_1 n_y$$

$$C_3 = S_1 o_x - C_1 o_y \tag{4.34}$$

Which lead to the solution for θ_3 as

$$\theta_3 = \text{Atan2}(S_1 n_x - C_1 n_y,\ S_1 o_x - C_1 o_y) \tag{4.35}$$

The inverse transform technique is to move one unknown to the left-hand side at a time and solve it. Therefore, it is also possible to achieve this by postmultiplying instead of premultiplying by the inverse transform matrix involving an unknown. This is illustrated here.

To solve for θ_3, that is, to move it to the left-hand side, postmultiplying both sides of matrix equation Eq. (4.28) by ${}^2\boldsymbol{T}_3^{-1}$ gives

$$\begin{bmatrix} n_x & o_x & a_x & 0 \\ n_y & o_y & a_y & 0 \\ n_z & o_z & a_z & 0 \\ 0 & 0 & 0 & 1 \end{bmatrix} \begin{bmatrix} C_3 & S_3 & 0 & 0 \\ -S_3 & C_3 & 0 & 0 \\ 0 & 0 & 1 & 0 \\ 0 & 0 & 0 & 1 \end{bmatrix} = \begin{bmatrix} C_1 & 0 & S_1 & 0 \\ S_1 & 0 & -C_1 & 0 \\ 0 & 1 & 0 & 0 \\ 0 & 0 & 0 & 1 \end{bmatrix} \begin{bmatrix} -S_2 & 0 & C_2 & 0 \\ C_2 & 0 & S_2 & 0 \\ 0 & 1 & 0 & 0 \\ 0 & 0 & 0 & 1 \end{bmatrix}$$

or

$$\begin{bmatrix} C_3 n_x - S_3 o_x & S_3 n_x + C_3 o_x & a_x & 0 \\ C_3 n_y - S_3 o_y & S_3 n_y + C_3 o_y & a_y & 0 \\ C_3 n_z - S_3 o_z & S_3 n_z + C_3 o_z & a_z & 0 \\ 0 & 0 & 0 & 1 \end{bmatrix} = \begin{bmatrix} -C_1 S_2 & S_1 & C_1 C_2 & 0 \\ -S_1 S_2 & -C_1 & S_1 C_2 & 0 \\ C_2 & 0 & S_2 & 0 \\ 0 & 0 & 0 & 1 \end{bmatrix} \tag{4.36}$$

Comparing elements of the matrices on both sides, the elements (3, 2) gives

$$S_3 n_z + C_3 o_z = 0$$

and, thus, the solution for θ_3 is

$$\theta_3 = \text{Atan2}\ (-o_z,\ n_y) \tag{4.37}$$

Similarly, from the elements (3,1) and (3, 3), θ_2 is obtained as

$$\theta_2 = \text{Atan2}\ (a_z,\ C_3 n_z - S_3 o_z) \tag{4.38}$$

and from the elements (1, 2) and (2, 2), θ_1 is obtained as

$$\theta_1 = \text{Atan2}(S_3\, n_x + C_3 o_x,\ -S_3\, n_y - C_3 o_y) \tag{4.39}$$

In this example, premulitplying or postmultiplying gives solution of similar complexity but this may not be always the case. The decision to premultiply or postmultiply is left to the discretion of the reader.

Example 4.3 ***SCARA manipulator inverse kinematics***

Analytically solve the inverse kinematic problem for the 4-DOF SCARA configuration manipulator given in Fig. 3.22, Example 3.6. Discuss the conditions for existence and multiplicity of solutions.

Solution For the SCARA manipulator of Example 3.6, equating ${}^0\boldsymbol{T}_4$ from Eq. (3.40) with $\boldsymbol{T}$ in Eq. (4.7) gives

$$\begin{bmatrix} C_{124} & S_{124} & 0 & L_2C_{12}+L_{11}C_1 \\ S_{124} & -C_{124} & 0 & L_2S_{12}+L_{11}S_1 \\ 0 & 0 & 1 & L_{12}+d_3-L_4 \\ 0 & 0 & 0 & 1 \end{bmatrix} = \begin{bmatrix} r_{11} & r_{12} & r_{13} & r_{14} \\ r_{21} & r_{22} & r_{23} & r_{24} \\ r_{31} & r_{32} & r_{33} & r_{34} \\ 0 & 0 & 0 & 1 \end{bmatrix} \tag{4.40}$$

The solution for joint displacement d_3 is directly obtained by equating the elements (3, 4) on both sides of Eq. (4.40),

$$L_{12} + d_3 - L_4 = r_{34}$$

or

$$d_3 = r_{34} + L_4 - L_{12} \tag{4.41}$$

Next, to solve for θ_1 elements (1, 4) and (2, 4) are compared. This gives

$$L_2C_{12} + L_{11}C_1 = r_{14} \tag{4.42}$$

$$L_2S_{12} + L_{11}S_1 = r_{24} \tag{4.43}$$

Squaring Eqs. (4.42) and (4.43), adding and simplifying using the trigonometric identity $\cos(\alpha+\beta) = \cos\alpha\cos\beta - \sin\alpha\sin\beta$, gives

$$L_{11}^2 + L_2^2 + 2L_{11}L_2C_2 = r_{14}^2 + r_{24}^2 \tag{4.44}$$

$$C_2 = \frac{r_{14}^2 + r_{24}^2 - L_{11}^2 - L_2^2}{2L_{11}L_2} \tag{4.45}$$

Since

$$S_2 = \pm\sqrt{1-C_2^2} \tag{4.46}$$

the solution for θ_2 is obtained from Eqs. (4.45) and (4.46) as

$$\theta_2 = \text{Atan2}(S_2,\ C_2) \tag{4.47}$$

Now that θ_2 being known, Eqs. (4.42) and (4.43) can be used to compute θ_1. These equations are written as

$$L_2(C_1C_2 - S_1S_2) + L_{11}C_1 = r_{14} \tag{4.48}$$

$$L_2(S_1C_2 + C_1S_2) + L_{11}S_1 = r_{24} \tag{4.49}$$

or

$$(L_{11} + L_2C_2)C_1 - (L_2S_2)S_1 = r_{14} \tag{4.50}$$

$$(L_{11} + L_2C_2)S_1 + (L_2S_2)C_1 = r_{24} \tag{4.51}$$

Let

$$(L_{11} + L_2C_2) = r\cos\phi \ \text{ and } \ (L_2S_2) = r\sin\phi \tag{4.52}$$

with

$$r = \sqrt{(L_{11} + L_2C_2)^2 + (L_2S_2)^2} \tag{4.53}$$

and

$$\phi = \text{Atan2}\left(\frac{L_2S_2}{r},\ \frac{L_{11} + L_2C_2}{r}\right) \tag{4.54}$$

Eqs. (4.50) and (4.51) reduce to

$$r\cos(\theta_1 + \phi) = r_{14} \tag{4.55}$$

$$r\sin(\theta_1 + \phi) = r_{24} \tag{4.56}$$

From Eqs. (4.55) and (4.56), θ_1 is obtained as

$$\theta_1 = \text{Atan2}\left(\frac{r_{24}}{r}, \frac{r_{14}}{r}\right) - \phi \tag{4.57}$$

$$\theta_1 = \text{Atan2}\left(\frac{r_{24}}{r}, \frac{r_{14}}{r}\right) - \text{Atan2}\left(\frac{L_2 S_2}{r}, \frac{L_{11} + L_2 C_2}{r}\right) \tag{4.58}$$

With θ_1, θ_2 and d_3 determined, only one variable, θ_4 is unknown. From elements (1,1) and (2,1), the equations are

$$C_{124} = r_{11} \tag{4.59}$$

$$S_{124} = r_{21} \tag{4.60}$$

Equations (4.59) and (4.60) give θ_4 as

$$\theta_1 + \theta_2 - \theta_4 = \text{Atan2}(r_{21}, r_{11})$$

or

$$\theta_4 = \theta_2 + \theta_1 - \text{Atan2}\,(r_{21}, r_{11}) \tag{4.61}$$

The complete closed form solution for the joint displacements θ_1, θ_2, d_3 and θ_4, of SCARA manipulator, is given by Eqs. (4.58), (4.47), (4.41) and (4.61), respectively, as explicit functions of the manipulator's tool position and orientation.

Existence of Solutions

Solutions to the inverse kinematic problem of a given arm exist, that is, the given Cartesian position and orientation of the tool is within the manipulator's workspace if the following condition is satisfied:

"Since *sin* and *cos* functions take values in the range [–1,1], right-hand side of the Eq. (4.45) must lie in the range [–1,1]."

Again, these solutions are for full 360 degrees of rotation for the revolute joints and limitless translation for the prismatic joint. The mechanical constraints, however, will permit only such solutions for which the joint variables take a value that lies in the range of motions allowed.

Multiplicity of Solutions

Due to the presence of the square root in Eq. (4.46), there are two solutions for θ_2 for a given position and orientation of the tool with respect to the base. From Eqs. (4.57) and (4.61), observe that there is one set of solution for θ_1 and θ_4 corresponding to each value of θ_2. Thus, the number of solutions to the inverse kinematics problem of the given SCARA arm is two. Note that multiple solutions exist due to the fact that revolute joint axes 1 and 2 are parallel.

Example 4.4 *Numerical solutions for a 3-DOF manipulator*

For the 3-DOF (RRP) configuration manipulator, shown in Fig. 4.7, the position and orientation of point *P* in Cartesian space is given by

$$\boldsymbol{T} = \begin{bmatrix} 0.354 & 0.866 & 0.354 & 0.106 \\ -0.612 & 0.500 & -0.612 & -0.184 \\ 0.707 & 0 & 0.707 & 0.212 \\ 0 & 0 & 0 & 1 \end{bmatrix} \quad (4.62)$$

Determine all values of all joint variables, that is, all solutions to the inverse kinematic problem. The joint displacements allowed (joint limits) for three joints are: $-100° < \theta_1 < 100°$, $-30° < \theta_2 < 70°$ and $0.05\text{m} < d_3 < 0.5$ m. Identify the feasible solutions.

Solution A manipulator with this configuration is another common structure widely used in industrial robots, as it is very effective in material handling and other applications, and gives a spherical workspace. The first two joints are revolute joints and provide motion in two perpendicular planes. Their sweep generates a constant radius sphere. The third prismatic joint provides the reach to the arm point where the wrist is attached. The three joint axes intersect at a point.

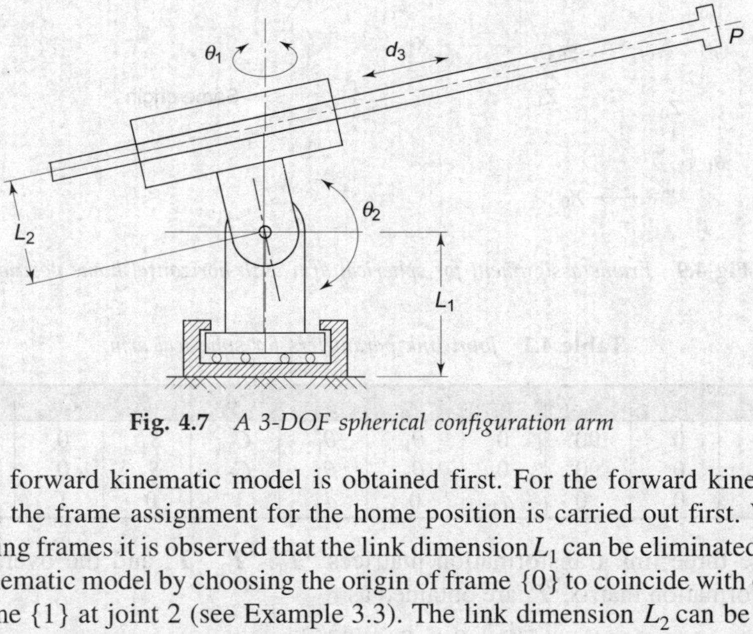

Fig. 4.7 *A 3-DOF spherical configuration arm*

The forward kinematic model is obtained first. For the forward kinematic model, the frame assignment for the home position is carried out first. While assigning frames it is observed that the link dimension L_1 can be eliminated from the kinematic model by choosing the origin of frame {0} to coincide with origin of frame {1} at joint 2 (see Example 3.3). The link dimension L_2 can be made zero by modifying the design slightly as shown in Fig. 4.8 such that the axis of prismatic link passes through the origin of frame {1}.

The final frame assignment with the origin of three frames, frame {0}, frame {1} and frame {2} at the same point is shown in Fig. 4.9. This minimizes the number of non-zero parameters as well as satisfies the necessary condition for existence of closed form solutions.

The joint-link parameters are tabulated in Table 4.1

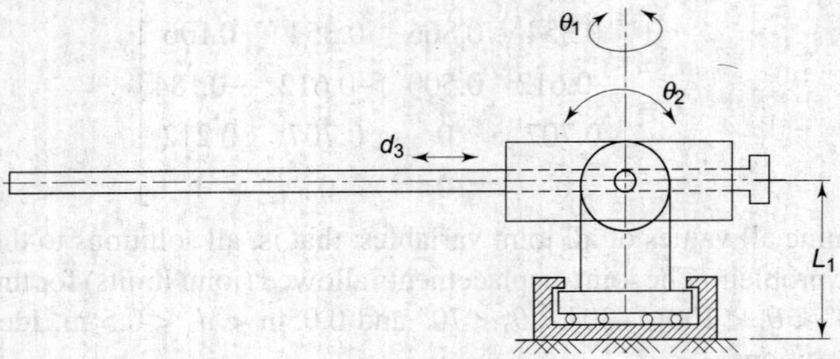

Fig. 4.8 *3-DOF spherical arm in home position: $\theta_1 = \theta_2 = 0$ and $d_3 = 0.05$*

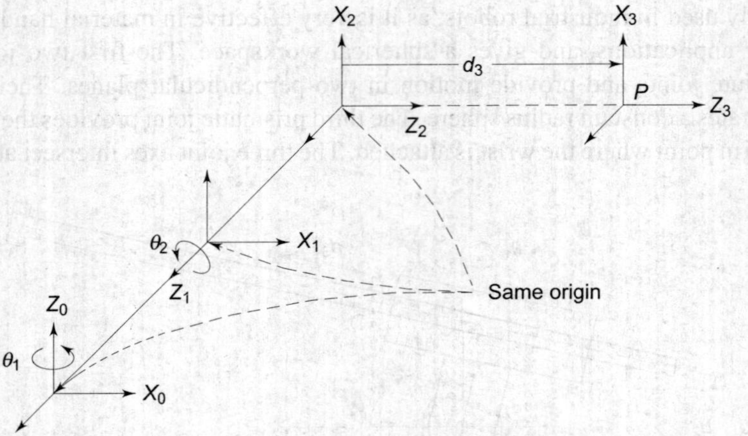

Fig 4.9 *Frame assignment for spherical arm with horizontal home position*

Table 4.1 *Joint-link parameters for spherical arm*

Link i	a_i	α_i	d_i	θ_i	q_i	$C\theta_i$	$S\theta_i$	$C\alpha_i$	$S\alpha_i$
1	0	90°	0	θ_1	θ_1	C_1	S_1	0	1
2	0	−90°	0	θ_2	θ_2	C_2	S_2	0	1
3	0	0	d_3	0	d_3	1	0	1	0

The three link transformation matrices ${}^0\boldsymbol{T}_1$, ${}^1\boldsymbol{T}_2$, ${}^2\boldsymbol{T}_3$ and the overall arm transformation matrix ${}^0\boldsymbol{T}_3$ are obtained as

$$ {}^0\boldsymbol{T}_1(\theta_1) = \begin{bmatrix} C_1 & 0 & S_1 & 0 \\ S_1 & 0 & -C_1 & 0 \\ 0 & 1 & 0 & 0 \\ 0 & 0 & 0 & 1 \end{bmatrix} \tag{4.63} $$

$$ {}^1\boldsymbol{T}_2(\theta_2) = \begin{bmatrix} C_2 & 0 & -S_2 & 0 \\ S_2 & 0 & C_2 & 0 \\ 0 & -1 & 0 & 0 \\ 0 & 0 & 0 & 1 \end{bmatrix} \tag{4.64} $$

$$ {}^2\boldsymbol{T}_3(d_3) = \begin{bmatrix} 1 & 0 & 0 & 0 \\ 0 & 1 & 0 & 0 \\ 0 & 0 & 1 & d_3 \\ 0 & 0 & 0 & 1 \end{bmatrix} \tag{4.65} $$

and, thus,

$$ {}^0\boldsymbol{T}_3 = {}^0\boldsymbol{T}_1\,{}^1\boldsymbol{T}_2\,{}^2\boldsymbol{T}_3 = \begin{bmatrix} C_1C_2 & -S_1 & -C_1S_2 & -d_3C_1S_2 \\ S_1C_2 & C_1 & -S_1S_2 & -d_3S_1S_2 \\ S_2 & 0 & C_2 & d_3C_2 \\ 0 & 0 & 0 & 1 \end{bmatrix} \tag{4.66} $$

First, a generalized solution is worked out, as in previous examples and then the values from Eq. (4.62) will be substituted to get specific solutions. Let the arm point position and orientation be specified as in Eq. (4.7).

The kinematic model equations are thus, obtained by equating Eq. (4.66) and Eq. (4.7) giving

$$ \begin{bmatrix} C_1C_2 & -S_1 & -C_1S_2 & -d_3C_1S_2 \\ S_1C_2 & C_1 & -S_1S_2 & -d_3S_1S_2 \\ S_2 & 0 & C_2 & d_3C_2 \\ 0 & 0 & 0 & 1 \end{bmatrix} = \begin{bmatrix} r_{11} & r_{12} & r_{13} & r_{14} \\ r_{21} & r_{22} & r_{23} & r_{24} \\ r_{31} & r_{32} & r_{33} & r_{34} \\ 0 & 0 & 0 & 1 \end{bmatrix} \tag{4.67} $$

The preferred solutions for joint displacements are obtained by comparing elements (1, 4), (2, 4) and (3, 4) in Eq. (4.67). The resulting equations are

$$ -d_3C_1S_2 = r_{14} \tag{4.68} $$

$$ -d_3S_1S_2 = r_{24} \tag{4.69} $$

$$ d_3C_2 = r_{34} \tag{4.70} $$

Dividing Eq. (4.69) by Eq. (4.68) the solution for θ_1 is

$$ \theta_1 = \text{Atan2}(-r_{24}, -r_{14}) \tag{4.71} $$

Squaring and adding Eq. (4.68) and Eq. (4.69) gives

$$ d_3^2\, S_2^2\,(C_1^2 + S_1^2) = r_{14}^2 + r_{24}^2 $$

or

$$ d_3S_2 = \pm\sqrt{r_{14}^2 + r_{24}^2} \tag{4.72} $$

Dividing Eq. (4.72) by Eq. (4.70) gives solution of θ_2 as

$$ \theta_2 = \text{Atan2}\left(\pm\sqrt{r_{14}^2 + r_{24}^2}\,,\, r_{34}\right) \tag{4.73} $$

The joint displacement d_3 for joint 3 is obtained by squaring and adding Eqs. (4.68), (4.69) and (4.70). Since the displacement d_3 cannot be negative, only positive sign is used. Thus,

$$ d_3 = +\sqrt{r_{14}^2 + r_{24}^2 + r_{34}^2} \tag{4.74} $$

The numerical values for joint displacements are obtained by substituting the values from given arm point position and orientation matrix, Eq. (4.62), in to Eqs. (4.71), (4.73) and (4.74). The specific solutions are:

$$\theta_1 = \text{Atan2}\,(0.184, -0.106) = -60°$$

$$\theta_2 = \text{Atan2}\left(\pm\sqrt{(0.106)^2 + (-0.184)^2},\ 0.212\right) = \pm 45° \tag{4.75}$$

$$\theta_3 = \text{Atan2}\,\sqrt{(0.106)^2 + (-0.184)^2 + (0.212)^2} = 0.30$$

The two possible solutions are tabulated in Table 4.2.

Table 4.2 *Two possible solutions for the specified arm point*

Solution No.	θ_1	θ_2	d_3
1	–60°	–45°	0.30
2	–60°	45°	0.30

The joint range specified for joint 2 is: $-30° < \theta_2 < 70°$. The solution 1 specifies angle θ_2 as $\theta_2 = -45°$. This violates the joint range constraint and, hence, solution 1 is not feasible.

Example 4.5 *Inverse Kinematics of 5-DOF Manipulator*

For the 5-DOF industrial manipulator discussed in Example 3.7, obtain the analytical solutions of joint variables.

Solution Let the end-effector tool point transformation matrix be given by

$$\boldsymbol{T}_E = \begin{bmatrix} n_x & o_x & a_x & d_x \\ n_y & o_y & a_y & d_y \\ n_z & o_z & a_z & d_z \\ 0 & 0 & 0 & 1 \end{bmatrix} \tag{4.76}$$

From the kinematic model obtained in Example 3.7, the overall transformation matrix for the end-effector tool point is (Eq. (3.48))

$$^0\boldsymbol{T}_5 = \begin{bmatrix} C_1S_{234}C_5 + S_1S_5 & -C_1S_{234}S_5 + S_1C_5 & C_1C_{234} & C_1(L_2C_2 + L_3C_{23} + L_5C_{234}) \\ S_1C_{234}C_5 - C_1S_5 & -S_1S_{234}S_5 - C_1C_5 & S_1C_{234} & S_1(L_2C_2 + L_3C_{23} + L_5C_{234}) \\ -C_{234}C_5 & C_{234}S_5 & -S_{234} & L_1 - L_2S_2 - L_3S_{23} - L_5S_{234} \\ 0 & 0 & 0 & 1 \end{bmatrix} \tag{4.77}$$

A close examination of Eqs.(4.76) and (4.77) clearly shows that no direct solution can be found for any of the joint-variables. Hence, the inverse matrix approach of guideline (e) is used here,

In order to solve for first joint variable θ_1, both matrices, Eq. (4.76) and Eq. (4.77) are premultiplied by inverse of $^0\boldsymbol{T}_1$ (that is $^0\boldsymbol{T}_1^{-1}$. This given left-hand side matrix of equation as

$$
{}^{0}\boldsymbol{T}_1^{-1}\,\boldsymbol{T}_E = \begin{bmatrix} C_1 n_x + S_1 n_y & C_1 o_x + S_1 o_y & C_1 a_x + S_1 a_y & C_1 d_x + S_1 d_y \\ -n_z & -o_z & -a_z & -d_z + L_1 \\ -S_1 n_x + C_1 n_y & -S_1 o_x + C_1 o_y & -S_1 a_x + C_1 a_y & -S_1 d_x + C_1 d_y \\ 0 & 0 & 0 & 1 \end{bmatrix} \tag{4.78}
$$

and the right-hand side of matrix equation is

$$
{}^{1}\boldsymbol{T}_5 = {}^{1}\boldsymbol{T}_2\,{}^{2}\boldsymbol{T}_3\,{}^{3}\boldsymbol{T}_4\,{}^{4}\boldsymbol{T}_5 = \begin{bmatrix} S_{234}C_5 & S_{234}S_5 & C_{234} & L_5 C_{234} + L_3 C_{23} + L_2 C_2 \\ -C_{234}C_5 & C_{234}S_5 & S_{234} & L_5 S_{234} + L_3 S_{23} + L_2 S_2 \\ -S_5 & -C_5 & 0 & 0 \\ 0 & 0 & 0 & 1 \end{bmatrix} \tag{4.79}
$$

or

$$
\begin{bmatrix} C_1 n_x + S_1 n_y & C_1 o_x + S_1 o_y & C_1 a_x + S_1 a_y & C_1 d_x + S_1 d_y \\ -n_z & -o_z & -a_z & -d_z + L_1 \\ -S_1 n_x + C_1 n_y & -S_1 o_x + C_1 o_y & -S_1 a_x + C_1 a_y & -S_1 d_x + C_1 d_y \\ 0 & 0 & 0 & 1 \end{bmatrix}
$$

$$
= \begin{bmatrix} S_{234}C_5 & S_{234}S_5 & C_{234} & L_5 C_{234} + L_3 C_{23} + L_2 C_2 \\ -C_{234}C_5 & C_{234}S_5 & S_{234} & L_5 S_{234} + L_3 S_{23} + L_2 S_2 \\ -S_5 & -C_5 & 0 & 0 \\ 0 & 0 & 0 & 1 \end{bmatrix} \tag{4.80}
$$

Comparing elements of Eq. (4.80), the single variable equations in θ_1 are obtained from elements (3, 3) and (3, 4) as:

$$
\begin{aligned} -S_1 a_x + C_1 a_y &= 0 \\ -S_1 d_x + C_1 d_y &= 0 \end{aligned} \tag{4.81}
$$

Using the later, the solution for θ_1 is

$$
\theta_1 = \text{Atan2}(d_y,\ d_x) \tag{4.82}
$$

From ratio of elements (3, 1) and (3, 2), solution for θ_5 is obtained as

$$
\begin{aligned} -S_5 &= -S_1 n_x + C_1 n_y \\ -C_5 &= -S_1 o_x + C_1 o_y \end{aligned} \tag{4.83}
$$

or

$$
\theta_5 = \text{Atan2}(-S_1 n_x + C_1 n_y, -S_1 o_x + C_1 o_y) \tag{4.84}
$$

The solution for remaining three variables θ_2, θ_3 and θ_4 is obtained by first solving for $\theta_2 + \theta_3 + \theta_4$. Equating elements (1, 3) and (2, 3) gives

$$
\begin{aligned} C_{234} &= C_1 a_x + S_1 a_y \\ S_{234} &= -a_z \end{aligned} \tag{4.85}
$$

which leads to

$$\theta_{234} = \theta_2 + \theta_3 + \theta_4 = A\tan 2(-a_z,\ C_1 a_x + S_1 a_y) \tag{4.86}$$

Premultiplying both sides of Eq. (4.80) by $({}^1\boldsymbol{T}_2)^{-1}$, gives

$$({}^1\boldsymbol{T}_2)^{-1}({}^0\boldsymbol{T}_1)^{-1}\boldsymbol{T}_E = {}^2\boldsymbol{T}_3\,{}^3\boldsymbol{T}_4\,{}^4\boldsymbol{T}_5 \tag{4.87}$$

$$\begin{bmatrix} C_2(C_1n_x + S_1n_y) - S_2n_z & C_2(C_1o_x + S_1o_y) - S_2o_z & C_2(C_1a_x + S_1a_y) - S_2a_z & C_2(C_1d_x + S_1d_y) - S_2(d_z - L_1) - L_2 \\ -S_2(C_1n_x + S_1n_y) - C_2n_z & -S_2(C_1o_x + S_1o_y) - C_2o_z & -S_2(C_1a_x + S_1a_y) - C_2a_z & -S_2(C_1d_x + S_1d_y) - C_2(d_z + L_1) \\ -S_1n_x + C_1n_y & -S_1o_x + C_1o_y & -S_1a_x + C_1a_y & -S_1d_x + C_1d_y \\ 0 & 0 & 0 & 1 \end{bmatrix}$$

$$= \begin{bmatrix} S_{34}C_5 & -S_{34}S_5 & C_{34} & L_5C_{34} + L_3C_3 \\ -C_{34}C_5 & C_{34}S_5 & S_{34} & L_5S_{34} + L_3C_3 \\ -S_5 & -C_5 & 0 & 0 \\ 0 & 0 & 0 & 1 \end{bmatrix} \tag{4.88}$$

From elements (1, 1) and (1, 2), two equations are obtained:

$$C_2\left(C_1n_x + S_1n_y\right) - S_2n_z = S_{34}C_5 \tag{4.89}$$

$$C_2\left(C_1o_x + S_1o_y\right) - S_2o_z = -S_{34}S_5 \tag{4.90}$$

Dividing Eq. (4.90) by Eq. (4.89) and rearranging gives

$$-(C_1o_x + S_1o_y)C_2 + S_2o_z = \tan\theta_5\left[\left(C_1n_x + S_1n_y\right)C_2 - S_2n_z\right] \tag{4.91}$$

or

$$\frac{S_2}{C_2} = \frac{\tan\theta_5\left(C_1n_x + S_1n_y\right) + (C_1o_x + S_1o_y)}{n_z\tan\theta_5 + o_z}$$

or

$$\theta_2 = A\tan 2\left[\tan\theta_5\left(C_1n_x + S_1n_y\right) + C_1o_x + S_1o_y,\ n_z\tan\theta_5 + o_z\right] \tag{4.92}$$

Similarly from elements (1, 4) and (2, 4) of Eq. (4.88), two equations are obtained after rearrangements as

$$\begin{aligned} L_5C_{34} + L_3C_3 &= C_1C_2d_x + S_1C_2d_y - S_2(d_z - L_1) - L_2 \\ L_5S_{34} + L_3S_3 &= -C_1S_2d_x + S_1S_2d_y - C_2(d_z + L_1) \end{aligned} \tag{4.93}$$

Substituting for C_{34} and S_{34} from elements (1, 3) and (2, 3), respectively form the left-hand matrix of Eq. (4.88) and simplifying gives

$$\begin{aligned} L_3C_3 &= C_1C_2d_x + S_1C_2d_y - S_2(d_z - L_1) - L_2 - L_5(C_1C_2a_x + S_1C_2a_y - S_2a_z) \\ L_3S_3 &= -C_1S_2d_x - S_1S_2d_y - C_2(d_z + L_1) + L_5(C_1S_2a_x + S_1S_2a_y + C_2a_z) \end{aligned} \tag{4.94}$$

Form Eq. (4.94) θ_3 is obtained as

$$\theta_3 = \text{Atan2}\begin{pmatrix} L_3 S_3 = -C_1 S_2 d_x - S_1 S_2 d_y - C_2(d_z + L_1) + L_5(C_1 S_2 a_x + S_1 S_2 a_y + C_2 a_z), \\ L_3 C_3 = C_1 C_2 d_x + S_1 C_2 d_y - S_2(d_z - L_1) - L_2 - L_5(C_1 C_2 a_x + S_1 C_2 a_y - S_2 a_z) \end{pmatrix} \tag{4.95}$$

The last unknown joint variable θ_4 is computed from Eq. (4.86).

$$\theta_4 = \theta_{234} - \theta_2 - \theta_3 \tag{4.96}$$

Example 4.6 *Inverse kinematics for 6-DOF Stanford manipulator*

Obtain all the joint displacements as explicit functions of the position and orientation of the end-effector for the 6-DOF manipulator of Example 3.8 by analytical method. Also, discuss the existence and multiplicity of solutions.

Solution The given 6-DOF-manipulator arm has a wrist whose joint axes 4, 5 and 6 intersect at a point. This satisfies the necessary condition for existence of closed form solutions. Hence, it is possible to solve the inverse problem in closed form.

Let the given position and orientation of the end-effector with respect to the base in Cartesian space be

$$\boldsymbol{T} = \begin{bmatrix} n_x & o_x & a_x & d_x \\ n_y & o_y & a_y & d_y \\ n_z & o_z & a_z & d_z \\ 0 & 0 & 0 & 1 \end{bmatrix} \tag{4.97}$$

From Example 3.8, the manipulator transformation matrix (the forward kinematic model) for the given manipulator is given by Eq. (3.58) that is

$$^0\boldsymbol{T}_6 = \begin{bmatrix} \begin{matrix} C_1C_2C_4C_5C_6 \\ -S_1S_4C_5C_6 \\ -C_1S_2S_5C_6 \\ -C_1C_2S_4S_6 \\ -S_1C_4S_6 \end{matrix} & \begin{matrix} C_1C_2C_4C_5S_6 \\ +S_1S_4C_5S_6 \\ +C_1S_2S_5S_6 \\ -C_1C_2S_4C_6 \\ -S_1C_4C_6 \end{matrix} & \begin{matrix} C_1C_2C_4S_5 \\ -S_1S_4S_5 \\ +C_1S_2C_5 \end{matrix} & \begin{matrix} C_1C_2C_4S_5L_6 \\ -S_1S_4S_5L_6 \\ +C_1S_2C_5L_6 \\ +C_1S_2d_3 - S_1L_2 \end{matrix} \\ \begin{matrix} S_1C_2C_4C_5C_6 \\ +C_1S_4C_5C_6 \\ -S_1S_2S_5C_6 \\ -S_1C_2S_4S_6 \\ +C_1C_4S_6 \end{matrix} & \begin{matrix} -S_1C_2C_4C_5S_6 \\ -C_1S_4C_5S_6 \\ +S_1S_2S_5S_6 \\ -S_1C_2S_4C_6 \\ +C_1C_4C_6 \end{matrix} & \begin{matrix} S_1C_2C_4S_5 \\ +C_1S_4S_5 \\ +S_1S_2C_5 \end{matrix} & \begin{matrix} S_1C_2C_4S_5L_6 \\ +C_1S_4S_5L_6 \\ +S_1S_2C_5L_6 \\ +S_1S_2d_3 + C_1L_2 \end{matrix} \\ \begin{matrix} S_2C_4C_5C_6 \\ -C_2S_5C_6 \\ +S_2S_4S_6 \end{matrix} & \begin{matrix} S_2C_4C_5S_6 \\ +C_2S_5S_6 \\ +S_2S_4C_6 \end{matrix} & \begin{matrix} -S_2C_4S_5 \\ +C_2C_5 \end{matrix} & \begin{matrix} -S_2C_4S_5L_6 \\ +C_2C_5L_6 + C_2d_3 \end{matrix} \\ 0 & 0 & 0 & 1 \end{bmatrix} \tag{4.98}$$

Observe that in Eq. (4.98) neither single variable terms are present nor simple algebra (like division of two elements) will give single variable isolation. Hence, to find a solution two alternatives are: (i) matrix premultiplication or postmultiplication to isolate one variable at a time, as was done in Example 4.1, and (ii) use more involved algebra and trigonometry. In this example latter is used.

Equating elements (1, 3), (1, 4), (2, 3), (2, 4), (3, 3) and (3, 4) in Eqs. (4.97) and (4.98), six equations are obtained.

$$C_1C_2C_4S_5 - S_1S_4S_5 + C_1S_2C_5 = a_x \tag{4.99}$$

$$L_6(C_1C_2C_4S_5 - S_1S_4S_5 + C_1S_2C_5) + C_1S_2d_3 - S_1L_2 = d_x \tag{4.100}$$

$$S_1C_2C_4S_5 + C_1S_4S_5 + S_1S_2C_5 = a_y \tag{4.101}$$

$$L_6(S_1C_2C_4S_5 + C_1S_4S_5 + S_1S_2C_5) + S_1S_2d_3 + C_1L_2 = d_y \tag{4.102}$$

$$-S_2C_4S_5 + C_2C_5 = a_z \tag{4.103}$$

$$L_6(-S_2C_4S_5 + C_2C_5) + C_2d_3 = d_z \tag{4.104}$$

Substituting Eq. (4.99) into Eq. (4.100) gives

$$L_6a_x + C_1S_2d_3 - S_1L_2 = d_x$$

or
$$C_1S_2d_3 - S_1L_2 = d_x - L_6a_x \tag{4.105}$$

Similarly, substituting Eq. (4.101) into Eq. (4.102) gives

$$L_6a_y + S_1S_2d_3 + C_1L_2 = d_y$$

or
$$S_1S_2d_3 + C_1L_2 = d_y - L_6a_y \tag{4.106}$$

Squaring Eqs. (4.105) and (4.106), adding and simplifying gives

$$S_2^2d_3^2 + L_2^2 = (d_x - L_6a_x)^2 + (d_y - L_6a_y)^2$$

or
$$S_2^2d_3^2 = (d_x - L_6a_x)^2 + (d_y - L_6a_y)^2 - L_2^2 \tag{4.107}$$

Also, combining Eqs. (4.103) and (4.104), rearranging and squaring gives

$$C_2d_3 = d_z - L_6a_z \tag{4.108}$$

or
$$C_2^2d_3^2 = (d_z - L_6a_z)^2 \tag{4.109}$$

Equations (4.107) and (4.109) are solved for d_3

$$d_3 = \pm\sqrt{(d_x - L_6a_x)^2 + (d_y - L_6a_y)^2 + (d_z - L_6a_z)^2 - L_2^2}$$

The displacement of prismatic joint d_3 is always positive; therefore, the negative solution is not valid. Thus

$$d_3 = \sqrt{(d_x - L_6a_x)^2 + (d_y - L_6a_y)^2 + (d_z - L_6a_z)^2 - L_2^2} \tag{4.110}$$

Thus, solution for the joint variable d_3, which gives the joint displacement for the prismatic joint, is found. Next, solve for joint displacement θ_2. From Eq. (4.107),

$$S_2 d_3 = \pm\sqrt{(d_x - L_6 a_x)^2 + (d_y - L_6 a_y)^2 - L_2^2} \tag{4.111}$$

Dividing Eq. (4.111) by Eq. (4.108), solution of joint variable θ_2 is obtained as

$$\theta_2 = A\tan2\left(\pm\sqrt{(d_x - L_6 a_x)^2 + (d_y - L_6 d_y)^2 - L_2^2}, (d_z - L_6 a_z)\right) \tag{4.112}$$

Next, to solve for θ_1, Eqs. (4.105) and (4.106) can be used as they have only θ_1 as unknown. First, let the constants K_1 and K_2 be defined as

$$K_1 = S_2 d_3 \quad \text{and} \quad K_2 = L_2$$

Substituting these constants in Eqs. (4.105) and (4.106):

$$K_1 C_1 - K_2 S_1 = d_x - L_6 a_x \tag{4.113}$$

$$K_1 S_1 + K_2 C_1 = d_y - L_6 a_y \tag{4.114}$$

The equations of this form can be solved by making trigonometric substitutions

$$K_1 = r \sin\phi \quad \text{and} \quad K_2 = r\cos\phi \tag{4.115}$$

where

$$r = +\sqrt{K_1^2 + K_2^2}$$

$$\phi = A\tan2\left(\frac{K_1}{r}, \frac{K_2}{r}\right) \tag{4.116}$$

Substituting K_1 and K_2 from Eq. (4.115) in Eqs. (4.113) and (4.114)

$$\sin(\phi - \theta_1) = \frac{d_x - L_6 a_x}{r} \tag{4.117}$$

$$\cos(\phi - \theta_1) = \frac{d_y - L_6 a_y}{r} \tag{4.118}$$

From Eqs. (4.117) and (4.118),

$$\phi - \theta_1 = A\tan2\left(\frac{d_x - L_6 a_x}{r}, \frac{d_y - L_6 a_y}{r}\right) \tag{4.119}$$

Substituting for ϕ, r, K_1, and K_2 and solving for θ_1 gives

$$\theta_1 = A\tan2\left(\frac{S_2 d_3}{\sqrt{S_2^2 d_3^2 + L_2^2}}, \frac{L_2}{\sqrt{S_2^2 d_3^2 + L_2^2}}\right)$$

$$- A\tan2\left(\frac{d_x - L_6 d_x}{\sqrt{S_2^2 d_3^2 + L_2^2}}, \frac{d_y - L_6 a_y}{\sqrt{S_2^2 d_3^2 + L_2^2}}\right) \tag{4.120}$$

Thus, Eqs. (4.110), (4.112) and (4.120) give solutions for the first three joint displacements θ_1, θ_2, and d_3, respectively.

To obtain the solutions for the remaining three joint displacements θ_4, θ_5, and θ_6, another approach will be used. It is possible to view the description of tool frame, frame {6}, with respect to frame {3}, the arm endpoint frame, along two different paths.

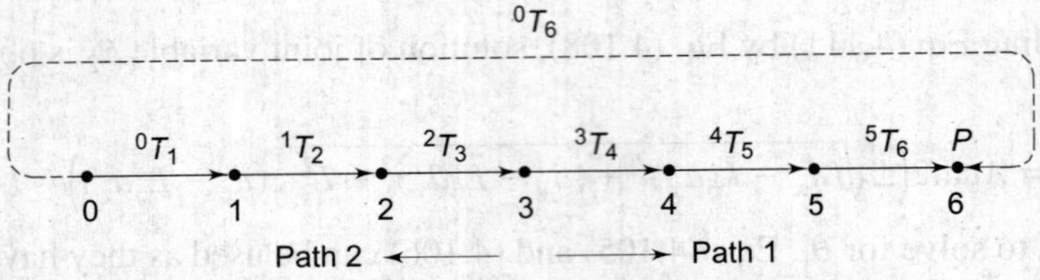

Fig. 4.10 *The transform graph for the manipulator*

The manipulator transformation matrix $^0\boldsymbol{T}_6$ is equal to the product of six link transformation matrices

$$^0\boldsymbol{T}_6 = {}^0\boldsymbol{T}_1\,{}^1\boldsymbol{T}_2\,{}^2\boldsymbol{T}_3\,{}^3\boldsymbol{T}_4\,{}^4\boldsymbol{T}_5\,{}^5\boldsymbol{T}_6 \tag{4.121}$$

This can be represented graphically as shown in Fig. 4.10 with nodes representing frames and edges representing the transform. From the graph in Fig. 4.10 observe that there are two paths to traverse from frame {3} to frame {6}, one is via frame {4} and {5} in the forward direction and other is via frame {0}, the base, in the reverse direction. The use of these two paths to get the solutions is discussed below.

Path 1 frame {3} → frame {4} → frame {5} → frame {6}

Along this path the transformation $^3\boldsymbol{T}_6$ can be obtained as

$$^3\boldsymbol{T}_6 = {}^3\boldsymbol{T}_4\,{}^4\boldsymbol{T}_5\,{}^5\boldsymbol{T}_6 \tag{4.122}$$

In Example 3.8, the transformation matrices $^3\boldsymbol{T}_4$, $^4\boldsymbol{T}_5$, and $^5\boldsymbol{T}_6$ were obtained in terms of θ_4, θ_5, and θ_6 respectively. On multiplying these matrices,

$$^3\boldsymbol{T}_6 = \begin{bmatrix} C_4C_5C_6 - S_4S_6 & -C_4C_5S_6 - S_4C_6 & -C_4S_5 & -L_6C_4S_5 \\ S_4C_5C_6 + C_4S_6 & -S_4C_5S_6 - C_4C_6 & -S_4S_5 & -L_6S_4S_5 \\ S_5C_6 & -S_5S_6 & C_5 & L_6C_5 + L_4 \\ 0 & 0 & 0 & 1 \end{bmatrix} \tag{4.123}$$

Path 2 frame {3} → frame {2} → frame {1} → frame {0} → frame {6}

This is a path via the base. The tool frame is defined with respect to the base and hence, it can be reached from frame {3} traversing the links of the arm. Thus,

$$^3\boldsymbol{T}_6 = {}^3\boldsymbol{T}_2\,{}^2\boldsymbol{T}_1\,{}^1\boldsymbol{T}_0\,{}^0\boldsymbol{T}_6$$

It is known that $^{i-1}\boldsymbol{T}_i = ({}^i\boldsymbol{T}_{i-1})^{-1}$ and $^0\boldsymbol{T}_6 = \boldsymbol{T}$ hence

$$^3\boldsymbol{T}_6 = ({}^2\boldsymbol{T}_3)^{-1}\,({}^1\boldsymbol{T}_2)^{-1}\,({}^0\boldsymbol{T}_1)^{-1}\,\boldsymbol{T} \tag{4.124}$$

Since θ_1, θ_2, and d_3 are known, the matrices $^2\boldsymbol{T}_3$, $^1\boldsymbol{T}_2$ and $^0\boldsymbol{T}_1$ and their inverse can be computed. Substituting these values and multiplying the matrices, $^3\boldsymbol{T}_6$ can be computed. Let it be denoted as

$$^3T_6 = \begin{bmatrix} r_{11} & r_{12} & r_{13} & r_{14} \\ r_{21} & r_{22} & r_{23} & r_{24} \\ r_{31} & r_{32} & r_{33} & r_{34} \\ 0 & 0 & 0 & 1 \end{bmatrix} \tag{4.125}$$

Since matrices in Eq. (4.123) and (4.125) represent the same point, the arm point, θ_4, θ_5, θ_6 can be found by equating the corresponding elements of matrices. First, for θ_5, the elements in row 3 are equated to give three equations as:

$$S_5C_6 = r_{31} \tag{4.126}$$

$$-S_5S_6 = r_{32} \tag{4.127}$$

$$C_5 = r_{33} \tag{4.128}$$

Squaring Eqs. (4.126) and (4.127), adding and dividing the result by Eq. (4.128) gives solution for θ_5 as

$$\theta_5 = \text{Atan2}(\pm\sqrt{(r_{31}^2 + r_{32}^2)}, r_{33}) \tag{4.129}$$

Next, joint angle θ_4 is obtained by equating the elements (2, 4) and (1, 4) as

$$-L_6S_4S_5 = r_{24} \tag{4.130}$$

$$-L_6S_4S_5 = r_{14} \tag{4.131}$$

Solving for θ_4 from Eqs. (4.130) and (4.131) gives

$$\theta_4 = \text{Atan2}\left(\frac{r_{24}}{S_5}, \frac{r_{14}}{S_5}\right) \tag{4.132}$$

Note that since S_5 is a variable and may have more than one value, it is going to influence solution of θ_4. Finally, to solve for θ_6, Eqs. (4.127) is divided by Eq. (4.126) to give

$$\theta_6 = \text{Atan2}\left(-\frac{r_{32}}{S_5}, \frac{r_{31}}{S_5}\right) \tag{4.133}$$

Thus, expressions for all the six joint displacements of 6-DOF manipulator are obtained as explicit functions of the desired position and orientation of end-effector. The values of joint displacements can be determined from these functions for the desired end-effector location data. Note that the inverse kinematics solutions could have been obtained by following alternate approaches. The issues of existence and multiplicity of solutions are discussed next.

Existence of Solutions

A close examination of the expressions of the joint variables reveals that the solutions to the inverse kinematics problem of the given 6-DOF arm exist only if the following conditions are satisfied:

(i) The term under the square root in Eq. (4.110) is non-negative, that is,

$$[(d_x - L_6a_x)^2 + (d_y - L_6a_y)^2 + (d_z - L_6a_z)^2 - L_2^2] \geq 0 \tag{4.134}$$

(ii) Similarly, the term under the square root in Eq. (4.111), is nonnegative, that is,

$$[(d_x - L_6 a_x)^2 + (d_y - L_6 a_y)^2 - L_2^2] \geq 0 \tag{4.135}$$

The above conditions are clearly geometric constraints on the dimensions of the links of the manipulator.

There are two ways to look at these. One, for given link dimensions (L_2 and L_6) the end-effector will not be able to attain the desired position (d_x, d_y, d_z) and orientation (a_x, a_y, a_z) if the above conditions are not satisfied, and second, the conditions can be used to design link dimensions L_2 and L_6 for the desired workspace.

Note that, if condition (ii) is satisfied, condition (i) is automatically satisfied. Also, note that here it is assumed that full 360° of rotation for all revolute joints and limitless translations of the prismatic joint are possible. But, due to mechanical constraints, joint motions are restricted and solutions exist only if, in addition to satisfying the conditions given above, each of the joint variables takes a value that lies within the range of motion allowed for that joint.

Multiplicity of Solutions

Equation (4.112) gives two solutions of θ_2 due to the presence of the square root. Since θ_1 depends on θ_2 (Eq. (4.120)), there is one value of θ_1 for each value of θ_2. Similarly, from Eq. (4.129), there are two solutions for θ_5 and hence, it gives one set of solutions for θ_4 and θ_6 for each value of θ_5.

Thus, the number of solutions to the inverse kinematics problem of the given 6-DOF-manipulator arm is four.

Example 4.7 ***Determination of joint variables for a 4-DOF RPPR manipulator***

For a 4-DOF, RPPR manipulator, the joint-link transformation matrices, with joint variables θ_1, d_2, d_3 and θ_4 are

$$^0\boldsymbol{T}_1(\theta_1) = \begin{bmatrix} C_1 & -S_1 & 0 & 0 \\ S_1 & C_1 & 0 & 0 \\ 0 & 0 & 1 & 0 \\ 0 & 0 & 0 & 1 \end{bmatrix} \tag{4.136}$$

$$^1\boldsymbol{T}_2(\theta_2) = \begin{bmatrix} 1 & 0 & 0 & 0 \\ 0 & 0 & 1 & 0 \\ 0 & -1 & 0 & d_2 \\ 0 & 0 & 0 & 1 \end{bmatrix} \tag{4.137}$$

$$^2\boldsymbol{T}_3(d_3) = \begin{bmatrix} 1 & 0 & 0 & 5 \\ 0 & 1 & 0 & 0 \\ 0 & 0 & 1 & d_3 \\ 0 & 0 & 0 & 1 \end{bmatrix} \tag{4.138}$$

$$^3T_4(\theta_4) = \begin{bmatrix} C_4 & -S_4 & 0 & 0 \\ S_4 & C_4 & 0 & 0 \\ 0 & 0 & 1 & 0 \\ 0 & 0 & 0 & 1 \end{bmatrix} \tag{4.139}$$

If the tool configuration matrix at a given instant is as given below, obtain the magnitude of each joint variable.

$$T_E = \begin{bmatrix} -0.250 & 0.433 & -0.866 & -89.10 \\ 0.433 & -0.750 & -0.500 & -45.67 \\ -0.866 & -0.500 & 0.000 & 50.00 \\ 0 & 0 & 0 & 1 \end{bmatrix} \tag{4.140}$$

Solution The kinematic model of the manipulator will be

$$^0T_1\,^1T_2\,^2T_3\,^3T_4 = T_E \tag{4.141}$$

Substituting from Eqs. (4.136) – (4.140) gives

$$\begin{bmatrix} C_1 & -S_1 & 0 & 0 \\ S_1 & C_1 & 0 & 0 \\ 0 & 0 & 1 & 0 \\ 0 & 0 & 0 & 1 \end{bmatrix} \begin{bmatrix} 1 & 0 & 0 & 0 \\ 0 & 0 & 1 & 0 \\ 0 & -1 & 0 & d_2 \\ 0 & 0 & 0 & 1 \end{bmatrix} \begin{bmatrix} 1 & 0 & 0 & 5 \\ 0 & 1 & 0 & 0 \\ 0 & 0 & 1 & d_3 \\ 0 & 0 & 0 & 1 \end{bmatrix} \begin{bmatrix} C_4 & -S_4 & 0 & 0 \\ S_4 & C_4 & 1 & 0 \\ 0 & 0 & 1 & 0 \\ 0 & 0 & 0 & 1 \end{bmatrix} = T_E \tag{4.142}$$

or

$$\begin{bmatrix} C_1C_4 & -C_1S_4 & -S_1 & -d_3S_1 + 5C_1 \\ S_1C_4 & -S_1S_4 & C_1 & d_3C_1 + 5S_1 \\ -S_4 & -C_4 & 0 & d_2 \\ 0 & 0 & 0 & 1 \end{bmatrix} = \begin{bmatrix} -0.250 & 0.433 & -0.866 & -89.10 \\ 0.433 & -0.750 & -0.500 & -45.67 \\ -0.866 & -0.500 & 0.000 & 50.00 \\ 0 & 0 & 0 & 1 \end{bmatrix} \tag{4.143}$$

The solutions for joint-variables are found by using the direct approach, guideline (a)–(d). The solution for first joint variable θ_1 is obtained by comparing elements (1, 3) and (2, 3)

$$-S_1 = -\sin\theta_1 = -0.866 \tag{4.144}$$
$$C_1 = \cos\theta_1 = -0.5$$

or $\theta_1 = A\tan 2(0.866, -0.5) = -60°$ (that is: $\theta_1 = A\tan 2\,(-a_x, a_y)$) (4.145)

The solution for second variable d_2 is obtained from element (3, 4) as:

$$d_2 = 50 \text{ (that is : } d_2 = d_z) \tag{4.146}$$

Similarly, the third joint variable d_3 is obtained from elements (1, 4) and elements (2, 4) by squaring, adding and simplifying

$$d_3 = \pm\sqrt{d_x^2 + d_y^2 - 25} \tag{4.147}$$

or $$d_3 = 100 \tag{4.148}$$

Note that d_3 can not be negative. The fourth joint variable θ_4 is computed from elements (3, 1) and (3, 2) as

$$\theta_4 = \text{Atan2}(0.866, 0.5) = 60^\circ \text{ (that is: } \theta_4 = \text{Atan2}(-n_z, -o_z)) \tag{4.149}$$

The given end-effector position and orientation, $\boldsymbol{T}_E$ will be achieved by setting the joint variable vector to

$$\boldsymbol{q} = \begin{bmatrix} -60^\circ & 50 & 100 & 60^\circ \end{bmatrix} \tag{4.150}$$

EXERCISES

4.1 For the two link planar manipulator in Fig. 4.3(a) the first link as is twice as long as the second link ($L_1 = 2L_2$). Sketch the reachable workspace of the manipulator if the joint range limits are

$$\begin{aligned} 0 < \theta_1 < 170^\circ, \\ -90^\circ < \theta_2 < 110^\circ. \end{aligned} \tag{4.151}$$

4.2 Sketch the approximate reachable workspace of the tip of a two-link planar arm with revolute joints. For this arm the first link is thrice as long as the second link, that is, $L_1 = 3L_2$ and the joint limits are $30° < \theta_1 < 180°$ and $-100° < \theta_2 < 160°$.

4.3 Sketch the approximate reachable workspace and the dexterous workspace of the 3-DOF planar manipulator shown in Fig. 4.5.

4.4 Show that for a 3R planar manipulator having link lengths as L_1, L_2 and L_3 with $(L_1 + L_2) > L_3$, the RWS is a circle with radius $r_{RWS} = (L_1 + L_2 + L_3)$ and DWS is a circle with radius $r_{DWS} = (L_1 + L_2 - L_3)$.

4.5 Explain why closed form analytical solutions are preferred over numerical iterative solutions.

4.6 Discuss the existence of multiple solutions for Example 4.1.

4.7 For a three-link planar manipulator (2-DOF for position and 1-DOF for orientation) two solutions are possible for a given position and orientation, as discussed in Section 4.3.2. If one more degree of freedom is added such that the manipulator is still planar, how many solutions will be possible for a given position and orientation when the added joint is
(a) revolute.
(b) prismatic.

4.8 How many solutions are possible, assuming no joint range limits for the following 3-DOF arm configuration?
(a) PPP manipulator shown in Fig. E4.8.
(b) RPP manipulator shown in Fig. 3.13.
(c) RRP manipulator shown in Fig. 4.7.
(d) RRR manipulator shown in Fig. 3.15.

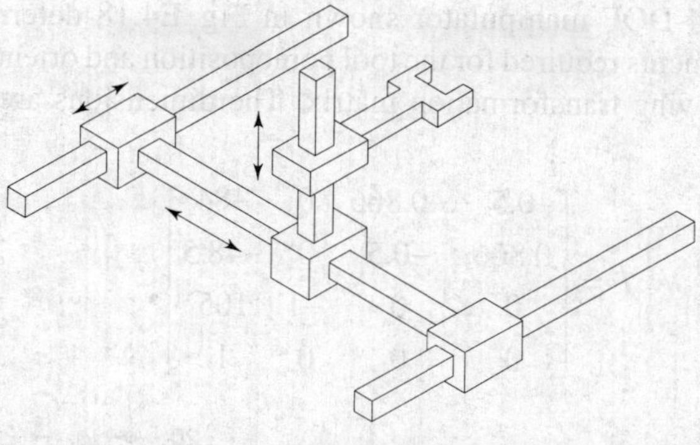

Fig. E4.8 *A three degree of freedom PPP configuration*

4.9 For the two degree of freedom planar RP configuration arm discussed in Example 3.1, how many solutions can be found for a given position and orientation. What will be the number of solutions if following alterations are made in the manipulator configuration?
(a) Add one revolute joint after the prismatic joint.
(b) Add one revolute joint before the prismatic joint.
(c) Add one degree of freedom (1R) wrist.
In each case, the manipulator remains planar.

4.10 For the 2-DOF manipulator shown in Fig. 3.11 determine the solution for all joint displacements $\boldsymbol{q}$ for a given tool point position and orientation.

4.11 Obtain the inverse kinematics solution for the 3-DOF planar manipulator shown in Fig. 4.5.

4.12 Workout Example 4.1 without using inverse transform approach.

4.13 Obtain the closed form solutions for the joint displacements of the cylindrical configuration arm described in Exercise 3.2.

4.14 For the manipulator arm consisting of 3-DOF described in Exercise 3.6, obtain the inverse kinematics solutions.

4.15 Obtain the inverse kinematics model for the 3-DOF articulated arm discussed in Example 3.3.

4.16 For the 3-DOF-RPY wrist shown in Fig. 3.18, obtain the conditions of singularities.

4.17 For the 3-DOF arm shown in Fig. E4.17, determine the forward kinematic model and, there from, obtain the general solution for inverse dinematics.

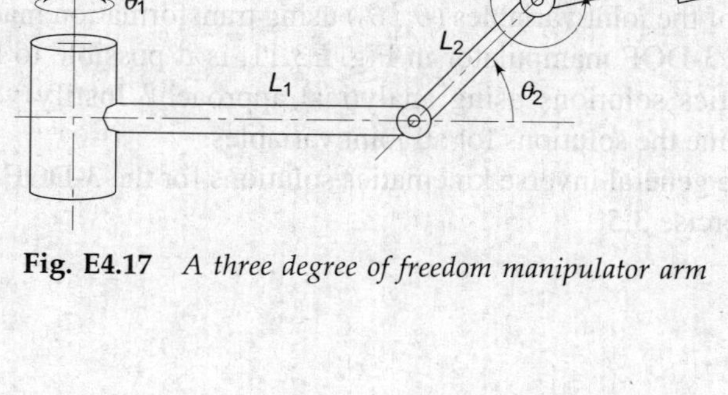

Fig. E4.17 *A three degree of freedom manipulator arm*

4.18 For the 4-DOF manipulator shown in Fig. E4.18 determine the joint displacements required for the tool point position and orientation given by the following transformation matrix. The dimensions are shown in the figure.

$$T = \begin{bmatrix} 0.5 & -0.866 & 0 & -84 \\ 0.866 & -0.5 & 0 & -48.5 \\ 0 & 0 & -1 & 105 \\ 0 & 0 & 0 & 1 \end{bmatrix} \tag{4.152}$$

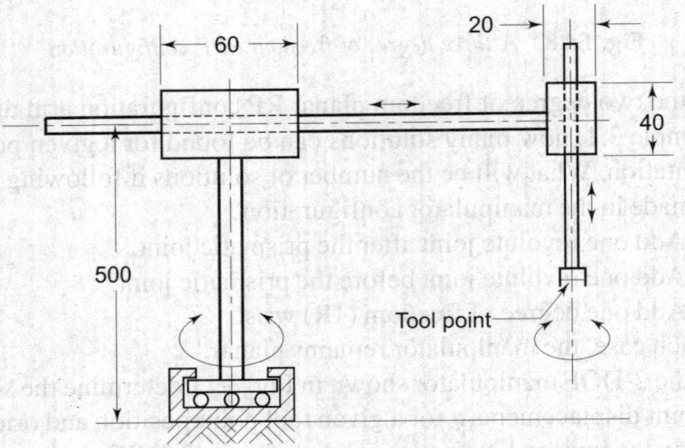

Note: Not to scale and all dimension in mm

Fig. E4.18 *A 4-DOF manipulator*

4.19 For a 5-DOF, RRR-RR articulated configuration manipulator shown in Fig. E3.14 obtain the inverse kinematics model.

4.20 Consider the 6-DOF manipulator in Exercise 3.18; find solutions for all the joint variables in terms of end-effector position and orientation by analytical method and inverse transform method.

4.21 For the SCARA robot discussed in Example 3.6, end-effector configuration (orientation and position) has been calculated in Exercise 3.13. Taking this as the desired end-effector configuration compute the joint displacement vector.

4.22 Consider the two-link planar manipulator (see Fig. 3.11) with its end-effector located at (4, 0). If the length of each link is 2 units, determine the values of the joint variables (θ_1, θ_2) using transformation matrices.

4.23 For the 3-DOF manipulator in Fig. E3.11, is it possible to find inverse kinematics solutions using analytical approach? Justify your answer. Determine the solutions for all joint variables.

4.24 Find the general inverse kinematics solutions for the 3-DOF Euler wrist, see Exercise 3.5.

4.25 The joint-link parameters of a 6-DOF freedom manipulator are described in Table E4.17. Find the inverse kinematics solutions for all the joint angles $\boldsymbol{q} = [\theta_1 \quad \theta_2 \quad \theta_3 \quad \theta_4 \quad \theta_5 \quad \theta_6]^T$. Assume that the position and orientation of the end-effector with respect to the base coordinates ${}^0\boldsymbol{T}_6$ is known and is given by

$$\boldsymbol{T} = \begin{bmatrix} n_x & o_x & a_x & d_x \\ n_y & o_y & a_y & d_y \\ n_z & o_z & a_z & d_z \\ 0 & 0 & 0 & 1 \end{bmatrix} \tag{4.153}$$

Table E4.17 *Joint-link parameters for a 6-DOF manipulator*

Link i	a_i	α_i	d_i	θ_i	q_i
1	0	−90°	d_1	θ_1	θ_1
2	a_2	0	0	θ_2	θ_2
3	a_3	0	0	θ_3	θ_3
4	a_4	−90°	0	θ_4	θ_4
5	0	90°	d_5	θ_5	θ_5
6	0	0	d_6	θ_6	θ_6

4.26 Describe the workspace of a manipulator. Make a list of factors on which the workspace, the dexterous and reachable workspace, of a given manipulator depends.

4.27 Solutions to inverse kinematics problem are generally difficult. Explain why.

4.28 Explain the factors on which the number of solutions to given inverse kinematics model depend.

4.29 How are the feasible solutions determined? What parameters have control on the number of feasible solutions to the given inverse kinematics problem?

4.30 Is it always possible to find analytical solutions to the inverse kinematics problem? Give a situation when the analytical solution to inverse kinematics problem cannot be found.

4.31 What are closed form solutions to inverse kinematics problem? Explain the methods for obtaining closed form solutions.

4.32 Why closed form solutions are preferred over numerical, iterative or other forms of solutions to the inverse kinematics problem?

SELECTED BIBLIOGRAPHY

1. D. Baker and C. Wampler, "On the Inverse Kinematics of Redundant Manipulators," *The International Journal of Robotics Research*, *7*(2), 1988.

2. A. A.Goldenberg, B.Benhabib and R. G. Fenton, "A complete generalized solution to the inverse kinematics of robots," *IEEE Journal of Robotics and Automation*, ***1***(1), 14–20, 1985.
3. A. A. Goldenberg, and D. L. Lawrence, "A generalized solution to the inverse kinematics of robotic manipulator," *Journal of Dynamic Systems, Measurement, and Control*, ***107***, 103–107, 1985.
4. K.C. Gupta and B. Roth, "Design Consideration for Manipulators Workspace," *J of Mechanical Design*, ***104***, 704–711, Oct 1992.
5. Wisama Khalil and Denis Creusot, "SYMORO+: A System for the Symbolic Modeling of Robots," *Robotica*, ***15***, 153–161, 1997.
6. C.L. Lai and C.H. Meng, "The Dexterous Workspace of Simple Manipulators," *IEEE Journal of Robotics and Automation*, ***107***, 99–103, Jun. 1988.
7. C.S.G. Lee, "A Geometric Approach in Solving Inverse Kinematics of PUMA Robots," *IEEE Tr on Aerospace and Electronic Systems*, ***20***(6), 696–706, Nov. 1984.
8. D. Manocha, and J. F. Canny, "Efficient inverse kinematics for general 6r manipulators," *IEEE Transactions on Robotics and Automation*, ***10***(5), 648–657, 1994.
9. Y. Nakamura and H. Hanafusa, "Inverse Kinematic Solutions with Singularity Robustness for Robot Manipulator Control," *Tr of ASME Journal of Dynamic Systems, Measurement, and Control*, ***108***, 163–171, 1986.
10. F.C. Park and R.W. Brockett, "Kinematic Dexterity of Robotic Mechanisms," *The International Journal of Robotics Research*, ***13***(1), 1–15, Feb. 1994.
11. R.P. Paul, B. Shimano, and G. Mayer, "Kinematic Control Equations for Simple Manipulator," *IEEE Tr on Systems, Man, and Cybernetics*, ***11***(6), 449–455, 1981.
12. R.P. Paul and H. Zhang, "Computationally Efficient Kinematics for Manipulators with Spherical Wrists Based on the Homogeneous Transformation Representation," *The International Journal of Robotics Research*, ***5***(2), 32–44, 1986.
13. R.G. Roberts, "The Dexterity and Singularities of an Under Actuated Robot," *J of Robotic Systems*, ***18***(4), 160–169, April 2001.
14. L. Sciavicco and B. Siciliano, "A Solution Algorithm to the Inverse Kinematic Problem for Redundant Manipulators," *IEEE Journal of Robotics and Automation*, ***4***, 403–410, 1988.
15. C.W. Wampler, "Manipulator Inverse Kinematic Solutions Based on Damped Least-Squares Solutions," *IEEE Tr on Systems, Man, and Cybernetics*, ***16***, 93–101, 1986.
16. T. Yoshikawa, "Manipulability of Robotic Manipulators," *The International Journal of Robotics Research*, ***4***, 3–9, 1985.

5

Manipulator Differential Motion and Statics

In the preceding chapters, the direct and inverse kinematic models were presented, which establish the relationship between the manipulator's joint displacements and position and orientation of its end-effector. These relationships permit the static control of the manipulator to place the end-effector at a specified location and make it traverse a specific path in space. However, for the manipulator, not only the final location of the end-effector is of concern, but also the velocities at which the end-effector would move to reach the final location is an equally important concern. This requires the coordination of the instantaneous end-effector velocity and joint velocities. One way to achieve this is to take the time derivative of kinematic equations of the manipulator.

The transformation from joint velocities to the end-effector velocity is described by a matrix, called the *Jacobian*. The Jacobian matrix, which is dependent on manipulator configuration is a linear mapping from velocities in joint space to velocities in Cartesian space. The inverse problem, where the joint velocities are to be determined for a given end-effector velocity is of practical importance and requires the inverse of the Jacobian.

The Jacobian is one of the most important tools for characterization of differential motions of the manipulator. At certain locations in joint space, the Jacobian matrix may lose rank and it may not be possible to find its inverse. These locations are referred to as *singular* configurations. In this chapter manipulator singularities are also discussed. These singularities are used to establish some indices of merit to judge a manipulator design, while it is still on the drawing board.

In addition, the Jacobian is also useful for describing the mapping between forces applied to the end-effector and resulting torques at joints. This is *statics* and is discussed in the later part of this chapter.

Before discussing the Jacobian, the requisite background will be prepared in the sections that follow.

5.1 LINEAR AND ANGULAR VELOCITY OF A RIGID BODY

A brief overview of the concepts of linear and angular velocity of a rigid body, as applicable to robot analysis, is carried out first in this section. Consider a link *i* of a manipulator in isolation, as shown in Fig. 5.1. The link joins with two other links at points O_{i-1} and O_i. Time-variation of position and orientation of the link in space produces linear and angular velocities $\boldsymbol{v}$ and ω, respectively. The concepts of translation and rotation, discussed in Chapter 2, are extended to the time-variation of location, that is, velocities. The motion of rigid body (the link), therefore, can be equivalently studied by attaching coordinate frames to the link and studying the motion of frames relative to one another.

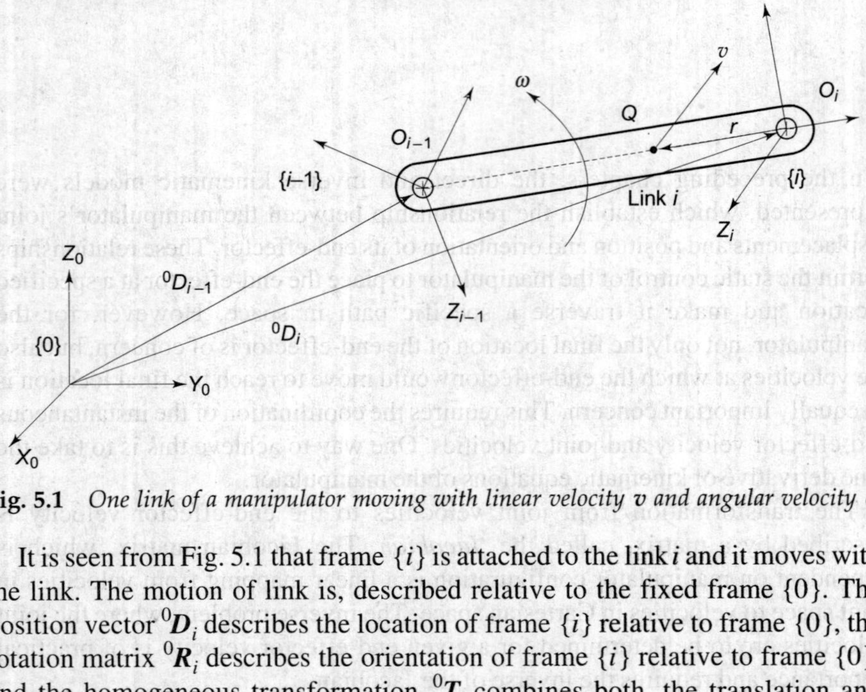

Fig. 5.1 *One link of a manipulator moving with linear velocity v and angular velocity ω*

It is seen from Fig. 5.1 that frame $\{i\}$ is attached to the link *i* and it moves with the link. The motion of link is, described relative to the fixed frame $\{0\}$. The position vector $^0\boldsymbol{D}_i$ describes the location of frame $\{i\}$ relative to frame $\{0\}$, the rotation matrix $^0\boldsymbol{R}_i$ describes the orientation of frame $\{i\}$ relative to frame $\{0\}$, and the homogeneous transformation $^0\boldsymbol{T}_i$ combines both, the translation and rotation of frame $\{i\}$ relative to frame $\{0\}$.

5.1.1 Linear Velocity

Suppose the link *i* is only translating relative to frame $\{0\}$. Every point on the link will have the same linear velocity. The linear velocity of link *i* or that of any point *Q* on it, relative to frame $\{0\}$ is given by

$$^0\boldsymbol{v}_i = {}^0\boldsymbol{v}_Q = \frac{d(^0\boldsymbol{D}_i)}{dt} = \lim_{\Delta t \to 0} \frac{^0\boldsymbol{D}_i(t+\Delta t) - {}^0\boldsymbol{D}_i(t)}{\Delta t} \tag{5.1}$$

For pure translation, linear link velocity can be obtained by taking the time derivative of the homogeneous transformation matrix because rotation matrix ${}^0\boldsymbol{R}_i$ is constant

$$
{}^0\boldsymbol{T}_i = \left[\begin{array}{ccc|c} & {}^0\boldsymbol{R}_i & & {}^0\boldsymbol{D}_i \\ & \text{(Constant)} & & \\ \hline 0 & 0 & 0 & 1 \end{array}\right] \tag{5.2}
$$

The link linearity velocity is thus given as

$$
{}^0\boldsymbol{v}_i = \lim_{\Delta t \to 0} \frac{{}^0\boldsymbol{T}_i(t+\Delta t) - {}^0\boldsymbol{T}_i(t)}{\Delta t} = {}^0\dot{\boldsymbol{T}}_i \tag{5.3}
$$

Note that in Eq. (5.3), ${}^0\boldsymbol{v}_i$ is a 3×1 vector with the components $[{}^0v_{ix}\ {}^0v_{iy}\ {}^0v_{iz}]^T$ each of which can be obtained by differentiating the corresponding components of the position vector ${}^0\boldsymbol{D}_i = [d_{ix}\ \ d_{iy}\ \ d_{iz}]^T$ with respect to time.

As with any other vector, a velocity vector can be mapped from one frame to another. But unlike position vectors, the velocity vectors are free vectors because they do not depend on their line of action and, hence, the mathematics of their mapping is different from that discussed in Chapter 2 as explained in the following subsection.

5.1.2 Angular Velocity

Consider now the rotation of the link in Fig. 5.1 with an angular velocity ω about some axis. In general, the axis of rotation itself may be changing with time. Therefore, there is an instantaneous axis of rotation and every point on the body as it rotates also has an instantaneous linear velocity, which is tangential to the arc of rotation at that instant. The angular velocity is discussed first.

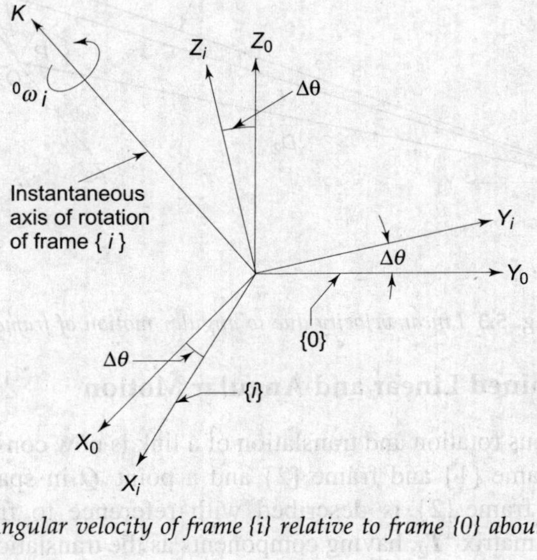

Fig. 5.2 *Angular velocity of frame {i} relative to frame {0} about axis k*

Let frame $\{i\}$ attached to the body be rotating with respect to a known stationary reference frame $\{0\}$, as shown in Fig. 5.2. The origins of the two frames are considered to be coincident such that only relative rotation is possible. The angular velocity of frame $\{i\}$, relative to frame $\{0\}$, is represented by angular velocity vector ${}^0\omega_i$ directed along the instantaneous axis of rotation of frame $\{i\}$, the axis k. The magnitude of angular velocity is equal to the speed of rotation. The instantaneous angular velocity ${}^0\omega_i$ is defined as

$$ {}^0\omega_i = \lim_{\Delta t \to 0} \frac{\Delta\theta}{\Delta t} = \frac{d\theta}{dt} = \dot{\theta} \tag{5.4} $$

5.1.3 Linear Velocity due to Angular Motion

To find the linear velocity arising due to angular motion consider frames $\{1\}$ and $\{2\}$ and a point Q on the link as shown in Fig. 5.3 (the link is not shown in figure for clarity). With respect to Fig. 5.1 for link $\boldsymbol{i}$, frame $\{1\}$ corresponds to frame $\{i-1\}$ and frame $\{2\}$ to frame $\{\boldsymbol{i}\}$. The frame $\{2\}$ along with the point Q is rotating about axis $\boldsymbol{k}$ with an instantaneous angular velocity ω. In small time interval Δt, the point Q describes an arc with radius $r = O_1Q$ and moves through an angle $\Delta\theta$, with respect to frame $\{1\}$. The linear distance translated in this time is $\Delta Q = r\Delta\theta$. Therefore, the linear velocity of Q is

$$ v = \lim_{\Delta t \to 0} \frac{\Delta Q}{\Delta t} = \lim_{\Delta t \to 0} \frac{r\Delta\theta}{\Delta t} = r\dot{\theta} = r\omega \tag{5.5} $$

The direction of this velocity is tangential to the arc described by Q.

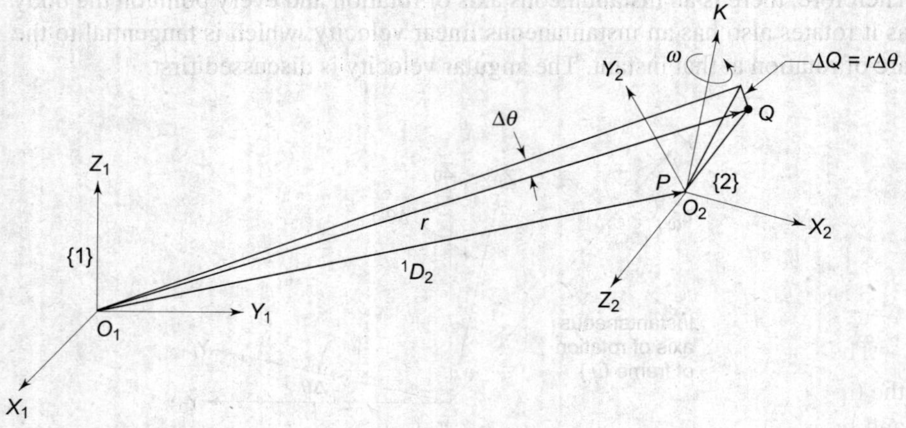

Fig. 5.3 *Linear velocity due to angular motion of frame {2}*

5.1.4 Combined Linear and Angular Motion

The simultaneous rotation and translation of a link is now considered. Consider two frames, frame $\{1\}$ and frame $\{2\}$ and a point Q in space, as shown in Fig. 5.3. The frame $\{2\}$ is described with reference to frame $\{1\}$ by the transformation matrix ${}^1\boldsymbol{T}_2$, having components as the translation vector ${}^1\boldsymbol{D}_2$ and rotation matrix ${}^1\boldsymbol{R}_2$. Small motion of frame {2} along with point Q, relative to frame {1}, results in a small displacement ΔQ of point Q along a circular arc of radius r in time Δt, subtending small angle $\Delta\theta$. Initially assume that the orientation of frame {2} is not changing with time, that is, ${}^1\boldsymbol{R}_2$ is constant. Then, the motion of point Q relative to frame {1} can be due to translation of origin of frame {2} (point P or origin O_2) and/or translation of point Q relative to frame {2}. The linear velocity of point Q with respect to frame {1} is, therefore,

$$ {}^1\boldsymbol{v}_Q = {}^1\boldsymbol{v}_P + {}^1T_2\ {}^2\boldsymbol{v}_Q \tag{5.6}$$

where ${}^2\boldsymbol{v}_Q$ is the velocity of point Q with respect to frame {2} and ${}^1\boldsymbol{v}_p$ is the velocity of frame {2}, with origin P, with respect to frame {1}.

Consider now the rotation of frame {2} about k-axis with an angular velocity ${}^1\omega_2$. Let Q be a fixed point in frame {2}, as shown in Fig. 5.4. Assume that frame {1} is translated to frame {1′} such that its origin coincides with origin of frame {2} and the axis of rotation (k-axis) is fixed with respect to frame {1′}. It means that vector $\overrightarrow{\boldsymbol{PQ}}$ rotates about k-axis at a fixed angular velocity ${}^1\omega_2$.

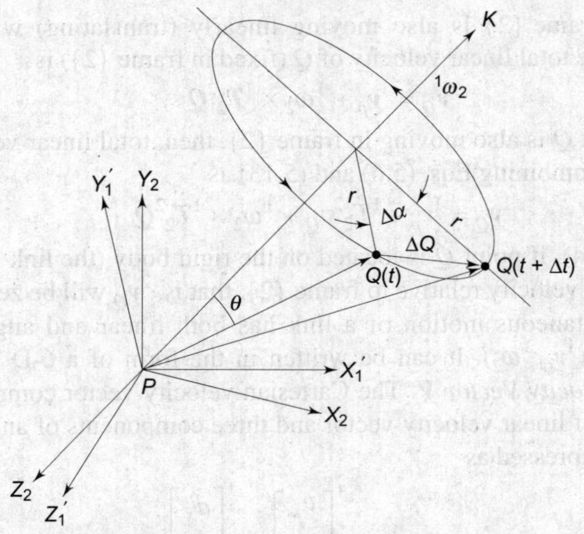

Fig. 5.4 *The velocity of a point due to angular velocity*

The figure shows the vector $\overrightarrow{\boldsymbol{PQ}}$ at any instant of time t. The rotation causes the tip Q of the vector $\overrightarrow{\boldsymbol{PQ}}$ to trace a circle in a plane perpendicular to the k-axis and line PQ describes the slanting surface of a cone. The radius of the circular path of Q is

$$ r = |\overrightarrow{\boldsymbol{PQ}}| \sin\theta \tag{5.7}$$

In time Δt, point Q turns by an angle $\Delta\alpha$ along the circle traversing a distance

$$\Delta Q = r\Delta\alpha$$

or

$$\Delta Q = |\overrightarrow{\boldsymbol{PQ}}| \sin\theta\, \Delta\alpha \tag{5.8}$$

Thus, the magnitude of linear velocity of Q is (in frame {1})

$$\lim_{\Delta t \to 0}\left|\frac{\Delta Q}{\Delta t}\right| = \left|\overrightarrow{PQ}\right| \sin\theta \lim_{t \to 0} \frac{\Delta \alpha}{\Delta t} \tag{5.9}$$

as

$$\lim_{\Delta t \to 0} \frac{\Delta \alpha}{\Delta t} = \left|{}^1\omega_2\right|$$

Therefore,

$$\left|{}^1\boldsymbol{v}_Q\right| = \left|{}^1\omega_2\right|\left|\overrightarrow{PQ}\right| \sin\theta \text{ or } {}^1\boldsymbol{v}_Q = {}^1\omega_2 \times {}^1\boldsymbol{Q} \tag{5.10}$$

Consider now the direction of the linear velocity of Q. It is tangential to the circle in the instantaneous direction of motion. The linear velocity direction at Q is, therefore, at 90° to both k-axis and $\overrightarrow{PQ}$ in right-handed direction. By virtue of these two observations, the linear velocity vector of point Q about the k-axis caused by rotation of $\overrightarrow{PQ}$ is the cross product

$${}^1\boldsymbol{v}_Q = {}^1\omega_2 \times {}^1\boldsymbol{Q} \tag{5.11}$$

Since the point Q is known in frame {2}, it can also be written as

$${}^1\boldsymbol{v}_Q = {}^1\omega_2 \times {}^1\boldsymbol{T}_2{}^2\boldsymbol{Q} \tag{5.12}$$

Now, if frame {2} is also moving linearly (translating) with respect to frame {1}, the total linear velocity of Q (fixed in frame {2}) is

$${}^1\boldsymbol{v}_Q = {}^1\boldsymbol{v}_p + {}^1\omega_2 \times {}^1\boldsymbol{T}_2{}^2\boldsymbol{Q} \tag{5.13}$$

If the point Q is also moving in frame {2}, then, total linear velocity of Q is obtained by combining Eqs. (5.6) and (5.13) as

$${}^1\boldsymbol{v}_Q = {}^1\boldsymbol{v}_p + {}^1\boldsymbol{T}_2{}^2v_Q + {}^1\omega_2 \times {}^1\boldsymbol{T}_2{}^2\boldsymbol{Q} \tag{5.14}$$

In Eq. (5.14), if point Q is located on the rigid body (the link in Fig. 5.1), it cannot have a velocity relative to frame {2}, that is, ${}^2\boldsymbol{v}_Q$ will be zero.

The instantaneous motion of a link has both linear and angular velocity components, $({}^1\boldsymbol{v}_Q, {}^1\omega_2)$. It can be written in the form of a 6-D vector, called *Cartesian Velocity Vector* $\boldsymbol{V}$. The Cartesian velocity vector comprises of three components of linear velocity vector and three components of angular velocity vector. It is expressed as

$${}^1\boldsymbol{V}_Q = {}^1\begin{bmatrix} \boldsymbol{v} \\ \omega \end{bmatrix}_Q = {}^1\begin{bmatrix} v_x \\ v_y \\ v_z \\ \omega_x \\ \omega_y \\ \omega_z \end{bmatrix}_Q = {}^1\begin{bmatrix} \dot{d}_x \\ \dot{d}_y \\ \dot{d}_z \\ \dot{\theta}_x \\ \dot{\theta}_y \\ \dot{\theta}_z \end{bmatrix}_Q \tag{5.15}$$

5.2 RELATIONSHIP BETWEEN TRANSFORMATION MATRIX AND ANGULAR VELOCITY

In Chapter 2, it was seen that finite rotations between two coordinate frames could be represented using a rotation matrix. The angular velocity vector also represents the rotational motion between frames. Hence, the homogeneous transformation matrix $\boldsymbol{T}$ and the angular velocity vector ω, at any instant of time, must be related. To analyze this relationship consider frame {2} attached to rigid body as it rotates with respect to frame {2′}, as shown in Fig. 5.5, in time t to $(t + \Delta t)$.

Let ${}^1\boldsymbol{T}_2$ and ${}^1\boldsymbol{T}_{2'}$ represent the homogeneous transformation matrix describing frame {2} and frame {2′} with respect to frame {1} at time instant t and $(t + \Delta t)$, respectively. The derivative of the transformation matrix ${}^1\boldsymbol{T}_2(t)$ with respect to time is

$$ {}^1\dot{\boldsymbol{T}}_2 = \lim_{\Delta t \to 0} \frac{{}^1\boldsymbol{T}_{2'} - {}^1\boldsymbol{T}_2}{\Delta t} \tag{5.16} $$

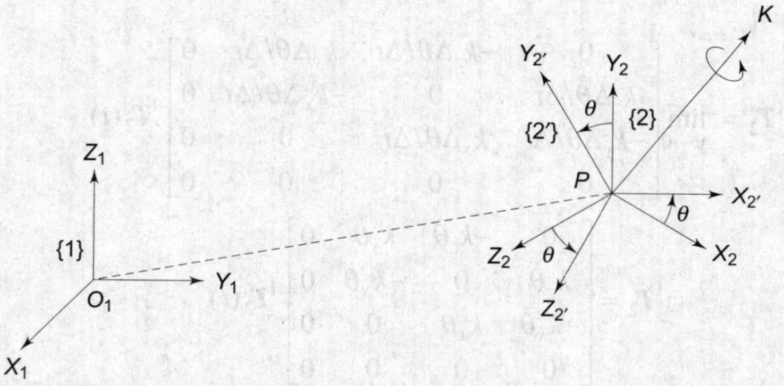

Fig. 5.5 *Relationship between transformation matrix and angular velocity vector*

Frame {2′} is obtained by rotating frame {2} about an axis K in the fixed reference frame {1} by a small angle θ during the time interval Δt. Here, axis K is nothing but the instantaneous axis of rotation at time t or the equivalent angle axis representation. From Section 2.6.4, Eq. (2.71), rotation of frame {2} to frame {2′} by angle θ about axis K is given by

$$ \boldsymbol{T}_k(\theta) = \begin{bmatrix} k_z^2 V\theta + C\theta & k_x k_y V\theta - k_z S\theta & k_x k_z V\theta + k_y S\theta & 0 \\ k_x k_y V\theta + k_z S\theta & k_y^2 V\theta + C\theta & k_y k_z V\theta - k_x S\theta & 0 \\ k_x k_z V\theta - k_y S\theta & k_y k_z V\theta + k_x S\theta & k_z^2 V\theta + C\theta & 0 \\ 0 & 0 & 0 & 1 \end{bmatrix} \tag{5.17} $$

where $\hat{\boldsymbol{k}} = [k_x \ k_y \ k_z]^T$, $C\theta = \cos\theta$, $S\theta = \sin\theta$ and $V\theta = 1 - \cos\theta$.

When θ is very small, $C\theta \approx 1$, $S\theta \approx \Delta\theta$, and $V\theta \approx 0$, such that, Eq. (5.17) reduces to

$$ \boldsymbol{T}_k(\Delta\theta) = \begin{bmatrix} 1 & -k_z\Delta\theta & k_y\Delta\theta & 0 \\ k_z\Delta\theta & 1 & -k_x\Delta\theta & 0 \\ -k_y\Delta\theta & k_x\Delta\theta & 1 & 0 \\ 0 & 0 & 0 & 1 \end{bmatrix} \tag{5.18} $$

The description of frame {2′} can be obtained by two successive transformations relative to the fixed frame {1} as

$$\text{frame}\{1\} \xrightarrow{T_2(t)} \text{frame}\{2\} \xrightarrow{T_k(\Delta\theta)} \text{frame}\{2'\}$$

That is

$$^1\boldsymbol{T}_{2'} = \boldsymbol{T}_k(\Delta\theta)\ ^1\boldsymbol{T}_2(t) \tag{5.19}$$

Equation (5.16) is simplified using Eq. (5.19) giving

$$^1\dot{\boldsymbol{T}}_2 = \lim_{\Delta t \to 0}\left(\frac{\boldsymbol{T}_k(\Delta\theta) - \boldsymbol{I}}{\Delta t}\right)\ ^1\boldsymbol{T}_2(t) \tag{5.20}$$

where $\boldsymbol{I}$ is 4×4 identity matrix. Substituting $\boldsymbol{T}_k(\Delta\theta)$ from Eq. (5.18) in Eq. (5.20) gives

$$^1\dot{\boldsymbol{T}}_2 = \lim_{\Delta t \to 0}\begin{bmatrix} 0 & -k_z\Delta\theta/\Delta t & k_y\Delta\theta/\Delta t & 0 \\ k_z\Delta\theta/\Delta t & 0 & -k_x\Delta\theta/\Delta t & 0 \\ -k_y\Delta\theta/\Delta t & k_x\Delta\theta/\Delta t & 0 & 0 \\ 0 & 0 & 0 & 0 \end{bmatrix}\ ^1\boldsymbol{T}_2(t)$$

or

$$^1\dot{\boldsymbol{T}}_2 = \begin{bmatrix} 0 & -k_z\dot{\theta} & k_y\dot{\theta} & 0 \\ k_z\dot{\theta} & 0 & -k_x\dot{\theta} & 0 \\ -k_y\dot{\theta} & k_x\dot{\theta} & 0 & 0 \\ 0 & 0 & 0 & 0 \end{bmatrix}\ ^1\boldsymbol{T}_2(t)$$

or

$$^1\dot{\boldsymbol{T}}_2 = \begin{bmatrix} 0 & -k_z & k_y & 0 \\ k_z & 0 & -k_x & 0 \\ -k_y & k_x & 0 & 0 \\ 0 & 0 & 0 & 0 \end{bmatrix}\dot{\theta}\ ^1\boldsymbol{T}_2(t) \tag{5.21}$$

where $\dot{\theta} = \lim_{\Delta t \to 0}\left(\frac{\Delta\theta}{\Delta t}\right)$ is the speed of angular rotation of frame {2} about axis K.

The angular rotation at speed $\dot{\theta}$ about axis K gives angular velocity of frame {2} relative to frame {1} as a vector $^1\omega_2$ directed along axis K. This is equivalent to three rotations of ω_x, ω_y, and ω_z, made in that order about the x-, y-, and z-axes, respectively, where

$$^1\omega_2 = \begin{bmatrix} \omega_x \\ \omega_y \\ \omega_z \end{bmatrix} = \begin{bmatrix} k_x \\ k_y \\ k_z \end{bmatrix}\dot{\theta} \tag{5.22}$$

Thus, Eq. (5.21) is rewritten as

$$^1\dot{\boldsymbol{T}}_2 = \begin{bmatrix} 0 & -\omega_z & \omega_y & 0 \\ \omega_z & 0 & -\omega_x & 0 \\ -\omega_y & \omega_x & 0 & 0 \\ 0 & 0 & 0 & 0 \end{bmatrix}\ ^1\boldsymbol{T}_2(t) \tag{5.23}$$

Equations (5.22) and (5.23) give the relationship between the angular velocity of the link and the transformation matrix. If a link undergoes only angular motion with no translation of the origin of frame {2}, in place of transformation matrix $\boldsymbol{T}$ the rotation matrix $\boldsymbol{R}$ may be used in the above derivation. For pure rotation, the relation between angular velocity of the link and rotation matrix follows from Eqs. (5.22) and (5.23) as

$$\dot{\boldsymbol{R}}(t) = \begin{bmatrix} 0 & -\omega_z & \omega_y \\ \omega_z & 0 & -\omega_x \\ -\omega_y & \omega_x & 0 \end{bmatrix} \boldsymbol{R}(t) = \begin{bmatrix} 0 & -k_z & k_y \\ k_z & 0 & -k_x \\ -k_y & k_x & 0 \end{bmatrix} \dot{\theta}\, \boldsymbol{R}(t) \tag{5.24}$$

The detailed derivation is left to the reader. The superscript and subscript have been dropped for simplicity. The derivative of the orthonormal rotation matrix has a very interesting property. Multiplying both sides of Eq. (5.24) by $\boldsymbol{R}^T$ ($= \boldsymbol{R}^{-1}$) gives

$$\dot{\boldsymbol{R}}(t)\boldsymbol{R}^T(t) = \begin{bmatrix} 0 & -\omega_z & \omega_y \\ \omega_z & 0 & -\omega_x \\ -\omega_y & \omega_x & 0 \end{bmatrix} = \begin{bmatrix} 0 & -k_z & k_y \\ k_z & 0 & -k_x \\ -k_y & k_x & 0 \end{bmatrix} \dot{\theta} \tag{5.25}$$

Let the left-hand term in Eq. (5.25) be denoted by a matrix $\boldsymbol{S}(t)$ and recall that $\boldsymbol{R}^T = \boldsymbol{R}^{-1}$ thus,

$$\boldsymbol{S}(t) = \dot{\boldsymbol{R}}(t)\,\boldsymbol{R}^T(t) = \dot{\boldsymbol{R}}\boldsymbol{R}^{-1}(t) \tag{5.26}$$

where
$$\boldsymbol{S}(t) = \begin{bmatrix} 0 & -\omega_z & \omega_y \\ \omega_z & 0 & -\omega_x \\ -\omega_y & \omega_x & 0 \end{bmatrix} \tag{5.27}$$

It is observed that $\boldsymbol{S}(t)$ is a skew-symmetric matrix. It immediately follows from the property of skew symmetric matrix that

$$\boldsymbol{S}(t) + \boldsymbol{S}^T(t) = \boldsymbol{0} \tag{5.28}$$

where $\boldsymbol{0}$ is a 3×3 zero matrix.

Equation (5.26) reveals an important property of the rotation matrix: the product of derivative of the orthonormal rotation matrix with its transpose or inverse yields a skew-symmetric matrix.

It is concluded from Eqs. (5.22) and (5.28), that at any instant the elements of the skew-symmetric matrix $\boldsymbol{S}(t)$ are indeed the components of the angular velocity vector ${}^1\omega_2$. Therefore, a change in orientation of a rotating frame can be viewed as a rotation about some axis K.

It follows from Eq. (5.26), that

$$\dot{\boldsymbol{R}}(t) = \boldsymbol{S}(t)\,\boldsymbol{R}(t) \tag{5.29}$$

Consider a constant vector ${}^2\boldsymbol{P}$ undergoing rotation. Then

$$^1\boldsymbol{P} = \boldsymbol{R}(t)\,{}^2\boldsymbol{P} \tag{5.30}$$

The time derivative of ${}^1\boldsymbol{P}(t)$ is

$$ {}^1\dot{\boldsymbol{P}} = \dot{\boldsymbol{R}}(t)\ {}^2\boldsymbol{P} \tag{5.31}$$

which, from Eq. (5.29) can be written as

$$ {}^1\dot{\boldsymbol{P}} = \boldsymbol{S}(t)\,\boldsymbol{R}(t)\ {}^2\boldsymbol{P} \tag{5.32}$$

The vector $\omega\,(t)$ of Eq. (5.22) denotes the angular velocity of frame {2} with respect to frame {1} at time t. Using $\omega\,(t)$ and $\boldsymbol{S}(t)$ of Eq. (5.27); it is easily verified that

$$ \boldsymbol{S}(t)\boldsymbol{P} = \omega(t) \times \boldsymbol{P} \tag{5.33}$$

Comparison of Eq. (5.32) and Eq. (5.33) gives

$$ {}^1\dot{\boldsymbol{P}}(t) = \omega(t) \times \boldsymbol{R}(t)\ {}^2\boldsymbol{P} \tag{5.34}$$

where $\times$ is the vector cross product and superscripts have been dropped for generalization discussed in next section.

5.3 MAPPING VELOCITY VECTORS

Velocity vectors belong to a class of vectors called *free vectors*. In the case of free vectors, all that needs to be preserved is the magnitude and direction. Thus, free vectors could be displaced anywhere in space without changing their meaning. A moment vector is another example of a free vector.

The mapping of the linear angular velocity vector, from frame {1} to frame {2}, can have reference to two frames, the frame of description and the frame of differentiation. In the study of motion of links of manipulators, the frame of differentiation for velocity vectors is always the base frame, that is, frame {0}. Hence, the leading superscript for the frame of differentiation can be dropped and a single leading superscript in velocity vector will denote the frame of description. Further, the absence of a superscript will indicate that the base frame serves as both the frame of description and the frame of differentiation. At any instant, each link of the manipulator in motion has both linear and angular velocities. The subscript describes the link or the link's frame. For example, for link i, $\boldsymbol{v}_i$ is linear velocity (of origin) of frame $\{i\}$, and ω_i is the angular velocity of frame $\{i\}$, with base frame {0} as the frame of description as well as the frame of differentiation.

5.4 VELOCITY PROPAGATION ALONG LINKS

Having discussed the concepts of linear and angular velocity of a rigid body, we now proceed to determine the linear and angular velocity of any link of a given manipulator.

The n-DOF manipulator is an open kinematic chain with n-links, each one capable of motion relative to its neighbour. The base, link 0, is the stationary reference and velocity of all other links is defined with respect to this. Therefore, the velocity of each link can be computed starting from the base. The velocity of link (i) will be the velocity of link (i – 1) plus velocity component added by joint (i).

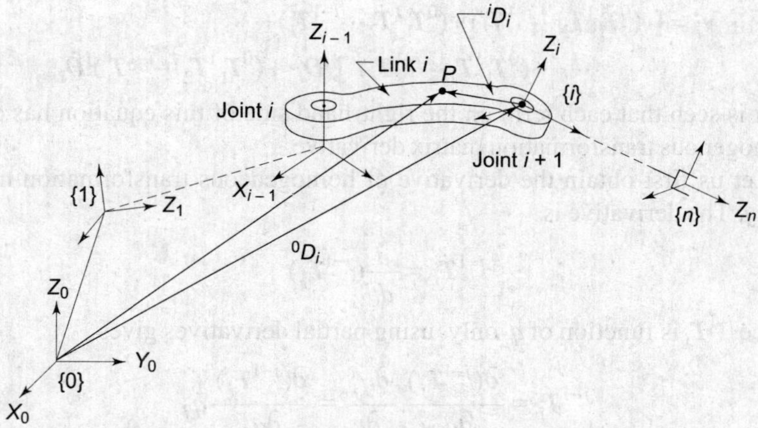

Fig. 5.6 *Velocity propagation along links of a n-DOF manipulator*

To determine the velocity of link i of a given n-DOF manipulator, consider link i with frame $\{i–1\}$ and frame $\{i\}$ attached to two ends of link i, as shown in Fig. 5.6. The origin of frame $\{i\}$, which is attached to the distal end of link will specify the velocity of link i.

5.4.1 Linear Velocity of a Link

To find the linear velocity of link i with respect to base, consider any point P on the link. The kinematic relationship between link i and link 0 (base) as given by Eq. (2.46) is

$$^0\boldsymbol{T}_i = {}^0\boldsymbol{T}_1\ {}^1\boldsymbol{T}_2 \cdots {}^{i-1}\boldsymbol{T}_i \tag{5.35}$$

The position vector $^i\boldsymbol{D}_i$ of this point P referred to frame $\{i\}$, can be referred to the base frame $\{0\}$ using Eq. (2.26). Thus,

$$^0\boldsymbol{D}_i = {}^0\boldsymbol{T}_i\ {}^i\boldsymbol{D}_i \tag{5.36}$$

The linear velocity of P and so of the link i, using Eq. (5.1) is given as

$$^0\boldsymbol{v}_i = \frac{d}{dt}({}^0\boldsymbol{D}_i) = \frac{d}{dt}({}^0\boldsymbol{T}_i\,{}^i\boldsymbol{D}_i) \tag{5.37}$$

In Eq. (5.37), as frame $\{0\}$ is the frame of differentiation as well as description, we can write $^0\boldsymbol{v}_i$ *as* $\boldsymbol{v}_i$. In each $^{j-1}\boldsymbol{T}_j$, only one variable exists, the jth joint displacement variable, q_j. It is useful to recall that

$$q_j = \begin{cases} \theta_j & \text{if joint } j \text{ is revolute.} \\ d_j & \text{if joint } j \text{ is prismatic.} \end{cases}$$

Substituting Eq. (5.35) in Eq. (5.37)

$$\boldsymbol{v}_i = \frac{d}{dt}\left[({}^0\boldsymbol{T}_1\ {}^1\boldsymbol{T}_2 \ldots {}^{i-1}\boldsymbol{T}_i)\,{}^i\boldsymbol{D}_i\right] \tag{5.38}$$

or
$$v_i = [\ ({}^0\dot{T}_1{}^1T_2\cdots{}^{i-1}T_i)+({}^0T_1{}^1\dot{T}_2\cdots{}^{i-1}T_i)+\cdots\ \cdots \\ \cdots\ \cdots\ +({}^0T_1{}^1T_2\cdots{}^{i-1}\dot{T}_i)\ ]\,{}^iD_i\ +({}^0T_1{}^1T_2\cdots{}^{i-1}T_i)\,{}^i\dot{D}_i \tag{5.39}$$

It is seen that each term on the right-hand side of this equation has only one homogenous transformation matrix derivative.

Let us first obtain the derivative of homogeneous transformation matrix of link j. The derivative is

$$ {}^{j-1}\dot{T}_j = \frac{d}{dt}({}^{j-1}T_j) \tag{5.40}$$

Since ${}^{j-1}T_j$ is function of q_j only, using partial derivatives gives

$$ {}^{j-1}\dot{T}_j = \frac{\partial({}^{j-1}T_j)}{\partial q_j}\frac{\partial q_j}{\partial t} = \frac{\partial({}^{j-1}T_j)}{\partial q_j}\dot{q}_j \tag{5.41}$$

Now consider the jth term in Eq. (5.39) containing term ${}^{j-1}\dot{T}_j$

$$ j^{\text{th}}\text{ term} = {}^0T_1{}^1T_2\cdots{}^{j-1}\dot{T}_j\cdots{}^{i-1}T_i$$

and substitute result of Eq. (5.41) in it. This gives

$$ j^{\text{th}}\text{ term} = {}^0T_1{}^1T_2\cdots\frac{\partial({}^{j-1}T_j)}{\partial q_j}\dot{q}_j\cdots{}^{i-1}T_i \tag{5.42}$$

In Eq. (5.42), the partial derivative can be taken out because no other transformation matrix except ${}^{j-1}T_j$ is a function of q_j. Thus,

$$ j^{\text{th}}\text{ term} = \frac{\partial}{\partial q_j}({}^0T_1{}^1T_2\cdots{}^{j-1}T_j\cdots{}^{i-1}T_i)\dot{q}_j \tag{5.43}$$

Combining transformation matrices with in brackets, Eq. (5.43) reduces to

$$ j^{\text{th}}\text{ term} = \frac{\partial({}^0T_i)}{\partial q_j}\dot{q}_j \tag{5.44}$$

where $\dot{q}_j$ is the time-rate of change of joint displacement q_j. The last term of Eq. (5.39) involves a derivative of ${}^iD_i$. Because the link is a rigid body, ${}^iD_i$ will be constant and, hence, its derivative will be zero. That is

$$ {}^i\dot{D}_i = 0 \tag{5. 45}$$

Substituting Eq. (5.44) and Eq. (5.45) in Eq. (5.39) and simplifying gives

$$ v_i = \sum_{j=1}^{i}\frac{\partial({}^0T_i)}{\partial q_j}\dot{q}_j\ {}^iD_i \tag{5.46}$$

The result of Eq. (5.46) may be extended to obtain the linear velocity of the end-effector of an n-DOF manipulator as

$$ v_{\text{end-effector}} = v_n = \sum_{j=1}^{n}\frac{\partial({}^0T_n)}{\partial q_j}\dot{q}_j\,{}^nD_n \tag{5.47}$$

Only the first three elements of 4 × 1 column vector are meaningful.

5.4.2 Angular Velocity of a Link

The angular velocity of link i is the vector sum of the angular velocity of link $(i-1)$ and the component generated by joint (i). In case joint i is prismatic, it allows only a translational motion between link $(i-1)$ and link i and, hence, it does not contribute to the angular velocity of link i. Thus for prismatic joint i

$$\omega_i = \omega_{i-1} \tag{5.48}$$

If the joint (i) is rotary, the relative rotation of link i with respect to link $(i-1)$ makes a contribution, so that

$$\omega_i = \omega_{i-1} + {}^{i-1}\omega_i \tag{5.49}$$

The speed of rotation of link i with respect to link $(i-1)$, is $\dot{\theta}_i$ about the z_{i-1}-axis. When referred to frame {0}

$${}^{i-1}\omega_i = {}^{0}\boldsymbol{R}_{i-1}\, z_{i-1}\, \dot{\theta}_i \tag{5.50}$$

Substituting in Eq. (5.49)

$$\omega_i = \omega_{i-1} + {}^{0}\boldsymbol{R}_{i-1}\, z_{i-1}\, \dot{\theta}_i = \omega_{i-1} + {}^{0}\boldsymbol{R}_{i-1}\, z_{i-1}\, \dot{q}_i \tag{5.51}$$

It is easily concluded from Eq. (5.51) that the rotary joint i contributes to the angular velocities of all links from link i to n. Further, it is to be noted that, both, the angular velocities of the manipulator links and the joint velocity are relative to the base frame {0}.

It is thus found from the above account that (i) linear velocities of all manipulator links are obtained using Eq. (5.46), and (ii) angular velocities of all the links are obtained using Eqs. (5.48) and (5.51).

5.5 MANIPULATOR JACOBIAN

From the results of the previous section, it is observed that the Cartesian velocities (linear as well as angular) of the end-effector are linearly related to the joint velocities. The relationship between infinitesimal (differential) joint motions with infinitesimal (differential) changes in end-effector position (and orientation) is investigated now. As these changes take place in infinitesimal (differential) time, we are really looking for a mapping between instantaneous end-effector velocity, $\boldsymbol{V}_e$ (in Cartesian space) to instantaneous joint velocities (in joint space). This mapping between differential changes is linear and can be expressed as

$$\boldsymbol{V}_e(t) = \boldsymbol{J}(\boldsymbol{q})\dot{\boldsymbol{q}} \tag{5.52}$$

where $\boldsymbol{V}_e(t)$ = 6 × 1 Cartesian velocity vector, [Eq. (5.15)],
$\boldsymbol{J}(\boldsymbol{q})$ = 6 × n *Manipulator Jacobian* or *Jacobian* matrix,
$\dot{\boldsymbol{q}}$ = n × 1 vector of n joint velocities.

Equation (5.52) can be written in column vectors of the Jacobian, that is,

$$\boldsymbol{V}_e(t) = [\boldsymbol{J}_1(\boldsymbol{q})\, \boldsymbol{J}_2(\boldsymbol{q}) \cdots\cdots \boldsymbol{J}_n(\boldsymbol{q})]\, \dot{\boldsymbol{q}}(t) \tag{5.53}$$

In the above equation, $J_i(q)$ is the ith column of the Jacobian matrix. From Eq. (5.15) and Eq. (5.53)

$$V_e(t) = \begin{bmatrix} v \\ \omega \end{bmatrix} = \begin{bmatrix} \dot{d} \\ \dot{\theta} \end{bmatrix} = J(q)\,\dot{q}(t) \tag{5.54}$$

Equation (5.54) represents the *forward differential motion model* or *differential kinematics model* presented schematically in Fig. 5.7, which is similar to the forward kinematic model. Note that the Jacobian $J(q)$ is a function of the joint variables.

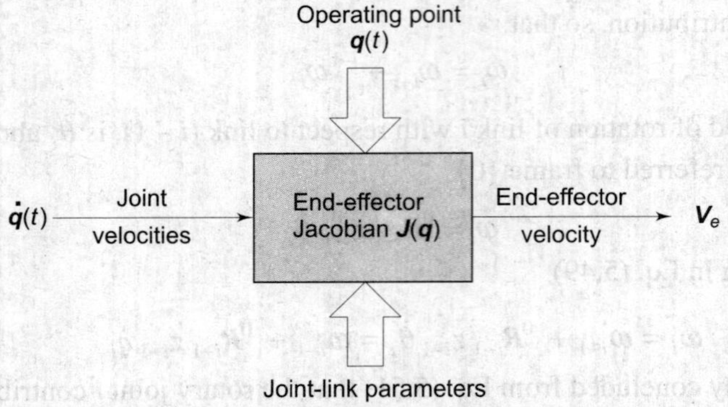

Fig. 5.7 *The forward differential motion model*

The first three rows of Jacobian $J(q)$ are associated with the linear velocity of end-effector v, while the last three rows correspond to the angular velocity ω of the end-effector. Each joint of the manipulator generates some linear and/or some angular velocity at the end-effector. Column i, of Jacobian matrix $J_i(q)$ is, thus, made up from three linear velocity components j_{vi} and three angular velocity components $j_{\omega i}$ and can be expressed as

$$J_i(q) = \begin{bmatrix} j_{vi} \\ j_{\omega i} \end{bmatrix} = \begin{bmatrix} j_{vxi} \\ j_{vyi} \\ j_{vzi} \\ j_{\omega xi} \\ j_{\omega yi} \\ j_{\omega zi} \end{bmatrix} \tag{5.55}$$

where j_{vki} and $j_{\omega ki}$ represents the component k of linear velocity and angular velocity, respectively, contributed by joint i with $k = x, y,$ or z and $i = 1, 2, 3, ..., n$. The contribution of joint i to the column J_i is computed depending on whether joint i is prismatic or rotary.

5.5.1 Jacobian Computation

The contribution of joint i to linear velocity of end-effector is $J_{vi}\,\dot{q}_i$ and to angular velocity of end-effector is $J_{\omega i}\,\dot{q}_i$. One simple method of computation of Jacobian is to determine J_{vi} and $J_{\omega i}$ for joint i. This is carried out in the following sections for both prismatic and revolute joints.

5.5.2 The Prismatic Joint Jacobian

Consider the prismatic joint i, as shown in Fig. 5.8. The link i translates along z_{i-1}-axis with a joint (scalar) velocity $\dot{d}_i$. To find the ith column $J_i(q)$ of the Jacobian matrix $J(q)$, assume that all joints except joint i are locked in position.

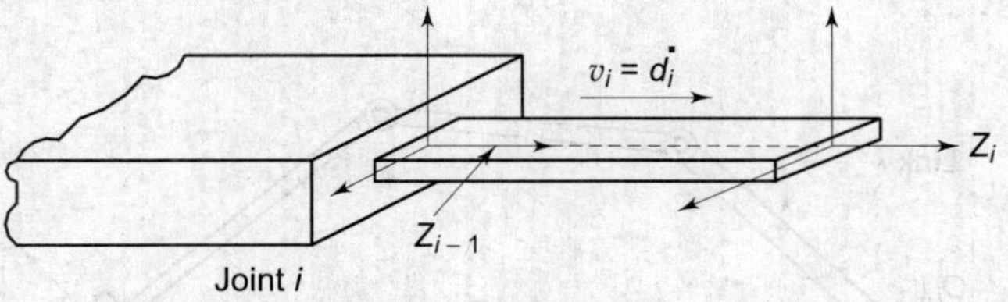

Fig. 5.8 *Prismatic joint contributing to end-effecter velocity*

In this situation, the translating prismatic joint i will produce a linear velocity at the end-effector in the same direction as the joint i axis, z_{i-1}. If a vector pointing along the direction z_{i-1}, with respect to base frame is denoted as P_{i-1}, the linear velocity produced at the end-effector by translating link i is

$$v_i = P_{i-1}\,\dot{d}_i \tag{5.56}$$

Note that for convenience P_{i-1} is written in place of ${}^0P_{i-1}$, that is, the leading superscript 0 is dropped. The prismatic joint cannot contribute any angular velocity at the end-effector, that is,

$$\omega_i = 0$$

From Eqs. (5.55) and (5.56), the $J_i(q)$ for a prismatic joint is, therefore

$$J_i(q) = \begin{bmatrix} J_{vi} \\ J_{\omega i} \end{bmatrix} = \begin{bmatrix} P_{i-1} \\ 0 \end{bmatrix} \tag{5.57}$$

Vector P_{i-1} in the above equation is computed using the coordinate transformations from base frame {0} to frame {i–1}. A unit vector in frame {i–1} along z_{i-1}-axis, with respect to frame {i–1} is given by $\hat{u} = [0 \quad 0 \quad 1]^T$.

The vector P_{i-1} is the transformation of this unit vector $\hat{u}$ with respect to base frame {0}. The homogeneous transformation matrix from frame {0} to frame {i–1} is

$${}^0T_1(q_1)\ \dots\ {}^{i-2}T_{i-1}(q_{i-1}) = {}^0T_{i-1} \tag{5.58}$$

The vector P_{i-1} is obtained as

$$P_{i-1} = {}^0R_{i-1}\,\hat{u} \tag{5.59}$$

where ${}^0R_{i-1}$ is the 3×3 orientation sub-matrix (the rotation matrix) of ${}^0T_{i-1}$. Note that P_{i-1} is the third column of the rotation matrix ${}^0R_{i-1}$.

5.5.3 The Rotary Joint Jacobian

Next, consider the joint i to be a rotary joint with angular velocity $\dot{\theta}_i$ about z_{i-1}-axis, as shown in Fig. 5.9. With all joints, except joint i, locked in position, the rotary joint rotates all the distal links from link i to link n at an angular velocity ω_i. For a rotary joint, the angular velocity will be given by

$$\omega_i = \boldsymbol{P}_{i-1}\,\dot{\theta}_i \tag{5.60}$$

where $\boldsymbol{P}_{i-1}$ is the vector in frame {0} describing the axis of rotation of joint i, that is z_{i-1}-axis.

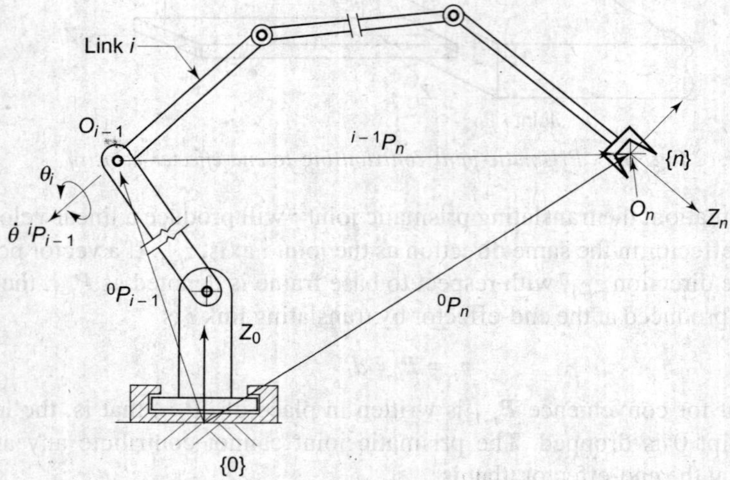

Fig. 5.9 *Motion of end-effecter generated by a rotary joint*

The angular velocity of link i also produces a linear velocity at the end-effector due to the rotation of all the distal links along with the end-effector about the origin O_{i-1} of frame $\{i-1\}$. From O_{i-1}, the end-effector frame, frame $\{n\}$ is defined by a position vector $^{i-1}\boldsymbol{P}_n$, as shown in Fig. 5.9. From Eqs. (5.11) and (5.60), the linear velocity generated by ω_i, is

$$\boldsymbol{v}_i = \omega_i \times {}^{i-1}\boldsymbol{P}_n = (\boldsymbol{P}_{i-1} \times {}^{i-1}\boldsymbol{P}_n)\,\dot{\theta}_i \tag{5.61}$$

It follows from Eqs. (5.60) and (5.61), that for a rotary joint,

$$\boldsymbol{J}_i = \begin{bmatrix} \boldsymbol{P}_{i-1} \times {}^{i-1}\boldsymbol{P}_n \\ \boldsymbol{P}_{i-1} \end{bmatrix} \tag{5.62}$$

For computing $\boldsymbol{J}_i$ in Eq. (5.62), we need vectors $\boldsymbol{P}_{i-1}$ and $^{i-1}\boldsymbol{P}_n$. The vector $\boldsymbol{P}_{i-1}$ is obtained from Eq. (5.59), while the vector $^{i-1}\boldsymbol{P}_n$ is found from Fig. (5.9) as

$$^{i-1}\boldsymbol{P}_n = {}^0\boldsymbol{P}_n - {}^0\boldsymbol{P}_{i-1}$$

The origin of frame $\{n\}$at the end-effector is $\boldsymbol{O}_n = [\,0\ \ 0\ \ 0\ \ 1\,]^T$. This applies for origin of any frame i.e. for any value of n. The above vector equation can then be written as

$$\text{or} \quad {}^{i-1}\boldsymbol{P}_n = [{}^0\boldsymbol{T}_1(q_1) \ldots {}^{n-1}\boldsymbol{T}_n(q_n)\ \boldsymbol{O}_n\ ] - [{}^0\boldsymbol{T}_1(q_1) \ldots {}^{i-2}\boldsymbol{T}_{i-1}(q_{i-1})\boldsymbol{O}_n]$$

$$\text{or} \quad {}^{i-1}\boldsymbol{P}_n = {}^0\boldsymbol{T}_n\boldsymbol{O}_n - {}^0\boldsymbol{T}_{i-1}\ \boldsymbol{O}_n \tag{5.63}$$

The computation of right-hand side of Eq. (5.63) would yield the three elements of position vector and the fourth term will be unity, the scale factor of $\boldsymbol{T}$ matrix. The vector ${}^{i-1}\boldsymbol{P}_i$ can be in homogeneous coordinates as 4×1 vector or in physical coordinates as 3×1 vector.

The manipulator Jacobian is obtained by determining all the elements using Eqs. (5.57), (5.59), (5.62) and (5.63). In summary, it is:

$$\boldsymbol{J}_i(q) = \begin{bmatrix} \boldsymbol{J}_{vi} \\ \boldsymbol{J}_{\omega i} \end{bmatrix} = \begin{cases} \begin{bmatrix} \boldsymbol{P}_{i-1} \\ 0 \end{bmatrix} & \text{for a prismatic joint} \\ \begin{bmatrix} \boldsymbol{P}_{i-1} \times {}^{i-1}\boldsymbol{P}_n \\ \boldsymbol{P}_{i-1} \end{bmatrix} & \text{for a revolute joint} \end{cases} \tag{5.64}$$

Equation (5.64) is a simple, systematic method of computing Jacobian of a manipulator. If, for any joint, the displacement axis is not *z*-axis, the above equations have to be appropriately modified. Note that the Jacobian matrix depends on the frame in which end-effector velocity is expressed.

5.6 JACOBIAN INVERSE

In practice, to make the manipulator's end-effector track a specified trajectory with a given velocity profile, it is required to coordinate individual joint motions. In other words, for a given end-effector velocity $\boldsymbol{V}_e$, the corresponding joint velocities ($\dot{\boldsymbol{q}}$) must be found that will cause the end-effector to move at the desired velocity.

In Section 5.5, it has been shown that the Jacobian of a manipulator defines a linear mapping between the vector $\dot{\boldsymbol{q}}$ of the joint velocities and end-effector velocity $\boldsymbol{V}_e$, from joint space to Cartesian space [Eq. (5.52)] as

$$\boldsymbol{V}_e = \boldsymbol{J}(\boldsymbol{q})\ \dot{\boldsymbol{q}} \tag{5.65}$$

From this, the reverse mapping from Cartesian space to joint space is

$$\dot{\boldsymbol{q}} = \boldsymbol{J}^{-1}(\boldsymbol{q})\ \boldsymbol{V}_e \tag{5.66}$$

It is known that for inverse of a matrix to exist, it must be a square matrix. The Jacobian, in general, for an n-DOF manipulator is a $6 \times n$ matrix. Therefore, only for a 6-DOF manipulator, the Jacobian $\boldsymbol{J}$ is a 6×6 square matrix. The inverse matrix $\boldsymbol{J}^{-1}$ exists, if $\boldsymbol{J}$ is nonsingular at the current configuration, in which case Eq. (5.66) determines the individual joint velocities required to obtain desired end-effector velocity.

Consequently at a configuration, where the Jacobian inverse exists, it is possible to move the end-effector in an arbitrary direction with an arbitrary velocity. For, a 6-DOF manipulator it means that all the six components of end-effector linear and angular velocities can be controlled by varying the six joint rates.

In earlier chapters, it has been shown that a manipulator must have at least 6-DOF to locate the end-effector in an arbitrary location. Similarly, 6-DOF are necessary for moving the end-effector in arbitrary direction with an arbitrary speed.

The above discussion is restricted to manipulator with exactly 6-DOF. Nevertheless, there are manipulators with less than 6-DOF for many applications. For example, to manipulate objects lying within a cylindrical workspace, the 4-DOF SCARA robot described in Examples 3.6 can be used. It can be shown that only four end-effector velocities, v_{x4}, v_{y4}, v_{z4}, and ω_{z4} can be controlled by varying the four joint velocities at any given configuration. Hence, the Cartesian velocity vector is $\boldsymbol{V}_4 = [\, v_{x4}\; v_{y4}\; v_{z4}\; \omega_{z4}\,]^T$ and Jacobian $\boldsymbol{J}(\boldsymbol{q})$ is a 4×4 matrix with $\boldsymbol{q} = [\theta_1\; \theta_2\; d_3\; \theta_4]^T$ and $\dot{\boldsymbol{q}} = [\dot{\theta}_1\;\; \dot{\theta}_2\;\; \dot{d}_3\;\; \dot{\theta}_4]^T$.

If a manipulator has less 6-DOF ($n < 6$), it has n controllable joint rates. As a result, there will be n controllable Cartesian velocity components. If this is acceptable for the jobs to be performed by the manipulator, its Jacobian is $n \times n$ square matrix, provided it is not in a singularity configuration. Jacobian singularities are discussed in the next section.

5.7 JACOBIAN SINGULARITIES

The manipulator Jacobian $\boldsymbol{J}$, a function of the configuration $\boldsymbol{q}$, may become rank-deficient or singular at certain configuration in Cartesian space. In such cases, the inverse Jacobian does not exist and Eq. (5.66) is not valid. Those manipulator configurations at which $\boldsymbol{J}$ becomes noninvertible are termed as Jacobian singularities and the configuration is itself called singular.

At a singular configuration, the Jacobian matrix $\boldsymbol{J}$ is not full rank; hence, its column vectors are linearly dependent. This means that there exists at least one direction in which the end-effector cannot be moved irrespective of values chosen for joint velocities $\dot{q}_1$ to $\dot{q}_n$.

The study of manipulator singularities is of great significance for the following reasons:

(a) It is not possible to give an arbitrary motion to end-effector; that is, singularities represent configurations at which structural mobility of the manipulator is reduced.

(b) At a singularity, no solution may exist for the inverse Jacobian problem.

(c) In the neighborhood of a singularity, small velocities in the Cartesian space require very high velocities in the joint space. This causes problems when the manipulator is required to track a trajectory that passes close to the singularity.

At a singular configuration, the manipulator loses one or more degrees of freedom. The singular configurations are classified into two categories based on the location of end-effector in the workspace.

(i) *Boundary singularities* occur when the end-effector is on the boundary of the workspace, that is, the manipulator is either fully stretched out or fully retracted. For example, consider the case of a two-link, 2-DOF-planar arm fully stretched out, as shown in Fig. 5.10. In this configuration, two links are in a straight line and the end-effector can be moved only in a direction perpendicular to the two links because it cannot move out of the workspace. Thus, the manipulator loses one degree of freedom. A similar situation will occur with θ_2 equal to 180°. Boundary singularities can be avoided by ensuring that the manipulator is not driven to boundaries of the reachable workspace during its work cycle.

(ii) *Interior singularities* occur when the end-effector is located inside the reachable workspace of the manipulator. These are caused when two or more joint axes become collinear or at specific end-effector configurations. For example, many spherical wrists cause interior singularities due to the lining up of two or more joint axes. These singularities cause serious problems such as path planning and control. They can occur anywhere in the reachable workspace. For certain manipulator configurations, the interior singularity points form a volume within the workspace called *void*.

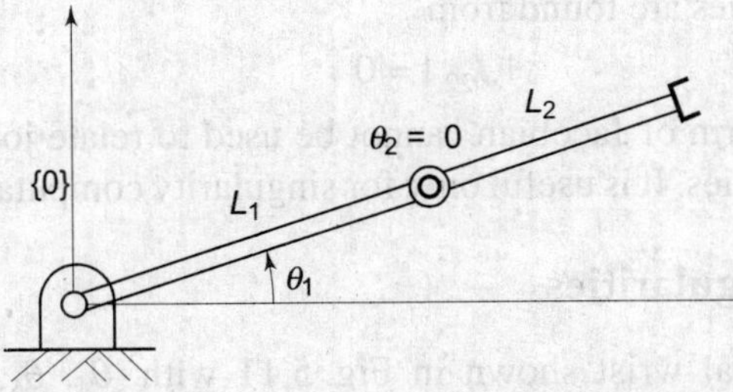

Fig. 5.10 *Two-link, 2-DOF planar manipulator fully stretched out*

In all the situations, it is essential that singularities are avoided. Therefore, one important criterion for a good design of manipulator configuration is to minimize the singularities.

5.7.1 Computation of Singularities

The computation of internal singularities can be carried out by analyzing the rank of the Jacobian matrix. The Jacobian matrix loses its rank and becomes illconditioned at values of joint variables $\boldsymbol{q}$ at which its determinant vanishes, that is,

$$|\boldsymbol{J}| = 0 \tag{5.67}$$

The solutions of Eq. (5.67) give the singular configurations of a manipulator. For manipulators that have 3-DOF arm and 3-DOF spherical wrists, it is possible to simplify the problem of singularity computation by dividing the problem into two separate problems:

- Computation of singularities resulting from motion of first three joints called *arm singularities*.
- Computation of singularities resulting from the motion of wrist joints called *wrist singularities*.

This is achieved by partitioning the Jacobian matrix into four 3 × 3 sub-matrices as

$$\boldsymbol{J} = \begin{bmatrix} \boldsymbol{J}_{11} & \boldsymbol{J}_{12} \\ \boldsymbol{J}_{21} & \boldsymbol{J}_{22} \end{bmatrix} \tag{5.68}$$

As the singularities are typical of a configuration and are not dependent on frames chosen for kinematic analysis, the origin of the end-effector frame can be chosen at the end-of-arm point. This will make

$$\boldsymbol{J}_{12} = 0 \tag{5.69}$$

In such a situation computation of determinant is greatly simplified, as,

$$|\boldsymbol{J}| = |\boldsymbol{J}_{11}|\,|\boldsymbol{J}_{22}| \tag{5.70}$$

Hence, for a manipulator with a spherical wrist, the arm singularities are found from

$$|\boldsymbol{J}_{11}| = 0 \tag{5.71}$$

and wrist singularities are found from

$$|\boldsymbol{J}_{22}| = 0 \tag{5.72}$$

Note that this form of Jacobian cannot be used to relate joint velocities and end-effector velocities. It is useful only for singularity computations.

5.7.2 Wrist Singularities

Consider a spherical wrist shown in Fig. 5.11 with θ_4, θ_5, and θ_6 as joint variables.

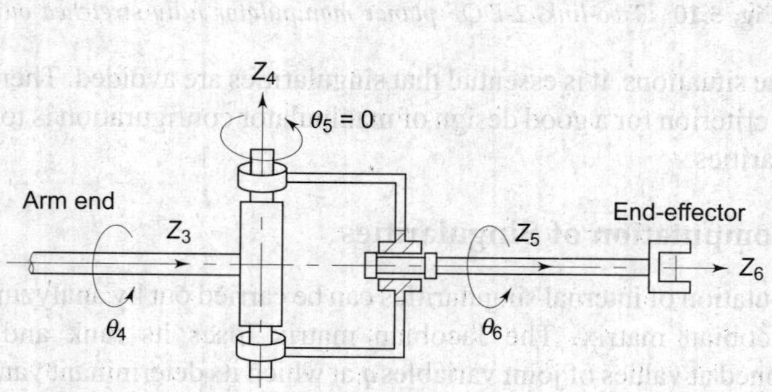

Fig. 5.11 *Spherical wrist at a singularity configuration*

It is known that a singularity occurs whenever two joint axes are aligned. The kinematic structure of the wrist reveals that only axis z_3 and axis z_5 can be aligned. For the configuration in Fig. 5.11 this occurs whenever

$$\theta_5 = 0 \text{ or } \theta_5 = \pi \tag{5.73}$$

The loss of mobility is caused by the fact that rotations of equal magnitude but in opposite direction about z_3- and z_5-axes do not produce any end-effector rotation. If axes z_3 and z_5 are aligned, it is not possible to rotate the wrist about the axis orthogonal to axes z_3 and z_4. This singularity can occur anywhere within the reachable workspace of the manipulator and consequently requires special care to be taken in programming wrist motions.

5.7.3 Arm Singularities

Arm singularities are characteristics of the manipulator's arm configuration and can be analyzed by considering the determinant of the upper left-hand corner 3×3 sub-matrix of Jacobian $\boldsymbol{J}$.

If arm singularities are identified in the workspace, the mechanical designs can be modified in such a manner that arm singularities are minimized. Alternately, they can be suitably avoided in the end-effector path planning. The specific steps for computation of arm singularities are illustrated through the examples at the end of this chapter.

5.8 STATIC ANALYSIS

There are instances in the work cycle when the manipulator is required to exert a force and/or moment on the environment. The end-effector makes a contact with the environment and all joints remain static. The contact between the end-effector and environment results in interactive forces and moments at the end-effector environment interface. The static problem of a manipulator is to determine the relationship between joint torques/forces—force for prismatic joint, torque for rotary joint and the force/moment exerted by its environment on the end-effector under static equilibrium conditions.

The force/torque at each joint of the manipulator are transmitted through the arm linkages to the end-effector, where the resultant force and moment acts on the environment. A manipulator carrying an object at its end point is equivalent to a force applied to end-effector by the environmental contact as discussed above. In both situations, the end point of the manipulator deflects by a magnitude determined by the stiffness of the manipulator. The end-point stiffness is an important characteristic of a manipulator that determines the accuracy of the manipulator and plays an important role in the control of mechanical interaction with the environment.

5.8.1 Force and Moment Balance

Consider an n-DOF manipulator with force and torque acting at each joint, which determine the torque/force to be exerted by the joint actuators, respectively. To find the force and torque acting on each joint requires the force and moment (torque) balance for the individual links. Figure 5.12 shows the free body diagram of link i with joint i and joint $(i+1)$ connecting link $(i–1)$ and link $(i+1)$, respectively. The corresponding coordinate frames at the two ends are frame $\{i–1\}$ and frame $\{i\}$ as shown.

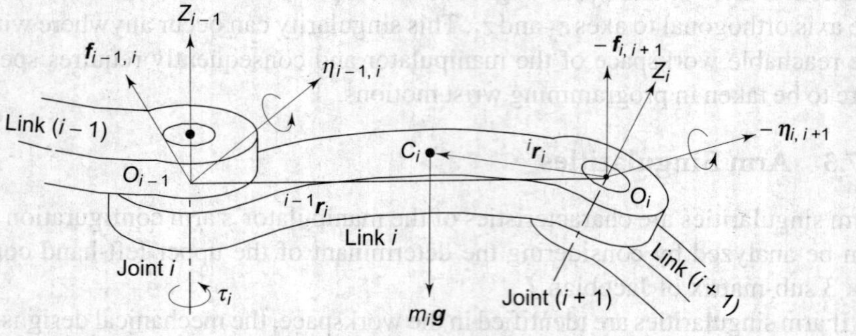

Fig. 5.12 *Forces and moments acting on link i in static equilibrium*

Let the vector $f_{i-1,\,i}$ denote the force acting on link i due to link $(i–1)$, where the first subscript denotes the link exerting the force, which acts on the link of second subscript. The vector $f_{i,\,i+1}$, therefore, represents the force applied by link i on link $(i+1)$. The force exerted by link $(i+1)$ on link i will be $-f_{i,\,i+1}$ because it is equal and opposite to force applied by link i on link $(i+1)$.

If m_i denotes the mass of link i and g is the 3×1 vector representing acceleration due to gravity, the gravity force acting at the centroid C_i of link i, is denoted by $m_i g$. Because the manipulator is at rest, each link is in static equilibrium, the force equilibrium gives

$$f_{i-1,\,i} - f_{i,\,i+1} + m_i g = 0 \tag{5.74}$$

All vectors in the above equation are with reference to the base frame $\{0\}$.

Also, for a link in static equilibrium, the vector sum of all moments acting on it about any arbitrary point is zero. Let $\eta_{i-1,\,i}$ represents the moment applied to link i by link $(i–1)$ with respect to frame $\{0\}$. Therefore, $-\eta_{i,\,i+1}$ is the moment applied to link i by link $(i+1)$. The moment balance about the centroid C_i of link i, see Fig. 5.12, gives

$$\eta_{i-1,\,i} + {}^0R_{i-1}({}^{i-1}r_i + {}^i r_i) \times f_{i-1,\,i} - \eta_{i,\,i+1} + {}^0R_i(-{}^i r_i) \times (-f_{i,\,i+1}) = 0$$
$$\text{for } i = 1, 2, \cdots, n \tag{5.75}$$

where ${}^{i-1}r_i$ is the position vector from origin O_{i-1} to origin O_i, ${}^i r_i$ is the position vector from origin O_i to centroid C_i and rotation matrix R is used to express all moment vectors in terms of frame $\{0\}$.

Equations (5.74) and (5.75) for each link give a set of n equations. Thus, there are $2n$ simultaneous equations involving $2(n+1)$ forces and moments. To solve these equations one force and one moment must be known. The force and moment applied by the manipulator on the environment can be specified to solve these equations. For $i = n$, $f_{n,\,n+1}$ and $\eta_{n,\,n+1}$ are the independent endpoint force and moment, respectively exerted by the manipulator on the environment.

The reaction force and moment from the environment to the manipulator, at its end point are $-f_{n,n+1}$ and $\eta_{n,\,n+1}$, respectively. The endpoint force and moment are combined into a 6-D vector, called *end-point force vector*, $\mathcal{F}$ as

$$\mathcal{F} = \begin{bmatrix} f_{n,\,n+1} \\ \eta_{n,\,n+1} \end{bmatrix} \tag{5.76}$$

Note that the negative sign has been dropped and will be appropriately taken care when used in a force/moment balance equations. Thus, for any n-DOF manipulator, the interactive forces and moments between links, $f_{i,\,i+1}$ and $\eta_{i,\,i+1}$ can be obtained from Eqs. (5.74) and (5.75) by performing inward iterations from $i = n$ down to 1.

Next, to determine the joint torques/forces to be exerted by the actuators to keep the manipulator in static equilibrium, let τ denote the $n \times 1$ vector of generalized joint torques

$$\tau = [\tau_1 \; \tau_2 \; \ldots \; \tau_n]^T \tag{5.77}$$

where τ_i is the generalized drive torque applied by the actuator i driving joint i and is defined as

$$\tau_i = \begin{cases} \text{Actuator force,} & \text{if joint } i \text{ is prismatic} \\ \text{Actuator torque,} & \text{if joint } i \text{ is revolute} \end{cases} \tag{5.78}$$

For a revolute joint i, the actuator at that joint will bear only that component of the moment $\eta_{i-1,\,i}$ which is in the direction of joint axis z_i, while the other components of the moment are borne by the joint structure. This gives, for a revolute joint, with ${}^{i-1}\hat{z}_{i-1}$ as the unit vector in the direction of the joint axis

$$\tau_i = [{}^0\boldsymbol{R}_{i-1} \; {}^{i-1}\hat{z}_{i-1}]^T \; \eta_{i-1,\,i} \tag{5.79}$$

Similarly, if joint i is prismatic, the actuator will bear that component of $f_{i-1,i}$ which is in the direction of joint axis, that is

$$\tau_i = [{}^0\boldsymbol{R}_{i-1} \; {}^{i-1}\hat{z}_{i-1}]^T \, f_{i-1,\,i} \tag{5.80}$$

5.8.2 The Jacobian in Statics

The relationship between the joint torques and the endpoint force/torque vector is derived using the *principle of virtual work*. This is used to determine the joint torques necessary to exert a given end-effector force and moment.

Consider an infinitesimal displacement of δq_i due to torque τ_i at joint i. Hence, $\boldsymbol{\delta q}$ is the $n \times 1$ vector of infinitesimal joint displacements. These displacements, also called virtual displacements, of a mechanical system need only satisfy the kinematic constraints. The end-point force and moment also act on the mechanical system producing infinitesimal endpoint displacement $\mathrm{d}x_e$ and rotation $\delta\phi_e$, respectively. Thus, the virtual work δW done by the forces and moments is given by

$$\delta W = \tau_1 \delta q_1 + \ldots + \tau_n \,\delta q_n - (f_{n,\,n+1})^T \,\delta x_e - (\eta_{n,\,n+1})^T \,\delta\phi_e$$

or $$\delta W = \tau^T\, \delta \boldsymbol{q} - \mathcal{F}^T\, \delta \boldsymbol{p} \tag{5.81}$$

where $\delta \boldsymbol{p}$ is the $n \times 1$ vector of infinitesimal end-effector displacements $(\delta x_e, \delta \phi_e)$ caused by the endpoint force $\mathcal{F}$. In the above, it is assumed that the joints of the mechanism are frictionless and the joint torques are the net torques that balance the endpoint force $\mathcal{F}$.

From Eq. (5.52), the Jacobian $\boldsymbol{J}$ relates infinitesimal joint displacement $\delta \boldsymbol{q}$ to infinitesimal end-effector displacement $\delta \boldsymbol{p}$ as:

$$\delta \boldsymbol{p} = \boldsymbol{J}(\boldsymbol{q})\, \delta \boldsymbol{q} \tag{5.82}$$

Substituting Eq. (5.82) into Eq. (5.81) and rearranging gives

$$\delta W = (\tau - \boldsymbol{J}(\boldsymbol{q})^T \mathcal{F})^T\, \delta \boldsymbol{q} \tag{5.83}$$

According to the principle of virtual work the manipulator mechanism is in static equilibrium, if and only if, the net virtual work is zero for arbitrary virtual displacements, that is,

$$\delta W = 0 \tag{5.84}$$

Substituing it in Eq. (5.83), leads to the result

$$\tau = \boldsymbol{J}(\boldsymbol{q})^T \mathcal{F} \tag{5.85}$$

Equation (5.85) states that the transpose of the Jacobian matrix transforms the end-effector torque to the corresponding joint torques.

SOLVED EXAMPLES

Example 5.1 ***Angular velocity of a frame***

A moving frame {1} is represented by the following rotation matrix R, where α is the angle of rotation of the frame {1} with respect to the base frame. If α is a function of time, find the angular velocity of frame {1}.

$$^0\boldsymbol{R}_1 = \begin{bmatrix} C\alpha & -S\alpha & 0 \\ S\alpha & C\alpha & 0 \\ 0 & 0 & 1 \end{bmatrix} \tag{5.86}$$

Solution The given rotation matrix is the fundamental rotation matrix of frame {1} about z-axis of frame {0}, see Eq. (2.54). From Eq. (5.26).

$$\boldsymbol{S}(t) = \dot{\boldsymbol{R}}(t)\, \boldsymbol{R}^T(t)$$

or $$\boldsymbol{S}(t) = \begin{bmatrix} -\dot{\alpha} S\alpha & -\dot{\alpha} C\alpha & 0 \\ \dot{\alpha} C\alpha & -\dot{\alpha} S\alpha & 0 \\ 0 & 0 & 0 \end{bmatrix} \begin{bmatrix} C\alpha & S\alpha & 0 \\ -S\alpha & C\alpha & 0 \\ 0 & 0 & 1 \end{bmatrix}$$

or $$\boldsymbol{S}(t) = \begin{bmatrix} 0 & -\dot{\alpha} & 0 \\ \dot{\alpha} & 0 & 0 \\ 0 & 0 & 0 \end{bmatrix} \tag{5.87}$$

Comparing Eq. (5.87) with Eq. (5.27) gives

$$\omega = [0 \quad 0 \quad \dot{\alpha}]^T \tag{5.88}$$

as the angular velocity of the frame about z-axis.

Example 5.2 *Velocity of RR-manipulator*

Calculate the velocity of the tip of the two-link, planar, RR-manipulator arm shown in Fig. 5.13.

Solution The frame assignments and joint-link parameters identification is carried out as discussed in Chapter 3 and are shown in Fig. 5.13 and Table 5.1. Frame {2} is attached to the end of manipulator and the velocity of origin of this frame is to be determined. This velocity can be expressed with respect to frame {2} as well as with respect to frame {0}. Assume length of each link is 1, that is

$$L_1 = L_2 = 1$$

The link transformation matrices are:

$$^0\boldsymbol{T}_1 = \begin{bmatrix} C_1 & -S_1 & 0 & C_1 \\ S_1 & C_1 & 0 & S_1 \\ 0 & 0 & 1 & 0 \\ 0 & 0 & 0 & 1 \end{bmatrix}; \quad ^1\boldsymbol{T}_2 \begin{bmatrix} C_2 & -S_2 & 0 & C_2 \\ S_2 & C_2 & 0 & S_2 \\ 0 & 0 & 1 & 0 \\ 0 & 0 & 0 & 1 \end{bmatrix} \tag{5.89}$$

and $$^0\boldsymbol{T}_2 = \begin{bmatrix} C_1C_2 - S_1S_2 & -C_1S_2 - S_1C_2 & 0 & C_1 + C_1C_2 - S_1S_2 \\ C_1S_2 + S_1C_2 & C_1C_2 - S_1S_2 & 0 & S_1 + C_1S_2 + S_1C_2 \\ 0 & 0 & 1 & 0 \\ 0 & 0 & 0 & 1 \end{bmatrix}$$

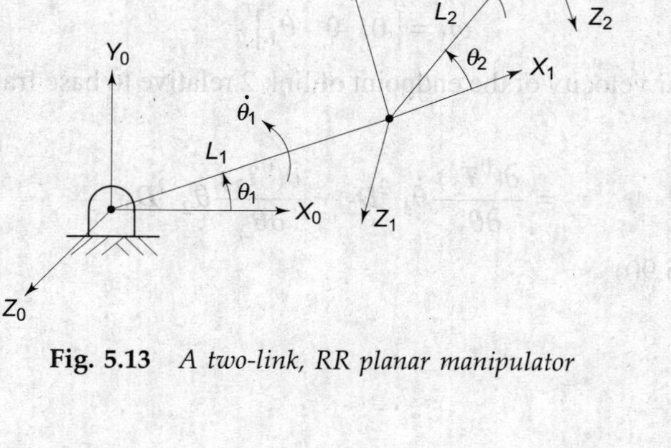

Fig. 5.13 *A two-link, RR planar manipulator*

or
$$
{}^0\boldsymbol{T}_2 = \begin{bmatrix} C_{12} & -S_{12} & 0 & C_1 + C_{12} \\ S_{12} & C_{12} & 0 & S_1 + S_{12} \\ 0 & 0 & 1 & 0 \\ 0 & 0 & 0 & 1 \end{bmatrix} \tag{5.90}
$$

Table 5.1 *Joint-link parameters for RR manipulator*

Link i	a_i	α_i	d_i	θ_i	$C\alpha_i$	$S\alpha_i$
1	1	0	0	θ_1	1	0
2	1	0	0	θ_2	1	0

Because both the joints of manipulator are rotary, according to Eq. (5.46) and Eq. (5.51), for endpoint of link 1, that is origin of frame {1}, linear velocity is computed from

$$
\boldsymbol{v}_1 = \frac{\partial({}^0\boldsymbol{T}_1)}{\partial\theta_1}\,\dot{\theta}_1\,{}^1\boldsymbol{D}_1
$$

with
$$
\frac{\partial({}^0\boldsymbol{T}_1)}{\partial\theta_1} = \begin{bmatrix} -S_1 & -C_1 & 0 & -S_1 \\ C_1 & -S_1 & 0 & C_1 \\ 0 & 0 & 0 & 0 \\ 0 & 0 & 0 & 0 \end{bmatrix}; \text{ and } {}^1\boldsymbol{D}_1 = \begin{bmatrix} 0 \\ 0 \\ 0 \\ 1 \end{bmatrix}
$$

Note that the origin of frame {1}is ${}^1\boldsymbol{D}_1$. The linear velocity of link 1 is, thus

$$
\boldsymbol{v}_1 = [-S_1 \quad -C_1 \quad 0 \quad 0]^T\,\dot{\theta} \tag{5.91}
$$

The angular velocity is computed from

$$
\omega_1 = \omega_0 + {}^0\boldsymbol{R}_0\,\hat{z}_0\,\dot{\theta}_1
$$

with
$$
\omega_0 = \begin{bmatrix} 0 \\ 0 \\ 0 \end{bmatrix}; \ \hat{z}_0 = \begin{bmatrix} 0 \\ 0 \\ 1 \end{bmatrix} \text{ and } {}^0\boldsymbol{R}_0 = \begin{bmatrix} 1 & 0 & 0 \\ 0 & 1 & 0 \\ 0 & 0 & 1 \end{bmatrix}
$$

The angular velocity of link 1 is found to be

$$
\omega_1 = \begin{bmatrix} 0 & 0 & \dot{\theta}_1 \end{bmatrix}^T \tag{5.92}
$$

The linear velocity of the endpoint of link 2 relative to base frame is obtained as:

$$
\boldsymbol{v}_2 = \frac{\partial({}^0\boldsymbol{T}_2)}{\partial\theta_1}\dot{\theta}_1\,{}^2\boldsymbol{D}_2 + \frac{\partial({}^0\boldsymbol{T}_2)}{\partial\theta_2}\dot{\theta}_2\,{}^2\boldsymbol{D}_2 \tag{5.93}
$$

From Eq. (5.90)

$$\frac{\partial({}^0\boldsymbol{T}_2)}{\partial\theta_1} = \begin{bmatrix} -S_1C_2 - C_1S_2 & S_1S_2 - C_1C_2 & 0 & -S_1 - S_1C_2 - C_1S_2 \\ -S_1S_2 + C_1C_2 & -S_1C_2 - C_1S_2 & 0 & C_1 - S_1S_2 + C_1S_2 \\ 0 & 0 & 0 & 0 \\ 0 & 0 & 0 & 0 \end{bmatrix} \tag{5.94}$$

and
$$\frac{\partial({}^0\boldsymbol{T}_2)}{\partial\theta_2} = \begin{bmatrix} -C_1S_2 - S_1C_2 & -C_1C_2 + S_1S_2 & 0 & -C_1S_2 - S_1C_2 \\ C_1C_2 - S_1S_2 & -C_1S_2 - S_1C_2 & 0 & C_1C_2 - S_1S_2 \\ 0 & 0 & 0 & 0 \\ 0 & 0 & 0 & 0 \end{bmatrix} \tag{5.95}$$

The position vector of endpoint of link 2 is origin of frame {2}, that is

$${}^2\boldsymbol{D}_2 = [0 \quad 0 \quad 0 \quad 1]^T$$

Substituting these gives,

$$\boldsymbol{v}_2 = \begin{bmatrix} -S_1\dot{\theta}_1 - S_{12}(\dot{\theta}_1 + \dot{\theta}_2) \\ C_1\dot{\theta}_1 + C_{12}(\dot{\theta}_1 + \dot{\theta}_2) \\ 0 \end{bmatrix} \tag{5.96}$$

And for the angular velocity of the endpoint,

$$\omega_2 = \omega_1 + {}^0\boldsymbol{T}_1\, z_1\, \dot{\theta}_2 \tag{5.97}$$

From Eq. (5.89) and Eq. (5.92) this simplifies to

$$\omega_2 = \left[0 \quad 0 \quad (\dot{\theta}_1 + \dot{\theta}_2)\right]^T \tag{5.98}$$

Example 5.3 *Jacobian of articulated arm*

Determine the manipulator Jacobian matrix for the 3-DOF articulated arm discussed in Example 3.3 and shown in Fig. 5.14.

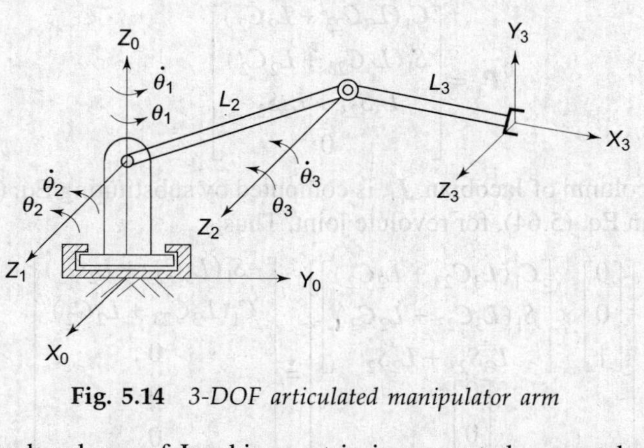

Fig. 5.14 *3-DOF articulated manipulator arm*

Solution Each column of Jacobian matrix is computed separately and all the columns are combined to form the total Jacobian matrix. The joint displacements

θ_1, θ_2, and θ_3 and joint velocities $\dot{\theta}_1$ $\dot{\theta}_2$, and $\dot{\theta}_3$ are shown in the figure and the transformation matrices are given in Eqs. (3.19)–(3.22).

The Jacobian matrix column $\boldsymbol{J}_1$ for joint 1, which is a rotary joint, is determined as follows:

From Eq. (5.59), the joint axis vector P_0 (P_{i-1} for $i = 1$) is

$$\boldsymbol{P}_0 = {}^0\boldsymbol{R}_0\, \hat{\boldsymbol{u}} \tag{5.99}$$

The transformation matrix ${}^0\boldsymbol{T}_0$ and rotation matrix ${}^0\boldsymbol{R}_0$ are the identity matrices. Thus,

$$\boldsymbol{P}_0 = \begin{bmatrix} 1 & 0 & 0 \\ 0 & 1 & 0 \\ 0 & 0 & 1 \end{bmatrix} \begin{bmatrix} 0 \\ 0 \\ 1 \end{bmatrix} = \begin{bmatrix} 0 \\ 0 \\ 1 \end{bmatrix} \tag{5.100}$$

The end-effector position vector (for $i = 1$ and $n = 3$) is determined from Eq. (5.63).

$${}^0\boldsymbol{P}_3 = {}^0\boldsymbol{T}_n \boldsymbol{O}_n - {}^0\boldsymbol{T}_0 \boldsymbol{O}_n$$

or

$${}^0\boldsymbol{P}_3 = {}^0\boldsymbol{T}_3 \begin{bmatrix} 0 \\ 0 \\ 0 \\ 1 \end{bmatrix} - {}^0\boldsymbol{T}_0 \begin{bmatrix} 0 \\ 0 \\ 0 \\ 1 \end{bmatrix}$$

Substituting ${}^0\boldsymbol{T}_3$ from Eq. (3.22), in above equation gives

$${}^0\boldsymbol{P}_3 = \begin{bmatrix} C_1C_{23} & -C_1S_{23} & S_1 & C_1(L_3C_{23} + L_2C_2) \\ S_1C_{23} & -S_1S_{23} & -C_1 & S_1(L_3C_{23} + L_2C_2) \\ S_{23} & C_{23} & 0 & L_3S_{23} + L_2S_2 \\ 0 & 0 & 0 & 1 \end{bmatrix} \begin{bmatrix} 0 \\ 0 \\ 0 \\ 1 \end{bmatrix} - \begin{bmatrix} 1 & 0 & 0 & 0 \\ 0 & 1 & 0 & 0 \\ 0 & 0 & 1 & 0 \\ 0 & 0 & 0 & 1 \end{bmatrix} \begin{bmatrix} 0 \\ 0 \\ 0 \\ 1 \end{bmatrix}$$

or

$${}^0\boldsymbol{P}_3 = \begin{bmatrix} C_1(L_3C_{23} + L_2C_2) \\ S_1(L_3C_{23} + L_2C_2) \\ L_3S_{23} + L_2S_2 \\ 0 \end{bmatrix} \tag{5.101}$$

The first column of Jacobian, $\boldsymbol{J}_1$, is computed by substituting Eq. (5.100) and Eq. (5.101) in Eq. (5.64), for revolute joint. Thus,

$$\boldsymbol{J}_1 = \begin{bmatrix} \begin{bmatrix} 0 \\ 0 \\ 1 \end{bmatrix} \times \begin{bmatrix} C_1(L_3C_{23} + L_2C_2) \\ S_1(L_3C_{23} + L_2C_2) \\ L_3S_{23} + L_2S_2 \end{bmatrix} \\ \begin{bmatrix} 0 \\ 0 \\ 1 \end{bmatrix} \end{bmatrix} = \begin{bmatrix} -S_1(L_3C_{23} + L_2C_2) \\ C_1(L_3C_{23} + L_2C_2) \\ 0 \\ 0 \\ 0 \\ 1 \end{bmatrix} \tag{5.102}$$

By following the similar steps for joint 2 and joint 3, $\boldsymbol{J}_2$ and $\boldsymbol{J}_3$ are obtained as

$$\boldsymbol{J}_2 = \begin{bmatrix} -C_1(L_3S_{23} + L_2S_2) \\ -S_1(L_3S_{23} + L_2S_2) \\ L_3C_{23} + L_2C_2 \\ S_1 \\ -C_1 \\ 0 \end{bmatrix} \tag{5.103}$$

and

$$\boldsymbol{J}_3 = \begin{bmatrix} -L_3C_1S_{23} \\ -L_3S_1S_{23} \\ L_3C_{23} \\ S_1 \\ -C_1 \\ 0 \end{bmatrix} \tag{5.104}$$

Combining the three columns, the total Jacobian matrix for the articulated arm is

$$\boldsymbol{J} = \begin{bmatrix} -S_1(L_3C_{23} + L_2C_2) & -C_1(L_3S_{23} + L_2S_2) & -L_3C_1S_{23} \\ C_1(L_3C_{23} + L_2C_2) & -S_1(L_3S_{23} + L_2S_2) & -L_3S_1S_{23} \\ 0 & L_3C_{23} + L_2C_2 & L_3C_{23} \\ 0 & S_1 & S_1 \\ 0 & -C_1 & -C_1 \\ 1 & 0 & 0 \end{bmatrix} \tag{5.105}$$

Example 5.4 ***Singularities and joint velocities of a 2-DOF arm***

For the 2-DOF-planar arm shown in Fig. 5.13, assuming both the links of unit length, determine (a) configuration singularities, and (b) joint velocities in terms of endpoint velocities.

Solution To determine the singularities of a given configuration manipulator Jacobian is required. The Jacobian can be computed using Eq. (5.57), which requires position vectors and unit vector of joint axes. From Eq. (5.89) and Eq. (5.90) the position vectors for various links are

$${}^0\boldsymbol{P}_0 = \begin{bmatrix} 0 \\ 0 \\ 0 \end{bmatrix}; \quad {}^0\boldsymbol{P}_1 = \begin{bmatrix} C_1 \\ S_1 \\ 0 \end{bmatrix}; \quad {}^0\boldsymbol{P}_2 = \begin{bmatrix} C_1 + C_{12} \\ S_1 + S_{12} \\ 0 \end{bmatrix} \tag{5.106}$$

The unit vectors along the revolute joint axes are

$$\hat{\boldsymbol{u}}_0 = \hat{\boldsymbol{u}}_1 = [0 \quad 0 \quad 1]^T \tag{5.107}$$

The Jacobian is given by (from Eq. 5.64)

$$J = \begin{bmatrix} \hat{u}_0 \times ({}^0P_2 - {}^0P_0) & \hat{u}_1 \times ({}^0P_2 - {}^0P_1) \\ \hat{u}_0 & \hat{u}_1 \end{bmatrix} \tag{5.108}$$

Substituting from Eqs. (5.106) and (5.107)

$$J = \begin{bmatrix} -S_1 - S_{12} & -S_{12} \\ C_1 + C_{12} & C_{12} \\ 0 & 0 \\ 0 & 0 \\ 0 & 0 \\ 1 & 1 \end{bmatrix} \tag{5.109}$$

(a) Configuration singularities In the Jacobian, Eq. (5.109) three rows are null and one row is unity. The arm has only 2-DOF, that is, it requires only two variables to be controlled. Hence, the upper 2×2 block of the Jacobian is only significant and sufficient for computing positions of two links. The significant Jacobian, to describe the relationship between the joint velocities and the arm-endpoint linear velocity, is therefore

$$J' = \begin{bmatrix} -S_1 - S_{12} & -S_{12} \\ C_1 + C_{12} & C_{12} \end{bmatrix} \tag{5.110}$$

It is instructive to note that Eq. (5.91) and Eq. (5.96) of Example 5.2, when written in the form

$$V_e = \begin{bmatrix} v \\ \omega \end{bmatrix} = J' \begin{bmatrix} \dot{\theta}_1 \\ \dot{\theta}_2 \end{bmatrix} \tag{5.111}$$

lead to the same result for J as in Eq. (5.109) from which J' of Eq. (5.110) follows.

Singularities of the configuration occur when determinant of J' vanishes. From Eq. (5.110), the determinant of J' is

$$|J'| = \begin{vmatrix} -S_1 - S_{12} & -S_{12} \\ C_1 + C_{12} & C_{12} \end{vmatrix} = S_2 \tag{5.112}$$

Note that the Jacobian determinant is independent of first joint variable θ_1. The determinant vanishes for non-zero link lengths whenever

$$\theta_2 = 0 \text{ or } \theta_2 = \pi$$

Hence, singular configurations occur when arm-endpoint is located either on the outer boundary ($\theta_2 = 0$) or on the inner boundary ($\theta_2 = \pi$) of the reachable workspace, that is, fully extended or fully retracted. At these values of θ_2, the two columns of J' become linearly dependent implying (i) the Jacobian loses rank (and the rank becomes one); (ii) the tip velocity components v_x and v_y are not independent, and (iii) the two links are collinear, as in Fig. 5.10.

(b) Joint velocities The joint velocities in terms of endpoint velocities v_x and v_y are obtained from Eq. (5.54). That is

$$\begin{bmatrix} v_x \\ v_y \end{bmatrix} = J' \begin{bmatrix} \dot{\theta}_1 \\ \dot{\theta}_2 \end{bmatrix}$$

or

$$\begin{bmatrix} \dot{\theta}_1 \\ \dot{\theta}_2 \end{bmatrix} = [J']^{-1} \begin{bmatrix} v_x \\ v_y \end{bmatrix} \tag{5.113}$$

The inverse of $\boldsymbol{J'}$ is

$$[J']^{-1} = \frac{1}{S_2} \begin{bmatrix} C_{12} & S_{12} \\ -C_1 - C_{12} & S_1 + S_{12} \end{bmatrix} \tag{5.114}$$

Thus, the joint velocities are

$$\dot{\theta}_1 = \frac{v_x C_{12} + v_y S_{12}}{S_2} \tag{5.115}$$

$$\dot{\theta}_2 = \frac{-v_x (C_1 + C_{12}) + v_y (S_1 + S_{12})}{S_2} \tag{5.116}$$

At regions near the singular points, the denominator in Eqs. (5.115) and (5.116) is almost zero, that is, to move the endpoint in the vicinity of the singularities excessively large joint velocities are required.

The joint velocity profiles for tracking a specified trajectory at specified speed (v_x, v_y) can be obtained by finding the angles that correspond to the endpoint position on the trajectory, that is, $\theta_1(t)$ and $\theta_2(t)$ substituting these into Eqs. (5.115) and (5.116) and plotting the joint velocities with respect to time.

Example 5.5 *Singularities of articulated arm configuration*

Determine the singularities of the 3-DOF articulated arm configuration discussed in Example 5.3.

Solution A close examination of the Jacobian of the articulated arm, Eq. (5.105) indicates that only three of the six rows of the Jacobian are linearly independent. It is also known that the arm has only three degrees of freedom, that is, it requires only three variables to be controlled. Hence, the upper 3 × 3 block of Jacobian is only significant and worth considering. The significant Jacobian is

$$J' = \begin{bmatrix} -S_1(L_3C_{23} + L_2C_2) & -C_1(L_3S_{23} + L_2S_2) & -L_3C_1S_{23} \\ C_1(L_3C_{23} + L_2C_2) & -S_1(L_3S_{23} + L_2S_2) & -L_3S_1S_{23} \\ 0 & L_3C_{23} + L_2C_2 & L_3C_{23} \end{bmatrix} \tag{5.117}$$

It is worth noting that for this 3-DOF arm endpoint angular velocity ω_x and ω_y are not independent, since $S_1\omega_y = -C_1\omega_x$ and this structure does not allow arbitrary endpoint angular velocity ω.

For determination of singularities, consider the Jacobian $\boldsymbol{J'}$, in Eq. (5.117). Its determinant is

$$|J'| = -L_2L_3S_3(L_2C_2 + L_3C_{23}) \tag{5.118}$$

It is seen from the determinant that it is independent of first joint variable. The determinant vanishes for nonzero link lengths when S_3=0 and/or $L_2C_2 + L_3C_3 = 0$.

Case 1 $S_3 = 0$ is possible for $\theta_3 = 0$ or $\theta_3 = \pi$.

At either of the values of θ_3, two links of arm are aligned, fully outstretched or retracted. This is similar to singularity condition of the RR planar manipulator discussed in Example 5.4.

Case 2 $L_2C_2 + L_3C_{23} = 0$.

This occurs when the endpoint lies on the axis of base frame, z_0. That is

$$p_x = p_y = 0$$

Example 5.6 *Singularities of RPY wrist*

Determine the singularities of the RPY wrist discussed in Example 3.4.

Solution To determine the singularities of the wrist, its Jacobian is determined first. For the 3-DOF RPY wrist, the link transformation matrices were obtained in Example 3.4, Eqs. (3.24) – (3.26), as

$$^0\boldsymbol{T}_1 = \begin{bmatrix} C_1 & 0 & S_1 & 0 \\ S_1 & 0 & -C_1 & 0 \\ 0 & 1 & 0 & 0 \\ 0 & 0 & 0 & 1 \end{bmatrix} \quad (5.119)$$

$$^1\boldsymbol{T}_2 = \begin{bmatrix} -S_2 & 0 & C_2 & 0 \\ C_2 & 0 & S_2 & 0 \\ 0 & 1 & 0 & 0 \\ 0 & 0 & 0 & 1 \end{bmatrix} \quad (5.120)$$

$$^2\boldsymbol{T}_3 = \begin{bmatrix} C_3 & -S_3 & 0 & 0 \\ S_3 & C_3 & 0 & 0 \\ 0 & 0 & 1 & 0 \\ 0 & 0 & 0 & 1 \end{bmatrix} \quad (5.121)$$

The rotational transformation matrices for the three links of the wrist are given by the upper-right-3 × 3 submatrix of the respective transformation matrices in Eqs. (5.119), (5.120) and (5.121), as

$$^0\boldsymbol{R}_1 = \begin{bmatrix} C_1 & 0 & S_1 \\ S_1 & 0 & -C_1 \\ 0 & 1 & 0 \end{bmatrix} \quad (5.122)$$

$$^1\boldsymbol{R}_2 = \begin{bmatrix} -S_2 & 0 & C_2 \\ C_2 & 0 & S_2 \\ 0 & 1 & 0 \end{bmatrix} \quad (5.123)$$

$$ {}^2\boldsymbol{R}_3 = \begin{bmatrix} C_3 & -S_3 & 0 \\ S_3 & C_3 & 0 \\ 0 & 0 & 1 \end{bmatrix} \tag{5.124} $$

The Jacobian for the 3-DOF RPY wrists is computed using Eq. (5.62) as all the three joints are revolute joints. That is

$$ \boldsymbol{J}_i = \begin{bmatrix} \boldsymbol{P}_{i-1} \times {}^{i-1}\boldsymbol{P}_n \\ \boldsymbol{P}_{i-1} \end{bmatrix} \tag{5.125} $$

with $\boldsymbol{P}_{i-1}$ and ${}^{i-1}\boldsymbol{P}_n$ given by Eq. (5.59) and Eq. (5.63), respectively.

For $i = 1$ and $n = 3$, the first coloumn of Jacobian $\boldsymbol{J}_1$ is given by

$$ \boldsymbol{J}_1 = \begin{bmatrix} \boldsymbol{P}_0 \times {}^0\boldsymbol{P}_3 \\ \boldsymbol{P}_0 \end{bmatrix} \tag{5.126} $$

where from Eq. (5.59)

$$ \boldsymbol{P}_0 = {}^0\boldsymbol{R}_0\, \hat{\boldsymbol{u}} = \begin{bmatrix} 1 & 0 & 0 \\ 0 & 1 & 0 \\ 0 & 0 & 1 \end{bmatrix} \begin{bmatrix} 0 \\ 0 \\ 1 \end{bmatrix} = \begin{bmatrix} 0 \\ 0 \\ 1 \end{bmatrix} \tag{5.127} $$

and from Eq. (5.63) and Eqs. (5.119), (5.120) and (5.121)

$$ {}^0\boldsymbol{P}_3 = {}^0\boldsymbol{T}_3\, \boldsymbol{O}_3 - {}^0\boldsymbol{T}_0 \boldsymbol{O}_3 $$

or

$$ {}^0\boldsymbol{P}_3 = \begin{bmatrix} -C_1S_2C_3 + S_1S_3 & C_1S_2S_3 + S_1C_3 & C_1C_2 & 0 \\ -S_1S_2C_3 - C_1S_3 & S_1S_2S_3 - C_1C_3 & S_1C_2 & 0 \\ C_2C_3 & -C_2S_3 & S_2 & 0 \\ 0 & 0 & 0 & 1 \end{bmatrix} \begin{bmatrix} 0 \\ 0 \\ 0 \\ 1 \end{bmatrix} - \begin{bmatrix} 1 & 0 & 0 & 0 \\ 0 & 1 & 0 & 0 \\ 0 & 0 & 1 & 0 \\ 0 & 0 & 0 & 1 \end{bmatrix} \begin{bmatrix} 0 \\ 0 \\ 0 \\ 1 \end{bmatrix} $$

$$ = \begin{bmatrix} 0 \\ 0 \\ 0 \\ 0 \end{bmatrix} \tag{5.128} $$

Thus,

$$ \boldsymbol{J}_1 = \begin{bmatrix} \begin{bmatrix} 0 \\ 0 \\ 1 \end{bmatrix} \times \begin{bmatrix} 0 \\ 0 \\ 0 \end{bmatrix} \\ \begin{bmatrix} 0 \\ 0 \\ 1 \end{bmatrix} \end{bmatrix} = \begin{bmatrix} 0 \\ 0 \\ 0 \\ 0 \\ 0 \\ 1 \end{bmatrix} \tag{5.129} $$

Similarly, for $i = 2$,

$$\boldsymbol{P}_1 = {}^0\boldsymbol{R}_1\,\hat{\boldsymbol{u}} = \begin{bmatrix} C_1 & 0 & S_1 \\ S_1 & 0 & -C_1 \\ 0 & 1 & 0 \end{bmatrix}\begin{bmatrix} 0 \\ 0 \\ 1 \end{bmatrix} = \begin{bmatrix} S_1 \\ -C_1 \\ 0 \end{bmatrix} \tag{5.130}$$

$${}^1\boldsymbol{P}_3 = {}^0\boldsymbol{T}_3\,\boldsymbol{O}_3 - {}^0\boldsymbol{T}_1\,\boldsymbol{O}_3 = \begin{bmatrix} 0 \\ 0 \\ 0 \\ 0 \end{bmatrix} \tag{5.131}$$

and

$$\boldsymbol{J}_2 = \begin{bmatrix} \boldsymbol{P}_1 \times {}^1\boldsymbol{P}_3 \\ \boldsymbol{P}_1 \end{bmatrix} = \begin{bmatrix} 0 \\ 0 \\ 0 \\ S_1 \\ -C_1 \\ 0 \end{bmatrix} \tag{5.132}$$

and for $i = 3$,

$$\boldsymbol{P}_2 = {}^0\boldsymbol{R}_2\,\hat{\boldsymbol{u}} = \begin{bmatrix} -C_1S_2 & S_1 & C_1C_2 \\ -S_1S_2 & -C_1 & S_1C_2 \\ C_2 & 0 & S_2 \end{bmatrix}\begin{bmatrix} 0 \\ 0 \\ 1 \end{bmatrix} = \begin{bmatrix} C_1C_2 \\ S_1C_2 \\ S_2 \end{bmatrix} \tag{5.133}$$

$${}^2\boldsymbol{P}_3 = {}^0\boldsymbol{T}_3\,\boldsymbol{O}_3 - {}^0\boldsymbol{T}_2\,\boldsymbol{O}_3 = \begin{bmatrix} 0 \\ 0 \\ 0 \\ 0 \end{bmatrix} \tag{5.134}$$

(*Note:* ${}^0\boldsymbol{R}_2$ and ${}^0\boldsymbol{T}_2$ are obtained from Eqs. (5.119), (5.120) and Eqs. (5.122), (5.123), respectively). Thus,

$$\boldsymbol{J}_3 = \begin{bmatrix} \boldsymbol{P}_2 \times {}^2\boldsymbol{P}_3 \\ \boldsymbol{P}_2 \end{bmatrix} = \begin{bmatrix} 0 \\ 0 \\ 0 \\ C_1C_2 \\ S_1C_2 \\ S_2 \end{bmatrix} \tag{5.135}$$

From Eqs. (5.129), (5.132) and (5.135) the Jacobian for the RPY wrist is

$$\boldsymbol{J} = [\boldsymbol{J}_1 \quad \boldsymbol{J}_2 \quad \boldsymbol{J}_3] = \begin{bmatrix} 0 & 0 & 0 \\ 0 & 0 & 0 \\ 0 & 0 & 0 \\ 0 & S_1 & C_1C_2 \\ 0 & -C_1 & S_1C_2 \\ 1 & 0 & S_2 \end{bmatrix} \tag{5.136}$$

Note that the first three rows of this Jacobian for the wrist are zero. This means that the wrist links have no linear velocities. This is expected because a wrist is used for orientation only.

The singularities of the RPY wrist are determined using the last three rows of the Jacobian in Eq. (5.136). The pseudo-Jacobian is

$$J' = \begin{bmatrix} 0 & S_1 & C_1C_2 \\ 0 & -C_1 & S_1C_2 \\ 1 & 0 & S_2 \end{bmatrix} \tag{5.137}$$

The wrist singularities will occur when inverse of J' cannot be found, or in other words J' becomes ill conditioned. This occurs whenever determinant of J' vanishes. Thus, the conditions of singularity are found from

$$|J'| = \begin{vmatrix} 0 & S_1 & C_1C_2 \\ 0 & -C_1 & S_1C_2 \\ 1 & 0 & S_2 \end{vmatrix} = 0 \tag{5.138}$$

or $$S_1(S_1C_2 - 0) + C_1C_2(0 + C_1) = 0$$

or $$C_2 = 0 \tag{5.139}$$

Hence, singularities of the RPY wrist are those positions of joints 2 where

$$\theta_2 = 90° \quad \text{or} \quad 270° \tag{5.140}$$

Example 5.7 *RPR arm Jacobian*

For the manipulator shown in Figure 5.15 below, obtain the Jacobian to express the Cartesian velocities in terms of the joint velocities. Obtain the singularities of the manipulator.

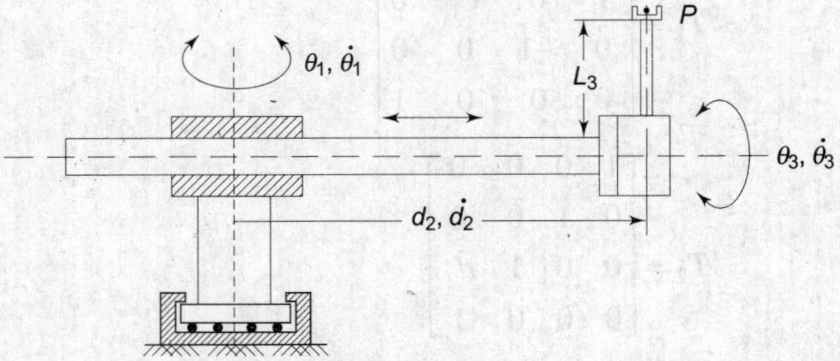

Fig. 5.15 *A 3-DOF RPR arm of a manipulator*

Solution In order to determine the Jacobian of the arm, first its forward kinematics model in terms of transformation matrices is required.

To determine the transformation matrices, frames are assigned as per Algorithm 3.1 and these are shown in Fig. 5.16. The joint-link parameters

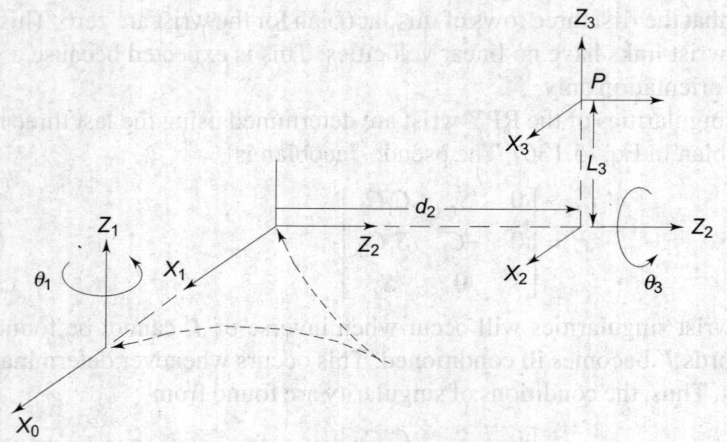

Fig. 5.16 *Frame assignment for RPR arm*

obtained there from are tabulated in Table 5.2 and the three transformation matrices obtained from these are given in Eqs. (5.141), (5.142) and (5.143).

Table 5.2 *Joint-link parameters for RPR arm*

i	θ_i	α_i	a_i	d_i	$C\theta$	$S\theta$	$C\alpha$	$S\alpha$
1	θ_1	–90°	0	0	C_1	S_1	0	–1
2	0	0	0	d_2	1	0	1	0
3	θ_3	90°	L_3	0	C_3	S_3	0	1

$$
{}^0\boldsymbol{T}_1 = \begin{bmatrix} C_1 & 0 & -S_1 & 0 \\ S_1 & 0 & C_1 & 0 \\ 0 & -1 & 0 & 0 \\ 0 & 0 & 0 & 1 \end{bmatrix} \tag{5.141}
$$

$$
{}^1\boldsymbol{T}_2 = \begin{bmatrix} 1 & 0 & 0 & 0 \\ 0 & 1 & 0 & 0 \\ 0 & 0 & 1 & d_2 \\ 0 & 0 & 0 & 1 \end{bmatrix} \tag{5.142}
$$

$$
{}^2\boldsymbol{T}_3 = \begin{bmatrix} C_3 & 0 & S_3 & L_3C_3 \\ S_3 & 0 & -C_3 & L_3S_3 \\ 0 & 1 & 0 & 0 \\ 0 & 0 & 0 & 1 \end{bmatrix} \tag{5.143}
$$

From the individual transformation matrices following products of transformation matrices are computed:

$$ {}^0\boldsymbol{T}_2 = {}^0\boldsymbol{T}_1\,{}^1\boldsymbol{T}_2 = \begin{bmatrix} C_1 & 0 & -S_1 & -S_1 d_2 \\ S_1 & 0 & C_1 & C_1 d_2 \\ 0 & -1 & 0 & 0 \\ 0 & 0 & 0 & 1 \end{bmatrix} \tag{5.144} $$

$$ {}^0\boldsymbol{T}_3 = {}^0\boldsymbol{T}_2\,{}^2\boldsymbol{T}_3 = \begin{bmatrix} C_1C_3 & -S_1 & C_1S_3 & L_3C_1C_3 - S_1d_2 \\ S_1C_3 & C_1 & S_1S_3 & L_3S_1C_3 + C_1d_2 \\ -S_3 & 0 & C_3 & -L_3S_3 \\ 0 & 0 & 0 & 1 \end{bmatrix} \tag{5.145} $$

The link 1 is a revolute link and Jacobian for link 1 is computed using Eq. (5.62) as:

$$ \boldsymbol{J}_1 = \begin{bmatrix} \boldsymbol{P}_0 \times {}^0\boldsymbol{P}_3 \\ \boldsymbol{P}_0 \end{bmatrix} \tag{5.146} $$

where from Eq. (5.59)

$$ \boldsymbol{P}_0 = {}^0\boldsymbol{R}_0\,\hat{\boldsymbol{u}} = \begin{bmatrix} 1 & 0 & 0 \\ 0 & 1 & 0 \\ 0 & 0 & 1 \end{bmatrix} \begin{bmatrix} 0 \\ 0 \\ 1 \end{bmatrix} = \begin{bmatrix} 0 \\ 0 \\ 1 \end{bmatrix} \tag{5.147} $$

and from Eq. (5.63)

$$ {}^0\boldsymbol{P}_3 = {}^0\boldsymbol{T}_3 \begin{bmatrix} 0 \\ 0 \\ 0 \\ 1 \end{bmatrix} - {}^0\boldsymbol{T}_0 \begin{bmatrix} 0 \\ 0 \\ 0 \\ 1 \end{bmatrix} = \begin{bmatrix} L_3C_1C_3 - S_1d_2 \\ L_3S_1C_3 + C_1d_2 \\ -L_3S_3 \\ 1 \end{bmatrix} - \begin{bmatrix} 0 \\ 0 \\ 0 \\ 1 \end{bmatrix} = \begin{bmatrix} L_3C_1C_3 - S_1d_2 \\ L_3S_1C_3 + C_1d_2 \\ -L_3S_3 \\ 0 \end{bmatrix} \tag{5.148} $$

Thus,

$$ \boldsymbol{J}_1 = \begin{bmatrix} \begin{bmatrix} 0 \\ 0 \\ 1 \end{bmatrix} \times \begin{bmatrix} L_3C_1C_3 - S_1d_2 \\ L_3S_1C_3 + C_1d_2 \\ -L_3S_3 \end{bmatrix} \\ \begin{bmatrix} 0 \\ 0 \\ 1 \end{bmatrix} \end{bmatrix} = \begin{bmatrix} -L_3S_1C_3 - C_1d_2 \\ L_3C_1C_3 - S_1d_2 \\ 0 \\ 0 \\ 0 \\ 1 \end{bmatrix} \tag{5.149} $$

For link 2, a prismatic link, the Jacobian is given by Eq. (5.57) as:

$$ \boldsymbol{J}_2 = \begin{bmatrix} \boldsymbol{P}_1 \\ 0 \end{bmatrix} \tag{5.150} $$

where

$$ \boldsymbol{P}_1 = {}^0\boldsymbol{R}_1\,\hat{\boldsymbol{u}} = \begin{bmatrix} C_1 & 0 & -S_1 \\ S_1 & 1 & C_1 \\ 0 & -1 & 0 \end{bmatrix} \begin{bmatrix} 0 \\ 0 \\ 1 \end{bmatrix} = \begin{bmatrix} -S_1 \\ C_1 \\ 0 \end{bmatrix} \tag{5.151} $$

Thus,

$$J_2 = \begin{bmatrix} -S_1 \\ C_1 \\ 0 \\ 0 \\ 0 \\ 0 \end{bmatrix} \tag{5.152}$$

For the revolute joint 3, the Jacobian is computed from

$$J_3 = \begin{bmatrix} P_2 \times {}^2P_3 \\ P_2 \end{bmatrix} \tag{5.153}$$

where

$$P_2 = {}^0R_2\,\hat{u} = \begin{bmatrix} C_1 & 0 & -S_1 \\ S_1 & 0 & C_1 \\ 0 & -1 & 0 \end{bmatrix} \begin{bmatrix} 0 \\ 0 \\ 1 \end{bmatrix} = \begin{bmatrix} -S_1 \\ C_1 \\ 0 \end{bmatrix} \tag{5.154}$$

and

$$ {}^2P_3 = {}^0T_3\,O_n - {}^0T_2\,O_n$$

or $$ {}^2P_3 = \begin{bmatrix} L_3C_1C_3 - S_1d_2 \\ L_3S_1C_3 + C_1d_2 \\ -L_3S_3 \\ 1 \end{bmatrix} - \begin{bmatrix} C_1 & 0 & -S_1 & -S_1d_2 \\ S_1 & 0 & C_1 & C_1d_2 \\ 0 & -1 & 0 & 0 \\ 0 & 0 & 0 & 1 \end{bmatrix} = \begin{bmatrix} L_3C_1C_3 \\ L_3S_1C_3 \\ -L_3S_3 \\ 0 \end{bmatrix} \tag{5.155}$$

Substituting Eqs. (5.154) and (5.155) in Eq. (5.153) gives

$$J_3 = \begin{bmatrix} \begin{bmatrix} -S_1 \\ C_1 \\ 0 \end{bmatrix} \times \begin{bmatrix} L_3C_1C_3 \\ L_3S_1C_3 \\ -L_3S_3 \end{bmatrix} \\ \begin{bmatrix} -S_1 \\ C_1 \\ 0 \end{bmatrix} \end{bmatrix} = \begin{bmatrix} -L_3C_1S_3 \\ -L_3S_1S_3 \\ -L_3C_3 \\ -S_1 \\ C_1 \\ 0 \end{bmatrix} \tag{5.156}$$

The Jacobian for the manipulator is, therefore, obtained by combining Eqs. (5.149), (5.152) and (5.156).

$$J = \begin{bmatrix} -L_3S_1C_3 - C_1d_2 & -S_1 & -L_3C_1S_3 \\ L_3C_1C_3 - S_2d_2 & C_1 & -L_3S_1S_3 \\ 0 & 0 & -L_3C_3 \\ 0 & 0 & -S_1 \\ 0 & 0 & C_1 \\ 1 & 0 & 0 \end{bmatrix} \tag{5.157}$$

For determination of singularities, consider first three rows of $\boldsymbol{J}$ in Eq. (5.157) as the pseudo Jacobian $\boldsymbol{J}'$:

$$\boldsymbol{J}' = \begin{bmatrix} -L_3S_1C_3 - d_2C_1 & -S_1 & -L_3C_1S_3 \\ L_3C_1C_3 - S_1d_2 & C_1 & L_3S_1S_3 \\ 0 & 0 & -L_3C_3 \end{bmatrix} \tag{5.158}$$

The determinant of $\boldsymbol{J}'$ is

$$|\boldsymbol{J}'| = L_3d_2S_3\left(C_1^2 + S_1^2\right) = L_3d_2S_3 \tag{5.159}$$

The singularity condition is

$$|\boldsymbol{J}'| = 0$$

or $$L_3\, d_2\, C_3 = 0 \tag{5.160}$$

Since L_3 and d_2 cannot be zero, singularities occurs at $C_3 = 0$

or $$\theta_3 = 90° \quad \text{or} \quad 270° \tag{5.161}$$

Example 5.8 ***Static forces in articulated arm***

The 3-DOF articulated arm of Example 5.3 applies a force $\mathcal{F}$ with its end-effector. Determine the torques required at the joints to maintain static equilibrium.

Solution Assume that the force $\mathcal{F}$ is acting at the origin of frame {3}(see Fig. 3.16) and is given by

$$\mathcal{F} = \begin{bmatrix} f_x & f_y & f_z & 0 & 0 & 0 \end{bmatrix}^T \tag{5.162}$$

Note that $\mathcal{F}$ is 6×1 Cartesian force-moment vector.

The joint torques are a function of the configuration and applied force and are given by Eq. (5.85) as

$$\tau = \boldsymbol{J}(\boldsymbol{q})^T\mathcal{F} \tag{5.163}$$

The Jacobian of the articulated arm was obtained in Example 5.3, Eq. (5.105) as

$$\boldsymbol{J} = \begin{bmatrix} -S_1(L_3C_{23} + L_2C_2) & -C_1(L_3S_{23} + L_2S_2) & -L_3C_1S_{23} \\ C_1(L_3C_{23} + L_2C_2) & -S_1(L_3S_{23} + L_2S_2) & -L_3S_1S_{23} \\ 0 & L_3C_{23} + L_2C_2 & L_3C_{23} \\ 0 & S_1 & S_1 \\ 0 & -C_1 & -C_1 \\ 1 & 0 & 0 \end{bmatrix} \tag{5.164}$$

The transpose of $\boldsymbol{J}$ is

$$\boldsymbol{J}^T = \begin{bmatrix} -S_1(L_3C_{23} + L_2C_2) & C_1(L_3C_{23} + L_2C_2) & 0 & 0 & 0 & 1 \\ -C_1(L_3S_{23} + L_2S_2) & -S_1(L_3S_{23} + L_2S_2) & L_3C_{23} + L_2C_2 & S_1 & -C_1 & 0 \\ -L_3C_1S_{23} & -L_3S_1S_{23} & L_3C_{23} & S_1 & -C_1 & 0 \end{bmatrix} \tag{5.165}$$

Substituting Eqs. (5.162) and (5.165) in Eq. (5.163) gives

$\tau =$

$$\begin{bmatrix} -S_1(L_3C_{23}+L_2C_2) & C_1(L_3C_{23}+L_2C_2) & 0 & 0 & 0 & 1 \\ -C_1(L_3S_{23}+L_2S_2) & -S_1(L_3S_{23}+L_2S_2) & L_3C_{23}+L_2C_2 & S_1 & -C_1 & 0 \\ -L_3C_1S_{23} & -L_3S_1S_{23} & L_3C_{23} & S_1 & -C_1 & 0 \end{bmatrix} \begin{bmatrix} f_x \\ f_y \\ f_z \\ 0 \\ 0 \\ 0 \end{bmatrix} \quad (5.166)$$

or

$$\begin{bmatrix} \tau_1 \\ \tau_2 \\ \tau_3 \end{bmatrix} = \begin{bmatrix} -S_1(L_3C_{23}+L_2C_2)f_x + C_1(L_3C_{23}+L_2C_2)f_y \\ -C_1(L_3S_{23}+L_2S_2)f_x - S_1(L_3S_{23}+L_2S_2)f_y + (L_3C_{23}+L_2C_2)f_z \\ -L_3C_1S_{23}f_x - L_3S_1S_{23}f_y + L_3C_{23}f_z \end{bmatrix} \quad (5.167)$$

EXERCISES

5.1 Show that the overall differential transformation due to three differential rotations of δ_x, δ_y, and δ_z about x-, y-, and x-axes, respectively, is independent of the order in which rotations are made. In the case of differential change $\sin\theta \to \delta$ and $\cos\theta \to 1$.

5.2 Show that the three differential rotations of δ_x, δ_y, and δ_z made in any order about the x-, y-, and z-axes, respectively are equivalent to a differential rotation of $d\theta$ about axis K.

5.3 Given a coordinate frame $\{A\}$

$$\boldsymbol{A} = \begin{bmatrix} 0 & 1 & 0 & 1 \\ 0 & 0 & 1 & 2 \\ 1 & 0 & 0 & 10 \\ 0 & 0 & 0 & 1 \end{bmatrix} \quad (5.168)$$

Find the differential transformation of $\boldsymbol{A}$ corresponding to a differential rotation $\Delta\theta = -0.1\hat{i} + 0\hat{j} + 0\hat{k}$ followed by a differential translation of $\Delta d = 0\hat{i} + 0.5\hat{j} + 1\hat{k}$ with respect to base frame.

5.4 A planar manipulator arm with one rotary and one prismatic joint is shown in Fig. E5.4.

(a) Compute the Jacobian matrix for the arm.

(b) What will be the joint velocities if the endpoint is required to move at a constant velocity along a flat surface?

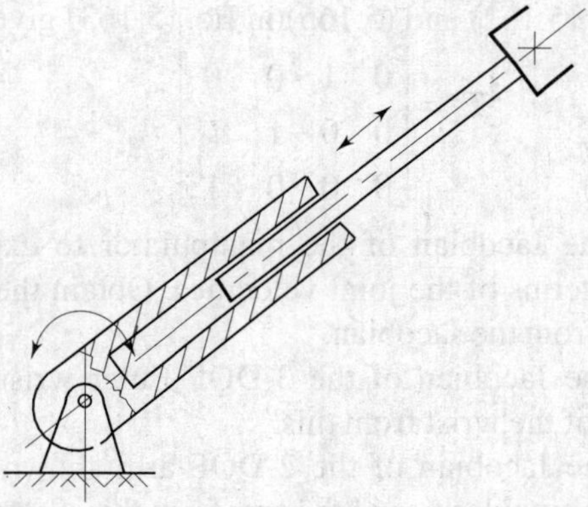

Fig. E5.4 *A 2-DOF planar arm with a rotary and a prismatic joint*

5.5 In Example 5.4, the Jacobian of a 2-DOF manipulator with $L_1 = L_2 = 1$ was obtained. Determine the Jacobian of the 2-DOF RR manipulator with $L_1 \neq L_2$ and show that the significant Jacobian is

$$\boldsymbol{J} = \begin{bmatrix} -L_1S_1 - L_2S_{12} & -L_2S_{12} \\ L_1C_1 + L_2C_{12} & L_2C_{12} \end{bmatrix} \tag{5.169}$$

Determine the singularities and joint velocities of this manipulator.

5.6 Determine (a) Jacobian (b) Singularities and (c) Joint velocities for a 3-DOF planar arm with the revolute joints.

5.7 Compute the Jacobian for the spherical configuration manipulator in Fig. 4.7, Example 4.4.

5.8 Compute the Jacobian for the SCARA manipulator of Fig. 3.22, Example 3.6. Find the singularities for the manipulator.

5.9 Determine the singularities of the spherical configuration manipulator of Fig. 4.7, Example 4.4.

5.10 Find the singularities of the 3R, roll-pitch-yaw wrist in Fig. 3.18.

5.11 For the 6-DOF Stanford manipulator in Fig. 3.26 obtain the Jacobian.

5.12 Determine the singularities of the six degrees of freedom manipulator in Example 3.8.

5.13 For the 5-DOF manipulator shown in Fig. E3.14 compute the Jacobian and determine the singularities, if any.

5.14 Show that the spherical configuration manipulator in Exercise 5.7 has singularities at $q = [\, q_1\, 0\, q_3\,]^T$. Which axes are collinear?

5.15 For a 3-DOF manipulator arm, the link transformation matrices are:

$$^0\boldsymbol{T}_1 = \begin{bmatrix} C_1 & -S_1 & 0 & C_1 \\ S_1 & C_1 & 0 & S_1 \\ 0 & 0 & 1 & 0 \\ 0 & 0 & 0 & 1 \end{bmatrix}; \quad ^1\boldsymbol{T}_2 = \begin{bmatrix} C_2 & 0 & -S_2 & C_2 \\ S_2 & 0 & C_2 & S_2 \\ 0 & 1 & 0 & 0 \\ 0 & 0 & 0 & 1 \end{bmatrix};$$

$$
{}^{2}T_{3} = \begin{bmatrix} 1 & 0 & 0 & 0 \\ 0 & 1 & 0 & 0 \\ 0 & 0 & 1 & d_3 \\ 0 & 0 & 0 & 1 \end{bmatrix} \tag{5.170}
$$

Determine the Jacobian of the manipulator to express the Cartesian velocities in terms of the joint velocities. Obtain the singularities of the manipulator from the Jacobian.

5.16 Determine the Jacobian of the 3-DOF Euler wrist and determine the singularities of the wrist from this.

5.17 Determine the Jacobian of the 2-DOF arm shown in Fig. E5.17, and determine the singularities of the arm from this. State assumptions made, if any.

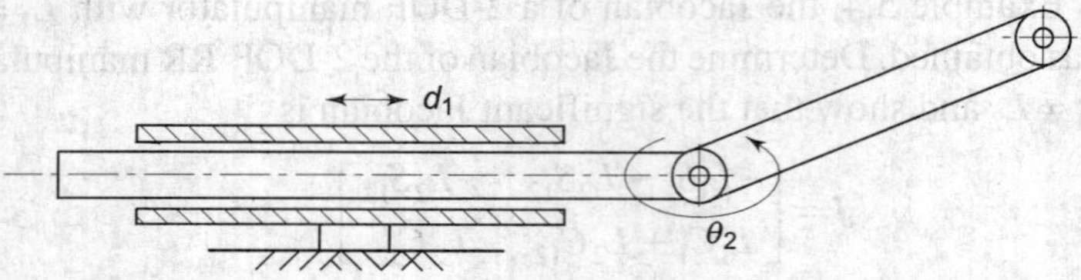

Fig. E5.17 *A 2-DOF planar arm with a rotary and a prismatic joint*

5.18 For the robotic arm shown in Fig. E5.18 obtain the Jacobian to express the Cartesian velocities in terms of joint velocities.

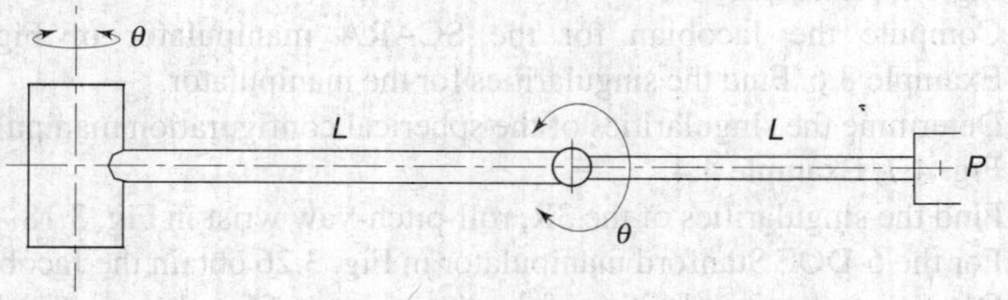

Fig. E5.18 *Two degree of freedom RR manipulator*

5.19 What are the singularities of a manipulator? How are they classified and determined? Explain briefly.

5.20 Jacobian of an n-DOF manipulator is a $6 \times n$ matrix. How can its inverse be determined to compute the kinematic singularities of the manipulator?

5.21 What is a skew symmetric matrix? How this matrix is related to the angular velocity of a link of an n-DOF manipulator?

5.22 How can a skew symmetric matrix be used to determine the velocity of a link if the position of the link, its angular velocity and rotational transformation matrix are known?

5.23 Determine the total derivative of the homogeneous transformation matrix of a link ${}^{i}T_{j}$.

5.24 For a given manipulator, are the velocity Jacobian and the static force Jacobian different? Explain your answer with an example.

5.25 A force $\mathcal{F} = [f_x \ f_y \ 0 \ 0 \ 0 \ 0]^T$ is applied by end-effector of the two link planar manipulator of Example 5.2. Find the required joint torques assuming a gravity free environment.

5.26 For the four axes SCARA manipulator discussed in Example 3.6 and whose Jacobian is computed in Exercise 5.8 above, determine the force/torques required at the joints to maintain a static equilibrium when end effector is exerting a force of

$$\mathcal{F} = [f_x \ f_y \ f_x \ \eta_x \ \eta_y \ \eta_z]^T \tag{5.171}$$

SELECTED BIBLIOGRAPHY

1. R. Featherstone, "Position and Velocity Transformations Between Robot End-effector Coordinates and Joint Angles," *Int. J. of Robotics Research*, *2*(2) 35–45, 1983.
2. Z.C. Lai and D.C.K. Yang, "A New Method for Singularity Analysis of Simple Six-Link Manipulator," *Int J. of Robotic Research*, ***5***(2), 66–74, 1986.
3. M. B.Leahy Jr, L. M. Nugent, G. N Saridis and K. P. Valavanis, "Efficient puma manipulator jacobian calculation and inversion," *Journal of Robotic Systems*, ***4***(4), 185–197, 1987.
4. S.K. Lin, "Singularity of a Nonlinear Feedback Control Scheme for Robots," *IEEE Tr on Systems, Man, and Cybernetics*, ***19***, 134–139, 1989.
5. Y. Nakamura, and H. Hanafusa, "Inverse kinematic solutions with singularity robustness for robot manipulator control," *Journal of Dynamic Systems, Measurement and Control*, ***108***(3), 163–171, 1986.
6. D. E. Orin, and W. W. Schrader, "Efficient computation of the Jacobian for robot manipulators," *Int. J. of Robotic Research*, ***3***(4), 66–75, 1984.
7. B. Siciliano, "Kinematic Control of Redundant Robot Manipulators: A Tutorial," *J. Intelligent and Robotic System*, ***3***, 201–212, 1990.

6

Dynamic Modeling

During the work cycle a manipulator must accelerate, move at constant speed, and decelerate. This time-varying position and orientation of the manipulator is termed as its dynamic behaviour. Time-varying torques are applied at the joints (by the joint actuators) to balance out the internal and external forces. The internal forces are caused by motion (velocity and acceleration) of links. Inertial, Coriolis, and frictional forces are some of the internal forces. The external forces are the forces exerted by the environment. These include the "load" and gravitational forces. As a result, links and joints have to withstand stresses caused by force/torque balance across these.

In this chapter, the mathematical model for the dynamic behaviour of the manipulator is developed. The mathematical equations, often referred as *manipulator dynamics*, are a set of *equations of motion* (EOM) that describe the dynamic response of the manipulator to input actuator torques.

The dynamic model of a manipulator is useful for computation of torque and forces required for execution of a typical work cycle, which is vital information for the design of links, joints, drives, and actuators. The dynamic behaviour of the manipulator provides relationship between joint actuator torques and motion of links for simulation and design of control algorithms. The manipulator control maintains the dynamic response of the manipulator to obtain the desired performance, which directly depends on the accuracy of the dynamic model and efficiency of the control algorithms. The control problem requires specifying the control strategies to achieve the desired response and performance. Simulations of manipulator motion permits testing of control strategies, motion planning, and performance studies without a physical prototype of the manipulator.

The serial link manipulator represents a complex dynamic system, which can be modeled by systematically using known physical laws of Lagrangian mechanics or Newtonian mechanics. Approaches such as *Lagrange-Euler* (LE), which is "energy-based", and *Newton-Euler* (NE), based on "force-balance", can be systematically applied to develop the manipulator EOM. Assuming rigid body motion, no backlash, no friction and neglecting effects of control component dynamics, the resulting EOM are a set of second order, coupled, nonlinear differential equations, consisting of inertia loading and coupling reaction forces between joints. Another method, the generalized d'Alembert principle, provides an "equivalent" dynamic model.

The Newton-Euler and Lagrange-Euler formulations of the dynamic model provide a closed-form solution, which are computationally intensive making real-time control based on such a dynamic model, inefficient, if not impossible. To improve the computational speed, recursive methods and approximate models based on simplifying assumptions have been developed. These approximate models, however, result in suboptimal dynamic performance and restrict arm movement to low speeds. The LE and NE models, presented in this chapter, provide a symbolic solution to manipulator dynamics and give insight into the control problem.

6.1 LAGRANGIAN MECHANICS

A scalar function called *Lagrange function* or *Lagrangian* $\mathcal{L}$ is defined as the difference between the total kinetic energy $\mathcal{K}$ and the total potential energy $\mathcal{P}$ of a mechanical system.

$$\mathcal{L} = \mathcal{K} - \mathcal{P} \tag{6.1}$$

The Lagrange-Euler dynamic formulation is based on a set of generalized coordinates to describe the system variables. In the generalized coordinates, generalized displacement '$\boldsymbol{q}$' is used as a joint variable, which describes a linear displacement $\boldsymbol{d}$ for a prismatic joint and angular displacement θ for a rotary joint, and $\dot{\boldsymbol{q}}$ describes linear velocity $\dot{\boldsymbol{d}}(=\boldsymbol{v})$ and angular velocity $\dot{\theta}(=\omega)$ for prismatic and rotary joints, respectively, as discussed in earlier chapters. Similarly, generalized torque 'τ' required at the joint to produce desired dynamics represents the force f for a prismatic joint and torque τ for a revolute joint, as seen in Eq. (5.78). Because kinetic and potential energies are function of $\boldsymbol{q}_i$ and $\dot{\boldsymbol{q}}_i$ ($i = 1, \ldots \boldsymbol{n}$), so is the Lagrangian $\mathcal{L}$.

The dynamic model based on Lagrange-Euler formulation is obtained from the Lagrangian, as a set of equations,

$$\frac{d}{dt}\left(\frac{\partial \mathcal{L}}{\partial \dot{q}_i}\right) - \frac{\partial \mathcal{L}}{\partial q_i} = \tau_i \quad \text{for } i = 1,\ 2,\ \ldots, \boldsymbol{n} \tag{6.2}$$

The left-hand side of dynamic equations can be interpreted as sum of the torques/forces due to kinetic and potential energy present in the system. The right-hand side τ_i is the joint torque for joint *i* that is provided by the actuator *i*. If $\tau_i = 0$, it means that joint *i* does not move and if $\tau_i \neq 0$, the manipulator movement is modified by the actuator at joint *i*.

6.2 TWO DEGREE OF FREEDOM MANIPULATOR—DYNAMIC MODEL

The dynamics of a simple manipulator is worked out to illustrate the Lagrange-Euler formulation and to clarify the problems involved in the dynamic modeling. A planar, 2-DOF manipulator with both rotary joints, as shown in Fig. 6.1, is considered and its dynamic model is obtained using direct geometric approach before discussing the general formulation. For the manipulator, coordinate frames {0} and {1}, joint variables θ_1 and θ_2, link lengths L_1 and L_2, and mass of links m_1 and m_2, respectively, are shown in the figure. The mass of each link is assumed to be a point mass located at the center of mass of each link and links are assumed to be slender members. The linear and angular velocities are v_1, v_2, $\dot{\theta}_1$ and $\dot{\theta}_2$, respectively.

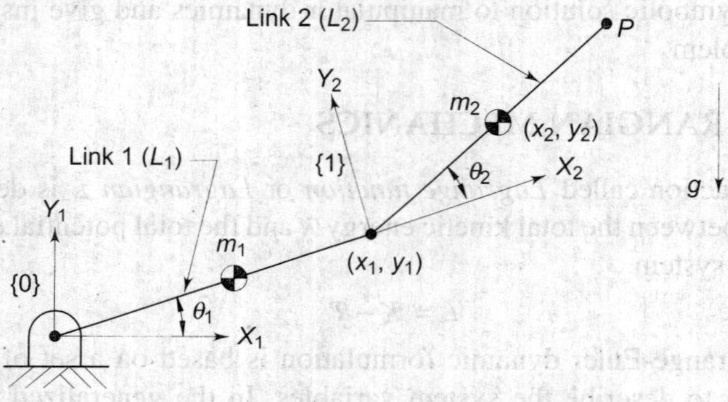

Fig. 6.1 *A 2-DOF planar articulated (RR) arm*

The Lagrangian requires kinetic and potential energies of the manipulator. The kinetic energy of a rigid body (a link), can be expressed as:

$$\mathcal{K} = \frac{1}{2} m v^2 + \frac{1}{2} I \omega^2 \tag{6.3}$$

where v is the linear velocity , ω is the angular velocity, m is the mass, and I is the moment of inertia of the rigid body at its center of mass.

Thus, the kinetic energy for the link 1 with the linear velocity $v_1 = \frac{1}{2} L_1 \dot{\theta}_1$, angular velocity $\omega_1 = \dot{\theta}_1$, moment of inertia $I_1 = \frac{1}{12} m_1 L_1^2$, and mass m_1 is

$$\mathcal{K}_1 = \frac{1}{2} m_1 v_1^2 + \frac{1}{2} I_1 \omega_1^2 = \frac{1}{8} m_1 L_1^2 \dot{\theta}_1^2 + \frac{1}{24} m_1 L_1^2 \dot{\theta}_1^2 = \frac{1}{6} m_1 L_1^2 \dot{\theta}_1^2 \tag{6.4}$$

and its potential energy is

$$\mathcal{P}_1 = \frac{1}{2} m_1 g L_1 \sin \theta_1 \tag{6.5}$$

where g is the magnitude of acceleration due to gravity in the negative y-axis direction.

For the second link, link 2, the Cartesian position coordinates (x_2, y_2) of the center of mass of link are:

$$
\begin{aligned}
x_2 &= L_1 \cos \theta_1 + \frac{1}{2} L_2 \cos (\theta_1 + \theta_2) \\
y_2 &= L_1 \sin \theta_1 + \frac{1}{2} L_2 \sin (\theta_1 + \theta_2)
\end{aligned}
\tag{6.6}
$$

Differentiating Eq. (6.6) gives the components of velocity of link 2 as

$$
\begin{aligned}
\dot{x}_2 &= - L_1 \sin \theta_1 \dot{\theta}_1 - \frac{1}{2} L_2 \sin (\theta_1 + \theta_2)(\dot{\theta}_1 + \dot{\theta}_2) \\
\dot{y}_2 &= L_1 \cos \theta_1 \dot{\theta}_1 + \frac{1}{2} L_2 \cos (\theta_1 + \theta_2)(\dot{\theta}_1 + \dot{\theta}_2)
\end{aligned}
\tag{6.7}
$$

From these components, the square of the magnitude of velocity of the end of link 2 is

$$v_2^2 = \dot{x}_2^2 + \dot{y}_2^2$$

or

$$
\begin{aligned}
v_2^2 = {} & L_1^2 S_1^2 \dot{\theta}_1^2 + \frac{1}{4} L_2^2 S_{12}^2 (\dot{\theta}_1 + \dot{\theta}_2)^2 + L_1 L_2 S_1 S_{12} (\dot{\theta}_1^2 + \dot{\theta}_1 \dot{\theta}_2) \\
& + L_1^2 C_1^2 \dot{\theta}_1^2 + \frac{1}{4} L_2^2 C_{12}^2 (\dot{\theta}_1 + \dot{\theta}_2)^2 + L_1 L_2 C_1 C_{12} (\dot{\theta}_1^2 + \dot{\theta}_1 \dot{\theta}_2)
\end{aligned}
$$

Simplifying

$$v_2^2 = L_1^2 \dot{\theta}_1^2 + \frac{1}{4} L_2^2 (\dot{\theta}_1 + \dot{\theta}_2)^2 + L_1 L_2 C_2 (\dot{\theta}_1^2 + \dot{\theta}_1 \dot{\theta}_2) \tag{6.8}$$

where $C_i = \cos \theta_i$, $S_i = \sin \theta_i$, $C_{12} = \cos (\theta_1 + \theta_2)$ and $S_{12} = \sin (\theta_1 + \theta_2)$.

Thus, the kinetic energy of link 2 with $\omega_2 = \dot{\theta}_1 + \dot{\theta}_2$ and $I_2 = \frac{1}{12} m_2 L_2^2$ is

$$
\begin{aligned}
\mathcal{K}_2 &= \frac{1}{2} m_2 v_2^2 + \frac{1}{2} I_2 \omega_2^2 \\
&= \frac{1}{2} m_2 [L_1^2 \dot{\theta}_1^2 + \frac{1}{4} L_2^2 (\dot{\theta}_1 + \dot{\theta}_2)^2 + L_1 L_2 C_2 (\dot{\theta}_1^2 + \dot{\theta}_1 \dot{\theta}_2)] \\
&\qquad + \frac{1}{24} m_2 L_2^2 \left(\dot{\theta}_1 + \dot{\theta}_2\right)^2 \qquad (6.9) \\
&= \frac{1}{2} m_2 L_1^2 \dot{\theta}_1^2 + \frac{1}{6} m_2 L_2^2 \left(\dot{\theta}_1^2 + \dot{\theta}_1^2 + 2\dot{\theta}_1 \dot{\theta}_2\right) + \frac{1}{2} m_2 L_1 L_2 C_2 (\dot{\theta}_1^2 + \dot{\theta}_1 \dot{\theta}_2)
\end{aligned}
$$

The potential energy of link 2, from Eq. (6.6), is

$$\mathcal{P}_2 = m_2 g L_1 S_1 + \frac{1}{2} m_2 g L_2 S_{12} \tag{6.10}$$

The Lagrangian $\mathcal{L} = \mathcal{K} - \mathcal{P} = \mathcal{K}_1 + \mathcal{K}_2 - \mathcal{P}_1 - \mathcal{P}_2$ is obtained from Eqs. (6.4), (6.5), (6.9), and (6.10). Rearranging and simplifying, the Lagrangian is

$$\mathcal{L} = \frac{1}{2}\left(\frac{1}{3} m_1 + m_2\right)L_1^2\dot{\theta}_1^2 + \frac{1}{6} m_2 L_2^2 (\dot{\theta}_1{}^2 + \dot{\theta}_2{}^2 + 2\dot{\theta}_1\dot{\theta}_2) + \frac{1}{2} m_2 L_1 L_2 C_2 (\dot{\theta}_1^2 + \dot{\theta}_1\dot{\theta}_2) - \left(\frac{1}{2} m_1 + m_2\right)gL_1S_1 - \frac{1}{2} m_2 g L_2 S_{12} \tag{6.11}$$

The Lagrange-Euler formulation for link 1, Eq. (6.2), gives the torque τ_1 at joint 1 as

$$\tau_1 = \frac{d}{dt}\left(\frac{\partial \mathcal{L}}{\partial \dot{\theta}_1}\right) - \frac{\partial \mathcal{L}}{\partial \theta_1} \tag{6.12}$$

The Lagrangian in Eq. (6.11) is differentiated wrt θ_1 and $\dot{\theta}_1$ to give

$$\frac{\partial \mathcal{L}}{\partial \theta_1} = -\left(\frac{1}{2} m_1 + m_2\right)g\, L_1 C_1 - \frac{1}{2} m_2 g L_2 C_{12} \tag{6.13}$$

and
$$\frac{\partial \mathcal{L}}{\partial \dot{\theta}_1} = \left(\frac{1}{3} m_1 + m_2\right)L_1^2\dot{\theta}_1 + \frac{1}{3} m_2 L_2^2 (\dot{\theta}_1 + \dot{\theta}_2) + \frac{1}{3} m_2 L_1 L_2 C_2 (2\dot{\theta}_1 + \dot{\theta}_2) \tag{6.14}$$

Differentiating Eq. (6.14) wrt time

$$\frac{d}{dt}\left(\frac{\partial \mathcal{L}}{\partial \dot{\theta}_1}\right) = \left[\left(\frac{1}{3} m_1 + m_2\right)L_1^2 + \frac{1}{3} m_2 L_2^2 + m_2 L_1 L_2 C_2\right]\ddot{\theta}_1 + m_2\left[\frac{1}{3} L_2^2 + \frac{1}{2} L_1 L_2 C_2\right]\ddot{\theta}_2 - m_2 L_1 L_2 S_2 \dot{\theta}_1\dot{\theta}_2 - \frac{1}{2} m_2 L_1 L_2 S_2 \dot{\theta}_2^2 \tag{6.15}$$

Substituting the results of Eqs. (6.13) and (6.15) into Eq. (6.12), the torque at joint 1 is obtained as

$$\tau_1 = \left[\left(\frac{1}{3} m_1 + m_2\right)L_1^2 + \frac{1}{3} m_2 L_2^2 + m_2 L_1 L_2 C_2\right]\ddot{\theta}_1 + m_2\left[\frac{1}{3} L_2^2 + \frac{1}{2} L_1 L_2 C_2\right]\ddot{\theta}_2 - m_2 L_1 L_2 S_2 \dot{\theta}_1\dot{\theta}_2 - \frac{1}{2} m_2 L_1 L_2 S_2 \dot{\theta}_2^2 + \left(\frac{1}{2} m_1 + m_2\right) g\, L_1 C_1 + \frac{1}{2} m_2 g\, L_2 C_{12} \tag{6.16}$$

Similarly, the derivatives of Lagrangian, Eq. (6.11), for joint 2 are

$$\frac{\partial \mathcal{L}}{\partial \theta_2} = -\frac{1}{2} m_2 L_1 L_2 S_2 (\dot{\theta}_1^2 + \dot{\theta}_1\dot{\theta}_2) - \frac{1}{2} m_2 g L_2 C_{12} \tag{6.17}$$

and
$$\frac{\partial \mathcal{L}}{\partial \dot{\theta}_2} = \frac{1}{3} m_2 L_2^2 (\dot{\theta}_1 + \dot{\theta}_2) + \frac{1}{2} m_2 L_1 L_2 C_2 \dot{\theta}_1 \tag{6.18}$$

and

$$\frac{d}{dt}\left(\frac{\partial \mathcal{L}}{\partial \dot{\theta}_2}\right) = \left[\frac{1}{3} m_2 L_2^2 + \frac{1}{2} m_2 L_1 L_2 C_2\right]\ddot{\theta}_1 + \frac{1}{3} m_2 L_2^2 \ddot{\theta}_2 - \frac{1}{2} m_2 L_1 L_2 S_2 \dot{\theta}_1\dot{\theta}_2 \tag{6.19}$$

Again from Eq (6.12),

$$\tau_2 = m_2\left[\frac{1}{3}L_2^2 + \frac{1}{2}L_1L_2C_2\right]\ddot{\theta}_1 + \frac{1}{3}m_2L_2^2\ddot{\theta}_2 + \frac{1}{2}m_2L_1L_2S_2\dot{\theta}_1^2 + \frac{1}{2}m_2gL_2C_{12} \tag{6.20}$$

Equations (6.16) and (6.20) are the EOM (dynamic model) of the 2-link planar manipulator. Because both the joints are revolute, the generalized torques τ_1 and τ_2 represent the actual joint torques.

Torque equation, Eqs. (6.16) and (6.20) can be written in the generalized form, which will be described in Section 6.4, as

$$\begin{aligned}\tau_1 &= M_{11}\ddot{\theta}_1 + M_{12}\ddot{\theta}_2 + H_1 + G_1\\ \tau_2 &= M_{21}\ddot{\theta}_1 + M_{22}\ddot{\theta}_2 + H_2 + G_2\end{aligned} \tag{6.21}$$

where

$$M_{11} = \left[\left(\frac{1}{3}m_1 + m_2\right)L_1^2 + \frac{1}{3}m_2L_2^2 + m_2L_1L_2C_2\right]$$

$$M_{12} = M_{21} = m_2\left[\frac{1}{3}L_2^2 + \frac{1}{2}L_1L_2C_2\right]$$

$$M_{22} = \frac{1}{3}m_2L_2^2$$

$$H_1 = -m_2L_1L_2S_2\dot{\theta}_1\dot{\theta}_2 - \frac{1}{2}m_2L_1L_2S_2\dot{\theta}_2^2$$

$$H_2 = \frac{1}{2}m_2L_1L_2S_2\dot{\theta}_1^2$$

$$G_1 = \left[\left(\frac{1}{2}m_1 + m_2\right)L_1C_1 + \frac{1}{2}m_2L_2C_{12}\right]g$$

$$G_2 = \frac{1}{2}m_2L_2C_{12}g$$

These coefficients are defined as

M_{ii} = effective inertia,

M_{ij} = effective coupling inertia,

H_i = centrifugal and Coriolis acceleration forces

This direct formulation approach becomes quite cumbersome when a manipulator with more than 2-DOF is analyzed. In the following sections, the derivation of EOM for an n-DOF manipulator, based on homogeneous coordinate transformation matrices is presented.

6.3 LAGRANGE-EULER FORMULATION

The Lagrange-Euler formulation is a systematic procedure for obtaining the dynamic model of an n-DOF manipulator. The n-DOF open kinematic chain serial link manipulator has n joint position or displacement variables, $\boldsymbol{q} = [q_1, \ldots, q_n]^T$. The LE formulation [Eq. (6.2)] establishes the relation between the joint positions, velocities, accelerations, and the generalized torques applied to the manipulator. The generalized torques are the nonconservative torques contributed by joint actuators, joint friction forces, and induced joint torques. The induced joint torques are the torques at the joints due to contact or interaction of the end-effector with the environment. In the present discussion only joint actuator torques are considered.

The derivation of EOM using LE formulation is carried out in the following subsections. It makes use of link transformation matrices $\boldsymbol{T}$, which are obtained from the kinematic modeling discussed in Chapter 3. First, the link velocity is computed and next the link inertia tensor is obtained. These are used to compute kinetic energy. Then potential energy is calculated and next, the Lagrangian is formed, which is substituted in Eq. (6.2) to get the dynamic model.

6.3.1 VELOCITY OF A POINT ON THE MANIPULATOR

For computing the kinetic energy of a link of an n-DOF manipulator, link velocity is required. The concepts of Chapter 5 are extended to compute the velocity of each link.

Consider a point P on link i of an n-link (n-DOF) manipulator, as shown in Fig. 6.2. The coordinate frames, frame {0}, frame {i–1}, and frame {i} are chosen as per convention. The position vector ${}^i\boldsymbol{r}_i$ describes the point P on the link with respect to frame {i}. In homogeneous coordinate notation,

$$ {}^i\boldsymbol{r}_i = [x_i \quad y_i \quad z_i \quad 1]^T \tag{6.22}$$

The position of point P with respect to base coordinate is

$$ {}^0\boldsymbol{r}_i = {}^0\boldsymbol{T}_i \, {}^i\boldsymbol{r}_i = \left({}^0\boldsymbol{T}_1 \, {}^1\boldsymbol{T}_2 \cdots {}^{i-1}\boldsymbol{T}_i\right) {}^i\boldsymbol{r}_i \tag{6.23}$$

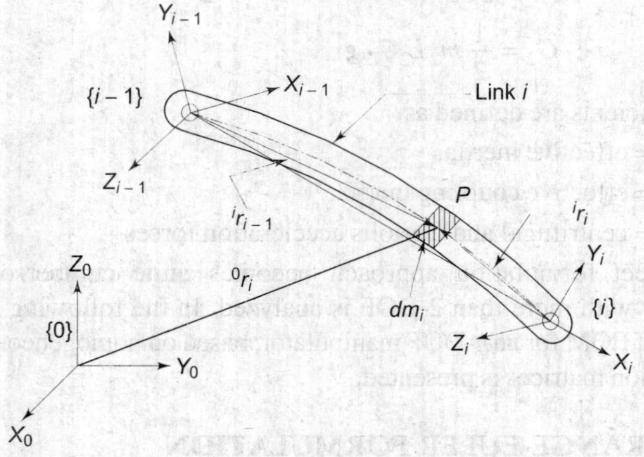

Fig. 6.2 *Velocity of a point on the link*

where ${}^{i-1}\boldsymbol{T}_i$ is given by Eq. (3.3) with $q_i = \theta_i$ for a rotary joint or $q_i = d_i$ for a prismatic joint. The velocity of point P with respect to base coordinates, frame {0}, is obtained from Eq. (5.46), as

$$ {}^{0}\boldsymbol{v}_i \equiv \boldsymbol{v}_i = {}^{o}\dot{\boldsymbol{r}}_i = \left[\sum_{j=1}^{i} \frac{\partial {}^{0}\boldsymbol{T}_i}{\partial q_j}\dot{q}_j\right] {}^{i}\boldsymbol{r}_i \tag{6.24}$$

where the fact that ${}^{i}\dot{\boldsymbol{r}}_i = 0$ is used.

The transformation matrix ${}^{0}\boldsymbol{T}_i$ involves complex trigonometric terms and its partial derivative with respect to q_j, required in Eq. (6.24), involves complex computation. The following steps simplify the computation of partial derivative of the homogeneous transformation matrix.

Consider the transformation matrix ${}^{j-1}\boldsymbol{T}_j$ for link j given by Eq. (3.4) for a rotary joint, that is,

$$ {}^{j-1}\boldsymbol{T}_j = \begin{bmatrix} C\theta_j & -S\theta_j C\alpha_j & S\theta_j S\alpha_j & a_j C\theta_j \\ S\theta_j & C\theta_j C\alpha_j & -C\theta_j S\alpha_j & a_j S\theta_j \\ 0 & S\alpha_j & C\alpha_j & d_j \\ 0 & 0 & 0 & 1 \end{bmatrix} \tag{6.25}$$

where a_j, d_j, α_j, and θ_j have the usual meaning. The partial derivative of Eq. (6.25) with respect to θ_j is

$$ \frac{\partial {}^{j-1}\boldsymbol{T}_j}{\partial \theta_j} = \begin{bmatrix} -S\theta_j & -C\theta_j C\alpha_j & C\theta_j S\alpha_j & -a_j S\theta_j \\ C\theta_j & -S\theta_j C\alpha_j & S\theta_j S\alpha_j & a_j C\theta_j \\ 0 & 0 & 0 & 0 \\ 0 & 0 & 0 & 0 \end{bmatrix} \tag{6.26}$$

Comparison of Eqs. (6.25) and (6.26) gives a pattern. It is observed that Eq. (6.26) can be obtained from Eq. (6.25) by

- interchanging row 1 with row 2,
- changing the sign of row 1, and
- making row 3 and row 4 zero.

Hence, partial derivative of ${}^{j-1}\boldsymbol{T}_j$ with respect to θ_j can be obtained using the above steps and without actually differentiating the terms. The same result can be obtained using matrix operations, which is more convenient in performing the operations using a computer. Mathematically these steps can be carried out using a 4×4 matrix $\boldsymbol{Q}_j$ defined as

$$ \boldsymbol{Q}_j = \begin{bmatrix} 0 & -1 & 0 & 0 \\ 1 & 0 & 0 & 0 \\ 0 & 0 & 0 & 0 \\ 0 & 0 & 0 & 0 \end{bmatrix} \quad \text{(for revolute joint)} \tag{6.27}$$

and premultiplying ${}^{j-1}\boldsymbol{T}_j$ with $\boldsymbol{Q}_j$, that is,

$$Q_j\,{}^{j-1}T_j = \begin{bmatrix} 0 & -1 & 0 & 0 \\ 1 & 0 & 0 & 0 \\ 0 & 0 & 0 & 0 \\ 0 & 0 & 0 & 0 \end{bmatrix} \begin{bmatrix} C\theta_j & -S\theta_j C\alpha_j & S\theta_j S\alpha_j & a_j C\theta_j \\ S\theta_j & C\theta_j C\alpha_j & -C\theta_j S\alpha_j & a_j S\theta_j \\ 0 & S\alpha_j & C\alpha_j & d_j \\ 0 & 0 & 0 & 1 \end{bmatrix}$$

or
$$Q_j\,{}^{j-1}T_j = \begin{bmatrix} -S\theta_j & -C\theta_j C\alpha_j & C\theta_j S\alpha_j & -a_j S\theta_j \\ C\theta_j & -S\theta_j C\alpha_j & S\theta_j S\alpha_j & a_j C\theta_j \\ 0 & 0 & 0 & 0 \\ 0 & 0 & 0 & 0 \end{bmatrix} \tag{6.28}$$

The result is same as Eq. (6.26). Hence,

$$\frac{\partial\,{}^{j-1}T_j}{\partial \theta_j} = Q_j\,{}^{j-1}T_j \tag{6.29}$$

The partial derivative equation, Eq. (6.29) also applies for a prismatic joint, with Q_j defined as

$$Q_j = \begin{bmatrix} 0 & 0 & 0 & 0 \\ 0 & 0 & 0 & 0 \\ 0 & 0 & 0 & 1 \\ 0 & 0 & 0 & 0 \end{bmatrix} \quad \text{(for prismatic joint)} \tag{6.30}$$

The reader may verify Eq. (6.29) with Q_j in Eq. (6.30) for a prismatic joint. Since ${}^0T_i = {}^0T_1\,{}^1T_2 \cdots {}^{i-1}T_i$, its partial derivative ${}^0T_i$ with respect to q_j is

$$\frac{\partial({}^0T_i)}{\partial q_j} = {}^0T_1\,{}^1T_2 \cdots {}^{j-2}T_{j-1}\,\frac{\partial({}^{j-1}T_j)}{\partial q_j}\,{}^jT_{j+1} \cdots {}^{i-1}T_i$$

Using Eq. (6.29), this simplifies to

$$\frac{\partial({}^0T_i)}{\partial q_j} = {}^0T_1\,{}^1T_2 \cdots {}^{j-2}T_{j-1}(Q_j\,{}^{j-1}T_j)\,{}^jT_{j+1} \cdots {}^{i-1}T_i = {}^0T_{j-1}\,Q_j\,{}^{j-1}T_i \tag{6.31}$$

This result is valid only for $j \le i$. Hence, for $i = 1, 2, \ldots, n$,

$$\frac{\partial({}^0T_i)}{\partial q_j} = \begin{cases} {}^0T_{j-1}\,Q_j\,{}^{j-1}T_i & \text{for } j \le i \\ 0 & \text{for } j > i \end{cases} \tag{6.32}$$

It is worth noting that the partial derivative of homogeneous transformation matrix ${}^{i-1}T_i$ with respect to q_j represents the effect of motion of joint j on link i. The link velocity v_i in Eq. (6.24) is, thus, simplified using Eq. (6.32) as

$$v_i = \sum_{j=1}^{i} {}^0T_{j-1}\,Q_j\,{}^{j-1}T_i\,\dot{q}_j\,{}^ir_i \tag{6.33}$$

6.3.2 The Inertia Tensor

The mass of the link contributes inertia forces during motion of the link. The mass properties, which reflect all the inertial loads with respect to rotations about the origin of frame of interest, are represented by a *moment of inertia tensor*. It is a 4×4 symmetric matrix, which characterizes the distribution of mass of a rigid body (the link i). The moment of inertia tensor is defined as

$$\boldsymbol{I}_i = \begin{bmatrix} \int x_i^2\, dm_i & \int x_i y_i\, dm_i & \int x_i z_i\, dm_i & \int x_i\, dm_i \\ \int x_i y_i\, dm_i & \int y_i^2\, dm_i & \int y_i z_i\, dm_i & \int y_i\, dm_i \\ \int x_i z_i\, dm_i & \int y_i z_i\, dm_i & \int z_i^2\, dm_i & \int z_i\, dm_i \\ \int x_i\, dm_i & \int y_i\, dm_i & \int z_i\, dm_i & \int dm_i \end{bmatrix} \tag{6.34}$$

where dm_i is the mass of the element on link i located at ${}^i\boldsymbol{r}_i = [x_i \quad y_i \quad z_i \quad 1]^T$, as shown in Fig. 6.2. Using the moment of inertia, I_{kk}, cross product of inertia, $\boldsymbol{I}_{kl}$ $(k \neq l)$ and first moments of body, the inertia tensor in Eq. (6.34) $\boldsymbol{I}_i$ is expressed as (see Appendix C for details).

$$\boldsymbol{I}_i = \begin{bmatrix} \frac{1}{2}(-I_{xx} + I_{yy} + I_{zz}) & I_{xy} & I_{xz} & m_i\bar{x}_i \\ I_{xy} & \frac{1}{2}(I_{xx} - I_{yy} + I_{zz}) & I_{yz} & m_i\bar{y}_i \\ I_{xz} & I_{yz} & \frac{1}{2}(I_{xx} + I_{yy} - I_{zz}) & m_i\bar{z}_i \\ m_i\bar{x}_i & m_i\bar{y}_i & m_i\bar{z}_i & m_i \end{bmatrix} \tag{6.35}$$

where m_i is the mass of the link i and ${}^i\bar{\boldsymbol{r}}_i = [\bar{x}_i \ \bar{y}_i \ \bar{z}_i \ 1]^T$ is its center of mass. The moment of inertia tensor $\boldsymbol{I}_i$ for link i depends on the mass distribution of the link and not on its position or rate of change of position.

6.3.3 The Kinetic Energy

The kinetic energy of the differential mass dm_i on link i, for $i = 1, 2, \ldots, n$, located at ${}^0\boldsymbol{r}_i$ and moving with velocity ${}^0\boldsymbol{v}_i$ $(= \boldsymbol{v}_i)$ with respect to the base frame $\{0\}$ is

$$d\mathcal{K}_i = \frac{1}{2} dm_i (\boldsymbol{v}_i)^2 \tag{6.36}$$

The trace operator $\left(\mathrm{Tr}(\boldsymbol{A}) = \sum_{i=1}^{n} a_{ii} \right)$ is used to obtain $(\boldsymbol{v}_i)^2$ as,

$$\boldsymbol{v}_i^2 = \boldsymbol{v}_i \cdot \boldsymbol{v}_i = {}^0\dot{\boldsymbol{r}}_i \cdot {}^0\dot{\boldsymbol{r}}_i = \mathrm{Tr}({}^0\dot{\boldsymbol{r}}_i \ {}^0\dot{\boldsymbol{r}}_i^{\,T}) = \mathrm{Tr}\,(\boldsymbol{v}_i \boldsymbol{v}_i^{\,T}) \tag{6.37}$$

Substituting $\boldsymbol{v}_i$ from Eq. (6.33) in Eq. (6.37) and the result in Eq. (6.36), the kinetic energy of the differential mass is obtained as

$$d\mathcal{K}_i = \frac{1}{2} \mathrm{Tr} \left[\left(\sum_{j=1}^{i} {}^0\boldsymbol{T}_{j-1} \boldsymbol{Q}_j \, {}^{j-1}\boldsymbol{T}_i \, \dot{\boldsymbol{q}}_j \, {}^i\boldsymbol{r}_i \right) \left(\sum_{k=1}^{i} {}^0\boldsymbol{T}_{k-1} \boldsymbol{Q}_k \, {}^{k-1}\boldsymbol{T}_i \, \dot{\boldsymbol{q}}_k \, {}^i\boldsymbol{r}_i \right)^T \right] dm_i$$

or $d\mathcal{K}_i = \dfrac{1}{2}\,\mathrm{Tr}\left[\sum_{j=1}^{i}\sum_{k=1}^{i}\left({}^0\boldsymbol{T}_{j-1}\boldsymbol{Q}_j\,{}^{j-1}\boldsymbol{T}_i\right){}^i\boldsymbol{r}_i\,{}^i\boldsymbol{r}_i^T dm_i\left({}^0\boldsymbol{T}_{k-1}\boldsymbol{Q}_k\,{}^{k-1}\boldsymbol{T}_i\right)^T\dot{\boldsymbol{q}}_j\dot{\boldsymbol{q}}_k\right]$

(6.38)

The total kinetic energy of link i is then

$$\mathcal{K}_i = \int d\mathcal{K}_i$$

$$\mathcal{K}_i = \frac{1}{2}\,\mathrm{Tr}\left[\sum_{j=1}^{i}\sum_{k=1}^{i}\left({}^0\boldsymbol{T}_{j-1}\boldsymbol{Q}_j\,{}^{j-1}\boldsymbol{T}_i\right)\int{}^i\boldsymbol{r}_i\,{}^i\boldsymbol{r}_i^T dm_i\left({}^0\boldsymbol{T}_{k-1}\boldsymbol{Q}_k\,{}^{k-1}\boldsymbol{T}_i\right)^T\dot{\boldsymbol{q}}_j\dot{\boldsymbol{q}}_k\right] \tag{6.39}$$

The integral term $\int{}^i\boldsymbol{r}_i\,{}^i\boldsymbol{r}_i^T dm_i$ in Eq. (6.39) is the moment of inertia tensor $\boldsymbol{I}_i$ given by Eq. (6.35). Therefore, $\mathcal{K}_i$ is

$$\mathcal{K}_i = \frac{1}{2}\,\mathrm{Tr}\left[\sum_{j=1}^{i}\sum_{k=1}^{i}\left({}^0\boldsymbol{T}_{j-1}\boldsymbol{Q}_j\,{}^{j-1}\boldsymbol{T}_i\right)\boldsymbol{I}_i\left({}^0\boldsymbol{T}_{k-1}\boldsymbol{Q}_k\,{}^{k-1}\boldsymbol{T}_i\right)^T\dot{\boldsymbol{q}}_j\dot{\boldsymbol{q}}_k\right] \tag{6.40}$$

Thus, for n-DOF manipulator, total kinetic energy of manipulator is

$$\mathcal{K} = \sum_{i=1}^{n}\mathcal{K}_i = \frac{1}{2}\sum_{i=1}^{n}\mathrm{Tr}\left[\sum_{j=1}^{i}\sum_{k=1}^{i}\left({}^0\boldsymbol{T}_{j-1}\boldsymbol{Q}_j\,{}^{j-1}\boldsymbol{T}_i\right)\boldsymbol{I}_i\left({}^0\boldsymbol{T}_{k-1}\boldsymbol{Q}_k\,{}^{k-1}\boldsymbol{T}_i\right)^T\dot{\boldsymbol{q}}_j\dot{\boldsymbol{q}}_k\right]$$

Exchanging the trace and sum operations

$$\mathcal{K} = \frac{1}{2}\sum_{i=1}^{n}\sum_{j=1}^{i}\sum_{k=1}^{i}\mathrm{Tr}\left[\left({}^0\boldsymbol{T}_{j-1}\boldsymbol{Q}_j\,{}^{j-1}\boldsymbol{T}_i\right)\boldsymbol{I}_i\left({}^0\boldsymbol{T}_{k-1}\boldsymbol{Q}_k\,{}^{k-1}\boldsymbol{T}_i\right)^T\right]\dot{\boldsymbol{q}}_j\dot{\boldsymbol{q}}_k \tag{6.41}$$

Note that the kinetic energy $\mathcal{K}$ of the manipulator is a scalar and is a function of joint position and velocity $(\boldsymbol{q}, \dot{\boldsymbol{q}})$.

6.3.4 The Potential Energy

The potential energy $\mathcal{P}_i$ of link i in a gravity field g is

$$\mathcal{P}_i = -\,m_i\,\boldsymbol{g}({}^0\bar{\boldsymbol{r}}_i) = -m_i\,\boldsymbol{g}\,{}^0\boldsymbol{T}_i\,{}^i\bar{\boldsymbol{r}}_i \tag{6.42}$$

where vectors ${}^0\bar{\boldsymbol{r}}_i$ and ${}^i\bar{\boldsymbol{r}}_i$ represent the center of mass of link i with respect to frame {0} and with respect to frame $\{i\}$, respectively, and the acceleration due to gravity $\boldsymbol{g} = \begin{bmatrix} g_x & g_y & g_z & 0\end{bmatrix}^T$ is the 4×1 gravity vector with respect to base frame {0}. The negative sign indicates that work is done on the system to raise link i against gravity. The total potential energy of the manipulator is sum of the potential energy of the links, that is,

$$\mathcal{P} = \sum_{i=1}^{n}\mathcal{P}_i = -\sum_{i=1}^{n} m_i\,\boldsymbol{g}\,{}^0\boldsymbol{T}_i\,{}^i\bar{\boldsymbol{r}}_i \tag{6.43}$$

Because ${}^0\bar{\boldsymbol{r}}_i$ is a function of q_i, the potential energy of a manipulator is descri' d by a scalar formula as a function of joint displacements $\boldsymbol{q}$.

6.3.5 Equations of Motion

The Lagrangian, $\mathcal{L} = \mathcal{K} - \mathcal{P}$, from Eqs. (6.41) and (6.43), is given by

$$\mathcal{L} = \frac{1}{2}\sum_{i=1}^{n}\sum_{j=1}^{i}\sum_{k=1}^{i}\mathrm{Tr}\left[\left({}^{0}\boldsymbol{T}_{j-1}\boldsymbol{Q}_{j}\,{}^{j-1}\boldsymbol{T}_{i}\right)\boldsymbol{I}_{i}\left({}^{0}\boldsymbol{T}_{k-1}\boldsymbol{Q}_{k}\,{}^{k-1}\boldsymbol{T}_{i}\right)^{T}\right]\dot{\boldsymbol{q}}_{j}\dot{\boldsymbol{q}}_{k} + \sum_{i=1}^{n} m_{i}\,\boldsymbol{g}\,{}^{0}\boldsymbol{T}_{i}\,{}^{i}\bar{\boldsymbol{r}}_{i} \tag{6.44}$$

According to the Lagrange-Euler dynamic formulation, the generalized torque τ_i of the actuator at joint i, to drive link i of the manipulator, is given by Eq. (6.2)

$$\tau_i = \frac{d}{dt}\left(\frac{\delta\mathcal{L}}{\delta\dot{q}_i}\right) - \frac{\delta\mathcal{L}}{\delta q_i}$$

By substituting $\mathcal{L}$ and carrying out the differentiation, the generalized torque τ_i applied to link i of n-DOF manipulator is obtained. The detailed derivations are carried out in Appendix D. The final EOM (dynamic model) is

$$\tau_i = \sum_{j=1}^{n} M_{ij}(\boldsymbol{q})\ddot{\boldsymbol{q}}_j + \sum_{j=1}^{n}\sum_{k=1}^{n} h_{ijk}\,\dot{\boldsymbol{q}}_j\dot{\boldsymbol{q}}_k + G_i \quad \text{for } i = 1, 2, \ldots, n \tag{6.45}$$

where

$$M_{ij} = \sum_{p=\max(i,j)}^{n}\mathrm{Tr}\left[\boldsymbol{d}_{pj}\boldsymbol{I}_p\boldsymbol{d}_{pi}^{T}\right] \tag{6.46}$$

$$h_{ijk} = \sum_{p=\max(i,j,k)}^{n}\mathrm{Tr}\left[\frac{\partial(\boldsymbol{d}_{pk})}{\partial q_p}\boldsymbol{I}_p\boldsymbol{d}_{pi}^{T}\right] \tag{6.47}$$

$$G_i = -\sum_{p=i}^{n} m_p\,\boldsymbol{g}\,\boldsymbol{d}_{pi}\,{}^{p}\bar{\boldsymbol{r}}_p \tag{6.48}$$

and

$$\boldsymbol{d}_{ij} = \begin{cases} {}^{0}\boldsymbol{T}_{j-1}\boldsymbol{Q}_j\,{}^{j-1}\boldsymbol{T}_i & \text{for } j \le i \\ 0 & \text{for } j > i \end{cases} \tag{6.49}$$

and

$$\frac{\partial \boldsymbol{d}_{ij}}{\partial q_k} = \begin{cases} {}^{0}\boldsymbol{T}_{j-1}\boldsymbol{Q}_j\,{}^{j-1}\boldsymbol{T}_{k-1}\boldsymbol{Q}_k\,{}^{k-1}\boldsymbol{T}_i & \text{for } i \ge k \ge j \\ {}^{0}\boldsymbol{T}_{k-1}\boldsymbol{Q}_k\,{}^{k-1}\boldsymbol{T}_{j-1}\boldsymbol{Q}_j\,{}^{j-1}\boldsymbol{T}_i & \text{for } i \ge j \ge k \\ 0 & \text{for } i < j \text{ or } i < k \end{cases} \tag{6.50}$$

Equation (6.45) is the dynamic model of the manipulator and gives a set of n nonlinear, coupled, second order ordinary differential equations for n-links of the n-DOF manipulator. These equations are the equations of motion or the dynamic equations of motion for the manipulator. Note that the mass of the payload would also contribute to the inertia and its position is continuously changing during the work cycle. This inertia is not included in the above EOM.

The physical meaning of various terms in Eqs. (6.45) to (6.49) is described as follows.

1. The coefficients of the $\ddot{q}_j$ terms in these equations represent inertia. It is known as effective inertia when acceleration of joint i causes a torque at joint i, and coupling inertia when acceleration at joint j causes a torque at joint i. In other words, the coefficient M_{ij} is related to acceleration of the joint and represent inertia loading of the actuator. In summary
   - M_{ii} = effective inertia at joint i where the driving torque τ_i acts.
   - M_{ij} = coupling inertia between joint i and joint j. It is the reaction torque $M_{ij}\ddot{q}_j$ at joint i induced by acceleration at joint j. Reverse applies equally with torque $M_{ji}\ddot{q}_j$ as torque at joint j due to acceleration of joint i.
   - Since $\text{Tr}(\boldsymbol{A}) = \text{Tr}(\boldsymbol{A}^T)$, it can be shown that $M_{ij} = M_{ji}$.
2. The coefficient h_{ijk} represents the velocity induced reaction torque at joint i, the first index. The indices j and k are related to velocities of joints j and joint k, whose dynamic interplay induces a reaction torque at joint i. In particular, a term of the form $h_{ijj}\dot{q}_j^2$ is the centrifugal force acting at joint i due to velocity at joint j and a term of the form $h_{ijk}\,\dot{q}_j\,\dot{q}_k$ is known as the Coriolis force acting at joint i due to velocities at joint j and k. In particular,
   - h_{ijk} = Coriolis force coefficient generated by the velocities of joint j and joint k and "felt" at joint i. Coriolis force acting at joint i due to velocities at joint j and joint k is a combination term of $h_{ijk}\,\dot{q}_j\,\dot{q}_k + h_{ijk}\,\dot{q}_k\,\dot{q}_j$.
   - h_{ijj} = Centrifugal force coefficient at joint i generated due to angular velocity at joint j. The centrifugal force acting at joint i due to velocity at joint j is given by $h_{ijj}\dot{q}_j^2$.
   - $h_{ijk} = h_{ikj}$ because the Coriolis force acts at joint i due to velocities of joints j and k and suffix order does not matter.
   - $h_{iii} = 0$, Coriolis force at joint i is not due to joint velocity itself
3. The terms involving gravity $\boldsymbol{g}$ represent the gravity generated moment at joint i. The coefficient G_i is the gravity loading force at joint i due to the links i to n. The gravity term is a function of the current position.
4. τ_i = generalized force applied at joint i due to motion of links.
5. $\boldsymbol{q}_i$ = joint displacement for joint i
6. $\dot{\boldsymbol{q}}_i$ = velocity of joint i
7. $\ddot{\boldsymbol{q}}_i$ = acceleration of joint i
8. The terms $\boldsymbol{d}_{ij}$ determine the rate of change of points ${}^i\boldsymbol{r}_i$ on link i, that is, the position and orientation of frame $\{i\}$ relative to the base coordinate frame as $\boldsymbol{q}_j$ changes.

In dynamic model equations [Eq. (6.45)], inertial and gravity terms are significant in manipulator control as they affect positioning accuracy and servo stability, which in turn determine the repeatability of the manipulator. The Coriolis and centrifugal forces are significant for high-speed motion of the manipulator.

The Lagrange-Euler formulation discussed above has the following characteristics:

- It is systematic and describes motion in real physical terms.
- The equations of motion obtained are analytical and compact.
- The matrix-vector form of equations (inertia matrix, centrifugal, and Coriolis force matrix, and gravitational force vector) is appealing for calculations and control system design.
- The control problem can be simplified by designing the structure of the manipulator with minimal joint couplings, that is, coefficients M_{ij} and h_{ijk} may be reduced or eliminated.
- The model is computationally intensive and is not amenable to online control.

An algorithm to get the dynamic model of any n-DOF manipulator using LE formulation is described in the next section.

6.3.6 The LE Dynamic Model Algorithm

The Lagrange-Euler formulation to derive the closed-form equations of motion (dynamic model) of a manipulator can be summarized in the form of an algorithm.

Algorithm 6.1 ➢ LE Formulation of Dynamic Equations

This algorithm carries out the complete dynamic formulation of an n-DOF manipulator that satisfies the condition for existence of closed-form geometric solutions. The various steps are

Step 1 Assign frames $\{0\}, \ldots, \{n\}$ using DH notation (Algorithm 3.1) such that frame $\{i\}$ is oriented (aligned) with principal axis of link i.

Step 2 Obtain the link transformation matrix ${}^{i-1}\boldsymbol{T}_i$ for each link and from these compute product matrices ${}^0\boldsymbol{T}_2$, ${}^0\boldsymbol{T}_3$ and so on, which are required for computing the coefficients $\boldsymbol{d}_{ij}$ and its derivatives, using

$$ {}^j\boldsymbol{T}_k = \prod_{p=j+1}^{k} {}^{p-1}\boldsymbol{T}_p \quad \text{for } j = 0, 1, 2, \ldots, n-2; \; k = j+2, \ldots, n \tag{6.51} $$

Step 3 Define $\boldsymbol{Q}_i$ for each link using Eq. (6.27) or Eq. (6.30), depending on whether the joint is revolute or prismatic.

Step 4 For each link i determine the inertia tensor I_i with respect to the frame $\{i\}$ (see Appendix C).

Step 5 Compute $\boldsymbol{d}_{ij}$ for $i, j = 1, 2, \ldots, n$ using Eq. (6.49).

Step 6 Compute the inertia coefficients M_{ij} for $i, j = 1, 2, \ldots n$ using Eq. (6.46).

Step 7 Compute the velocity coupling coefficients h_{ijk} for $i, j, k = 1, 2, \ldots, n$ using Eqs. (6.47) and (6.50).

Step 8 Compute gravity loading terms G_i for each link, $i = 1, 2, \ldots, n$ using Eq. (6.48).

Step 9 Substitute all the coefficients in Eq. (6.45) to formulate the ith equation for torque τ_i.

An example is worked out using Algorithm 6.1 to get the dynamic model of a two-link planar manipulator.

Example 6.1 ***EOM for 2-DOF planar manipulator using LE formulation***

Determine the equations of motion (dynamic model) for the 2-DOF RR-planar manipulator arm using the Lagrange-Euler formulation. Assume both links have equal length ($L_1 = L_2 = L$) and have equal mass ($m_1 = m_2 = m$). Assume further that the links are slender members with a uniform mass distribution, that is, the center of mass of each link is located at the mid point of the link.

Solution The dynamic model of the two link, 2-DOF planar manipulator was developed in Section 6.3 using a direct approach. The manipulator is shown in Fig. 6.1.

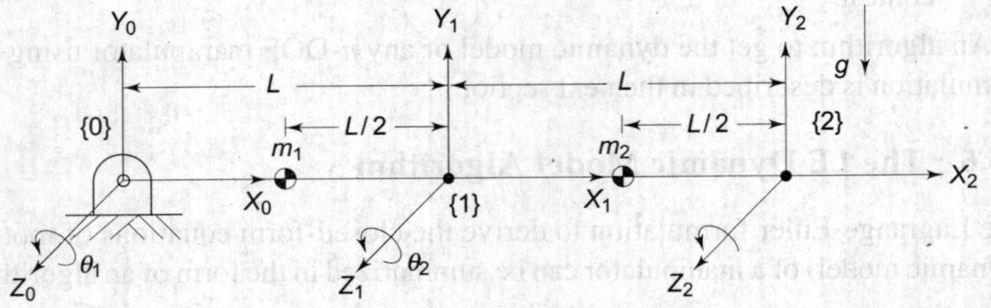

Fig. 6.3 *Frame assignments for two-link, planar RR manipulator*

The LE formulation begins with the determination of kinematic model. Applying step 1 of Algorithm 6.1, the frame assignment is carried out as shown in Fig. 6.3 and the joint-link parameters are tabulated in Table 6.1.

Table 6.1 *Joint-link parameters for RR planar manipulator*

Link i	a_i	α_i	d_i	θ_i	$C\alpha_i$	$S\alpha_i$
1	L	0	0	θ_1	1	0
2	L	0	0	θ_2	1	0

The link-transformation matrices and the required products of transformation matrices are obtained as per step 2. These are given by Eqs. (6.52) and (6.53).

$$
{}^0\boldsymbol{T}_1 = \begin{bmatrix} C_1 & -S_1 & 0 & LC_1 \\ S_1 & C_1 & 0 & LS_1 \\ 0 & 0 & 1 & 0 \\ 0 & 0 & 0 & 1 \end{bmatrix}, \quad {}^1\boldsymbol{T}_2 = \begin{bmatrix} C_2 & -S_2 & 0 & LC_2 \\ S_2 & C_2 & 0 & LS_2 \\ 0 & 0 & 1 & 0 \\ 0 & 0 & 0 & 1 \end{bmatrix} \tag{6.52}
$$

$$ {}^0\boldsymbol{T}_2 = \begin{bmatrix} C_{12} & -S_{12} & 0 & L(C_{12}+C_1) \\ S_{12} & C_{12} & 0 & L(S_{12}+S_1) \\ 0 & 0 & 1 & 0 \\ 0 & 0 & 0 & 1 \end{bmatrix} \tag{6.53} $$

where $C_{12} = \cos(\theta_1 + \theta_2)$ and $S_{12} = \sin(\theta_1 + \theta_2)$. Applying step 3, $\boldsymbol{Q}_i$ matrices for rotary joints 1 and 2, from Eq. (6.27), are

$$ \boldsymbol{Q}_1 = \boldsymbol{Q}_2 = \begin{bmatrix} 0 & -1 & 0 & 0 \\ 1 & 0 & 0 & 0 \\ 0 & 0 & 0 & 0 \\ 0 & 0 & 0 & 0 \end{bmatrix} \tag{6.54} $$

Next step (step 4) requires computation of inertia tensors. The inertia tensors $\boldsymbol{I}_1$ and $\boldsymbol{I}_2$ for two slender links of length with mass $m_1 = m_2 = m$ at the centroid of the link and $L_1 = L_2 = L$, with respect to frame $\{i\}$, $i = 1, 2$, are computed using Table C.1 in Appendix C as:

$$ \boldsymbol{I}_1 = \boldsymbol{I}_2 = \begin{bmatrix} \frac{1}{3}mL^2 & 0 & 0 & -\frac{1}{2}mL \\ 0 & 0 & 0 & 0 \\ 0 & 0 & 0 & 0 \\ -\frac{1}{2}mL & 0 & 0 & m \end{bmatrix} \tag{6.55} $$

Step 5 of Algorithm 6.1 requires the computation of matrices $\boldsymbol{d}_{ij}$, which are required to compute all other coefficients in Eqs (6.46) – (6.48). According to Eq. (6.49)

$$ \boldsymbol{d}_{ij} = \begin{cases} {}^0\boldsymbol{T}_{j-1}\,\boldsymbol{Q}_j\,{}^{j-1}\boldsymbol{T}_i & \text{for } j \le i \\ 0 & \text{for } j > i \end{cases} $$

For $i, j = 1, 2$, the four $\boldsymbol{d}_{ij}$ matries are: $\boldsymbol{d}_{11}$, $\boldsymbol{d}_{12}$, $\boldsymbol{d}_{21}$, and $\boldsymbol{d}_{22}$. These are computed now. For $i = j = 1$, $\boldsymbol{d}_{11}$ is computed as:

$$ \boldsymbol{d}_{11} = {}^0\boldsymbol{T}_0\,\boldsymbol{Q}_1\,{}^0\boldsymbol{T}_1 = \boldsymbol{Q}_1\,{}^0\boldsymbol{T}_1 = \begin{bmatrix} 0 & -1 & 0 & 0 \\ 1 & 0 & 0 & 0 \\ 0 & 0 & 0 & 0 \\ 0 & 0 & 0 & 0 \end{bmatrix} \begin{bmatrix} C_1 & -S_1 & 0 & LC_1 \\ S_1 & C_1 & 0 & LS_1 \\ 0 & 0 & 1 & 0 \\ 0 & 0 & 0 & 1 \end{bmatrix} $$

Note that ${}^0\boldsymbol{T}_0$ is the identity matrix.

or
$$ \boldsymbol{d}_{11} = \begin{bmatrix} -S_1 & -C_1 & 0 & -LS_1 \\ C_1 & -S_1 & 0 & LC_1 \\ 0 & 0 & 0 & 0 \\ 0 & 0 & 0 & 0 \end{bmatrix} \tag{6.56} $$

Similarly, $\boldsymbol{d}_{12} = 0$ (because $j > i$) (6.57)

$$\boldsymbol{d}_{21} = {}^0\boldsymbol{T}_0\,\boldsymbol{Q}_1\,{}^0\boldsymbol{T}_2 = \begin{bmatrix} -S_{12} & -C_{12} & 0 & -L(S_{12}+S_1) \\ C_{12} & -S_{12} & 0 & L(C_{12}+C_1) \\ 0 & 0 & 0 & 0 \\ 0 & 0 & 0 & 0 \end{bmatrix} \tag{6.58}$$

Finally,

$$\boldsymbol{d}_{22} = {}^0\boldsymbol{T}_1\,\boldsymbol{Q}_2\,{}^1\boldsymbol{T}_2 = \begin{bmatrix} -S_{12} & -C_{12} & 0 & -LS_{12} \\ C_{12} & -S_{12} & 0 & LC_{12} \\ 0 & 0 & 0 & 0 \\ 0 & 0 & 0 & 0 \end{bmatrix} \tag{6.59}$$

Step 6 is applied to compute the elements of inertia matrix $\boldsymbol{M}(M_{ij})$ using Eq. (6.46), that is

$$M_{ij} = \sum_{p=\max(i,j)}^{n} \mathrm{Tr}\left[\boldsymbol{d}_{pj}\boldsymbol{I}_p\boldsymbol{d}_{pi}^T\right]$$

Using the $\boldsymbol{d}_{ij}$ and $\boldsymbol{I}_i$ from Eqs (6.55) to (6.59), the effective inertia coefficients M_{11} and M_{12} are computed, first, as:

$$M_{11} = \mathrm{Tr}(\boldsymbol{d}_{11}\,\boldsymbol{I}_1\,\boldsymbol{d}_{11}^T) + \mathrm{Tr}(\boldsymbol{d}_{21}\,\boldsymbol{I}_2\,\boldsymbol{d}_{21}^T) \tag{6.60}$$

Now,

$$\boldsymbol{d}_{11}\boldsymbol{I}_1\boldsymbol{d}_{11}^T = \begin{bmatrix} -S_1 & -C_1 & 0 & -LS_1 \\ C_1 & -S_1 & 0 & LC_1 \\ 0 & 0 & 0 & 0 \\ 0 & 0 & 0 & 0 \end{bmatrix} \begin{bmatrix} \frac{1}{3}mL^2 & 0 & 0 & -\frac{1}{2}mL \\ 0 & 0 & 0 & 0 \\ 0 & 0 & 0 & 0 \\ -\frac{1}{2}mL & 0 & 0 & m \end{bmatrix} \begin{bmatrix} -S_1 & C_1 & 0 & 0 \\ -C_1 & -S_1 & 0 & 0 \\ 0 & 0 & 0 & 0 \\ -LS_1 & LC_1 & 0 & 0 \end{bmatrix}$$

or

$$\boldsymbol{d}_{11}\,\boldsymbol{I}_1\,\boldsymbol{d}_{11}^T = \begin{bmatrix} \frac{1}{3}mL^2S_1^2 & -\frac{1}{3}mL^2S_1C_1 & 0 & 0 \\ -\frac{1}{3}mL^2S_1C_1 & \frac{1}{3}mL^2C_1^2 & 0 & 0 \\ 0 & 0 & 0 & 0 \\ 0 & 0 & 0 & 0 \end{bmatrix} \tag{6.61}$$

Thus,

$$\mathrm{Tr}\,(\boldsymbol{d}_{11}\,\boldsymbol{I}_1\,\boldsymbol{d}_{11}^T) = \frac{1}{3}mL^2(S_1^2 + C_1^2) = \frac{1}{3}mL^2 \tag{6.62}$$

and

$$\boldsymbol{d}_{21}\,\boldsymbol{I}_2\,\boldsymbol{d}_{21}^T = \begin{bmatrix} -S_{12} & -C_{12} & 0 & -L(S_{12}+S_1) \\ C_{12} & -S_{12} & 0 & L(C_{12}+C_1) \\ 0 & 0 & 0 & 0 \\ 0 & 0 & 0 & 0 \end{bmatrix} \begin{bmatrix} \frac{1}{3}mL^2 & 0 & 0 & -\frac{1}{2}mL \\ 0 & 0 & 0 & 0 \\ 0 & 0 & 0 & 0 \\ -\frac{1}{2}mL & 0 & 0 & m \end{bmatrix} \boldsymbol{d}_{21}^T$$

$$= \begin{bmatrix} mL^2(\frac{1}{6}S_{12} + \frac{1}{2}S_1) & 0 & 0 & -mL(\frac{1}{2}S_{12} + S_1) \\ -mL^2(\frac{1}{6}C_{12} + \frac{1}{6}C_1) & 0 & 0 & mL(\frac{1}{2}C_{12} + C_1) \\ 0 & 0 & 0 & 0 \\ 0 & 0 & 0 & 0 \end{bmatrix}$$

$$\begin{bmatrix} -S_{12} & C_{12} & 0 & 0 \\ -C_{12} & -S_{12} & 0 & 0 \\ 0 & 0 & 0 & 0 \\ -L(S_{12}+S_1) & L(C_{12}+C_1) & 0 & 0 \end{bmatrix}$$

$$= mL^2 \begin{bmatrix} \frac{1}{3}S_{12}^2 + S_1S_{12} + S_1^2 & \begin{matrix} -(\frac{1}{2}S_{12}C_{12} + \frac{1}{2}C_1S_{12} \\ +\frac{1}{2}S_1C_{12} + S_1C_1) \end{matrix} & 0 & 0 \\ \begin{matrix} -(\frac{1}{3}S_{12}C_{12} + \frac{1}{2}C_1S_{12} \\ +\frac{1}{2}S_1C_{12} + S_1C_1) \end{matrix} & \frac{1}{3}C_{12}^2 + C_1C_{12} + C_1^2 & 0 & 0 \\ 0 & 0 & 0 & 0 \\ 0 & 0 & 0 & 0 \end{bmatrix} \tag{6.63}$$

Thus,
$$\mathrm{Tr}(\boldsymbol{d}_{21}\,\boldsymbol{I}_2\,\boldsymbol{d}_{21}^T) = mL^2\left(\frac{4}{3} + C_2\right) \tag{6.64}$$

Substituting Eq. (6.62) and Eq. (6.64) in Eq. (6.60) and simplifying gives

$$M_{11} = \frac{5}{3}mL^2 + mL^2C_2 \tag{6.65}$$

Similarly,
$$M_{22} = \mathrm{Tr}(\boldsymbol{d}_{22}\,\boldsymbol{I}_2\,\boldsymbol{d}_{22}^T) = \frac{1}{3}mL^2 \tag{6.66}$$

The coupling inertia coefficients are computed as

$$M_{12} = M_{21} = \mathrm{Tr}(\boldsymbol{d}_{22}\,\boldsymbol{I}_2\,\boldsymbol{d}_{21}^T) = \frac{1}{3}mL^2 + \frac{1}{2}mL^2C_2 \tag{6.67}$$

From the inertia coefficients obtained above, the inertia matrix for the manipulator is

$$M(\ddot{\theta}) = \begin{bmatrix} mL^2(\frac{5}{3} + C_2) & mL^2(\frac{1}{3} + \frac{1}{2}C_2) \\ mL^2(\frac{1}{3} + \frac{1}{2}C_2) & \frac{1}{3}mL^2 \end{bmatrix} \tag{6.68}$$

In step 7, the Coriolis and centrifugal force coefficients (or velocity coupling coefficients), h_{ijk} for $i, j, k = 1, 2$ are obtained using Eq. (6.47), that is

$$h_{ijk} = \sum_{p=\max(i,j,k)}^{n} \mathrm{Tr}\left[\frac{\partial(\boldsymbol{d}_{pk})}{\partial \boldsymbol{q}_p}\boldsymbol{I}_p\,\boldsymbol{d}_{pi}^T\right]$$

From Eqs. (6.50) and (6.55), the centrifugal acceleration coefficients are

$$h_{111} = 0$$

$$h_{122} = \mathrm{Tr}\left[\frac{\partial(\boldsymbol{d}_{22})}{\partial \boldsymbol{q}_2}\boldsymbol{I}_2\,\boldsymbol{d}_{21}^T\right]$$

$$= \mathrm{Tr}\,({}^0\boldsymbol{T}_1\,\boldsymbol{Q}_2\,{}^1\boldsymbol{T}_1\,\boldsymbol{Q}_2\,{}^1\boldsymbol{T}_2\,\boldsymbol{I}_2\,\boldsymbol{d}_{21}^T) = -\frac{1}{2}mL^2S_2 \tag{6.69}$$

$$h_{211} = \mathrm{Tr}\,({}^0\boldsymbol{T}_0\,\boldsymbol{Q}_1\,{}^0\boldsymbol{T}_0\,\boldsymbol{Q}_1\,{}^0\boldsymbol{T}_2\,\boldsymbol{I}_2\,\boldsymbol{d}_{22}^T) = \frac{1}{2}mL^2S_2$$

$$h_{222} = 0$$

and the Coriolis acceleration coefficients are

$$h_{112} = h_{121} = \mathrm{Tr}\,({}^0\boldsymbol{T}_0\, \boldsymbol{Q}_1\, {}^0\boldsymbol{T}_1\, \boldsymbol{Q}_2\, {}^1\boldsymbol{T}_2\, \boldsymbol{I}_2\, \boldsymbol{d}_{21}^T) = -\frac{1}{2} mL^2 S_2$$
$$h_{212} = h_{221} = \mathrm{Tr}\,({}^0\boldsymbol{T}_0\, \boldsymbol{Q}_1\, {}^0\boldsymbol{T}_1\, \boldsymbol{Q}_2\, {}^1\boldsymbol{T}_2\, \boldsymbol{I}_2\, \boldsymbol{d}_{22}^T) = 0 \tag{6.70}$$

The Coriolis and centrifugal coefficient terms are computed using the series summation.

For $i = 1$, $H_1 = \sum_{j=1}^{2}\sum_{k=1}^{2} h_{1jk}\dot{\theta}_j\dot{\theta}_k = h_{111}\dot{\theta}_1^2 + h_{112}\dot{\theta}_1\dot{\theta}_2 + h_{121}\dot{\theta}_1\dot{\theta}_2 + h_{122}\dot{\theta}_2^2$

$$= -mL^2 S_2\dot{\theta}_1\dot{\theta}_2 - \frac{1}{2} mL^2 S_2\dot{\theta}_2^2 \tag{6.71}$$

and for $i = 2$, $H_2 = \sum_{j=1}^{2}\sum_{k=1}^{2} h_{2jk}\dot{\theta}_j\dot{\theta}_k = h_{211}\dot{\theta}_1^2 + h_{212}\dot{\theta}_1\dot{\theta}_2 + h_{221}\dot{\theta}_1\dot{\theta}_2 + h_{222}\dot{\theta}_2^2$

$$= \frac{1}{2} mL^2 S_2\dot{\theta}_1^2 \tag{6.72}$$

Thus, the Coriolis and centrifugal coefficient matrix $\boldsymbol{H}$ is, therefore,

$$\boldsymbol{H}(\theta,\dot{\theta}) = \begin{bmatrix} H_1 \\ H_2 \end{bmatrix} = \begin{bmatrix} -mL^2 S_2\dot{\theta}_1\dot{\theta}_2 - \frac{1}{2} mL^2 S_2\dot{\theta}_2^2 \\ \frac{1}{2} mL^2 S_2\dot{\theta}_1^2 \end{bmatrix} \tag{6.73}$$

Step 8 is applied to compute the gravity loading at the two joints. From Eqs. (6.48) and (6.56) – (6.59),

$$G_1 = -(m\boldsymbol{g}\,\boldsymbol{d}_{11}\, {}^1\bar{\boldsymbol{r}}_1 + m\boldsymbol{g}\,\boldsymbol{d}_{21}\, {}^2\bar{\boldsymbol{r}}_2) \tag{6.74}$$

The mass of the links has been assumed to be concentrated at the centroid of the links. The centroid is located from the origins of frame {1} and frame {2}

$${}^1\bar{\boldsymbol{r}}_1 = {}^2\bar{\boldsymbol{r}}_2 = \begin{bmatrix} -\frac{1}{2}L & 0 & 0 & 1 \end{bmatrix}^T \tag{6.75}$$

and the gravity is in the negative y-direction giving $\boldsymbol{g} = [0 \quad -g \quad 0 \quad 0]$
Hence,

$$m\boldsymbol{g}\boldsymbol{d}_{11}\, {}^1\bar{\boldsymbol{r}}_1 = m[0 \quad -g \quad 0 \quad 0] \begin{bmatrix} -S_1 & -C_1 & 0 & -LS_1 \\ C_1 & -S_1 & 0 & LC_1 \\ 0 & 0 & 0 & 0 \\ 0 & 0 & 0 & 0 \end{bmatrix} \begin{bmatrix} -\frac{1}{2}L \\ 0 \\ 0 \\ 1 \end{bmatrix}$$

$$= -\frac{1}{2} mgLC_1 \tag{6.76}$$

$$\boldsymbol{mgd}_{12}{}^2\bar{\boldsymbol{r}}_2 = m[0 \ -g \ 0 \ 0]\begin{bmatrix} -S_{12} & -C_{12} & 0 & -L(S_{12}+S_1) \\ C_{12} & -S_{12} & 0 & L(C_{12}+C_1) \\ 0 & 0 & 0 & 0 \\ 0 & 0 & 0 & 0 \end{bmatrix}\begin{bmatrix} -\frac{1}{2}L \\ 0 \\ 0 \\ 1 \end{bmatrix}$$

$$= -mgL\left(\frac{1}{2}C_{12} + C_1\right) \tag{6.77}$$

Finally, $$G_1 = \frac{1}{2}mgLC_{12} + \frac{3}{2}mgLC_1 \tag{6.78}$$

Similarly, $$G_2 = -mg\,\boldsymbol{d}_{22}{}^2\bar{\boldsymbol{r}}_2 = \frac{1}{2}mgLC_{12} \tag{6.79}$$

The complete dynamic model is obtained in the final step, step 9, by substituting the above results, Eqs (6.68), (6.73), (6.78), and (6.79) into Eq. (6.45). The equations of motion in the vector-matrix form are

$$\begin{bmatrix} \tau_1 \\ \tau_2 \end{bmatrix} = \begin{bmatrix} mL^2(\frac{5}{3}+C_2) & mL^2(\frac{1}{3}+\frac{1}{2}C_2) \\ mL^2(\frac{1}{3}+\frac{1}{2}C_2) & \frac{1}{3}mL^2 \end{bmatrix}\begin{bmatrix} \ddot{\theta}_1 \\ \ddot{\theta}_2 \end{bmatrix}$$

$$+\begin{bmatrix} -mL^2S_2\dot{\theta}_1\dot{\theta}_2 - \frac{1}{2}mL^2S_2\dot{\theta}_2^2 \\ \frac{1}{2}mL^2S_2\dot{\theta}_1^2 \end{bmatrix} + \begin{bmatrix} \frac{1}{2}mgL\,C_{12} + \frac{3}{2}mgLC_1 \\ \frac{1}{2}mgL\,C_{12} \end{bmatrix} \tag{6.80}$$

The reader should verify that the dynamic model in Eq. (6.21) is identical to dynamic model in Eq. (6.80) for $m_1 = m_2 = m$ and $L_1 = L_2 = L$. Equations (6.80) are rather complex for a simple manipulator. Obviously, the dynamic equations for a 6-DOF manipulator would be quite complex.

6.4 NEWTON-EULER FORMULATION

The second approach to dynamic modeling of robotic manipulators is discussed now. The Newton-Euler (NE) formulation is based on the Newton's second law and d'Alembert principle. The balance of all forces acting on a link of the manipulator leads to a set of equations, whose structure allows a recursive solution. A forward recursion, which describes the kinematic relations of a moving coordinate frame, is performed for propagating velocities and accelerations, followed by a backward recursion for propagating forces and moments. Initially, it is assumed that the position, velocity, and acceleration of each joint, $(\boldsymbol{q}, \dot{\boldsymbol{q}}, \ddot{\boldsymbol{q}})$ are known. The joint torques required to cause these time-dependent motions to realize a trajectory are computed using the recursive NE dynamic equations of motion. To understand the Newton-Euler formulation, some basic concepts of kinetics are reviewed first.

6.4.1 Newton's Equation, Euler's Equation

The robotic manipulator is an open kinematic chain of rigid links connected with 1-DOF joints. The mass distribution of a link is completely characterized by the location of the centre of mass and the inertia tensor of the link. The forces or torques required for moving the links, and accelerating or decelerating them, are a function of the mass distribution and inertia tensor of the links.

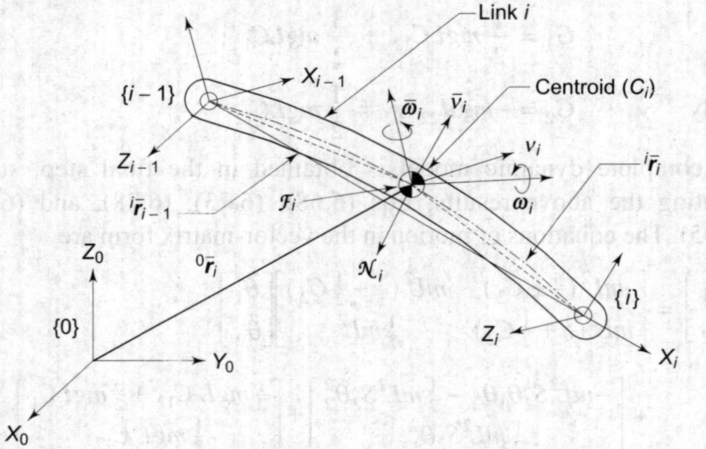

Fig. 6.4 *The geometry and kinematics of link i for Newton-Euler formulation*

Consider the rigid link i of the manipulator kinematic chain connected between joints i and $(i + 1)$. The frames at the two ends are frame $\{i–1\}$ and frame $\{i\}$ as shown in Fig. 6.4, with reference to inertial frame $\{0\}$. The centre of mass of the link is C_i and the parameters listed below characterize the geometry and kinematics:

${}^{i-1}\bar{r}_i$ = position vector of C_i from frame $\{i–1\}$,
${}^{i}\bar{r}_i$ = position vector of C_i from frame $\{i\}$,
m_i = mass of link,
$\boldsymbol{I}_i$ = inertia tensor of link with respect to a frame $\{C_i\}$, whose origin is located at the center of mass of the link C_i and orientation of frame $\{C_i\}$ is same as the orientation of the base frame $\{0\}$,
$\bar{v}_i$ = linear velocity of center of mass of link,
$\dot{\bar{v}}_i$ = linear acceleration of center of mass,
ω_i = angular velocity of link,
$\dot{\omega}_i$ = angular acceleration of link,
$\mathcal{F}_i$ = total external force acting at the centre of mass of link,
$\mathcal{N}_i$ = total external moment acting on link at the centre of mass of link.

The translational motion of the link in terms of the balance of forces is described by the *Newton's equation*. The force $\mathcal{F}_i$, acting at the centre of mass of the link is given by

$$\mathcal{F}_i = m_i \dot{\bar{v}}_i \tag{6.81}$$

where $\bar{v}_i$ is the linear acceleration of the link.

The Euler equation for the rotational motion of the link describes the moment balance about the centre of mass of the link. The angular velocity of the link ω_i and the moment of inertia tensor $\boldsymbol{I}_i$ relate to the total moment $\mathcal{N}_i$ acting on link as

$$\mathcal{N}_i = \frac{d}{dt}(\boldsymbol{I}_i\omega_i) = \boldsymbol{I}_i\dot{\omega}_i + \omega_i \times (\boldsymbol{I}_i\omega_i) \tag{6.82}$$

where the second term is the gyroscopic torque induced by the dependence of I_i on link's orientation with respect to base frame.

Equations (6.81) and (6.82) are the Newton-Euler equations that are recursively applied to compute the inertia force and torque acting at the center of mass of each link of the manipulator.

6.4.2 Kinematics of Links

To compute the inertial forces and torques acting on the links it is necessary to compute the linear and rotational velocity and acceleration of the centre of mass of each link of the manipulator at any given instant. The kinematic relationship of the moving-rotating links with respect to the base coordinate system is first described as a set of mathematical equations.

Two adjacent links *i* and (*i*–1), forming the joint *i*, are shown in Fig. 6.5. The orthogonal coordinate frames are established as described in Section 3.4 with

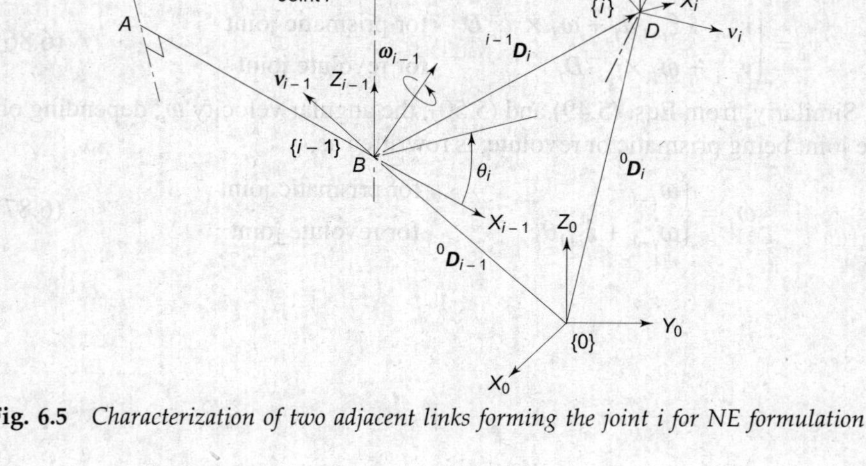

Fig. 6.5 *Characterization of two adjacent links forming the joint i for NE formulation*

frame{0} as the base coordinate frame; frame $\{i-1\}$ at joint i attached to link $(i-1)$ with point B as the origin, and frame $\{i\}$ at joint $(i+1)$ attached to link i with origin D. The origin D and origin B are located by position vectors ${}^0\boldsymbol{D}_i$ and ${}^0\boldsymbol{D}_{i-1}$, with respect to the base frame {0}, respectively, and the position vector ${}^{i-1}\boldsymbol{D}_i$ locates the origin D from the origin B with respect to the base coordinate system.

Let the linear and angular velocities of frame $\{i-1\}$, with respect to the base frame {0}, be $\boldsymbol{v}_{i-1}$ and ω_{i-1}, respectively, and ω_i be the angular velocity of frame $\{i\}$ with respect to the base frame {0}. Let ${}^{i-1}\omega_i$ be the relative angular velocity of frame $\{i\}$ with respect to frame $\{i-1\}$ referred to base frame {0}. *Note that superscript '0' is omitted for quantities expressed in the base frame* {0}.

The linear and angular velocities of frame $\{i\}$ (or origin D), as a function of translational and rotational velocities of link $\{i-1\}$, with respect to base frame {0} were determined in Chapter 5. [Eqs. (5.14) and (5.51). The linear velocity $\boldsymbol{v}_i$ and angular velocity ω_i of the frame $\{i\}$ with respect to base frame {0} are, respectively,

$$\boldsymbol{v}_i = \boldsymbol{v}_{i-1} + {}^{i-1}\dot{\boldsymbol{D}}_i + \omega_{i-1} \times {}^{i-1}\boldsymbol{D}_i \tag{6.83}$$

$$\omega_i = \omega_{i-1} + {}^{i-1}\omega_i \tag{6.84}$$

where ${}^{i-1}\dot{\boldsymbol{D}}_i$ denotes the velocity of frame $\{i\}$ with respect to origin of frame $\{i-1\}$ expressed in base frame {0}.

Recalling the behavior of prismatic and revolute joints discussed in Chapter 5, the linear and angular velocities are computed. If joint i is prismatic, link i travels in the direction of axes z_{i-1} with a linear joint velocity $\dot{d}_i$ relative to link $(i-1)$, that is, ${}^{i-1}\dot{\boldsymbol{D}}_i = \hat{z}_{i-1}\dot{d}_i$, with $\hat{z}_{i-1}$ as the unit vector along z_{i-1}-axis. The angular velocity of prismatic link i is same as that of link $(i-1)$, that is, $\omega_{i-1} = \omega_i$. Similarly, if the joint is revolute, the link i has an angular motion about the z_{i-1} axis with the angular velocity $\hat{z}_{i-1}\dot{\theta}_i$ and zero linear velocity, that is ${}^{i-1}\dot{\boldsymbol{D}}_i = 0$.

$$ {}^{i-1}\omega_i = \begin{cases} 0 & \text{for prismatic joint} \\ \hat{z}_{i-1}\dot{\omega}_i & \text{for revolute joint} \end{cases} \tag{6.85}$$

Substituting these linear and angular velocities in Eq. (6.83), the linear velocity of link i with respect to reference frame is

$$\boldsymbol{v}_i = \begin{cases} \boldsymbol{v}_{i-1} + \hat{z}_{i-1}\dot{\boldsymbol{d}}_i + \omega_i \times {}^{i-1}\boldsymbol{D}_i & \text{for prismatic joint} \\ \boldsymbol{v}_{i-1} + \omega_i \times {}^{i-1}\boldsymbol{D}_i & \text{for revolute joint} \end{cases} \tag{6.86}$$

Similarly, from Eqs. (5.49) and (5.50), the angular velocity ω_i, depending on the joint being prismatic or revolute, is rewritten as

$$\omega_i = \begin{cases} \omega_{i-1} & \text{for prismatic joint} \\ \omega_{i-1} + \hat{z}_{i-1}\dot{\theta}_i & \text{for revolute joint} \end{cases} \tag{6.87}$$

Remember that the linear velocity is associated with a point and angular velocity is associated with a body. Hence, $\boldsymbol{v}_i$, the linear velocity of link i is the velocity of the origin of the frame $\{i\}$, and the angular velocity of the frame $\{i\}$, ω_i is the angular velocity of the whole link i.

6.4.3 Link Acceleration

The linear acceleration of link i (i.e. frame $\{i\}$) is obtained by differentiating Eq. (6.86) with respect to time. The required complex derivational details are not given here. Only qualitative arrangements are advanced. For a prismatic joint the following terms contribute to the linear acceleration of link i

1. $\dot{\boldsymbol{v}}_{i-1}$
2. $\hat{z}_{i-1}\, \ddot{\boldsymbol{d}}_i$
3. Angular velocity ω_{i-1} interacting with linear velocity $\hat{z}_{i-1}\, \dot{\boldsymbol{d}}_i$ contributes $\omega_{i-1} \times \left(\dot{z}_{i-1}\, \dot{\boldsymbol{d}}_i\right)$.
4. Angular acceleration $\dot{\omega}_i$ interacting with distance ${}^{i-1}\boldsymbol{D}_i$ contributes $\dot{\omega}_i \times {}^{i-1}\boldsymbol{D}_i$.
5. Angular velocity $\dot{\omega}_i$ interacting with linear velocity $\hat{z}_{i-1}\dot{\boldsymbol{d}}_i$ contributes $\dot{\omega}_i \times \left(\hat{z}_{i-1}\dot{\boldsymbol{d}}_i\right)$.
6. Angular velocity ω_{i-1} interact with ${}^{i-1}\boldsymbol{D}_i$ to cause linear velocity $\left(\dot{\omega}_i \times {}^{i-1}\boldsymbol{D}_i\right)$ which interacting with angular velocity ω_i cause the linear acceleration
$$\omega_i \times \left(\dot{\omega}_i \times {}^{i-1}\boldsymbol{D}_i\right)$$

Note that all quantities are referred to frame $\{0\}$ but superscript '0' is omitted for symbolic simplification. Hence, the linear acceleration of a link prismatically jointed to the preceding link is given by

$$\begin{aligned}\dot{\boldsymbol{v}}_i = \dot{\boldsymbol{v}}_{i-1} &+ \hat{z}_{i-1}\ddot{\boldsymbol{d}}_i + \omega_{i-1} \times (\hat{z}_{i-1}\dot{\boldsymbol{d}}_i) + \dot{\omega}_i \times {}^{i-1}\boldsymbol{D}_i \\ &+ \omega_i \times \hat{z}_{i-1}\dot{\boldsymbol{d}}_i + \omega_i \times \left(\omega_{i-1} \times {}^{i-1}\boldsymbol{D}_i\right)\end{aligned} \tag{6.88}$$

For prismatic joint $\omega_{i-1} = \omega_i$, Eq. (6.88) is simplified to

$$\begin{aligned}\dot{\boldsymbol{v}}_i = \dot{\boldsymbol{v}}_{i-1} &+ \hat{z}_{i-1}\ddot{\boldsymbol{d}}_i + 2\omega_i \times (\hat{z}_{i-1}\dot{\boldsymbol{d}}_i) \\ &+ \dot{\omega}_i \times {}^{i-1}\boldsymbol{D}_i + \omega_i \times \left(\omega_i \times {}^{i-1}\boldsymbol{D}_i\right)\end{aligned} \tag{6.89}$$

For the revolute joint, differentiating second part of Eq. (6.86) and simplifying gives the linear acceleration of revolute link as

$$\dot{v}_i = \dot{v}_{i-1} + \dot{\omega}_i \times {}^{i-1}D_i + \omega_i \times \left(\omega_i \times {}^{i-1}D_i\right) \tag{6.90}$$

Combining Eqs. (6.89) and (6.90), the linear acceleration of link i is given by

$$\dot{v}_i = \begin{cases} \dot{v}_{i-1} + \hat{z}_{i-1}\ddot{d}_i + 2\omega_i \times (\hat{z}_{i-1}\dot{d}_i) + & \text{for prismatic joint} \\ \dot{\omega}_i \times {}^{i-1}D_i + \omega_i \times \left(\omega_i \times {}^{i-1}D_i\right) & \\ v_{i-1} + \omega_i \times {}^{i-1}D_i + \omega_i \times \left(\omega_i \times {}^{i-1}D_i\right) & \text{for revolute joint} \end{cases} \tag{6.91}$$

Similarly, Eq.(6.87) is differentiated with respect to time to get the angular accelerations for prismatic and revolute links. The resulting angular accelerations are:

$$\dot{\omega}_i = \begin{cases} \dot{\omega}_{i-1} & \text{for prismatic joint} \\ \dot{\omega}_{i-1} + \hat{z}_{i-1}\ddot{\theta}_i + \omega_{i-1} \times \hat{z}_{i-1}\dot{\theta}_i & \text{for revolute joint} \end{cases} \tag{6.92}$$

The linear velocity and acceleration of the center of mass of link i are, respectively,

$$\bar{v}_i = v_i + \omega_i \times {}^i\bar{r}_i \tag{6.93}$$

and

$$\dot{\bar{v}}_i = \dot{v}_i + \dot{\omega}_i \times {}^i\bar{r}_i + \omega_i \times \left(\omega_i \times {}^i\bar{r}_i\right) \tag{6.94}$$

where ${}^i\bar{r}_i$ is the position vector of centre of mass of link i from frame $\{i\}$ with respect to base frame $\{0\}$.

6.4.4 Recursive Newton-Euler Formulation

The recursive formulation of dynamic equations based on NE equations is now carried out from the above kinematic information of each link. In the recursive formulation the serial open kinematic chain structure of a manipulator is exploited. The NE formulation requires two passes over the links of the manipulator, one for computing the velocities and accelerations of the links and, second, to compute joint forces and torques, as shown in Fig. 6.6.

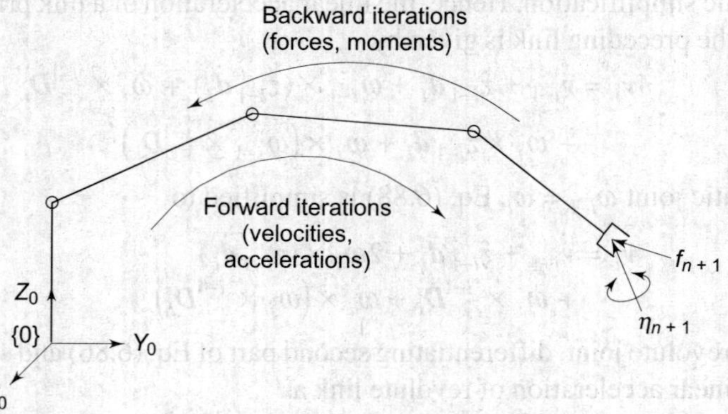

Fig. 6.6 *Two-pass recursive Newton-Euler formulation of dynamic equations*

The *forward iteration* or *outward iteration* is carried out to compute the velocities and accelerations of each link recursively, starting at the base and propagating forward to the end-effector. The boundary conditions are base velocity, linear and angular, which are zero, and the boundary acceleration is the gravitational acceleration.

In the *backward* or *inward iteration*, the forces and moments acting on each link are computed using the Newton's and Euler's equations.

First, the force and moment that must be applied to the centre of mass of each link are computed, and from these, the force and moment that must be applied at each joint are computed starting from the end-effector and moving inward (working backward) to the base. The end-effector force f_{n+1} and moment η_{n+1} exerted by the environment or the grasped object provide the necessary boundary conditions.

6.4.5 Forward Iteration

In the equations for velocity and acceleration, Eqs. (6.86), (6.87), (6.91), and (6.92), all the physical parameters are referenced to the base frame {0}. Because the parameters referred to the base frame change, as the manipulator is moving, the computations are complex. The computations are much simplified by referring all velocities, accelerations, inertia tensors, and location of centre of mass of each link to their own link-coordinate frames. The reference to link frame $\{i\}$ results in constant inertia tensor $\boldsymbol{I}_i$ and constant vectors appear in the equations, simplifying the numerical computations. The change of frames is accomplished by using the 3 × 3 rotational transformation matrices, which give the kinematic relationship between the links.

The 3 × 3 rotation matrix ${}^{i-1}\boldsymbol{R}_i$ transforms rotation of any vector with reference to frame $\{i\}$ to the frame $\{i-1\}$. The rotation matrix ${}^{i-1}\boldsymbol{R}_i$ is the upper left 3 × 3 sub-matrix of ${}^{i-1}\boldsymbol{T}_i$. That is,

$$
{}^{i-1}\boldsymbol{T}_i = \left[\begin{array}{ccc|c} C\theta_i & -S\theta_i C\alpha_i & S\theta_i S\alpha_i & a_i C\theta_i \\ S\theta_i & C\theta_i C\alpha_i & -C\theta_i S\alpha_i & a_i S\theta_i \\ 0 & S\alpha_i & C\alpha_i & d_i \\ \hline 0 & 0 & 0 & 1 \end{array}\right] = \left[\begin{array}{c|c} {}^{i-1}\boldsymbol{R}_i & {}^{i-1}\boldsymbol{D}_i \\ \hline 0 \;\; 0 \;\; 0 & 1 \end{array}\right] \tag{6.95}
$$

with

$$
{}^{i-1}\boldsymbol{R}_i = \begin{bmatrix} C\theta_i & -S\theta_i C\alpha_i & S\theta_i S\alpha_i \\ S\theta_i & C\theta_i C\alpha_i & -C\theta_i S\alpha_i \\ 0 & S\alpha_i & C\alpha_i \end{bmatrix}; \text{ and } {}^{i-1}\boldsymbol{D}_i = \begin{bmatrix} a_i C\theta_i \\ a_i S\theta_i \\ d_i \end{bmatrix} \tag{6.96}
$$

It has been shown earlier, [Eq. (2.15)] that

$$
({}^{i-1}\boldsymbol{R}_i)^{-1} = {}^{i}\boldsymbol{R}_{i-1} = ({}^{i-1}\boldsymbol{R}_i)^T \tag{6.97}
$$

Thus, using the rotational transformation matrices, it is possible to express the vectors related to link i with respect to link frame $\{i\}$ instead of base frame $\{0\}$*. This gives constant vectors instead of variable vectors, simplifying the numerical computations. Applying this the velocity and acceleration relationships of Eqs. (6.87), (6.91), and (6.94) are modified for the outward iteration as below. All the variables in these equations are now referred to their own frame.

$$ {}^{i}\omega_i = \begin{cases} {}^{i}\boldsymbol{R}_{i-1}\,{}^{i-1}\omega_{i-1} & \text{for prismatic joint} \\ {}^{i}\boldsymbol{R}_{i-1}[\,{}^{i-1}\omega_{i-1} + \hat{z}_0\dot{\theta}_i\,] & \text{for revolute joint} \end{cases} \tag{6.98} $$

$$ {}^{i}\dot{\omega}_i = \begin{cases} {}^{i}\boldsymbol{R}_{i-1}\,{}^{i-1}\dot{\omega}_{i-1} & \text{for prismatic joint} \\ {}^{i}\boldsymbol{R}_{i-1}[\,{}^{i-1}\dot{\omega}_{i-1} + \hat{z}_0\ddot{\theta}_i + {}^{i-1}\omega_{i-1}\times\hat{z}_0\dot{\theta}_i\,] & \text{for revolute joint} \end{cases} \tag{6.99} $$

---

* *Transforming vectors to link's own frame*

This is indicated here for some typical terms in the expressions of Eqs. (6.98), (6.99) and (6.100).

(1) ω_i, ω_{i-1} terms:

$$ \omega_i = {}^{i}\omega_i,\ {}^{i}R_0\,\omega_{i-1} = {}^{i}R_{i-1}\,{}^{i-1}R_0\omega_{i-1} = {}^{i}R_{i-1}{}^{i-1}\omega_{i-1} $$

These results apply to linear velocities $\boldsymbol{v}_i$ and $\boldsymbol{v}_{i-1}$ and also to angular and linear acceleration $\dot{\omega}_i$, $\dot{\omega}_{i-1}$; $\dot{\boldsymbol{v}}_i$ and $\dot{\boldsymbol{v}}_{i-1}$.

(2) $\hat{z}_{i-1}\dot{d}_i$, $\hat{z}_{i-1}\ddot{d}_i$, $\hat{z}_{i-1}\dot{\theta}_i$, $\hat{z}_{i-1}\ddot{\theta}_i$ terms:

These terms are of one kind, therefore

$$ {}^{i}\boldsymbol{R}_0\hat{z}_{i-1}\dot{d}_i = {}^{i}\boldsymbol{R}_{i-1}\,{}^{i-1}\boldsymbol{R}_0\hat{z}_{i-1}\dot{d}_i = {}^{i}\boldsymbol{R}_{i-1}\hat{z}_0\dot{d}_i $$

where $\hat{z}_0$ is $\hat{z}_{i-1}$-axis referred to its own frame. These apply for all i and $\hat{z}_0 = [0 \quad 0 \quad 1]^T$ always.

(3) (i) $\omega_i \times (\hat{z}_{i-1}\dot{d}_i)$ term:

$$ {}^{i}\boldsymbol{R}_0\,\omega_i \times ({}^{i}\boldsymbol{R}_0\hat{z}_{i-1}\dot{d}_i) = {}^{i}\omega_i \times ({}^{i}\boldsymbol{R}_{i-1}\hat{z}_0 d_i) $$

(ii) $\omega_i \times {}^{i-1}\boldsymbol{D}_i$ term:

$$ {}^{i}\boldsymbol{R}_0\,\omega_i \times ({}^{i}\boldsymbol{R}_0\,{}^{i-1}\boldsymbol{D}_i) = {}^{i}\omega_i \times ({}^{i}\boldsymbol{R}_{i-1}\,{}^{i-1}\boldsymbol{D}_i) $$

(iii) $\omega_i \times (\omega_i \times {}^{i-1}\boldsymbol{D}_i)$:

$$ {}^{i}\boldsymbol{R}_0\,\omega_i \times \left[{}^{i}\boldsymbol{R}_0\,\omega_i \times ({}^{i}\boldsymbol{R}_0\,{}^{i-1}\boldsymbol{D}_i)\right] = {}^{i}\omega_i \times {}^{i}\omega_i \times ({}^{i-1}\boldsymbol{R}_0\,{}^{i-1}\boldsymbol{D}_i) $$

$$
{}^{i}\dot{\boldsymbol{v}}_i = \begin{cases} {}^{i}\boldsymbol{R}_{i-1}[{}^{i-1}\dot{\boldsymbol{v}}_{i-1} + \hat{z}_0\ddot{d}_i] + 2\,{}^{i}\omega_i \times ({}^{i}\boldsymbol{R}_{i-1}\,\hat{z}_0\dot{d}_i) \\ +\,{}^{i}\omega_i \times ({}^{i}\boldsymbol{R}_0\,{}^{i-1}\boldsymbol{D}_i) + {}^{i}\omega_i \times \left[{}^{i}\omega_i \times ({}^{i}\boldsymbol{R}_0\,{}^{i-1}\boldsymbol{D}_i)\right] & \text{for prismatic joint} \\ {}^{i}\boldsymbol{R}_{i-1}\,{}^{i-1}\dot{\boldsymbol{v}}_{i-1} + {}^{i}\dot{\omega}_i \times ({}^{i}\boldsymbol{R}_0\,{}^{i-1}\boldsymbol{D}_i) + \\ {}^{i}\omega_i \times \left[{}^{i}\omega_i \times ({}^{i}\boldsymbol{R}_0\,{}^{i-1}\boldsymbol{D}_i)\right] & \text{for revolute joint} \end{cases} \tag{6.100}
$$

The linear acceleration of the center of mass is given by

$$
{}^{i}\dot{\bar{\boldsymbol{v}}}_i = {}^{i}\dot{\boldsymbol{v}}_i + {}^{i}\omega_i \times {}^{i}\bar{\boldsymbol{r}}_i + {}^{i}\omega_i \times \left({}^{i}\omega_i \times {}^{i}\bar{\boldsymbol{r}}_i\right) \tag{6.101}
$$

where $\hat{z}_0 = [0 \quad 0 \quad 1]^T$ is the unit vector in the direction of z_0-axis and the matrix products $({}^{i}\boldsymbol{R}_0\,{}^{i-1}\boldsymbol{D}_i)$ in Eq. (6.100) are simplified, using Eq. (6.96) and (6.97), to yield

$$
{}^{i}\boldsymbol{R}_{i-1}\,{}^{i-1}\boldsymbol{D}_i = \begin{bmatrix} C_i & S_i & 0 \\ -S_i C\alpha_i & C_i C\alpha_i & S\alpha_i \\ S_i S\alpha_i & -C_i C\alpha_i & C\alpha_i \end{bmatrix} \begin{bmatrix} a_i C_i \\ a_i S_i \\ d_i \end{bmatrix} = \begin{bmatrix} a_i \\ d_i S\alpha_i \\ d_i C\alpha_i \end{bmatrix} \tag{6.102}
$$

with $C_i = C\theta_i = \cos\theta_i$ and $S_i = S\theta_i = \sin\theta_i$.

Equations (6.98) to (6.101) give forward Newton-Euler equations. The forward iteration starts at the base, that is $i = 0$. The initial conditions for forward NE recursion for a fixed (inertial) base of manipulator are

$$
\begin{aligned} {}^{0}\boldsymbol{v}_0 &= 0 \\ {}^{0}\dot{\boldsymbol{v}}_0 &= 0 \\ {}^{0}\omega_0 &= 0 \\ {}^{0}\dot{\omega}_0 &= 0 \end{aligned} \tag{6.103}
$$

The effect of gravity can be included by considering the linear accelerations of base frame as

$$
{}^{0}\dot{\boldsymbol{v}}_0 = \boldsymbol{g} = \begin{bmatrix} g_x & g_y & g_z \end{bmatrix}^T \tag{6.104}
$$

The gravity loading on each link is then automatically propagated through the links by the forward recursion.

6.4.6 Backward Iteration

Once the velocities and accelerations of each link are known, the forces and moments acting on each link are computed by starting at the end-effector and working backwards.

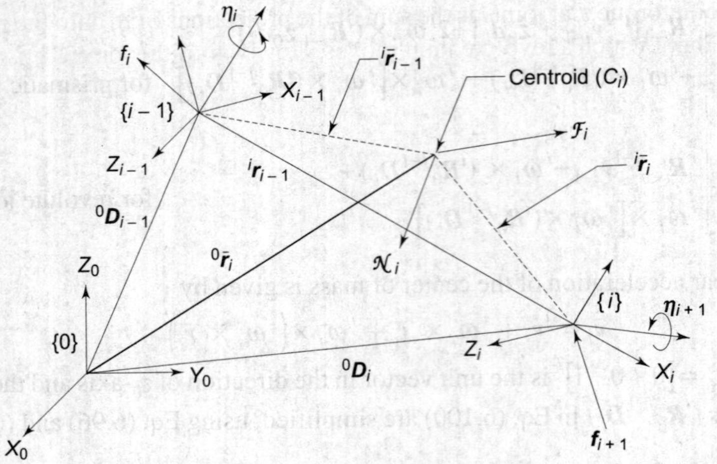

Fig. 6.7 *Forces and moments acting on a link and its centroid*

Let f_i and η_i be the force and moment exerted by link $(i–1)$ on link i at the origin of frame $\{i–1\}$ with respect to base frame $\{0\}$ as shown in Fig. 6.7. Further, link i exerts force f_{i+1} and moment η_{i+1} on link $(i+1)$ at the origin of frame $\{i\}$. The reaction force f_{i+1} and moment η_{i+1} act on link i in the opposite direction. This force and moment would, therefore, have a negative sign attached to them. The total external force $\mathcal{F}_i$ and $\mathcal{N}_i$ acting on link i are given by Eqs. (6.81) and (6.82) with these observations, the force and moment balance equations for link i from Fig. 6.7 are

$$\mathcal{F}_i = f_i - f_{i+1} \tag{6.105}$$

$$\mathcal{N}_i = \eta_i - \eta_{i+1} + ({}^0D_{i-1} - {}^0\bar{r}_i) \times f_i - ({}^0D_i - {}^0\bar{r}_i) \times f_{i+1} \tag{6.106}$$

Substituting f_i from Eq. (6.105) in Eq. (6.106)

$$\mathcal{N}_i = \eta_i - \eta_{i+1} + ({}^0D_{i-1} - {}^0\bar{r}_i) \times (\mathcal{F}_i + f_{i+1}) - ({}^0D_i - {}^0\bar{r}_i) \times f_{i+1} \tag{6.107}$$

or $$\mathcal{N}_i = \eta_i - \eta_{i+1} + ({}^0D_{i-1} - {}^0D_i) \times f_{i+1} + ({}^0D_{i-1} - {}^0\bar{r}_i) \times \mathcal{F}_i \tag{6.108}$$

From Fig. 6.7, the position vectors are related as

$$({}^0D_{i-1} - {}^0\bar{r}_i) = -({}^{i-1}D_i + {}^i\bar{r}_i) \text{ and } ({}^0D_{i-1} - {}^0D_i) = -{}^{i-1}D_i \tag{6.109}$$

Therefore,

$$\mathcal{N}_i = \eta_i - \eta_{i+1} - \left({}^{i-1}D_i + {}^i\bar{r}_i\right) \times \mathcal{F}_i - {}^{i-1}D_i \times f_{i+1} \tag{6.110}$$

These equations can be rewritten as recursive backward iteration equations from Eqs. (6.105) and (6.110) as

$$f_i = \mathcal{F}_i + f_{i+1} \tag{6.111}$$

$$\eta_i = \eta_{i+1} + {}^{i-1}D_i \times f_{i+1} + \left({}^{i-1}D_i + {}^i\bar{r}_i\right) \times \mathcal{F}_i + \mathcal{N}_i \tag{6.112}$$

The joint torque τ_i at joint i is the sum of the projections of η_i onto the z_{i-1}-axis, the axis about which a revolute joint moves. However, if the joint is prismatic, it translates along z_{i-1}-axis, and the joint torque τ_i is the sum of the projections of $\boldsymbol{f}_i$ onto the z_{i-1}-axis. Hence, the generalized torque for joint i in base frame {0} is given by

$$\tau_i = \begin{cases} \boldsymbol{f}_{i-1}^T \; \hat{z}_{i-1} & \text{for prismatic joint} \\ \eta_{i-1}^T \; \hat{z}_{i-1} & \text{for revolute joint} \end{cases} \tag{6.113}$$

These equations [Eqs. (6.81), (6.82), (6.111), (6.112), and (6.113)] are transformed to joint-link coordinates using the rotation transformation matrix $\boldsymbol{R}$ to give final backward recursive equations as

$${}^i\mathcal{F}_i = m_i \, {}^i\dot{\bar{\boldsymbol{v}}}_i \tag{6.114}$$

$${}^i\mathcal{N}_i = \boldsymbol{I}_i \, {}^i\dot{\omega}_i + {}^i\omega_i \times \left(\boldsymbol{I}_i \, {}^i\omega_i\right) \tag{6.115}$$

$${}^i\boldsymbol{f}_i = {}^i\mathcal{F}_i + {}^i\boldsymbol{R}_{i+1} \; {}^{i+1}\boldsymbol{f}_{i+1} \tag{6.116}$$

$$\begin{aligned} {}^i\eta_i = {}^i\boldsymbol{R}_{i+1} \; {}^{i+1}\eta_{i+1} + ({}^i\boldsymbol{R}_0 \; {}^{i-1}\boldsymbol{D}_i) \times {}^i\boldsymbol{R}_{i+1} \; {}^{i+1}\boldsymbol{f}_{i+1} \\ + ({}^i\boldsymbol{R}_0 \; {}^{i-1}\boldsymbol{D}_i + {}^i\boldsymbol{R}_0 \; {}^i\bar{\boldsymbol{r}}_i) \times {}^i\mathcal{F}_i + {}^i\mathcal{N}_i \end{aligned} \tag{6.117}$$

and

$$\tau_i = \begin{cases} {}^i\boldsymbol{f}_i^T \; {}^i\boldsymbol{R}_{i-1} \; \hat{z}_0 & \text{for prismatic joint} \\ {}^i\eta_i^T \; {}^i\boldsymbol{R}_{i-1} \; \hat{z}_0 & \text{for revolute joint} \end{cases} \tag{6.118}$$

The above five equations, Eq.(6.114) to (6.118), are recursive and can be used to compute the forces and moments ($\boldsymbol{f}_i$, η_i) at the joints for $i = 1, 2, \ldots, n$ for a n–DOF manipulator when the force and moment exerted by the end-effector on the external object or environment $\boldsymbol{f}_{n+1}$ and η_{n+1} are known.

In summary, recursive Newton-Euler equation of motion are a set of forward and backward recursive equations with the kinematics and dynamics of each link referred to link coordinate system. These equations are systematically reproduced in Table 6.2.

The recursive equations are amenable to an algorithmic approach for obtaining the dynamic model of the manipulator. The algorithm for applying NE recursive approach to obtain dynamic model follows.

Algorithm 6.2 ➢ Newton-Euler formulation

This algorithm generates the joint torque equations for all the joint actuators of an n-DOF manipulator. The steps of algorithm are:

Step 0. Define the variables

n = number of degrees of freedom = number of links = number of joints

Joint variables: $\boldsymbol{q}_i$, $\dot{\boldsymbol{q}}_i$, $\ddot{\boldsymbol{q}}_i$, for $i = 1, 2, \ldots, n$

Link variables: $\mathcal{F}_i$, $\boldsymbol{f}_i$, $\mathcal{N}_i$, η_i, τ_i

Table 6.2 *Recursive Newton-Euler equations of motion*

Initial Conditions:

$$ {}^0\omega_0 = {}^0\dot{\omega}_0 = {}^0v_0 = 0 $$

$$ {}^0\dot{v}_0 = g = \begin{bmatrix} g_x & g_y & g_z \end{bmatrix}^T $$

set f_{n+1}, η_{n+1} to given end-effector force and moment.

Forward Iteration: for i = 1, 2, …, n

For revolute joint $\quad {}^i\omega_i = {}^iR_{i-1}\left[{}^{i-1}\omega_{i-1} + \hat{z}_0\dot{\theta}_i\right]$

$$ {}^i\dot{\omega}_i = {}^iR_{i-1}\left[{}^{i-1}\dot{\omega}_{i-1} + \hat{z}_0\ddot{\theta}_i + {}^{i-1}\omega_{i-1} \times \hat{z}_0\dot{\theta}_i\right] $$

$$ {}^i\dot{v}_i = {}^iR_{i-1}\,{}^{i-1}\dot{v}_{i-1} + {}^i\dot{\omega}_i \times ({}^iR_0\,{}^{i-1}D_i) + {}^i\omega_i \times \left[{}^i\omega_i \times ({}^iR_0\,{}^{i-1}D_i)\right] $$

For prismatic joint $\quad {}^i\omega_i = {}^iR_{i-1}\,{}^{i-1}\omega_{i-1}$

$$ {}^i\dot{\omega}_i = {}^iR_{i-1}\,{}^{i-1}\dot{\omega}_{i-1} $$

$$ {}^i\dot{v}_i = {}^iR_{i-1}\left[{}^{i-1}\dot{v}_{i-1} + \hat{z}_0\ddot{d}_i\right] + 2\,{}^i\omega_i \times ({}^iR_{i-1}\hat{z}_0\dot{d}_i) + {}^i\dot{\omega}_i \times \left({}^iR_0\,{}^{i-1}D_i\right) + {}^i\omega_i \times \left[{}^i\omega_i \times ({}^iR_0\,{}^{i-1}D_i)\right] $$

$$ {}^i\dot{\bar{v}}_i = {}^i\dot{v}_i + {}^i\dot{\omega}_i \times {}^i\bar{r}_i + {}^i\omega_i \times \left[{}^i\omega_i \times {}^i\bar{r}_i\right] $$

where $\hat{z}_0 = \begin{bmatrix} 0 & 0 & 1 \end{bmatrix}^T$,

${}^{i-1}D_i$ is referred to frame {0}, when referred to frame {i}, it is

$$ ({}^iR_0\,{}^{i-1}D_i) = \begin{bmatrix} a_i \\ d_i S\alpha_i \\ d_i C\alpha_i \end{bmatrix} $$

${}^i\bar{r}_i$ is the center of mass of link i referred to link frame {i}.

Backward Iteration: for i = 1, 2, …, n

$$ {}^i\mathcal{F}_i = m_i\,{}^i\dot{\bar{v}}_i $$

$$ {}^i\mathcal{N}_i = I_i\,{}^i\dot{\omega}_i + {}^i\omega_i \times \left(I_i\,{}^i\omega_i\right) $$

$$ {}^if_i = {}^i\mathcal{F}_i + {}^iR_{i+1}\,{}^{i+1}f_{i+1} $$

$$ {}^i\eta_i = {}^iR_{i+1}\,{}^{i+1}\eta_{i+1} + ({}^iR_0\,{}^{i-1}D_i) \times {}^iR_{i+1}\,{}^{i+1}f_{i+1} + ({}^iR_0\,{}^{i-1}D_i + {}^iR_0\,{}^i\bar{r}_i) \times {}^iF_i + {}^i\mathcal{N}_i $$

where I_i is the top left 3 × 3 submatrix of inertia tensor of link i about its center of mass with respect to frame {i}.

The joint torque: i = 1, 2, 3 ⋯, n

$$ \tau_i = \begin{cases} {}^if_i^T\,{}^iR_{i-1}\,\hat{z}_0 & \text{for prismatic joint} \\ {}^i\eta_i^T\,{}^iR_{i-1}\,\hat{z}_0 & \text{for revolute joint} \end{cases} $$

Step 1. Set the initial conditions

$$^0\omega_0 = {}^0\dot{\omega}_0 = {}^0v_0 = 0$$

$$^0\dot{v}_0 = g = \begin{bmatrix} g_x & g_y & g_z \end{bmatrix}^T$$

Step 2. Assign frames {0}, ..., {n}using DH notation (Algorithm 3.1) such that frame {i} is oriented with principle axes of link i. Obtain rotational transformation matrices $^{i-1}\boldsymbol{R}_i$, their products and inverses $^i\boldsymbol{R}_{i-1}$, centre of mass of links, inertia tensors $\boldsymbol{I}_i$ of links at centre of mass with respect to frame {i}.

Forward Iteration

Step 3. Set $i = 1$

Step 4. Compute: $^i\omega_i$, $^i\dot{\omega}_i$, ^iv_i and $^i\dot{\bar{v}}_i$ using equations in Table 6.2.

Step 5. If $i = n$, then go to step 6; otherwise set $i = i+1$ and go to step 4.

Backward Iteration

Step 6. Set f_{n+1} = required end-effector force and η_{n+1} = required end-effector moment. Note that if end-effector is free to move in space, it has no load and required force and moment are zero.

Step 7. Compute $\mathcal{F}_i$, $\mathcal{N}_i$, f_i, η_i and τ_i using equations in Table 6.2.

Step 8. If $i = 1$, then stop; otherwise set $i = i-1$ and go to step 7.

Example 6.2 ***NE formulation of EOM for 2-DOF planar manipulator***

The NE formulation to obtain the EOM is illustrated by considering the 2-DOF planar manipulator discussed in Example 6.1, Fig. 6.3.

Solution The Algorithm 6.2 is applied, step-by-step, to obtain the EOM using the recursive Newton-Euler formulation. The base of the robot is fixed and the gravity is assumed to be acting in negative y_0-direction. Thus, the initial conditions are set as per step 1 for the 2-DOF manipulator as

$$^0v_0 = 0; \quad ^0\omega_0 = 0; \quad ^0\dot{\omega}_0 = 0; \quad ^0\dot{v}_0 = \begin{bmatrix} 0 & -g & 0 \end{bmatrix}^T \tag{6.119}$$

Next step (step 2) requires assignment of frames and based on it determination of geometrical parameters, rotational transformation matrices and inertia tensors for the two links. Because, the manipulator is the same as in Example 6.1, the rotational transformation matrices from Eqs. (6.52) and (6.53) are:

$$^{i-1}\boldsymbol{R}_i = \begin{bmatrix} C_i & -S_i & 0 \\ S_i & C_i & 0 \\ 0 & 0 & 1 \end{bmatrix}, \; i = 1, 2; \text{ and } {}^0\boldsymbol{R}_2 = \begin{bmatrix} C_{12} & -S_{12} & 0 \\ S_{12} & C_{12} & 0 \\ 0 & 0 & 1 \end{bmatrix} \tag{6.120}$$

and their inverses are

$$^i\boldsymbol{R}_{i-1} = \begin{bmatrix} C_i & S_i & 0 \\ -S_i & C_i & 0 \\ 0 & 0 & 1 \end{bmatrix}, \; i = 1, 2; \text{ and } {}^2\boldsymbol{R}_0 = \begin{bmatrix} C_{12} & S_{12} & 0 \\ -S_{12} & C_{12} & 0 \\ 0 & 0 & 1 \end{bmatrix} \tag{6.121}$$

The location of center of mass for each length i of length L with respect to its frame $\{i\}$ is given by

$$^{i}\bar{r}_i = \begin{bmatrix} -\frac{1}{2}L \\ 0 \\ 0 \end{bmatrix}, \; i = 1, 2 \tag{6.122}$$

and from Eq. (6.52), the relative positions of origin of frames are

$$^{0}\boldsymbol{D}_1 = \begin{bmatrix} LC_1 \\ LS_1 \\ 0 \end{bmatrix}; \; ^{1}\boldsymbol{D}_2 = \begin{bmatrix} LC_2 \\ LS_2 \\ 0 \end{bmatrix} \tag{6.123}$$

and from Eq. (6.102)

$$^{1}\boldsymbol{R}_0 \, ^{0}\boldsymbol{D}_1 = \begin{bmatrix} L \\ 0 \\ 0 \end{bmatrix}; \; ^{2}\boldsymbol{R}_0 \, ^{1}\boldsymbol{D}_2 = \begin{bmatrix} L \\ 0 \\ 0 \end{bmatrix} \tag{6.124}$$

The inertia tensors of each link with mass m at the center of mass with respect to the link frame from Eq. (6.55) is

$$\boldsymbol{I}_1 = \boldsymbol{I}_2 = \begin{bmatrix} \frac{1}{3}mL^2 & 0 & 0 & -\frac{1}{2}mL \\ 0 & 0 & 0 & 0 \\ 0 & 0 & 0 & 0 \\ -\frac{1}{2}mL & 0 & 0 & m \end{bmatrix} \tag{6.125}$$

Now, forward iterations are carried out as per step 3–5 for $i = 1, 2$.

For i = 1

The angular velocity of revolute joint is calculated from Eq. (6.98) with ω_0 and $^{1}\boldsymbol{R}_0$ from Eqs. (6.119) and (6.121), respectively

$$^{1}\omega_1 = {}^{1}\boldsymbol{R}_0[{}^{0}\omega_0 + \hat{z}_0\dot{\theta}_1] = \begin{bmatrix} C_1 & S_1 & 0 \\ -S_1 & C_1 & 0 \\ 0 & 0 & 1 \end{bmatrix}\begin{bmatrix} 0 \\ 0 \\ 1 \end{bmatrix}\dot{\theta}_1 = \begin{bmatrix} 0 \\ 0 \\ 1 \end{bmatrix}\dot{\theta}_1 \tag{6.126}$$

The angular acceleration is computed using Eq. (6.99) for revolute joint with $^{0}\omega_0 = {}^{0}\dot{\omega}_0 = 0$ as:

$$^{1}\dot{\omega}_1 = {}^{1}\boldsymbol{R}_0\left[{}^{0}\dot{\omega}_0 + \hat{z}_0\ddot{\theta}_1 + {}^{0}\omega_0 \times \hat{z}_0\dot{\theta}_1\right]$$

or
$$^{1}\dot{\omega}_1 = {}^{1}\boldsymbol{R}_0\hat{z}_0\ddot{\theta}_1 = \begin{bmatrix} C_1 & S_1 & 0 \\ -S_1 & C_1 & 0 \\ 0 & 0 & 1 \end{bmatrix}\begin{bmatrix} 0 \\ 0 \\ 1 \end{bmatrix}\ddot{\theta}_1 = \begin{bmatrix} 0 \\ 0 \\ 1 \end{bmatrix}\ddot{\theta}_1 \tag{6.127}$$

The linear acceleration for revolute joint is computed from Eq. (6.100), as

$$^{1}\dot{v}_1 = {}^{1}\boldsymbol{R}_0 \, {}^{0}\dot{v}_0 + {}^{1}\dot{\omega}_1 \times {}^{1}\boldsymbol{R}_0 \, {}^{0}\boldsymbol{D}_1 + {}^{1}\omega_1 \times [{}^{1}\omega_1 \times ({}^{1}\boldsymbol{R}_0 \, {}^{0}\boldsymbol{D}_1)]$$

Substituting values from Eqs. (6.119), (6.121), (6.124), (6.126) and (6.127)

$$
{}^1\dot{v}_1 = \begin{bmatrix} C_1 & S_1 & 0 \\ -S_1 & C_1 & 0 \\ 0 & 0 & 1 \end{bmatrix}\begin{bmatrix} 0 \\ -g \\ 0 \end{bmatrix} + \begin{bmatrix} 0 \\ 0 \\ 1 \end{bmatrix}\ddot{\theta}_1 \times \begin{bmatrix} L \\ 0 \\ 0 \end{bmatrix} + \begin{bmatrix} 0 \\ 0 \\ 1 \end{bmatrix}\dot{\theta}_1 \times \left\{ \begin{bmatrix} 0 \\ 0 \\ 1 \end{bmatrix}\dot{\theta}_1 \times \begin{bmatrix} L \\ 0 \\ 0 \end{bmatrix} \right\}
$$

$$
= \begin{bmatrix} -gS_1 - L\dot{\theta}_1^2 \\ -gC_1 + L\ddot{\theta}_1 \\ 0 \end{bmatrix} \tag{6.128}
$$

and from Eq. (6.101), the linear acceleration of the centre of mass of link 1 is

$$
{}^1\dot{\bar{v}}_1 = {}^1\dot{v}_1 + {}^1\dot{\omega}_1 \times {}^1\bar{r}_1 + {}^1\omega_1 \times \left({}^1\omega_1 \times {}^1\bar{r}_1 \right)
$$

Substituting ${}^1\dot{v}_1$, ${}^1\omega_1$, ${}^1\dot{\omega}_1$ and ${}^1\bar{r}_1$ from Eqs. (6.122), (6.126), (6.127) and (6.128), gives:

$$
{}^1\dot{\bar{v}}_1 = \begin{bmatrix} -gS_1 - L\dot{}_1^2 \\ -gC_1 + L\ddot{}_1 \\ 0 \end{bmatrix} + \begin{bmatrix} 0 \\ 0 \\ 1 \end{bmatrix}\ddot{\theta}_1 \times \begin{bmatrix} -\frac{1}{2}L \\ 0 \\ 0 \end{bmatrix} + \begin{bmatrix} 0 \\ 0 \\ 1 \end{bmatrix}\dot{\theta}_1 \times \left\{ \begin{bmatrix} 0 \\ 0 \\ 1 \end{bmatrix}\dot{\theta}_1 \times \begin{bmatrix} -\frac{1}{2}L \\ 0 \\ 0 \end{bmatrix} \right\}
$$

$$
= \begin{bmatrix} -gS_1 - \frac{1}{2}L\dot{\theta}_1^2 \\ -gC_1 + \frac{1}{2}L\ddot{\theta}_1 \\ 0 \end{bmatrix} \tag{6.129}
$$

For i = 2

The angular velocity is

$$
{}^2\omega_2 = {}^2\boldsymbol{R}_1\left[{}^1\omega_1 + \hat{z}_0\dot{\theta}_2 \right] = \begin{bmatrix} C_2 & S_2 & 0 \\ -S_2 & C_2 & 0 \\ 0 & 0 & 1 \end{bmatrix}\left\{ \begin{bmatrix} 0 \\ 0 \\ 1 \end{bmatrix}\dot{\theta}_1 + \begin{bmatrix} 0 \\ 0 \\ 1 \end{bmatrix}\dot{\theta}_2 \right\}
$$

$$
= \begin{bmatrix} 0 \\ 0 \\ 1 \end{bmatrix}(\dot{\theta}_1 + \dot{\theta}_2) \tag{6.130}
$$

and angular acceleration is

$$
{}^2\dot{\omega}_2 = {}^2\boldsymbol{R}_1\left[{}^1\dot{\omega}_1 + \hat{z}_0\ddot{\theta}_2 + {}^1\omega_1 \times \hat{z}_0\dot{\theta}_2 \right]
$$

or

$$
{}^2\dot{\omega}_2 = \begin{bmatrix} C_2 & S_2 & 0 \\ -S_2 & C_2 & 0 \\ 0 & 0 & 1 \end{bmatrix}\left\{ \begin{bmatrix} 0 \\ 0 \\ 1 \end{bmatrix}\ddot{\theta}_1 + \begin{bmatrix} 0 \\ 0 \\ 1 \end{bmatrix}\ddot{\theta}_2 + \begin{bmatrix} 0 \\ 0 \\ 1 \end{bmatrix}\dot{\theta}_1 \times \begin{bmatrix} 0 \\ 0 \\ 1 \end{bmatrix}\dot{\theta}_2 \right\}
$$

$$
= \begin{bmatrix} 0 \\ 0 \\ 1 \end{bmatrix}(\ddot{\theta}_1 + \ddot{\theta}_2) \tag{6.131}
$$

The linear velocity of link 2 is

$$ {}^2\dot{\boldsymbol{v}}_2 = {}^2\boldsymbol{R}_1 {}^1\dot{\boldsymbol{v}}_1 + {}^2\dot{\omega}_2 \times ({}^2\boldsymbol{R}_0 {}^1\boldsymbol{D}_2) + {}^2\omega_2 \times \left[{}^2\omega_2 \times ({}^2\boldsymbol{R}_0 {}^1\boldsymbol{D}_2)\right] $$

Substituting values from Eqs. (6.121), (6.124), (6.128), (6.130), and (6.131),

$$ {}^2\dot{\boldsymbol{v}}_2 = \begin{bmatrix} C_2 & S_2 & 0 \\ -S_2 & C_2 & 0 \\ 0 & 0 & 1 \end{bmatrix} \begin{bmatrix} -gS_1 - L\dot{\theta}_1^2 \\ -gC_1 + L\ddot{\theta}_1 \\ 0 \end{bmatrix} + \begin{bmatrix} 0 \\ 0 \\ 1 \end{bmatrix} (\ddot{\theta}_1 + \ddot{\theta}_2) \times \begin{bmatrix} L \\ 0 \\ 0 \end{bmatrix} $$

$$ + \begin{bmatrix} 0 \\ 0 \\ 1 \end{bmatrix} (\dot{\theta}_1 + \dot{\theta}_2) \times \left\{ \begin{bmatrix} 0 \\ 0 \\ 1 \end{bmatrix} (\dot{\theta}_1 + \dot{\theta}_2) \times \begin{bmatrix} L \\ 0 \\ 0 \end{bmatrix} \right\} $$

or
$$ {}^2\dot{\boldsymbol{v}}_2 = \begin{bmatrix} L\ddot{\theta}_1 S_2 - L\dot{\theta}_1^2 C_2 - L\dot{\theta}_1^2 - L\dot{\theta}_2^2 - 2L\dot{\theta}_1\dot{\theta}_2 - gS_{12} \\ L\ddot{\theta}_1 C_2 + L\ddot{\theta}_1 + L\ddot{\theta}_2 + L\dot{\theta}_1^2 S_2 - gC_{12} \\ 0 \end{bmatrix} \tag{6.132} $$

and the linear acceleration of link 2 is

$$ {}^2\dot{\bar{\boldsymbol{v}}}_2 = {}^2\dot{\boldsymbol{v}}_2 + {}^2\dot{\omega}_2 \times {}^2\bar{\boldsymbol{r}}_2 + {}^2\omega_2 \times \left[{}^2\omega_2 \times {}^2\bar{\boldsymbol{r}}_2\right] $$

Substituting values obtained above

$$ {}^2\dot{\bar{\boldsymbol{v}}}_2 = \begin{bmatrix} L\ddot{\theta}_1 S_2 - L\dot{\theta}_1^2 C_2 - L\dot{\theta}_1^2 - L\dot{\theta}_2^2 - 2L\dot{\theta}_1\dot{\theta}_2 - gS_{12} \\ L\ddot{\theta}_1 C_2 + L\ddot{\theta}_1 + L\ddot{\theta}_2 + L\dot{\theta}_1^2 S_2 - gC_{12} \\ 0 \end{bmatrix} + \begin{bmatrix} 0 \\ 0 \\ 1 \end{bmatrix} (\ddot{\theta}_1 + \ddot{\theta}_2) $$

$$ \times \begin{bmatrix} -\frac{1}{2}L \\ 0 \\ 0 \end{bmatrix} + \begin{bmatrix} 0 \\ 0 \\ 1 \end{bmatrix} (\dot{\theta}_1 + \dot{\theta}_2) \times \left\{ \begin{bmatrix} 0 \\ 0 \\ 1 \end{bmatrix} (\dot{\theta}_1 + \dot{\theta}_2) \times \begin{bmatrix} -\frac{1}{2}L \\ 0 \\ 0 \end{bmatrix} \right\} $$

or
$$ {}^2\dot{\bar{\boldsymbol{v}}}_2 = \begin{bmatrix} L\ddot{\theta}_1 S_2 - L\dot{\theta}_1^2 C_2 - \frac{1}{2}L\dot{\theta}_1^2 - \frac{1}{2}L\dot{\theta}_2^2 - L\dot{\theta}_1\dot{\theta}_2 - gS_{12} \\ L\ddot{\theta}_1 C_2 + \frac{1}{2}L\ddot{\theta}_1 + \frac{1}{2}L\ddot{\theta}_2 + L\dot{\theta}_1^2 S_2 - gC_{12} \\ 0 \end{bmatrix} \tag{6.133} $$

Backward iterations

Backward iterations are now carried out for $i = 2, 1$, according to step 6–8. The end of link 2 is assumed to move freely, that is, no-load condition, giving

$$ {}^3f_3 = {}^3\eta_3 = 0 \tag{6.134} $$

For i = 2

From Eqs. (6.114) to (6.117)

$$ {}^2\mathcal{F}_2 = m\,{}^2\dot{\bar{\boldsymbol{v}}}_2 $$

$$\text{or}\quad {}^2\mathcal{F}_2 = \begin{bmatrix} mL\ddot{\theta}_1 S_2 - mL\dot{\theta}_2^2 C_2 - \frac{1}{2} mL\dot{\theta}_1^2 - \frac{1}{2} mL\dot{\theta}_2^2 - mL\dot{\theta}_1\dot{\theta}_2 - mgS_{12} \\ mL\ddot{\theta}_1 C_2 + \frac{1}{2} mL\ddot{\theta}_1 + \frac{1}{2} mL\ddot{\theta}_2 + mL_1\dot{\theta}_1^2 S_2 - mgC_{12} \\ 0 \end{bmatrix} \tag{6.135}$$

$${}^2\mathcal{N}_2 = \boldsymbol{I}_2\ {}^2\dot{\omega}_2 + {}^2\omega_2 \times \left(\boldsymbol{I}_2\ {}^2\omega_2\right) = [0\ 0\ 0]^T$$

$${}^2\boldsymbol{f}_2 = {}^2\mathcal{F}_2 + {}^2\boldsymbol{R}_3\ {}^3\boldsymbol{f}_3 = {}^2\mathcal{F}_2$$

$${}^2\eta_2 = {}^2\boldsymbol{R}_3\ {}^3\eta_3 + ({}^2\boldsymbol{R}_0\ {}^1\boldsymbol{D}_2) \times {}^2\boldsymbol{R}_3\ {}^3\boldsymbol{f}_3 + ({}^2\boldsymbol{R}_0\ {}^1\boldsymbol{D}_2 + {}^2\boldsymbol{R}_0\ {}^2\bar{\boldsymbol{r}}_2) \times {}^2\mathcal{F}_2 + {}^2\mathcal{N}_2$$

$$\text{or}\quad {}^2\eta_2 = \begin{bmatrix} 0 \\ 0 \\ \frac{1}{2} mL^2 C_2\ddot{\theta}_1 + \frac{1}{3} mL^2\ddot{\theta}_1 + \frac{1}{3} mL^2\ddot{\theta}_2 + \frac{1}{2} mL^2 S_2\dot{\theta}_2^2 - \frac{1}{2} mLgC_{12} \end{bmatrix} \tag{6.136}$$

For i = 1

$${}^1\mathcal{F}_1 = m\ {}^1\dot{\bar{v}}_1 = \begin{bmatrix} -mgS_1 - \frac{1}{2} mL\dot{\theta}_1^2 \\ -mgC_1 + \frac{1}{2} mL\ddot{\theta}_1 \\ 0 \end{bmatrix} \tag{6.137}$$

$${}^1\mathcal{N}_1 = \boldsymbol{I}_1\ {}^1\dot{\omega}_1 + {}^1\omega_1 \times (\boldsymbol{I}_1\ {}^1\omega_1) = [0\ 0\ 0]^T$$

$${}^1\boldsymbol{f}_1 = {}^1\mathcal{F}_1 + {}^1\boldsymbol{R}_2\ {}^2\boldsymbol{f}_2$$

or

$${}^1\boldsymbol{f}_1 = \begin{bmatrix} -mgS_1 - \frac{1}{2} mL\dot{\theta}_1^2 \\ -mgC_1 + \frac{1}{2} mL\ddot{\theta}_1 \\ 0 \end{bmatrix} +$$

$$\begin{bmatrix} C_2 & -S_2 & 0 \\ S_2 & C_2 & 0 \\ 0 & 0 & 1 \end{bmatrix} \begin{bmatrix} mL\ddot{\theta}_1 S_2 - mL\dot{\theta}_2^2 C_2 - \frac{1}{2} mL\dot{\theta}_1^2 - \frac{1}{2} mL\dot{\theta}_2^2 - mL\dot{\theta}_1\dot{\theta}_2 - mgS_{12} \\ mL\ddot{\theta}_1 C_2 + \frac{1}{2} mL\ddot{\theta}_1 + \frac{1}{2} mL\ddot{\theta}_2 + mL_1\dot{\theta}_1^2 S_2 - mgC_{12} \\ 0 \end{bmatrix}$$

Simplifying,

$${}^1\boldsymbol{f}_1 = \begin{bmatrix} -\frac{1}{2} mLS_2\ddot{\theta}_1 - \frac{1}{2} mLS_2\ddot{\theta}_2 - \frac{3}{2} mL\dot{\theta}_1^2 - \frac{1}{2} mLC_2\dot{\theta}_1^2 - \frac{1}{2} mLC_2\dot{\theta}_2^2 - mLC_2\dot{\theta}_1\dot{\theta}_2 - 2mgS_1 \\ \frac{3}{2} mL\ddot{\theta}_1 + \frac{1}{2} mLC_2\ddot{\theta}_1 + \frac{1}{2} mLC_2\ddot{\theta}_2 - \frac{1}{2} mLS_2\dot{\theta}_1^2 - \frac{1}{2} mLS_2\dot{\theta}_2^2 - mLS_2\dot{\theta}_1\dot{\theta}_2 - 2mgC_1 \\ 0 \end{bmatrix} \tag{6.138}$$

Similarly,

$${}^1\eta_1 = {}^1\boldsymbol{R}_2\ {}^2\eta_2 + ({}^1\boldsymbol{R}_0\ {}^0\boldsymbol{D}_1) \times {}^1\boldsymbol{R}_2\ {}^2\boldsymbol{f}_2 + ({}^1\boldsymbol{R}_0\ {}^0\boldsymbol{D}_1 + {}^1\boldsymbol{R}_0\ {}^1\bar{\boldsymbol{r}}_1) \times {}^1\boldsymbol{F}_1 + {}^1\mathcal{N}_1$$

$$ {}^1\eta_1 = \begin{bmatrix} 0 \\ 0 \\ mL^2C_2\ddot{\theta}_1 - mL^2S_2\dot{\theta}_2^2 + mLgC_{12} + mL^2(\ddot{\theta}_1 + \ddot{\theta}_2) \end{bmatrix} + \begin{bmatrix} 0 \\ 0 \\ m_1L^2\ddot{\theta}_1 + m_1LgC_1 \end{bmatrix} $$

$$ + \begin{bmatrix} 0 \\ 0 \\ mL^2\ddot{\theta}_1 - mL^2S_2(\dot{\theta}_1 + \dot{\theta}_2)^2 + mLgS_2S_{12} + mL^2C_2(\ddot{\theta}_1 + \ddot{\theta}_2) + mLgC_2C_{12} \end{bmatrix} \tag{6.139} $$

The joint torques for two joints are finally, obtained using Eq. (6.118).

$$ \tau_1 = {}^1\eta_1^{T\,1}\boldsymbol{R}_0\,\hat{z}_0 $$

or

$$ \tau_1 = \frac{5}{3}mL^2\ddot{\theta}_1 + \frac{1}{3}mL^2\ddot{\theta}_2 + mL^2C_2\ddot{\theta}_1 + \frac{1}{2}mL^2C_2\ddot{\theta}_2 $$

$$ -\frac{1}{2}mL^2S_2\dot{\theta}_2^2 - mL^2S_2\dot{\theta}_1\dot{\theta}_2 + \frac{1}{2}mLgC_{12} + \frac{3}{2}mLgC_1 \tag{6.140} $$

and

$$ \tau_2 = \frac{1}{2}mL^2C_2\ddot{\theta}_1 + \frac{1}{3}mL^2\ddot{\theta}_1 + \frac{1}{3}mL^2\ddot{\theta}_2 + \frac{1}{2}mL^2S_2\dot{\theta}_1^2 $$

$$ + \frac{1}{2}mLgC_{12} \tag{6.141} $$

The above equations, Eqs. (6.140) and (6.141), are written in matrix form as:

$$ \begin{bmatrix} \tau_1 \\ \tau_2 \end{bmatrix} = \begin{bmatrix} mL^2(\frac{5}{3} + C_2) & mL^2(\frac{1}{3} + \frac{1}{2}C_2) \\ mL^2(\frac{1}{3} + \frac{1}{2}C_2) & \frac{1}{3}mL^2 \end{bmatrix} \begin{bmatrix} \ddot{\theta}_1 \\ \ddot{\theta}_2 \end{bmatrix} $$

$$ + \begin{bmatrix} -mL^2S_2\dot{\theta}_1\dot{\theta}_2 - \frac{1}{2}mL^2S_2\dot{\theta}_2^2 \\ \frac{1}{2}mL^2S_2\dot{\theta}_1^2 \end{bmatrix} + \begin{bmatrix} \frac{1}{2}mgLC_{12} + \frac{3}{2}mgLC_1 \\ \frac{1}{2}mgLC_{12} \end{bmatrix} \tag{6.142} $$

This is same as equation (6.80), obtained from the Lagrange-Euler formulation in Example 6.1.

6.5 COMPARISON OF LAGRANGE-EULER AND NEWTON-EULER FORMULATIONS

Two approaches to dynamic modeling of robotic manipulators have been discussed above. Both Lagrange-Euler and Newton-Euler formulations give the relationship between the joint and end-effector forces/torques and the manipulator joint-link position, velocities, and accelerations. These formulations can be compared on the basis of their computational requirements.

The mathematical operations required to compute τ for an n-DOF manipulator in Lagrange-Euler formulation is estimated to be (see Appendix D)

$$\begin{aligned} &\text{Multiplications:} \quad \frac{128}{3}n^4 + \frac{512}{3}n^3 + \frac{844}{3}n^2 + \frac{76}{3}n \\ &\text{Additions:} \quad \frac{98}{3}n^4 + \frac{786}{3}n^3 + \frac{673}{3}n^2 + \frac{107}{3}n \end{aligned} \tag{6.143}$$

In the Newton-Euler formulation, the computations are dependent on the configuration, in addition to the number of degrees of freedom n. For a typical manipulator the numbers of computations for NE formulation are estimated to be

$$\begin{aligned} &\text{Multiplications} : \quad 117n - 24 \\ &\text{Additions} \quad\quad : \quad 103n - 21 \end{aligned} \tag{6.144}$$

The recursive Newton-Euler formulation has a computational complexity of order $O(n)$, that is, mathematical operations of multiplication and addition are proportional to n, the number of degrees of freedom of the manipulator. This is markedly less compared to the Lagrange-Euler formulation that has a complexity of order $O(n^4)$. The drawback of recursive formulation is that it is not as amenable to simple physical interpretation because the closed-form Lagrange-Euler formulation is. However, recursive formulation is much more efficient in terms of computational effort, particularly as the number of degrees of freedom n increases, they are ideally suited for computer implementation.

Example 6.3 ***Application of dynamic model to study the torque variations***

Determine the variation of joint torques from the dynamic model of 2-DOF manipulator discussed in Example 6.1. Take the following data for the manipulator:

$$m_1 = m_2 = 6 \text{ kg} \quad \text{and} \quad L_1 = L_2 = 1 \text{ m}$$

The specified motion cycle is: Both joints move from rest with equal velocity, attain maximum velocity then decelerate and come to rest, traversing $\pi/2$ rad in 1 s.

Solution The given values of m and L are substituted in Eq. (6.142). The resulting joint torque equations for the 2-DOF manipulator are:

$$\begin{bmatrix} \tau_1 \\ \tau_2 \end{bmatrix} = \begin{bmatrix} 10 + 6C_2 & 2 + 2C_2 \\ 2 + 2C_2 & 2 \end{bmatrix} \begin{bmatrix} \ddot{\theta}_1 \\ \ddot{\theta}_2 \end{bmatrix} + \begin{bmatrix} -6S_2\dot{\theta}_1\dot{\theta}_2 - 3S_2\dot{\theta}_2^2 \\ 3S_2\dot{\theta}_1^2 \end{bmatrix} + \begin{bmatrix} 2gC_{12} + 9gC_1 \\ 3gC_{12} \end{bmatrix} \tag{6.145}$$

For the specified motion the velocity profile for both joints is a triangular velocity profile, as shown in Fig. 6.8(a). The variation of position of each link is obtained by integrating the velocity profile and values of all other terms in Eq. (6.145) are computed as a function of time. The time history of position, acceleration, components of torque (M_{ij}), and net torque for this velocity profile for each joint is shown in Fig. 6.8(b) to Fig. 6.8(h).

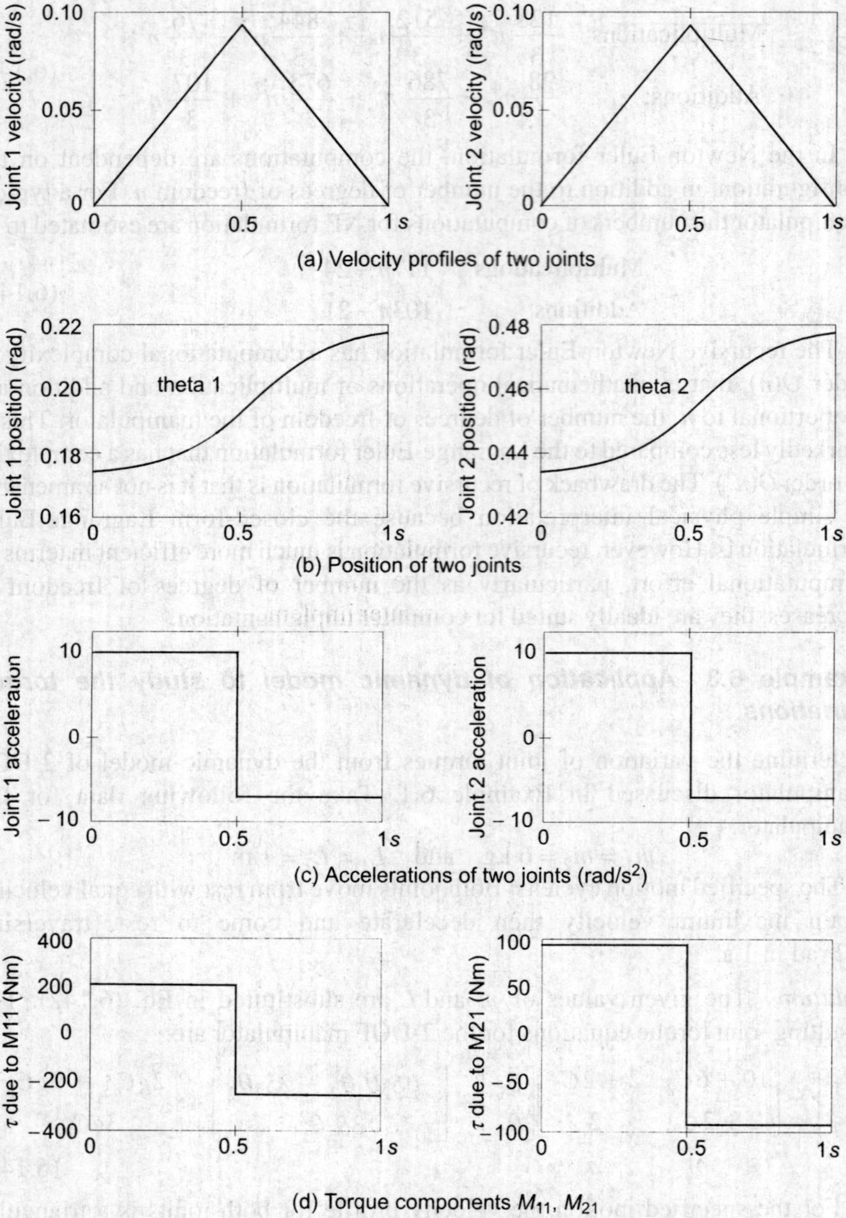

(a) Velocity profiles of two joints

(b) Position of two joints

(c) Accelerations of two joints (rad/s^2)

(d) Torque components M_{11}, M_{21}

Fig. 6.8 *Time histories of position, velocity, acceleration and torques for the two joints of the 2-DOF manipulator (Contd….)*

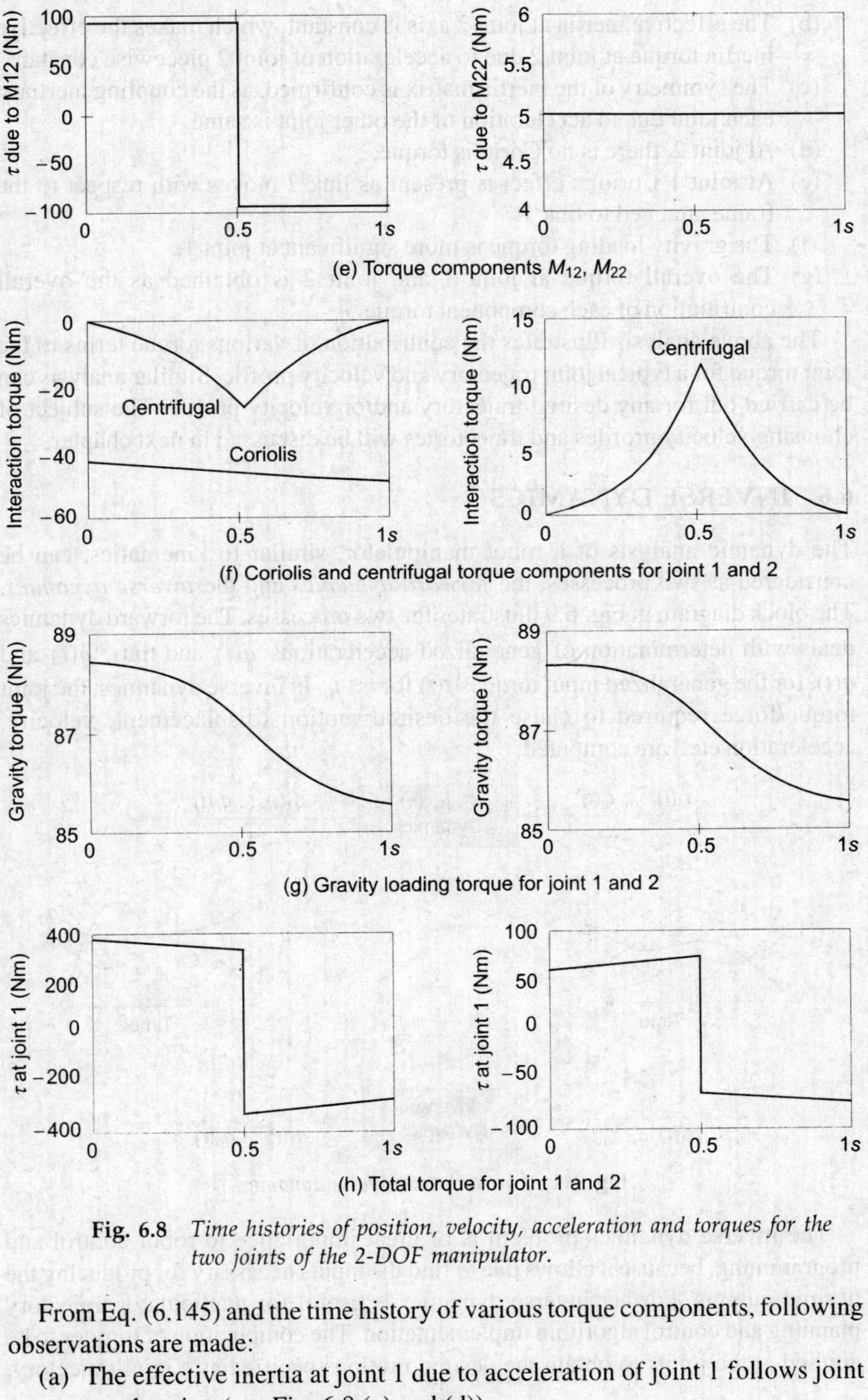

Fig. 6.8 *Time histories of position, velocity, acceleration and torques for the two joints of the 2-DOF manipulator.*

From Eq. (6.145) and the time history of various torque components, following observations are made:

(a) The effective inertia at joint 1 due to acceleration of joint 1 follows joint acceleration (see Fig. 6.8 (c) and (d)).

(b) The effective inertia at joint 2 axis is constant, which makes the effective inertia torque at joint 2 due to acceleration of joint 2 piecewise constant.
(c) The symmetry of the inertia matrix is confirmed, as the coupling inertia at each joint due to acceleration of the other joint is same.
(d) At joint 2, there is no Coriolis torque.
(e) At joint 1 Coriolis effect is present as link 2 moves with respect to the frame attached to link 1.
(f) The gravity loading torque is more significant at joint 1.
(g) The overall torque at joint 1 and joint 2 is obtained as the overall contribution of each component torque.

The above analysis illustrates the contribution of various torque terms to the joint torque for a typical joint trajectory and velocity profile. Similar analysis can be carried out for any desired trajectory and/or velocity profile. The subject of choosing velocity profiles and trajectories will be discussed in next chapter.

6.6 INVERSE DYNAMICS

The dynamic analysis of a robot manipulator, similar to kinematics, can be considered as two processes, the *forward dynamics* and the *inverse dynamics.* The block diagram in Fig. 6.9 illustrates the two processes. The forward dynamics deals with determination of generalized accelerations $\ddot{\boldsymbol{q}}(t)$ and thus, $\dot{\boldsymbol{q}}(t)$ and $\boldsymbol{q}(t)$, for the generalized input torques $\tau(t)$ for $t > t_0$. In inverse dynamics, the joint torque/force required to cause the desired motion (displacement, velocity, acceleration etc.) are computed.

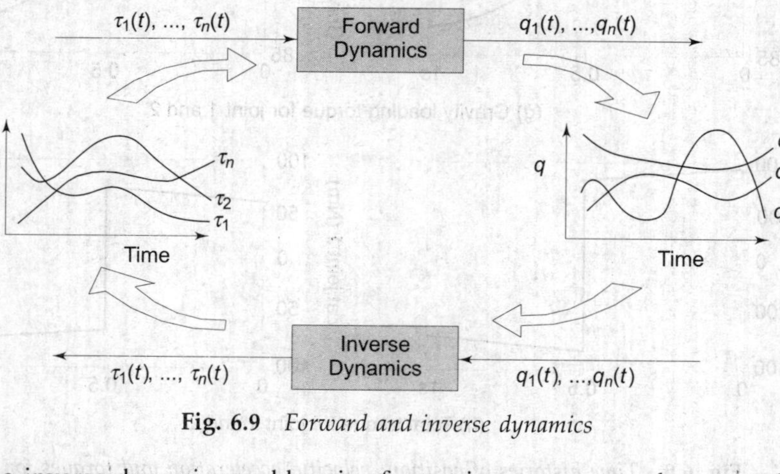

Fig. 6.9 *Forward and inverse dynamics*

The inverse dynamics problem is of great importance to robot control and programming, because it allows one to find the inputs necessary for producing the desired outputs. Further inverse dynamics is useful for manipulator trajectory planning and control algorithm implementation. The computation of torques to be applied at the joints to obtain the desired motion specified as a joint trajectory, which is specified in terms of positions, velocities, and accelerations, is useful both for verifying feasibility of the desired trajectory and compensating nonlinear terms in the linear control models. Both of these problems are discussed in the next chapter.

Example 6.4 *Lagrange-Euler formulation for 2-DOF RR non-planar manipulator arm*

Determine the equations of motions (dynamic model) for the two-link, 2-DOF, RR non-planar manipulator arm using the Lagrange-Euler formulation. The manipulator is shown in Fig. 6.10. Assume that both links are slender with mass m_1 and m_2 at distal end.

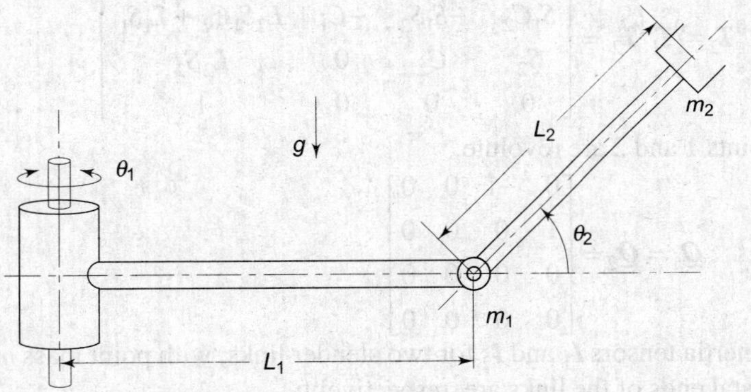

Fig. 6.10 *2-DOF RR non-planar manipulator arm*

Solution The LE formulation is carried out to obtain the EOM as per Algorithm 6.1. The frames assignment is shown in Fig. 6.11 and Table 6.3 gives the joint-link parameters.

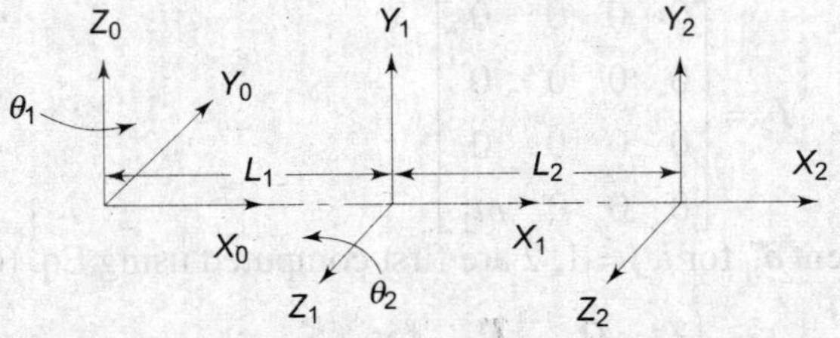

Fig. 6.11 *Frame assignment for 2-DOF RR non-planar arm*

Table 6.3 *Joint link parameters for 2-DOF RR non-planar arm*

i	a_i	α_i	d_i	θ_i	$C\theta_i$	$S\theta_i$	$C\alpha_i$	$S\alpha_i$
1	L_1	90°	0	θ_1	C_1	S_1	0	1
2	L_2	0	0	θ_2	C_2	S_2	1	0

The link transformation matrices and the overall transformation matrix are:

$$^0\boldsymbol{T}_1 = \begin{bmatrix} C_1 & 0 & S_1 & L_1C_1 \\ S_1 & 0 & -C_1 & L_1S_1 \\ 0 & 1 & 0 & 0 \\ 0 & 0 & 0 & 1 \end{bmatrix} \tag{6.146}$$

$$^1\boldsymbol{T}_2 = \begin{bmatrix} C_2 & -S_2 & 0 & L_2C_2 \\ S_2 & C_2 & 0 & L_2S_2 \\ 0 & 0 & 1 & 0 \\ 0 & 0 & 0 & 1 \end{bmatrix} \tag{6.147}$$

$$^0\boldsymbol{T}_2 = {^0\boldsymbol{T}_1}\,{^1\boldsymbol{T}_2} = \begin{bmatrix} C_1C_2 & -C_1S_2 & S_1 & L_2C_1C_2 + L_1C_1 \\ S_1C_2 & -S_1S_2 & -C_1 & L_2S_1C_2 + L_1S_1 \\ S_2 & C_2 & 0 & L_2S_2 \\ 0 & 0 & 0 & 1 \end{bmatrix} \tag{6.148}$$

Since joints 1 and 2 are revolute,

$$\boldsymbol{Q}_1 = \boldsymbol{Q}_2 = \begin{bmatrix} 0 & -1 & 0 & 0 \\ 1 & 0 & 0 & 0 \\ 0 & 0 & 0 & 0 \\ 0 & 0 & 0 & 0 \end{bmatrix} \tag{6.149}$$

The inertia tensors $\boldsymbol{I}_1$ and $\boldsymbol{I}_2$ for two slender links, with point mass m_1 and m_2 at the distal ends of the links are, respectively:

$$\boldsymbol{I}_1 = \begin{bmatrix} 0 & 0 & 0 & 0 \\ 0 & 0 & 0 & 0 \\ 0 & 0 & 0 & 0 \\ 0 & 0 & 0 & m_1 \end{bmatrix} \tag{6.150}$$

$$\boldsymbol{I}_2 = \begin{bmatrix} 0 & 0 & 0 & 0 \\ 0 & 0 & 0 & 0 \\ 0 & 0 & 0 & 0 \\ 0 & 0 & 0 & m_2 \end{bmatrix} \tag{6.151}$$

The coefficient $\boldsymbol{d}_{ij}$ for $i, j = 1, 2$ are first computed using Eq. (6.49), that is

$$\boldsymbol{d}_{ij} = \begin{cases} \boldsymbol{T}_{j-1}\,\boldsymbol{Q}_j\,{^{j-1}\boldsymbol{T}_i} & \text{for } j \le i \\ 0 & \text{for } j > i \end{cases} \tag{6.152}$$

which gives $\boldsymbol{d}_{11}$, $\boldsymbol{d}_{12}$, $\boldsymbol{d}_{21}$ and $\boldsymbol{d}_{22}$ as:

$$\boldsymbol{d}_{11} = {^0\boldsymbol{T}_0}\boldsymbol{Q}_1\,{^0\boldsymbol{T}_1} = \begin{bmatrix} -S_1 & 0 & C_1 & -L_1S_1 \\ C_1 & 0 & S_1 & L_1C_1 \\ 0 & 0 & 0 & 0 \\ 0 & 0 & 0 & 0 \end{bmatrix}$$

$d_{12} = 0$ (because $j > i$)

$$d_{21} = {}^0T_1 Q_2 {}^1T_2 = \begin{bmatrix} -S_1C_2 & S_1S_2 & C_1 & -L_2S_1C_2 - L_1S_1 \\ C_1C_2 & -C_1S_2 & S_1 & L_2C_1C_2 + L_1C_1 \\ 0 & 0 & 0 & 0 \\ 0 & 0 & 0 & 0 \end{bmatrix} \quad (6.153)$$

$$d_{22} = {}^0T_1 Q_2 {}^1T_2 = \begin{bmatrix} -C_1S_2 & -C_1C_2 & 0 & -L_2C_1S_2 \\ -S_1S_2 & -S_1C_2 & 0 & -L_2S_1S_2 \\ C_2 & -S_2 & 0 & L_2C_2 \\ 0 & 0 & 0 & 0 \end{bmatrix}$$

The elements of inertia matrix M are computed using Eq. (6.46).

$$M_{ij} = \sum_{p=\max(i,j)}^{n} \mathrm{Tr}\left[d_{pj} I_p d_{pi}^T\right] \quad \text{for } i, j = 1, 2 \quad (6.154)$$

This gives,

$$M_{11} = \mathrm{Tr}(d_{11} I_1 d_{11}^T) + \mathrm{Tr}(d_{21} I_2 d_{21}^T) \quad (6.155)$$

or $$M_{11} = m_1 L_1^2 + (L_2C_2 + L_1)^2 m_2$$

and $$M_{22} = \mathrm{Tr}(d_{22} I_2 d_{22}^T) = m_2 L_2^2$$

$$M_{12} = M_{21} = \mathrm{Tr}(d_{22} I_2 d_{21}^T) = 0$$

The Coriolis and centrifugal force coefficients (or velocity coupling coefficients), h_{ijk} for $i, j, k = 1, 2$ are computed using Eq. (6.47).

$$h_{ijk} = \sum_{\mathrm{p}=\max(i,j,k)}^{\mathrm{n}} \left[\frac{\partial (d_{pk})}{\partial q_p} I_p d_{pi}^T \right] \quad (6.156)$$

The partial derivatives in Eq. (6.156) are obtained from Eq. (6.50). Thus,

$$h_{111} = 0$$
$$h_{222} = 0$$
$$h_{122} = \mathrm{Tr}\left[{}^0T_1 Q_2 {}^1T_1 Q_2 {}^1T_2 I_2 d_{21}^T \right]$$
$$= -L_2C_1C_2m_2(-L_2S_1C_2 - L_1S_1) - L_2S_1C_2m_2(L_2C_1C_2 + L_1C_1) = 0$$
$$h_{211} = \mathrm{Tr}\left[{}^0T_0 Q_1 {}^0T_0 Q_1 {}^0T_2 I_2 d_{22}^T \right]$$
$$= -L_2C_1S_2m_2(-L_2C_1C_2 - L_1C_1) - L_2S_1S_2m_2(-L_2S_1C_2 - L_1S_1) \quad (6.157)$$
$$= m_2S_2(L_2^2C_2 + L_1L_2)$$
$$h_{121} = h_{112} = \mathrm{Tr}\left[{}^0T_0 Q_1 {}^0T_1 Q_2 {}^1T_2 I_2 d_{21}^T \right]$$
$$= L_2S_1S_2m_2(-L_2S_1C_2 - L_1S_1) - L_2C_1S_2m_2(L_2C_1C_2 + L_1C_1)$$
$$= -m_2S_2(L_2^2C_2 + L_1L_2)$$
$$h_{221} = h_{212} = \mathrm{Tr}\left[{}^0T_0 Q_1 {}^0T_1 Q_2 {}^1T_2 I_2 d_{22}^T \right] = 0$$

The Coriolis and centrifugal torque terms are computed using the series summation:

$$H_i = \sum_{j=1}^{n}\sum_{k=1}^{n} h_{ijk}\,\dot{\theta}_j\dot{\theta}_k \tag{6.158}$$

$$H_1 = \sum_{j=1}^{2}\sum_{k=1}^{2} h_{ijk}\dot{\theta}_j\dot{\theta}_k = h_{111}\dot{\theta}_1^2 + h_{112}\dot{\theta}_1\dot{\theta}_2 + h_{121}\dot{\theta}_1\dot{\theta}_2 + h_{122}\dot{\theta}_2^2 \tag{6.159}$$

Substituting values of h_{ijk} from Eq. (6.157) and simplifying gives

$$H_1 = -2m_2S_2(L_2^2C_2 + L_1L_2)\dot{\theta}_1\dot{\theta}_2 \tag{6.160}$$

Similarly,

$$H_2 = h_{211}\dot{\theta}_1^2 + h_{212}\dot{\theta}_1\dot{\theta}_2 + h_{221}\dot{\theta}_1\dot{\theta}_2 + h_{222}\dot{\theta}_2^2$$

or

$$H_2 = m_2S_2(L_2^2C_2 + L_1L_2)\dot{\theta}_1^2 \tag{6.161}$$

Thus,

$$\boldsymbol{H}(\theta,\dot{\theta}) = \begin{bmatrix} H_1 \\ H_2 \end{bmatrix} = \begin{bmatrix} -2m_2S_2(L_2^2C_2 + L_1L_2)\dot{\theta}_1\dot{\theta}_2 \\ m_2S_2(L_2^2C_2 + L_1L_2)\dot{\theta}_1^2 \end{bmatrix} \tag{6.162}$$

The mass of the links is at the distal end of the links, that is, at origin of frame {1} and frame {2}. Thus,

$${}^1\bar{\boldsymbol{r}}_1 = {}^2\bar{\boldsymbol{r}}_2 = [0 \quad 0 \quad 0 \quad 1]^T \tag{6.163}$$

and the gravity is in the –ve direction of z-axis of frame {0}, that is,

$$\boldsymbol{g} = [0 \quad 0 \quad -g] \tag{6.164}$$

where $g = 9.8062$ m/s^2. Therefore, the gravity loading at the joint 1 is

$$G_1 = (m_1\boldsymbol{g}\boldsymbol{d}_{11}\,{}^1\bar{\boldsymbol{r}}_1 + m_2\boldsymbol{g}\boldsymbol{d}_{21}\,{}^2\bar{\boldsymbol{r}}_2) = 0 \tag{6.165}$$

Similarly, at joint 2

$$G_2 = -m_2\boldsymbol{g}\boldsymbol{d}_{22}\,{}^2\bar{\boldsymbol{r}}_2 = m_2gL_2C_2 \tag{6.166}$$

Hence, equations of motion in the vector matrix form are obtained from Eqs. (6.155), (6.162) (6.165) and (6.166) as

$$\begin{bmatrix} \tau_1 \\ \tau_2 \end{bmatrix} = \begin{bmatrix} m_1L_1^2 + (L_2C_2 + L_1)^2m_2 & 0 \\ 0 & m_2L_2^2 \end{bmatrix}\begin{bmatrix} \ddot{\theta}_1 \\ \ddot{\theta}_2 \end{bmatrix} + \begin{bmatrix} -2m_2S_2(L_2^2C_2 + L_1L_2)\dot{\theta}_1\dot{\theta}_2 \\ m_2S_2(L_2^2C_2 + L_1L_2)\dot{\theta}_1^2 \end{bmatrix} + \begin{bmatrix} 0 \\ m_2gL_2C_2 \end{bmatrix} \tag{6.167}$$

Example 6.5 ***NE formulation of EOM for two link, 2-DOF nonplanar manipulator***

As the last example, consider the NE formulation to obtain the EOM for the two-link, 2-DOF nonplanar manipulator arm discussed in Example 6.4 and shown in Fig. 6.10. The assumption about the link masses is same as in Example 6.4, that is, the link masses m_1 and m_2 are at the distal end of the respective links.

Solution The recursive NE formulation is carried out using equations in Table 6.2 (Algorithm 6.2) to obtain the EOM. The base of the arm is fixed and the gravity is assumed to be acting in negative z_0-direction. Thus, the initial conditions are set as:

$$^0\boldsymbol{v}_0 = {}^0\omega_0 = {}^0\dot{\omega}_0 = 0; \quad {}^0\dot{\boldsymbol{v}}_0 = [0 \quad 0 \quad -g]^T \tag{6.168}$$

where $g = 9.8062\ m/s^2$.

The assignment of frames was carried out in Example 6.4, Fig. 6.11 and transformation matrices were determined in, Eqs. (6.146)-(6.148). The rotational transformation matrices from these equations are:

$$^0\boldsymbol{R}_1 = \begin{bmatrix} C_1 & 0 & S_1 \\ S_1 & 0 & -C_1 \\ 0 & 1 & 0 \end{bmatrix}, \ ^1\boldsymbol{R}_2 = \begin{bmatrix} C_2 & -S_2 & 0 \\ S_2 & C_2 & 0 \\ 0 & 0 & 1 \end{bmatrix}, \ ^0\boldsymbol{R}_2 = \begin{bmatrix} C_1C_2 & -C_1S_2 & S_1 \\ S_1C_2 & -S_1S_2 & -C_1 \\ S_2 & C_2 & 1 \end{bmatrix} \tag{6.169}$$

and their inverses are

$$^0\boldsymbol{R}_1 = \begin{bmatrix} C_1 & -S_1 & 0 \\ 0 & 0 & 1 \\ S_1 & -C_1 & 0 \end{bmatrix}, \ ^2\boldsymbol{R}_1 = \begin{bmatrix} C_2 & -S_2 & 0 \\ -S_2 & C_2 & 0 \\ 0 & 0 & 1 \end{bmatrix}, \ ^2\boldsymbol{R}_0 = \begin{bmatrix} C_{12} & S_{12} & S_2 \\ -S_{12} & C_{12} & C_2 \\ -S_1 & -C_1 & 1 \end{bmatrix} \tag{6.170}$$

The location of center of mass for each link with respect to its frame $\{\boldsymbol{i}\}$ is given by

$$^1\bar{\boldsymbol{r}}_1 = \begin{bmatrix} 0 \\ 0 \\ 0 \end{bmatrix}, \quad ^2\bar{\boldsymbol{r}}_2 = \begin{bmatrix} 0 \\ 0 \\ 0 \end{bmatrix} \tag{6.171}$$

and from Eq. (6.52), the relative positions of origin of frames are

$$^0\boldsymbol{D}_1 = \begin{bmatrix} L_1C_1 \\ L_1S_1 \\ 0 \end{bmatrix}; \quad ^1\boldsymbol{D}_2 = \begin{bmatrix} L_2C_2 \\ L_2S_2 \\ 0 \end{bmatrix} \tag{6.172}$$

and from Eq. (6.102)

$$^1\boldsymbol{R}_0\,{}^0\boldsymbol{D}_1 = \begin{bmatrix} L_1 \\ 0 \\ 0 \end{bmatrix}; \quad ^2\boldsymbol{R}_0\,{}^1\boldsymbol{D}_2 = \begin{bmatrix} L_2 \\ 0 \\ 0 \end{bmatrix} \tag{6.173}$$

The inertia tensors of the two links with mass m_1 and m_2 at the distal end with respect to the link frame are as given by upper 3×3 submatrix of Eqs. (6.150) and (6.151) of Example 6.4.

Now, forward iterations are carried out for $i = 1, 2$.

For i = 1

The angular velocity of first joint is calculated from Eq. (6.98) as:

$$^1\omega_1 = {}^1\boldsymbol{R}_0[{}^0\omega_0 + \hat{z}_0\dot{\theta}_1] = \begin{bmatrix} C_1 & S_1 & 0 \\ 0 & 0 & 1 \\ S_1 & -C_1 & 0 \end{bmatrix}\begin{bmatrix} 0 \\ 0 \\ 1 \end{bmatrix}\dot{\theta}_1 = \begin{bmatrix} 0 \\ 1 \\ 0 \end{bmatrix}\dot{\theta}_1 \tag{6.174}$$

The angular acceleration is computed using Eq. (6.99) for revolute joint with as:

$$^1\dot{\omega}_1 = {}^1\boldsymbol{R}_0\left[{}^0\dot{\omega}_0 + \hat{z}_0\ddot{\theta}_1 + {}^0\omega_0 \times \hat{z}_0\dot{\theta}_1\right]$$

or

$$^1\dot{\omega}_1 = {}^1\boldsymbol{R}_0\hat{z}_0\ddot{\theta}_1 = \begin{bmatrix} C_1 & S_1 & 0 \\ 0 & 0 & 1 \\ S_1 & -C_1 & 0 \end{bmatrix}\begin{bmatrix} 0 \\ 0 \\ 1 \end{bmatrix}\ddot{\theta}_1 = \begin{bmatrix} 0 \\ 1 \\ 0 \end{bmatrix}\ddot{\theta}_1 \tag{6.175}$$

The linear acceleration for revolute joint is computed from Eq. (6.100), as

$$^1\dot{\boldsymbol{v}}_1 = {}^1\boldsymbol{R}_0\,{}^0\dot{\boldsymbol{v}}_0 + {}^1\dot{\omega}_1 \times {}^1\boldsymbol{R}_0\,{}^0\boldsymbol{D}_1 + {}^1\omega_1 \times [{}^1\omega_1 \times ({}^1\boldsymbol{R}_0\,{}^0\boldsymbol{D}_1)]$$

Substituting values from Eqs. (6.119), (6.121), (6.124), (6.126) and (6.127)

$$^1\dot{\boldsymbol{v}}_1 = \begin{bmatrix} C_1 & S_1 & 0 \\ 0 & 0 & 1 \\ S_1 & -C_1 & 0 \end{bmatrix}\begin{bmatrix} 0 \\ 0 \\ -g \end{bmatrix} + \begin{bmatrix} 0 \\ 1 \\ 0 \end{bmatrix}\ddot{\theta}_1 \times \begin{bmatrix} L_1 \\ 0 \\ 0 \end{bmatrix} + \begin{bmatrix} 0 \\ 1 \\ 0 \end{bmatrix}\dot{\theta}_1 \times \left\{\begin{bmatrix} 0 \\ 1 \\ 0 \end{bmatrix}\dot{\theta}_1 \times \begin{bmatrix} L_1 \\ 0 \\ 0 \end{bmatrix}\right\}$$

$$= \begin{bmatrix} -L_1\dot{\theta}_1^2 \\ g \\ -L_1\ddot{\theta}_1 \end{bmatrix} \tag{6.176}$$

and from Eq. (6.101), the linear acceleration of the centre of mass of link 1 is

$$^1\dot{\bar{v}}_1 = {}^1\dot{v}_1 + {}^1\dot{\omega}_1 \times {}^1\bar{r}_1 + {}^1\omega_1 \times \left({}^1\omega_1 \times {}^1\bar{r}_1\right)$$

Substituting $^1\dot{v}_1$, $^1\omega_1$, $^1\dot{\omega}_1$ and $^1\bar{r}_1$ from Eqs. (6.122), (6.126), (6.127) and (6.128), gives:

$$^1\dot{\bar{v}}_1 = \begin{bmatrix} -L\dot{\theta}_1^2 \\ g \\ L_1\ddot{\theta}_1 \end{bmatrix} + \begin{bmatrix} 0 \\ 1 \\ 0 \end{bmatrix}\ddot{\theta}_1 \times \begin{bmatrix} 0 \\ 0 \\ 0 \end{bmatrix} + \begin{bmatrix} 0 \\ 1 \\ 0 \end{bmatrix}\dot{\theta}_1 \times \left\{\begin{bmatrix} 0 \\ 1 \\ 0 \end{bmatrix}\dot{\theta}_1 \times \begin{bmatrix} 0 \\ 0 \\ 0 \end{bmatrix}\right\} = \begin{bmatrix} -L_1\dot{\theta}_1^2 \\ g \\ -L_1\ddot{\theta}_1 \end{bmatrix} \tag{6.177}$$

For i = 2

The angular velocity is

$$^2\omega_2 = {}^2\boldsymbol{R}_1\left[{}^1\omega_1 + \hat{z}_0\dot{\theta}_2\right] = \begin{bmatrix} C_2 & S_2 & 0 \\ -S_2 & C_2 & 0 \\ 0 & 0 & 1 \end{bmatrix}\left\{\begin{bmatrix} 0 \\ 1 \\ 0 \end{bmatrix}\dot{\theta}_1 + \begin{bmatrix} 0 \\ 0 \\ 1 \end{bmatrix}\dot{\theta}_2\right\} = \begin{bmatrix} S_2\dot{\theta}_1 \\ C_2\dot{\theta}_1 \\ \dot{\theta}_2 \end{bmatrix} \tag{6.178}$$

and angular acceleration is

$$^2\dot{\omega}_2 = {}^2\boldsymbol{R}_1\left[{}^1\dot{\omega}_1 + \hat{z}_0\ddot{\theta}_2 + {}^1\omega_1 \times \hat{z}_0\dot{\theta}_2\right]$$

$$\text{or } {}^2\dot{\omega}_2 = \begin{bmatrix} C_2 & S_2 & 0 \\ -S_2 & C_2 & 0 \\ 0 & 0 & 1 \end{bmatrix} \left\{ \begin{bmatrix} 0 \\ 1 \\ 0 \end{bmatrix} \ddot{\theta}_1 + \begin{bmatrix} 0 \\ 0 \\ 1 \end{bmatrix} \ddot{\theta}_2 + \begin{bmatrix} 0 \\ 1 \\ 0 \end{bmatrix} \dot{\theta}_1 \times \begin{bmatrix} 0 \\ 0 \\ 1 \end{bmatrix} \dot{\theta}_2 \right\} = \begin{bmatrix} C_2\dot{\theta}_1\dot{\theta}_2 + S_2\ddot{\theta}_1 \\ -S_2\dot{\theta}_1\dot{\theta}_2 + C_2\ddot{\theta}_1 \\ \ddot{\theta}_2 \end{bmatrix} \tag{6.179}$$

The linear velocity of link 2 is

$${}^2\dot{\boldsymbol{v}}_2 = {}^2\boldsymbol{R}_1\,{}^1\dot{\boldsymbol{v}}_1 + {}^2\dot{\boldsymbol{\omega}}_2 \times ({}^2\boldsymbol{R}_0\,{}^1\boldsymbol{D}_2) + {}^2\omega_2 \times \left[{}^2\omega_2 \times ({}^2\boldsymbol{R}_0\,{}^1\boldsymbol{D}_2)\right]$$

Substituting values from Eqs. (6.121), (6.124), (6.128), (6.130) and (6.131),

$${}^2\dot{\boldsymbol{v}}_2 = \begin{bmatrix} C_2 & S_2 & 0 \\ -S_2 & C_2 & 0 \\ 0 & 0 & 1 \end{bmatrix} \begin{bmatrix} -L_1\dot{\theta}_1^2 \\ g \\ -L_1\ddot{\theta}_1 \end{bmatrix} + \begin{bmatrix} C_2\dot{\theta}_1\dot{\theta}_2 + S_2\ddot{\theta}_1 \\ -S_2\dot{\theta}_1\dot{\theta}_2 + C_2\ddot{\theta}_1 \\ \ddot{\theta}_2 \end{bmatrix} \times \begin{bmatrix} L_2 \\ 0 \\ 0 \end{bmatrix} + \begin{bmatrix} S_2\dot{\theta}_1 \\ C_2\dot{\theta}_1 \\ \dot{\theta}_2 \end{bmatrix} \times \left\{ \begin{bmatrix} S_2\dot{\theta}_1 \\ C_2\dot{\theta}_1 \\ \dot{\theta}_2 \end{bmatrix} \times \begin{bmatrix} L_2 \\ 0 \\ 0 \end{bmatrix} \right\}$$

$$\text{or} \quad {}^2\dot{\boldsymbol{v}}_2 = \begin{bmatrix} -L_1\dot{\theta}_1^2 C_2 - C_2^2 L_2\dot{\theta}_1^2 - L_2\dot{\theta}_2^2 + gS_2 \\ L_2\ddot{\theta}_2 + L_1\dot{\theta}_1^2 S_2 + S_2C_2L_2\dot{\theta}_1^2 + gC_2 \\ -L_1\ddot{\theta}_1 - L_2C_2\ddot{\theta}_1 + 2S_2L_2\dot{\theta}_1\dot{\theta}_2 \end{bmatrix} \tag{6.180}$$

and the linear acceleration of link 2 is

$${}^2\dot{\boldsymbol{v}}_2 = {}^2\dot{\boldsymbol{v}}_2 + {}^2\dot{\boldsymbol{\omega}}_2 \times {}^2\bar{\boldsymbol{r}}_2 + {}^2\boldsymbol{\omega}_2 \times \left[{}^2\boldsymbol{\omega}_2 \times {}^2\bar{\boldsymbol{r}}_2 1\right]$$

Substituting values obtained above

$${}^2\dot{\bar{\boldsymbol{v}}}_2 = \begin{bmatrix} -L_1\dot{\theta}_1^2 C_2 - C_2^2 L_2\dot{\theta}_1^2 - L_2\dot{\theta}_2^2 + gS_2 \\ L_2\ddot{\theta}_2 + L_1\dot{\theta}_1^2 S_2 + S_2C_2L_2\dot{\theta}_1^2 + gC_2 \\ -L_1\ddot{\theta}_1 - L_2C_2\ddot{\theta}_1 + 2S_2L_2\dot{\theta}_1\dot{\theta}_2 \end{bmatrix} + \begin{bmatrix} C_2\dot{\theta}_1\dot{\theta}_2 + S_2\ddot{\theta}_1 \\ -S_2\dot{\theta}_1\dot{\theta}_2 + C_2\ddot{\theta}_1 \\ \ddot{\theta}_2 \end{bmatrix} \times \begin{bmatrix} 0 \\ 0 \\ 0 \end{bmatrix} + \begin{bmatrix} S_2\dot{\theta}_1 \\ C_2\dot{\theta}_1 \\ \dot{\theta}_2 \end{bmatrix} \times \left\{ \begin{bmatrix} S_2\dot{\theta}_1 \\ C_2\dot{\theta}_1 \\ \dot{\theta}_2 \end{bmatrix} \times \begin{bmatrix} L_2 \\ 0 \\ 0 \end{bmatrix} \right\}$$

$$\text{or} \quad {}^2\dot{\bar{\boldsymbol{v}}}_2 = \begin{bmatrix} -L_1\dot{\theta}_1^2 C_2 - C_2^2 L_2\dot{\theta}_1^2 - L_2\dot{\theta}_2^2 + gS_2 \\ L_2\ddot{\theta}_2 + L_1\dot{\theta}_1^2 S_2 + S_2C_2L_2\dot{\theta}_1^2 + gC_2 \\ -L_1\ddot{\theta}_1 - L_2C_2\ddot{\theta}_1 + 2S_2L_2\dot{\theta}_1\dot{\theta}_2 \end{bmatrix} \tag{6.181}$$

Backward iterations

Backward iterations are now carried out for $i = 2, 1$. The end of link 2 is assumed to move freely, that is, no-load condition, giving

$${}^3\boldsymbol{f}_3 = {}^3\boldsymbol{\eta}_3 = \boldsymbol{0} \tag{6.182}$$

For i = 2

From Eqs. (6.114) to (6.117)

$${}^2\mathcal{F}_2 = \boldsymbol{m}_2\,{}^2\dot{\bar{\boldsymbol{v}}}_2$$

$$\text{or} \quad {}^2\mathcal{F}_2 = m_2 \begin{bmatrix} -L_1\dot{\theta}_1^2 C_2 - C_2^2 L_2\dot{\theta}_1^2 - L_2\dot{\theta}_2^2 + gS_2 \\ L_2\ddot{\theta}_2 + L_1\dot{\theta}_1^2 S_2 + S_2C_2L_2\dot{\theta}_1^2 + gC_2 \\ -L_1\ddot{\theta}_1 - L_2C_2\ddot{\theta}_1 + 2S_2L_2\dot{\theta}_1\dot{\theta}_2 \end{bmatrix} \tag{6.183}$$

$$
\begin{aligned}
{}^{2}\mathcal{N}_2 &= \boldsymbol{I}_2\ {}^{2}\dot{\boldsymbol{\omega}}_2 + {}^{2}\boldsymbol{\omega}_2 \times \left(\boldsymbol{I}_2\ {}^{2}\boldsymbol{\omega}_2\right) = [0\ 0\ 0]^T \\
{}^{2}\boldsymbol{f}_2 &= {}^{2}\mathcal{F}_2 + {}^{2}\boldsymbol{R}_3\ {}^{3}\boldsymbol{f}_3 = {}^{2}\mathcal{F}_2 \qquad (6.184) \\
{}^{2}\boldsymbol{\eta}_2 &= {}^{2}\boldsymbol{R}_3\ {}^{3}\boldsymbol{\eta}_3 + ({}^{2}\boldsymbol{R}_0\ {}^{1}\boldsymbol{D}_2) \times {}^{2}\boldsymbol{R}_3\ {}^{3}\boldsymbol{f}_3 + ({}^{2}\boldsymbol{R}_0\ {}^{1}\boldsymbol{D}_2 + {}^{2}\boldsymbol{R}_0\ {}^{2}\bar{\boldsymbol{r}}_2) \times {}^{2}\mathcal{F}_2 + {}^{2}\mathcal{N}_2
\end{aligned}
$$

or

$$
{}^{2}\boldsymbol{\eta}_2 = m_2 L_2 \begin{bmatrix} 0 \\ L_1\ddot{\theta}_1 + C_2 L_2 \ddot{\theta}_1 - 2 S_2 L_2 \dot{\theta}_1 \dot{\theta}_2 \\ L_2 \ddot{\theta}_2 + L_1 S_2 \dot{\theta}_1^2 + S_2 C_2 L_2 \dot{\theta}_1^2 + g C_2 \end{bmatrix} \qquad (6.185)
$$

For i = 1

$$
{}^{1}\mathcal{F}_1 = m_1\ {}^{1}\dot{\bar{\boldsymbol{v}}}_1 = m_1 \begin{bmatrix} -L_1\dot{\theta}_1^2 \\ g \\ -L_1\ddot{\theta}_1 \end{bmatrix} \qquad (6.186)
$$

$$
\begin{aligned}
{}^{1}\mathcal{N}_1 &= \boldsymbol{I}_1\ {}^{1}\dot{\boldsymbol{\omega}}_1 + {}^{1}\boldsymbol{\omega}_1 \times (\boldsymbol{I}_1\ {}^{1}\boldsymbol{\omega}_1) = [0\ 0\ 0]^T \\
{}^{1}\boldsymbol{f}_1 &= {}^{1}\mathcal{F}_1 + {}^{1}\boldsymbol{R}_2\ {}^{2}\boldsymbol{f}_2
\end{aligned}
$$

Substituting from Eqs (6. 169), (6.184) and (6.186) gives

$$
{}^{1}\boldsymbol{f}_1 = \begin{bmatrix} -m_1 L\dot{\theta}_1^2 \\ m_1 g \\ -m_1 L\ddot{\theta}_1 \end{bmatrix} + \begin{bmatrix} C_2 & -S_2 & 0 \\ S_2 & C_2 & 0 \\ 0 & 0 & 1 \end{bmatrix} m_2 \begin{bmatrix} -L_1\dot{\theta}_1^2 C_2 - L_2 C_2^2 \dot{\theta}_1^2 - L_2\dot{\theta}_2^2 + g S_2 \\ L_2\ddot{\theta}_2 + L_1\dot{\theta}_1^2 S_2 + S_2 C_2 L_2 \dot{\theta}_1^2 + g C_2 \\ -L_1\ddot{\theta}_1 - L_2 C_2 \ddot{\theta}_1 + 2 S_2 L_2 \dot{\theta}_1 \dot{\theta}_2 \end{bmatrix}
$$

Simplifying,

$$
{}^{1}\boldsymbol{f}_1 = \begin{bmatrix} -m_2 S_2 L_2 \ddot{\theta}_2 - m_1 L_1 \dot{\theta}_1^2 - m_2 (L_1 + L_2 C_2) \dot{\theta}_1^2 - m_2 C_2 L_2 \dot{\theta}_2^2 \\ (m_1 + m_2) g + m_2 L_2 \dot{\theta}_2^2 + m_2 C_2 L_2 \ddot{\theta}_2 \\ -m_1 L_1 \ddot{\theta}_1 - m_2 L_1 \ddot{\theta}_1 - m_2 L_2 C_2 \ddot{\theta}_1 + 2 S_2 L_2 \dot{\theta}_1 \dot{\theta}_2 \end{bmatrix} \qquad (6.187)
$$

Similarly,

$$
{}^{1}\boldsymbol{\eta}_1 = {}^{1}\boldsymbol{R}_2\ {}^{2}\boldsymbol{\eta}_2 + ({}^{1}\boldsymbol{R}_0\ {}^{0}\boldsymbol{D}_1) \times {}^{1}\boldsymbol{R}_2\ {}^{2}\boldsymbol{f}_2 + ({}^{1}\boldsymbol{R}_0\ {}^{0}\boldsymbol{D}_1 + {}^{1}\boldsymbol{R}_0\ {}^{1}\bar{\boldsymbol{r}}_1) \times {}^{1}\mathcal{F}_1 + {}^{1}\mathcal{N}_1
$$

$$
\begin{aligned}
{}^{1}\boldsymbol{\eta}_1 = &\begin{bmatrix} -m_2 S_2 L_2 (L_1 + L_2 C_2) \ddot{\theta}_1 + 2 m_2 S_2 L_2 \dot{\theta}_1 \dot{\theta}_2 \\ m_2 C_2 L_2 (L_1 + L_2 C_2) \ddot{\theta}_1 + 2 m_2 C_2 L_2 \dot{\theta}_1 \dot{\theta}_2 \\ m_2 S_2 L_2 (L_1 + L_2 C_2) \dot{\theta}_1^2 + m_2 L_2^2 \ddot{\theta}_2 + m_2 L_2 C_2 g \end{bmatrix} + \begin{bmatrix} 0 \\ m_1 L_1^2 \ddot{\theta}_1 \\ -m_1 L_1 g \end{bmatrix} \\
&+ \begin{bmatrix} 0 \\ m_2 L_1 (L_1 + L_2 C_2) \ddot{\theta}_1 - 2 m_2 S_2 L_1 L_2 \dot{\theta}_1 \dot{\theta}_2 \\ m_2 L_1 g - m_2 S_2 L_1 L_2 \dot{\theta}_1^2 + m_2 C_2 L_1 L_2 \ddot{\theta}_2 \end{bmatrix}
\end{aligned} \qquad (6.188)
$$

Joint torques for the two joints are, finally, obtained using Eq. (6.118) as

$$
\tau_1 = {}^{1}\boldsymbol{\eta}_1^T\ {}^{1}\boldsymbol{R}_0\ \hat{z}_0
$$

or $\tau_1 = m_1 L_1^2 \ddot{\theta}_1 + m_2(L_1^2 + L_2^2 C_2^2 + 2L_1 L_2 C_2)\ddot{\theta}_1 - m_2 S_2 (L_2^2 C_2 + L_1 L_2)\dot{\theta}_1\dot{\theta}_2$ (6.189)\

and $\tau_2 = L_2 m_2 (L_2 \ddot{\theta}_2 + L_1 S_2 \dot{\theta}_1^2 + L_2 S_2 C_2 \dot{\theta}_1^2 + g C_2$ (6.190)

The above equations, Eqs. (6.189) and (6.190), are written in matrix form as:

$$\begin{bmatrix}\tau_1\\ \tau_2\end{bmatrix} = \begin{bmatrix} m_1 L_1^2 + (L_2 C_2 + L_1)^2 m_2 & 0 \\ 0 & m_2 L_2^2 \end{bmatrix} \begin{bmatrix}\ddot{\theta}_1 \\ \ddot{\theta}_2\end{bmatrix} + \begin{bmatrix} -2m_2 S_2 (L_2^2 C_2 + L_1 L_2)\dot{\theta}_1\dot{\theta}_2 \\ m_2 S_2 (L_2^2 C_2 + L_1 C_2)\dot{\theta}_1^2 \end{bmatrix} + \begin{bmatrix} 0 \\ m_2 g L_2 C_2 \end{bmatrix} \quad (6.191)$$

This EOM is same as equation (6.167), obtained from the Lagrange-Euler formulation in Example 6.4.

EXERCISES

6.1 Show that the velocity coupling vector for a one-axis robot is always zero, that is

$$h(\dot{q}, q) = 0 \quad (6.192)$$

6.2 Prove that $\dfrac{\partial ({}^0T_i)}{\partial q_j} = 0$ for $j > i$ for all $i, j = 1, 2, \ldots, n$.

6.3 Determine the partial derivative of d_{ij} with respect to q_k where d_{ij} is

$$d_{ij} = \begin{cases} {}^0T_{j-1}\, Q_j\, {}^{j-1}T_i & \text{for } j \le i \\ 0 & \text{for } j > i \end{cases} \quad (6.193)$$

and show that

$$\frac{\partial d_{ij}}{\partial q_k} = \begin{cases} {}^0T_{j-1}\, Q_j\, {}^{j-1}T_{k-1}\, Q_k\, {}^{k-1}T_i & \text{for } i \ge k \ge j \\ {}^0T_{k-1}\, Q_k\, {}^{k-1}T_{j-1}\, Q_j\, {}^{j-1}T_i & \text{for } i \ge j \ge k \\ 0 & \text{for } i < j \text{ or } i < k \end{cases} \quad (6.194)$$

6.4 Determine the dynamic model of a 1-DOF, 1-axis planar manipulator with one rotary joint (the inverted pendulum). Assume the link to be a thin cylinder (slender member) with length L and mass m acting at the centroid of the link. Obtain direct solution and solution using Lagrange-Euler formulation and compare the two.

6.5 Find the dynamic model of a 2-DOF Cartesian manipulator shown in Fig. E6.5 using Newton-Euler formulation. Assume that the mass of the links and the actuators to drive them are m_1, and m_2, respectively, and are concentrated at the centroid of each link. Take link lengths to be L_1 and L_2.

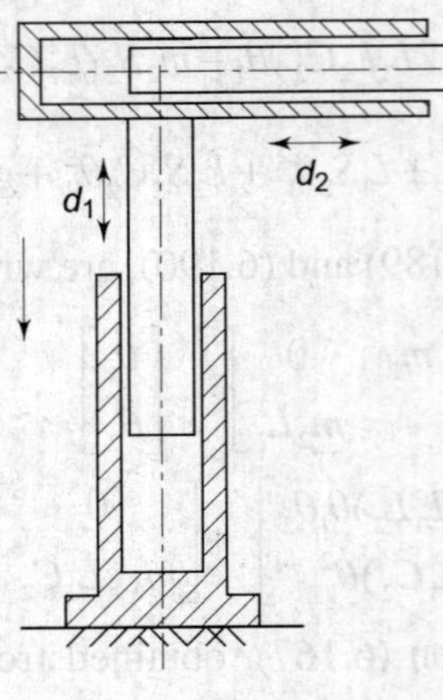

Fig. E6.5 *A 2-DOF Cartesian manipulator*

6.6 If the axis of second joint makes an angle of 60° with the axis of first joint for the two-link Cartesian manipulator instead of 90° in Fig. 6.12, find the dynamic model. Compare the resulting model with one obtained in Exercise 6.5.

6.7 Derive the equation of motion for the 1-DOF manipulator in Exercise 6.4 using recursive Newton-Euler formulation. Verify that the result is equivalent to that of Exercise 6.4.

6.8 Obtain the dynamic model of the two-link RP manipulator shown in Fig. E6.8 by determining total kinetic and potential energy of the manipulator. Assume that the links are thin homogeneous cylinders with radii b, mass m_1, and m_2, and lengths L_1 and L_2.

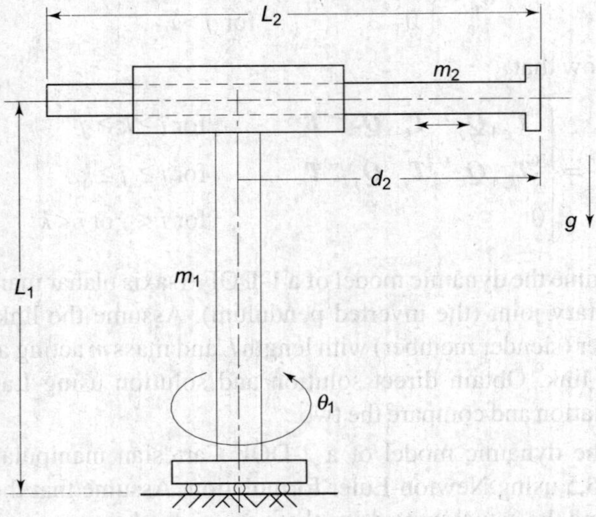

Fig. E6.8 *A two-link RP manipulator*

6.9 Determine the dynamic model of the manipulator in Exercise 6.8 using Newton-Euler formulation.

6.10Find the dynamic equations of motion for the planar 2-DOF manipulator with one prismatic joint and one rotary joint, as shown in Fig. E6.10, using Lagrange-Euler formulation. The link parameters are shown in the figure.

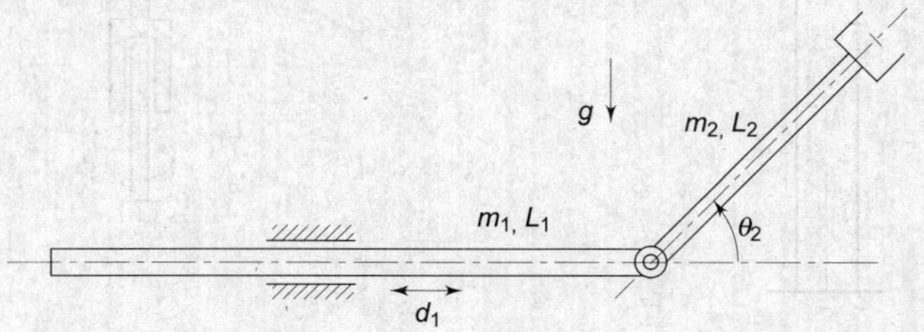

Fig. E6.10 *A 2-DOF planar manipulator with one prismatic and one rotary join*

6.11 Compute the joint torques for the two-link planar manipulator discussed in Example 6.1 for $\theta_1 = 45°$ and $\theta_2 = 60°$. Take other link parameters from Example 6.3.

6.12 Obtain the dynamic equations for a three-axes Cartesian arm configuration in Fig. 1.11 using the Newton-Euler and Lagrange-Euler formulations.

6.13 For the nonplanar 2-DOF manipulator with two rotary joints, shown in Fig. 6E.13, find the dynamic equations of motion using Lagrange-Euler formulation. Assume link masses, m_1 and m_2 to be unity and concentrated at the distal ends of the respective links.

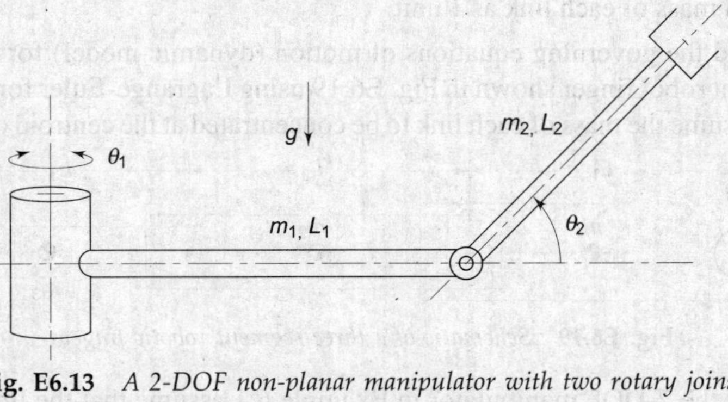

Fig. E6.13 *A 2-DOF non-planar manipulator with two rotary joins*

6.14 Determine the effective inertia, coupling inertia, and gravity terms for the 4-DOF SCARA manipulator shown in Fig. E6.14. Take the mass of four links as m_1, m_2, m_3, and m_4, and lengths as L_1, L_2, L_3, and L_4, respectively, acting at the centroid of each link. Assume each link to be a slender member. The SCARA manipulator shown in Fig. E6.14 is a simplified model of SCARA manipulator discussed in Example 3.5, Fig. 3.20, with $L_{11} = L_1$ and $L_{12} = 0$.

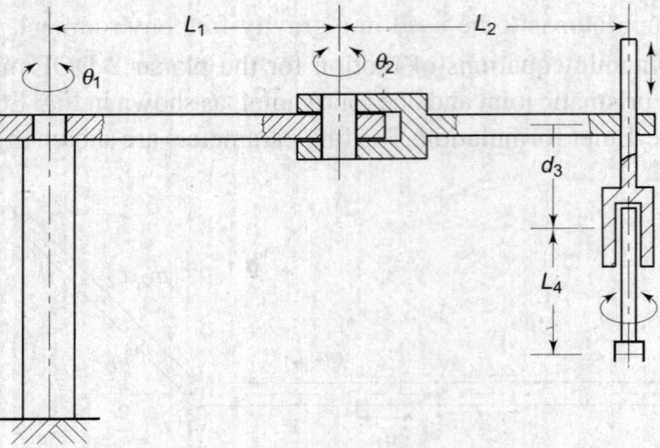

Fig. E6.14 *Schematic diagram of SCARA manipulator*

6.15 Obtain the equations to determine the joint torques (τ_1, τ_2, τ_3, and τ_4) using Lagrange-Euler formulation for the SCARA manipulator described in Exercise 6.14.

6.16 Formulate the dynamic model of the 2-link manipulator in Fig. E6.10, Exercise 6.10 using the Newton-Euler formulation. Identify the centrifugal force, Coriolis force, and gravity loading for each joint.

6.17 Derive the dynamic model for the 3-DOF cylindrical arm (see Fig. 1.12) using Lagrange-Euler formulation.

6.18 For the SCARA manipulator in Exercise 6.14, plot the time history of position and torque for each joint assuming length of each link as 1 unit and mass of each link as 1 unit.

6.19 Find the governing equations of motion (dynamic model) for the three-joint robot finger shown in Fig. E6.19 using Lagrange-Euler formulation. Assume the mass of each link to be concentrated at the centroid of the link.

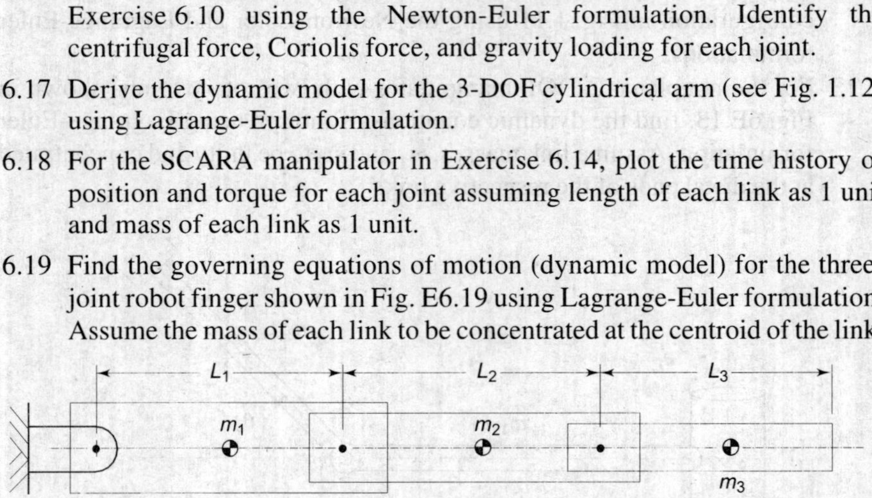

Fig. E6.19 *Schematic of a three-segment robotic finger*

6.20 For the 2-DOF manipulator in Example 6.1 assume that the links have a length of 2 units and a mass of 1 unit, each. Calculate the linear velocity and acceleration and angular velocity and acceleration of the centroids of the two links as a function of time, if the joint angle of both joints varies with the time function:

$$\theta_i(t) = 2t + t^2 \tag{6.195}$$

6.21 A n-DOF manipulator is required to work at very low velocities only. What simplifications will result from this condition in the Lagrange-Euler formulation? Explain.

6.22 If a manipulator is to be used in a gravity-free environment, say space, what will be the effect of this on the dynamic model. Which forces will be significant under these conditions?

6.23 A two link, 2-DOF RP manipulator is shown in Fig. E6.23. Assuming the centroid inertia tensors for both links as

$$\boldsymbol{I}_i = \begin{bmatrix} I_{x_i} & 0 & 0 & 0 \\ 0 & I_{y_i} & 0 & 0 \\ 0 & 0 & I_{z_i} & 0 \\ 0 & 0 & 0 & m_i \end{bmatrix}; i = 1,2 \tag{6.196}$$

Determine the dynamic model of the manipulator using (i) Newton-Euler formulation (ii) Lagrange-Euler formulation.

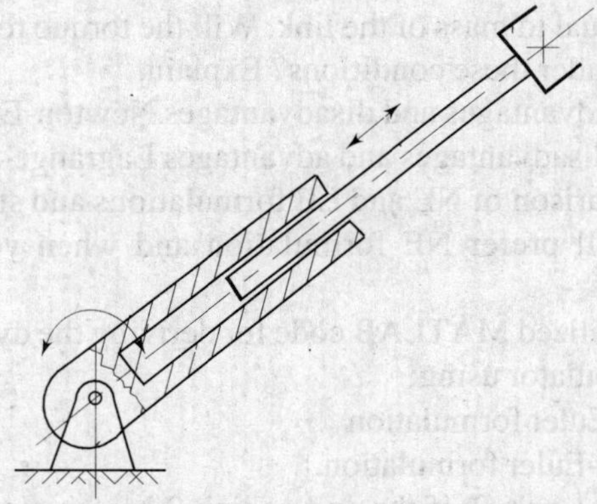

Fig. E6.23 *Two degree of freedom RP manipulator*

6.24 Derive the dynamic equations of for the 3-DOF RPR manipulator shown in Fig. E6.24. The centroidal inertia tensors of link 1 and 3 is given by

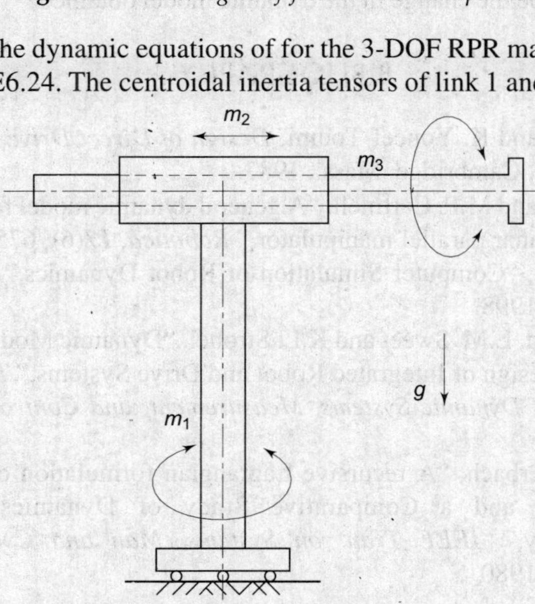

Fig. E6.24 *A 3-DOF RPR arm*

$$I_i = \begin{bmatrix} I_{x_i} & 0 & 0 & 0 \\ 0 & I_{y_i} & 0 & 0 \\ 0 & 0 & I_{z_i} & 0 \\ 0 & 0 & 0 & m_i \end{bmatrix}; i = 1,3 \quad (6.197)$$

The link 2 has a point mass m_2 located at the origin of its frame. Use, both, Newton-Euler and Lagrange-Euler formulations and compare the results.

6.25 Write a generalized MATLAB code for deriving the inertia tensor of a link. It should take input about the link geometry, the reference frame about which inertia tensor is required and assumptions about location of center of mass, if any.

6.26 Under the assumption that the mass of a slender link is concentrated at its distal end, the inertia tensor becomes a zero matrix with only element (4,4) being equal to mass of the link. Will the torque required to move the link be zero under these conditions? Explain.

6.27 Describe the advantages and disadvantages Newton-Euler formulation.

6.28 Describe the disadvantages and advantages Lagrange-Euler formulation.

6.29 Make a comparison of NE and LE formulations and state the situation(s) when you will prefer NE formulation and when you will prefer LE formulation.

6.30 Write a generalized MATLAB code for deriving the dynamic model of an ***n***-DOF manipulator using:

(a) Newton-Euler formulation

(b) Lagrange-Euler formulation.

6.31 In Example 6.1 or 6.2, if the end of link 2 has an external load of 1 kg, what will be the change in the dynamic model obtained?

BIBLIOGRAPHY

1. H. Asada and K. Youcef-Toumi, *Design of Direct Drive Manipulators*, MIT Press, Cambridge, Mass., 1987.
2. K.A. Atia and M.P. Cartmell, "A general dynamic model for a large scale 2-DOF Planar parallel manipulator," *Robotica*, **17**(6), 675–684, 1999.
3. M. Eroglu, "Computer Simulation of Robot Dynamics," *Robotica*, **16**, 615–621, 1998.
4. M.C. Good, L.M. Sweet and K.L. Strobel, "Dynamic Models for Control System Design of Integrated Robot and Drive Systems," *Trans of ASME Journal of Dynamic Systems, Measurement, and Control*, **107**, 53–59, 1985.
5. J.M. Hollerbach, "A recursive Lagrangian formulation of Manipulator Dynamics and a Comparative Study of Dynamics Formulation Complexity," *IEEE Tran. on Systems, Man and Cybernetics*, **10**, 730–736, 1980.

6. M.B. Leahy and G.N. Saridis, "Compensation of Industrial Manipulator Dynamics," *The International Journal of Robotics Research*, ***8***(4), 73–84, 1989.
7. C.S.G. Lee, "Robot Arm Kinematics, Dynamics and Control," *Computer*, ***15***(12), 62–80, 1982.
8. D. Naderi, A. Meghdari and M.Durali, "Dynamic Modeling and Analysis of Two DOF Mobile Manipulator," *Robotica*, ***19***(2), 177–186, 2001.
9. M. Rackovic, M. Vukobratovic and D. Surla, "Generation of Dynamic Models of Complex Robotic Mechanisms in Symbolic Form," *Robotica*, ***16***, .23–36, 1998.
10. L. Sciavicco, B. Siciliano and L. Villani, "On Dynamic Modeling of Gear-Driven Rigid Robot Manipulators," *Postpr. 4th IFAC Symp. Robot Control*, Capri, Italy, 477–483, 1994.
11. W. Silver, "On the Equivalence of Lagrangian and Newton-Euler Dynamics for Manipulators," *The International Journal of Robotics Research*, ***1***(2), 60–70, 1982.
12. M.W. Spong and M. Vidyasagar, *Robot Dynamics and Control*, Wiley, 1989.
13. M. Walker and D. Orin, "Efficient Dynamic Computer Simulation of Robotic Mechanisms," *Trans of ASME Journal of Dynamic Systems, Measurement, and Control*, ***104***, 1982.

7

Trajectory Planning

The end-effector of the manipulator is required to move in a particular fashion to accomplish a specified task. The previous chapters discussed the mathematical models for geometric, kinematic, static, and dynamic behaviour of mechanical manipulators, which form the background for this chapter. The execution of the specific task requires the manipulator to follow a preplanned path, which is the larger problem of motion or trajectory planning and motion control for the manipulator. The goal of trajectory planning is to describe the requisite motion of the manipulator as a time sequence of joint/link/end-effector locations and derivatives of locations, which are generated by "interpolating" or "approximating" the desired path by a polynomial function. These time base sequence locations, also called time law or time history, obtained from the trajectory planning serve as reference input or "control set points" to the manipulator's control system. The control system, in turn, assures that the manipulator executes the planned trajectories. The motion control is discussed in the next chapter.

This chapter focuses on various trajectory-planning issues. The chapter begins with describing the terminology involved in trajectory planning. Some techniques for trajectory generation for two types of motion, point-to-point motion, and continuous-path motion, are presented. First, the problem of trajectory planning in joint space is considered and then Cartesian space planning is discussed. The chapter concludes with examples illustrating various trajectory-planning techniques.

The user usually defines the desired trajectory by a number of parameters. A polynomial function is identified to "approximate" the trajectory. The trajectory planning consists of generating a time sequence of values of different parameters attained by the polynomial interpolating the trajectory.

Every manipulator has at least capability to move from an initial location (position and orientation) to a final desired location. The kinematics and dynamics of the manipulator govern the transition of posture with actuators exerting the desired torques. As no limits must be violated and no unmodelled behaviour should be executed, it is necessary to devise planning algorithms that generate smooth trajectories.

One of the main objectives of any trajectory-planning algorithm is to achieve a smooth motion of the manipulator. A smooth function is one that is continuous and has a continuous first derivative. The smooth motion has the important advantage of reducing the vibrations and wear of the mechanical system.

7.1 DEFINITIONS AND PLANNING TASKS

To begin with, a trajectory planner may be thought of as a black box to which the user inputs a list of parameters and constraints describing the desired trajectory. The trajectory planner generates a time sequence of intermediate configurations expressed in either joint or Cartesian coordinate frames. The block diagram in Fig. 7.1 shows the parameters involved in the trajectory-planning problem. To solve the trajectory-planning problem the terminology used in trajectory planning is discussed first.

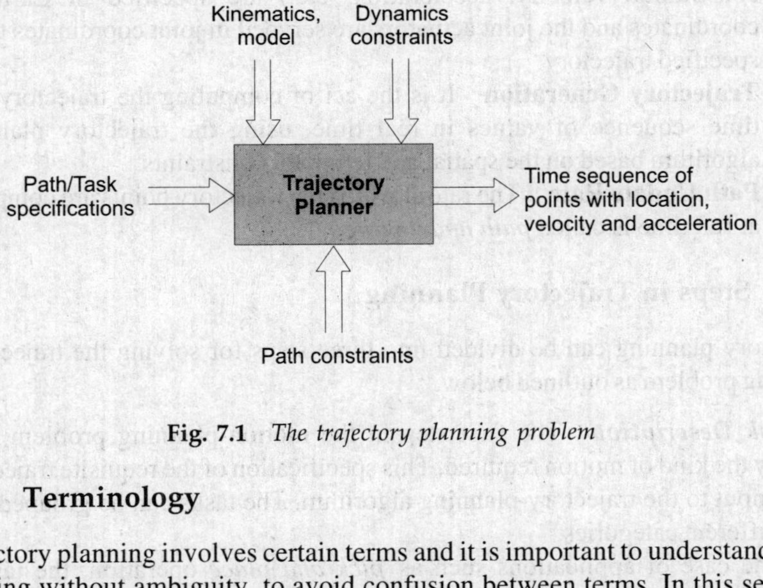

Fig. 7.1 *The trajectory planning problem*

7.1.1 Terminology

Trajectory planning involves certain terms and it is important to understand their meaning without ambiguity, to avoid confusion between terms. In this section, some of the frequently used terms in trajectory planning are defined.

(i) **Path** A path is the locus of points (either in the joint space or in the Cartesian space) to be traversed by the manipulator to execute the specified task. A path is a purely geometric (spatial) description of the motion.

(ii) **Trajectory** A trajectory is a path with specified qualities of motion, that is, a path on which a time law is specified in terms of velocities and/or accelerations at each point. In other words, a trajectory is the time sequence (or time history) of position, velocity, and acceleration for each joint or end-effector of the manipulator. A trajectory is both a spatial and temporal representation of motion. It can be specified either in joint space or in Cartesian space.

(iii) **Knot Points or Via Points** Knot points or via points or interpolation points are the set of intermediate locations between the start and goal points on the trajectory through which the manipulator must pass enroute to the destination.

(iv) **Spline** It is the smooth function that passes through the set of via points.

(v) **Joint Space Trajectory Planning** In joint space trajectory planning each path point is specified in terms of a desired position and orientation of the end-effector frame relative to the base frame. Each of these points is converted into a set of desired joint positions by application of inverse kinematics. A smooth function is then found for each of the joints, which passes through these points.

(vi) **Cartesian Space Trajectory Planning** In Cartesian space trajectory planning, the path is explicitly specified in the Cartesian space. The path constraints (velocity, acceleration, etc.) are specified in Cartesian coordinates and the joint actuators are servoed in joint coordinates to the specified trajectory.

(vii) **Trajectory Generation** It is the act of computing the trajectory as a time sequence of values in real time, using the trajectory planning algorithm based on the spatial and temporal constraints.

(viii) **Path Update Rate** The rate at which the trajectory points are computed at run time is called *path update rate*.

7.1.2 Steps in Trajectory Planning

Trajectory planning can be divided into three steps for solving the trajectory-planning problem as outlined below.

(i) Task Description The first step in the motion-planning problem is to identify the kind of motion required. This specification of the requisite trajectory is the input to the trajectory-planning algorithm. The tasks can be grouped into three different categories.

In the case of applications such as *pick and place* operation, the task is specified as initial and final end-effector location. This is called *point-to-point* (PTP) motion. In a point-to-point motion, except for the constraint that the motion be smooth, no particular specification about the intermediate locations of the end-effector is given and the planner is free to formulate any convenient path. In such cases, the user specifies only the goal point (location) in Cartesian space for a known initial point.

If in addition to start and finish points, a specific path between them is required to be traced by the end-effector in Cartesian space, this is called *continuous path* (CP) motion and trajectory. Operations such as arc welding and plotting are examples of continuous path motion. In such cases, user specifies the type and parameters of the path to be tracked.

There is a third type of task description, where more than two points of the path are specified. This is done to ensure better monitoring of the executed trajectory in case of applications similar to those discussed for point-to-point motion. A typical example is a pick-n-place operation where obstacles are present in the path. Here, the first point on the path is called the "initial" point, while the last point is called the "goal" point. The intermediate locations are the via points. Together, these are referred to as *path points*.

Apart from the required task attributes, the user can also include temporal attributes such as the travel time in any of the task specifications discussed above. The task definition must not be misunderstood to mean that end-effector or manipulator is only required to traverse a path or trajectory. The task performed by the tool or gripper at a specific point in PTP motion or along the path in CP motion is not a part of trajectory planning. It is an additional task for the "tool" which may or may not be controlled by the robot controller.

(ii) Selecting and Employing a Trajectory Planning Technique The various trajectory-planning techniques fall into one of the two categories: *Joint space techniques* or *Cartesian space techniques*. Next step in trajectory planning is to select a particular technique based on the task required.

In the case of point-to-point motion (with or without via points), joint space techniques are employed in which motion planning is done at the joint level. The joint-space planning schemes generate time-dependent functions (time history) of all joint variables and their first two derivatives to describe the desired motion of the manipulator.

For applications requiring continuous path motion, Cartesian space techniques are used. The Cartesian space-planning techniques provide time history of the location, velocity, and acceleration of the end-effector with respect to the base. The corresponding joint variables and their derivatives are computed, using inverse kinematics.

(iii) Computing the Trajectory The final step is to compute the time sequence of values attained by the functions generated from the trajectory planning technique. These values are computed at a particular path update or sampling rate. The path update rate in real time, lies between 20 Hz and 200 Hz in typical industrial manipulator systems.

7.2 JOINT SPACE TECHNIQUES

To plan a trajectory, the values of joint variables have to be determined from the end-effector location specified by the user. It is useful to recall that joint-space techniques are used for point-to-point motion with or without via points and each path point is specified by the end-effector position and orientation with respect to the base in Cartesian space.

The first step in these techniques is to obtain the corresponding set of joint variable values for each specified path point, either by employing the inverse kinematics algorithm or the variable can be directly recorded if trajectory planning is performed by *teaching-by-showing technique.*

The next step is to find a smooth function $q(t)$ for each of the n joints of an n-DOF manipulator, which interpolates the joint-variable vector for each given path point, with respect to imposed constraints. In general, following features are required in a joint space trajectory-planning algorithm:

- An important temporal constraint is that the travelling time between any two path points is the same for each joint so as to make all the joints reach their corresponding path-point locations simultaneously. This guarantees the attainment of the desired location by tool with respect to the base at each path point.
- Joint locations and velocities be continuous functions of time (continuity of accelerations may be imposed, too) so as to ensure smooth motion.
- The interpolating function should not be computationally intensive.
- Nonsmooth trajectories and other similar undesirable effects be minimized.

For moving the end-effector from starting (initial) position q^s to a goal (target) position q^g in a certain amount of time, it is seen that there are innumerable functions that satisfy the aforesaid criteria as shown in Fig. 7.2. The selection of appropriate polynomial function plays a pivotal role in joint-space planning. For making a choice of polynomial, two of the common approaches are:

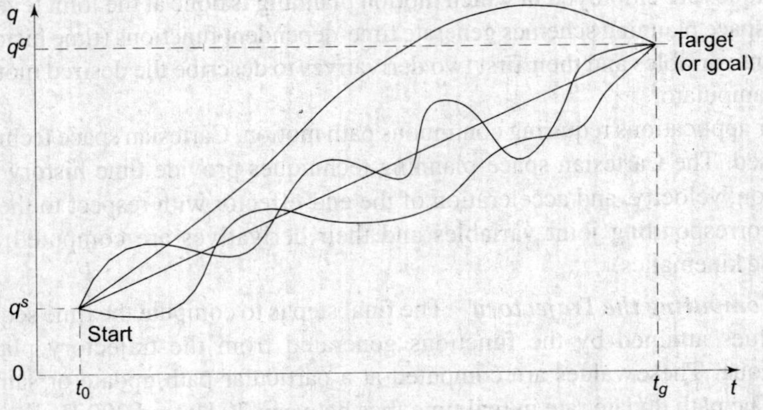

Fig. 7.2 *Smooth functions for motion to goal position*

(a) Choose the highest degree polynomial with at least as many coefficients as are the specified constraints.

(b) Split the trajectory into segments and use lower degree polynomial to interpolate each segment with smooth transition from one segment to next segment.

In the first approach, the higher degree polynomials tend to be more oscillatory and numerically less accurate. The second approach imposes additional constraints required for blending two segments for smooth motion.

Two polynomial functions are examined here, one in each of above approaches:

(i) a *cubic polynomial* in which cubics connect the path points of each joint in a smooth way, and

(ii) a *linear function with parabolic blends* in which each segment between two successive path points contains a linear function with parabolic blends near the path points.

Both these functions are widely used in joint-space techniques and are not very demanding from computational viewpoint. Before taking up these specific functions, general discussion on a p-degree polynomial function is carried out to illustrate the concepts of constraint identification and their use to compute polynomial coefficients.

7.2.1 Use of a *p*-Degree Polynomial as Interpolation Function

The selection of a single polynomial for the entire joint path depends on the number of constraints imposed and the type of motion desired. The minimum number of constraints for a smooth motion between two points are

(i) Initial position,
(ii) Initial velocity,
(iii) Final position, and
(iv) Final velocity.

With four constraints, a third-degree polynomial (cubic spline) with four coefficients can be used, that is,

$$q(t) = a_0 + a_1 t + a_2 t^2 + a_3 t^3 \tag{7.1}$$

If, in addition, the initial and final accelerations are also specified, then one may use a fifth-degree polynomial,

$$q(t) = a_0 + a_1 t + a_2 t^2 + a_3 t^3 + a_4 t^4 + a_5 t^5 \tag{7.2}$$

However, for practical purposes of simple pick-n-place task, just specifying a start and finish point is not enough for satisfactory performance. The reason is that while picking up an object the motion of the end-effector must be directed away from the supporting surface so that the gripper does not crash into the supporting surface. Thus, a "lift-off position" along the outward normal to the surface, from the initial position, is selected as additional specification. On the basis of similar considerations, a "set-down position" is selected to specify the correct approach direction. This gives, in all, four position constraints as

(i) Initial position,
(ii) Lift-off position,
(iii) Set-down position, and
(iv) Final position.

In a typical pick-n-place task, the manipulator begins moving at constant acceleration, away from the initial position in appropriate direction (for example, in vertically upward direction from a horizontal surface). After it has moved a safe distance, lift-off is complete. From then onwards, it moves at constant velocity, that is, with zero acceleration. The constant velocity motion continues until the set-down position from where it begins to decelerate ending at the final position with zero velocity.

A typical trajectory of this kind is shown in Fig. 7.3, which has six constraints, four position constraints given above and two acceleration constraints—constant acceleration in the initial phase and the constant deceleration in the final phase. The two velocity constraints are in fact subsumed by the two acceleration constraints.

These concepts for the end-effector motion can be extended to position, velocity, and acceleration of the joints. Further, additional constraints are required for continuity of each joint variable and its derivatives at every path point. In summary, the constraints that can be applied to a joint for joint interpolated trajectory planning can be grouped into four categories: initial point constraints, intermediate constraints, final point constraints, and time constraints. There are a number of possible constraints in each of these categories.

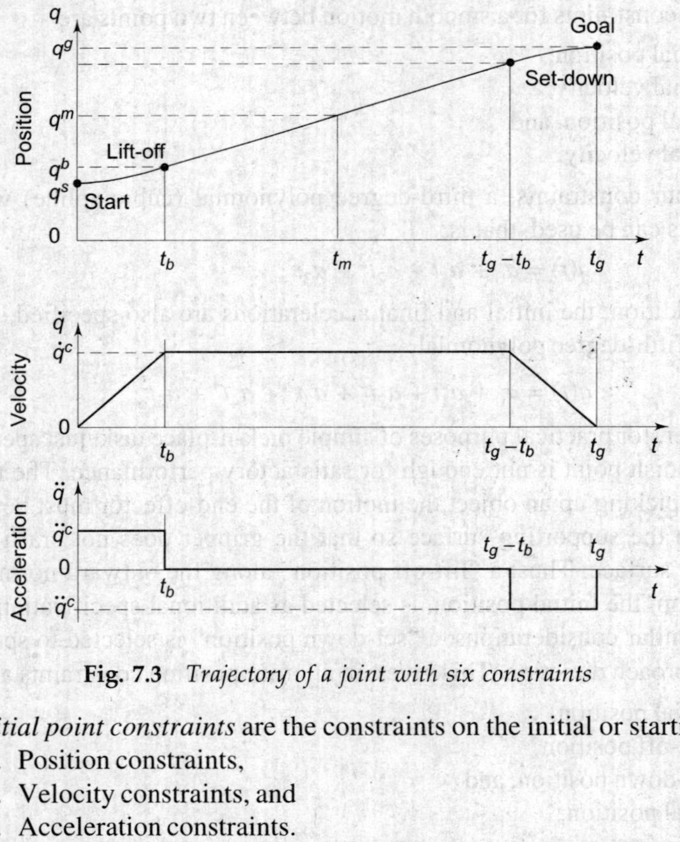

Fig. 7.3 *Trajectory of a joint with six constraints*

(a) *Initial point constraints* are the constraints on the initial or starting point.
   1. Position constraints,
   2. Velocity constraints, and
   3. Acceleration constraints.

(b) *Intermediate constraints* are the constraints on the intermediate points of the trajectory.

1. Lift-off position,
2. Continuity at lift-off position,
3. Continuity of velocity at lift-off point,
4. Continuity of acceleration at lift-off point,
5. Set-down position,
6. Continuity at set-down position,
7. Continuity of velocity at set-down point, and
8. Continuity of acceleration at set-down point.

Additional via points or knot points may be specified and at each of such point velocity/acceleration may be specified with additional constraint of continuity for smooth transition.

(c) *Final point constraints* are the constraints on the final or goal or target point of the trajectory.
1. Position,
2. Velocity, and
3. Acceleration.

(d) *Time constraints are* the constraints imposed by the cycle time for the task.
1. Total traversal time,
2. Traversal time for each segment, and
3. Duration of a specific polynomial interpolating a trajectory in case of blended trajectories.

A suitable polynomial function is selected such that the required constraints are satisfied and the trajectory is smooth. It is known that a p-degree polynomial has $(p+1)$ coefficients and can therefore satisfy $(p+1)$ constraints. Thus, for a trajectory with $(p+1)$ constraints, a p-degree polynomial can be the general choice for trajectory planning. However, if the degree of polynomial is high (say > 5), it becomes computationally intensive and it tends to cause extraneous motion.

The complexity can be reduced by splitting the trajectory into two or more segments so that in each segment polynomials of lower degree can be used to interpolate trajectory. For example, if a trajectory is split with two intermediate points, there will be three segments. For each of these three segments, lower degree polynomials can be used to interpolate the joint positions.

The type of polynomial used for interpolation of the trajectory also has options like

1. Linear position or constant velocity trajectory
2. Parabolic position or linear velocity or constant acceleration trajectory

Some other complex possibilities of polynomials are:

(a) *Linear trajectory with parabolic blends* — The first segment is interpolated with a quadratic polynomial (parabola) specifying trajectory from initial position to the "blend" point. The second segment is interpolated as linear or constant velocity followed by a quadratic polynomial in the last (third) segment, identical to first segment.

(b) *4-3-4 Trajectory* — A fourth-degree polynomial in the first segment followed by a cubic polynomial in the second segment followed by a fourth-degree polynomial in the last segment.

(c) *3-5-3 Trajectory* — A third-degree polynomial or cubic spline for the first segment, a fifth-degree polynomial for the second segment, and a third-degree polynomial in the last segment.

In segmented trajectories, the trajectories must blend smoothly at the intermediate points. These intermediate points may, sometimes, be only "virtual points" and the specification of the location of these points may not be required. Only the continuity constraints are imposed on them to solve the coefficients of the polynomials. An estimation of the computations involved with segmented trajectories is made as follows.

Suppose the trajectory is split into segments such that there are k path points. Thus, there will be (k–2) intermediate or via points and (k–1) polynomials will be required to connect the k path points. For each of the (k–2) via points, there will be (p+1) constraints for the p-degree polynomial and, say, there are six constraints on the initial and final points. Hence, in all there will be $(p+1)(k-2)+6$ constraints that must be solved to determine the trajectory consisting of (k–1) p-degree polynomials.

A set of cubic polynomials (p=3) can be used for interpolation of trajectory, preserving the continuity in the first and second derivatives at the interpolation points. Such cubic polynomials are known as *cubic-spline functions*. The cubic-spline trajectories are most preferred among the interpolating functions as the cubic polynomial is the lowest degree polynomial for which both first and second derivatives exist, allowing for continuity in velocity and acceleration, and it is computationally more efficient. The number of constraint equations for a k path point trajectory interpolated with cubic polynomials will be $4k$–2. Cubic polynomial and other trajectories are discussed in the following sections. For example, a five-cubic polynomials interpolation will have five segments and six path or interpolation points and the number of constraint equations will be 22.

7.2.2 Cubic Polynomial Trajectories

The problem of obtaining cubic polynomial trajectories is considered first, for a point-to-point motion without via points and then it is extended to the case with via points.

Point-to-Point motion without via points In this scheme, the goal point and travel time are specified by the user. The set of joint-variable values for the given goal point are obtained from the inverse kinematics. It is assumed that the joint-variable values corresponding to the initial location of the end-effector are known.

A cubic trajectory is obtained for each joint variable, $q(t)$, of the n-DOF manipulator. The determination of the interpolation cubic polynomial is carried out for each of the n-joint variables.

Let q^s and q^g be the starting and goal-point values, respectively, of the joint variable q. Note that the generalized joint variable q_i denotes the joint variable (θ_i or d_i) for joint i of the n-DOF manipulator. Therefore, to denote time-dependent values of q_i, at different path points, a superscript is used, for example, q_i^j implies the value of $q_i(t)$ at path point j. The subscript i for joint i is dropped whenever only one joint variable is considered, for the sake of clarity. The manipulator motion begins at time $t = 0$ ($t_0 = 0$) and ends at $t = t_g$, that is, the travel time or time to reach goal is t_g.

For a smooth motion between the initial and goal points, the functions $q(t)$ and $\dot{q}(t)$ have to be smooth. This requirement imposes two constraints each on the joint position and velocity functions. Further, $q(t)$ is zero for $t < 0$ and $t > t_g$. This, along with the continuity requirements gives the four constraints as

$$\begin{aligned} q(0) &= q^s \\ q(t_g) &= q^g \\ \dot{q}(0) &= 0 \\ \dot{q}(t_g) &= 0 \end{aligned} \tag{7.3}$$

To satisfy these four constraints, the polynomial $q(t)$ must be of order three, that is, a cubic polynomial with four coefficients. This justifies our assumption that a cubic polynomial would give a smooth motion profile for joint variable q. Therefore, to describe the joint motion, assume that the cubic polynomial is

$$q(t) = a_0 + a_1 t + a_2 t^2 + a_3 t^3 \tag{7.4}$$

which gives a parabolic velocity profile

$$\dot{q}(t) = a_1 + 2a_2 t + 3a_3 t^2 \tag{7.5}$$

and a linear acceleration profile

$$\ddot{q}(t) = 2a_2 + 6a_3 t \tag{7.6}$$

Applying the constraints of Eq. (7.3) to Eqs. (7.4) and (7.5), gives the following set of four equations in four unknowns:

$$\begin{aligned} a_0 &= q^s \\ a_0 + a_1 t_g + a_2 t_g^2 + a_3 t_g^3 &= q^g \\ a_1 &= 0 \\ a_1 + 2a_2 t_g + 3a_3 t_g^2 &= 0 \end{aligned} \tag{7.7}$$

The solution of Eq. (7.7) gives the coefficients of cubic polynomial, Eq. (7.4), as

$$\begin{aligned} a_0 &= q^s \\ a_1 &= 0 \\ a_2 &= \frac{3}{t_g^2}(q^g - q^s) \\ a_3 &= -\frac{2}{t_g^3}(q^g - q^s) \end{aligned} \tag{7.8}$$

Thus, the cubic polynomial to interpolate the path connecting the initial joint position to final joint position is

$$q(t) = q^s + \frac{3}{t_g^2}(q^g - q^s)t^2 - \frac{2}{t_g^3}(q^g - q^s)t^3 \tag{7.9}$$

It may be noted that this solution is for the case when the velocity is zero both at the start and finish points.

The acceleration profile, though linear, has discontinuity at initial and final positions and the motion will be jerky at the start and at the goal point due to this. The variation of the joint variable, velocity, and acceleration are shown in Fig.7.4. The acceleration at the start and goal points is difficult to realize with physical actuators.

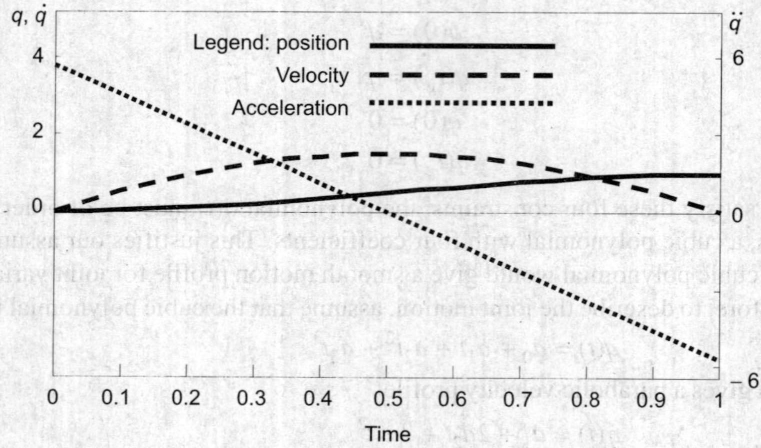

Fig. 7.4 *Cubic polynomials trajectory for PTP motion*

Point-to-point motion with via points Now consider the problem of determining the cubics that connect the given "k" path points smoothly. If the manipulator comes to rest at each via point, then the cubic solution of above section can be applied between each pair of successive via points. A different approach is needed when the manipulator is required to pass the via points without stopping or coming to rest at the via points. For a path with $\boldsymbol{k}$ path points, there are (k–2) intermediate or via points, and (k–1) polynomials are required to connect the k path points in a smooth way.

Usually, each path point is specified in terms of the desired end-effector position and orientation. Similar to the single goal point case, each of these path points is transformed into a set of joint-variable values by inverse kinematics. Assume that q_j denotes the value taken by joint variable q corresponding to path point j and the cubic polynomial connecting path points j and (j+1) is $P_j(t)$. This cubic is defined in the time interval $t_{j-1} < t < t_j$, with $T_j = t_j - t_{j-1}$ denoting the travel time between path points j and (j+1). These time intervals for k path point with (k–2) via points are shown in Fig. 7.5. The total travel time for the path is $T(=t_g)$ with

$$T = \sum_{m=1}^{k-1} T_m \tag{7.10}$$

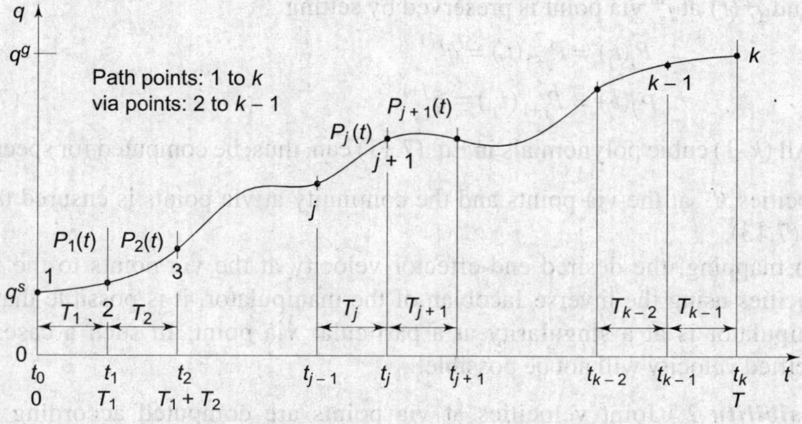

Fig. 7.5 *Time intervals for k path points and (k–2) via points*

Thus, the joint position function, $q(t)$ for the entire path over the time interval $0 < t < T$ is defined by $(k–1)$ cubic polynomials as

$$q(t) = \begin{cases} P_1(t) & 0 < t < T_1 \text{ or } t_0 < t < t_1 \\ P_2(t - T_1) & T_1 < t < T_1 + T_2 \\ \vdots & \\ P_j(t - \sum_{m=0}^{j-1} T_m) & \sum_{m=0}^{j-1} T_m < t < \sum_{m=0}^{j} T_m \text{ or } T_{j-1} < t < T_j \\ \vdots & \\ P_{k-1}(t - \sum_{m=0}^{k-2} T_m) & \sum_{m=0}^{k-2} T_m < t < \sum_{m=0}^{k-1} T_m \text{ or } T_{K-2} < t < T_{K-1} \end{cases} \tag{7.11}$$

For smooth trajectory at each via point, additional constraints are required. Once these constraints are specified, the $(k–1)$ cubic polynomials are determined as before. There are many possible ways of specifying the additional constraints at via points. Three possibilities are presented here.

Possibility 1 The desired velocity at each via point is specified in terms of Cartesian linear and angular velocity of the tool frame at that instant. These desired velocities are mapped to the joint velocities using the inverse Jacobian. The initial and goal point joint velocities are set to zero to ensure the continuity of $q(t)$ at the end points, as before. Thus, the set of constraints imposed on j^{th} cubic $P_j(t)$ between via points j and $(j+1)$ is given by

$$\begin{aligned} P_j(t_{j-1}) &= q^j \\ P_j(t_j) &= q^{j+1} \\ \dot{P}_j(t_{j-1}) &= \dot{q}^j \\ \dot{P}_j(t_j) &= \dot{q}^{j+1} \end{aligned} \tag{7.12}$$

The four constraints in Eq. (7.12) represent a set of four equations from which the four unknown coefficients of cubic $P_j(t)$ can be computed. The continuity of q^j and $\dot{q}^j(t)$ at j^{th} via point is preserved by setting

$$P_j(t_j) = P_{j+1}(t_j) = q^{j+1}$$

and
$$\dot{P}_j(t_j) = \dot{P}_{j+1}(t_j) = \dot{q}^{j+1} \tag{7.13}$$

All (k–1) cubic polynomials in Eq. (7.11) can, thus, be computed for specified velocities $\dot{q}^j$ at the via points and the continuity at via points is ensured using Eq. (7.13).

In mapping, the desired end-effector velocity at the via points to the joint velocities using the inverse Jacobian of the manipulator, it is possible that the manipulator is at a singularity at a particular via point. In such a case, the specified velocity will not be possible.

***Possibility* 2** Joint velocities at via points are computed according to a heuristic criterion. In this case, the constraints imposed on the cubics are same as those used in the previous case but the method of specifying via-point joint velocity is different.

The joint variables $q(t)$ at the path points along with the travel time T, are specified by the user. The joint velocities at via points are assigned using a criterion, which preserves the continuity of $q(t)$. One such criterion is to use the average velocity as

$$\dot{q}^1 = 0$$

$$\dot{q}^j = \begin{cases} 0 & \text{if } \operatorname{sign}(\dot{\bar{q}}^{j-1}) \neq \operatorname{sign}(\dot{\bar{q}}^j) \\ \frac{1}{2}(\dot{q}^{j-1} + \dot{q}^j) & \text{if } \operatorname{sign}(\dot{\bar{q}}^{j-1}) = \operatorname{sign}(\dot{\bar{q}}^j) \end{cases} \quad \text{for } j = 2, \cdots, (k-1) \tag{7.14}$$

$$\dot{q}^k = 0$$

where $\dot{\bar{q}}^j = \dfrac{q^{j+1} - q^j}{T_j}$ is the average velocity for segment j.

These values of $\dot{q}^j$ given by Eq. (7.14) are used in Eq. (7.11) to obtain the unknown coefficients of the cubics.

***Possibility* 3** Another alternative to generate cubics, interpolating more than two path points, is to cause the joint accelerations to be continuous at the via points. This implies that the velocities at via points are automatically chosen by the system to satisfy acceleration continuity, which also ensures velocity continuity at via points.

Under these conditions, constraints at the end-points are same as the case of single goal point, that is, joint velocities are continuous at the end-points but accelerations are discontinuous. The velocity and acceleration continuity restrictions at the via points give the following constraints

$$
\begin{aligned}
&P_1(t_0) = q^1 = q^s \\
&\dot{P}_1(t_0) = \dot{q}^1 = \dot{q}^s = 0 \\
&\cdots \\
&\left.\begin{aligned}
&P_{j-1}(t_j) = q^j \; P_{j-1}(t_j) = q^j \\
&P_{j-1}(t_j) = P_j(t_j) \\
&\dot{P}_{j-1}(t_j) = \dot{P}_j(t_j) \\
&\ddot{P}_{j-1}(t_j) = \ddot{P}_j(t_j)
\end{aligned}\right\} \quad \text{for } j = 2, \cdots, (k-1) \qquad (7.15) \\
&\cdots \\
&P_k(t_k) = q^k = q^g \\
&\dot{P}_k(t_k) = \dot{q}^k = \dot{q}^g = 0
\end{aligned}
$$

Equation (7.15) represents a set of $4(k-1)$ equations from which the four unknown coefficients for each of the $(k-1)$ cubics can be obtained.

It is important to note that Eq. (7.15) guarantees that the path points are reached as per the temporal constraints specified as well as ensures the continuity of joint velocity and acceleration functions at the via points.

The three possibilities discussed above to connect the cubics smoothly at the via points require different amount of computations. The first requires the least computations, whereas the last one is most computationally intensive. The possibility 3 ensures the velocity and acceleration continuity at the via points giving a smooth motion throughout, while in possibility 2, jerks are possible at via points because acceleration continuity is not enforced. The heuristics used would also influence the trajectory. Possibility 1 is useful when it is desired to have specific velocities at the via points. This obviously requires the knowledge of desired velocities at via points, which may be a burden.

7.2.3 Linear Function with Parabolic Blends

Another possible trajectory is to use linear interpolation from start to goal point. This would cause the velocity to be discontinuous at the beginning and end of motion and would require infinite acceleration. Also, in the case of cubic trajectory for PTP motion without via points (section 7.2.2), the discontinuity of acceleration at the end-points is observed (see Fig.7.4) and the acceleration is not physically realizable. One method to avoid this difficulty is to use higher degree polynomial with more constraints, requiring more complex computations, as discussed in section 7.2.1. Alternately, it is possible to use a blended polynomial, which allows for verification whether the resulting velocities and accelerations can be supported by the physical mechanical manipulator.

A trapezoidal velocity profile, for example, imposes a constant acceleration in start phase, a cruise velocity, and a constant deceleration in the final phase, (see Fig. 7.3). The resulting trajectory has a linear segment with parabolic segments at both ends. In general, a blended trajectory is constructed using polynomial interpolation function with blending functions at both ends, which are splined together so that the entire path has continuity of position and velocity, including the smooth motion at resulting via points.

The trapezoidal velocity profile trajectory planning is discussed, first for point-to-point motion without via points and later, it is extended to the case of point-to-point motion with via points.

Blended trajectory for point-to-point motion without via points For the given start and target points, a smooth trajectory, which consists of a linear segment with parabolic blends near the end-points, is used. During the blend portion of the trajectory, constant acceleration is used to achieve a smooth velocity transition. The two parabolic blends are assumed to be identical, that is, they have the same constant acceleration. Therefore, parabolic blends near the path points are of same duration (t_b) and the whole trajectory is symmetric about the halfway point in time and position (t_m, q^m). Thus, the trajectory gives continuity of position, velocity, and acceleration throughout. The time variation of position for this trajectory is shown in Fig. 7.6.

For this type of trajectory planning, the user input is the target point position q^g, travel time t_g with known initial point q^s and the constant blend acceleration $\ddot{q}^c$ in two-end segments.

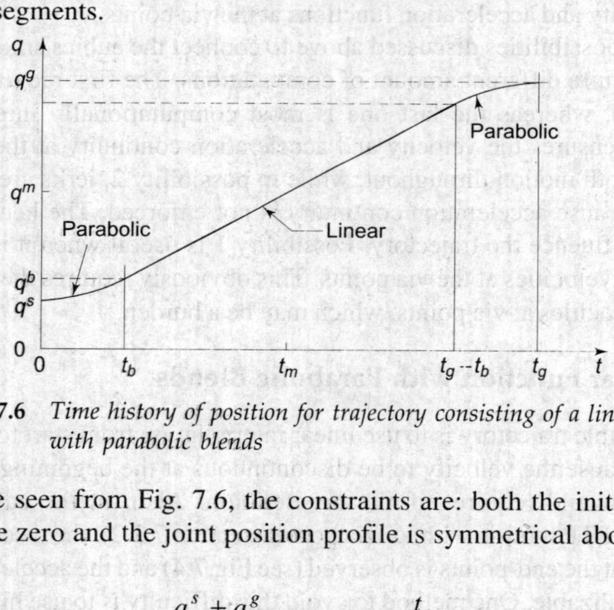

Fig. 7.6 *Time history of position for trajectory consisting of a linear segment with parabolic blends*

As can be seen from Fig. 7.6, the constraints are: both the initial and final velocities are zero and the joint position profile is symmetrical about the point (q^m, t_m) with

$$q^m = \frac{q^s + q^g}{2} \quad \text{and} \quad t_m = \frac{t_g}{2} \tag{7.16}$$

The trajectory has to satisfy the additional constraint for smooth transition from parabolic blend to linear segment. For joint velocity to be continuous, the velocity at the end of the first blend (or beginning of second blend) must be equal to the velocity of the linear segment, that is, at the blend point (q^b, t_b)

$$\ddot{q}^c t_b = \frac{q^m - q^b}{t_m - t_b} \tag{7.17}$$

where q^b is the value of the joint variable at the end of the parabolic segment, at time t_b. For constant acceleration $\ddot{q}^c$, the joint position q^b, at the end of the first blend is given by

$$q^b = q^s + \frac{1}{2}\ddot{q}^c t_b^2 \tag{7.18}$$

Combining Eqs (7.16), (7.17), and (7.18) gives quadratic equation in t_b as

$$\ddot{q}^c t_b^2 - \ddot{q}^c t_g t_b + (q^g - q^s) = 0 \tag{7.19}$$

To predict the time history of q, $\dot{q}$, and $\ddot{q}$, with given q^s, q^g, and t_g, the only unknown parameter is t_b, the blend duration. Solving Eq. (7.19) for t_b, gives

$$t_b = \frac{1}{2}t_g - \frac{1}{2}\sqrt{\frac{t_g^2\ddot{q}^c - 4(q^g - q^s)}{\ddot{q}^c}} \tag{7.20}$$

Note that there are many solutions to Eq. (7.20), depending on the value of $\ddot{q}^c$ but the answer is always symmetric about the point (q^m, t_m). The constraint on the choice of acceleration during the blend, $\ddot{q}^c$, is

$$\left|\ddot{q}^c\right| \geq \frac{4\left|q^g - q^s\right|}{t_g^2}; \text{ and } \ddot{q}^c \neq 0 \tag{7.21}$$

Solution to Eq. (7.19) will not exist if the above condition is not satisfied. When equality occurs in Eq. (7.21), the two blends meet at t_m and there is no linear or constant velocity segment and only acceleration and deceleration segments are present, giving a triangular velocity profile.

On the other hand, as the acceleration becomes larger and larger, t_b becomes shorter and shorter. The trajectory tends to linear interpolation with acceleration tending to infinity or t_b tending to zero.

The trapezoidal velocity profile (linear interpolation with parabolic blends) trajectory using Eq. (7.20) generates the following sequence of polynomials for the time history of joint position, velocity and acceleration.

$$q(t) = \begin{cases} q^s + \frac{1}{2}\ddot{q}^c t^2 & 0 \leq t \leq t_b \\ q^s - \frac{1}{2}\ddot{q}^c t_b^2 + \ddot{q}^c t_b t & t_b < t \leq t_g - t_b \\ q^g - \frac{1}{2}\ddot{q}^c (t_g - t)^2 & t_g - t_b < t \leq t_g \end{cases} \tag{7.22}$$

$$\dot{q}(t) = \begin{cases} \ddot{q}^c t & 0 \leq t \leq t_b \\ \ddot{q}^c t_b & t_b < t \leq t_g - t_b \\ \ddot{q}^c (t_g - t) & t_g - t_b < t \leq t_g \end{cases} \tag{7.23}$$

$$\ddot{q}(t) = \begin{cases} \ddot{q}^c & 0 \leq t \leq t_b \\ 0 & t_b < t \leq t_g - t_b \\ -\ddot{q}^c & t_g - t_b < t \leq t_g \end{cases} \tag{7.24}$$

These polynomials confirm with Fig. 7.3 and Fig. 7.6. Alternative trajectory solutions are possible if cruise velocity $\dot{q}^c$ is specified instead of constant blend acceleration $\ddot{q}^c$.

Blended trajectory for point-to-point motion with via points Polynomial with linear-parabolic-blend splines can also be used to interpolate the given path points in a smooth way. As shown in Fig. 7.7, a smooth trajectory for a joint variable, q, is constructed with linear function connecting via points, and parabolic blends are added to the linear segments near the via points.

To maintain the continuity at via points, the parabolic blends of two segments must also have a smooth transition, This is achieved by choosing a continuous parabola for the two-blend segments at a via point. The result of this is that the polynomial $q(t)$ does not touch the via points and the via points become virtual or pseudo via points. This means that different methods to compute the blend duration and trajectory for end-points and for via points have to be employed.

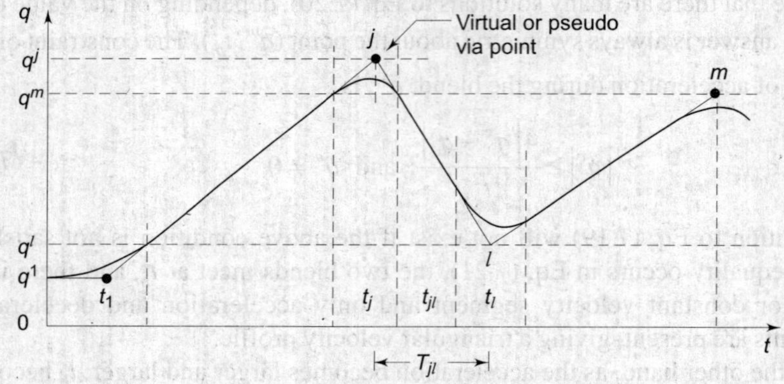

Fig. 7.7 *Multisegment linear-parabolic-blend splines*

The path points, travel time between successive path points and constant blend acceleration at path points constitute the input for the problem. The only missing parameter required to determine the time history of q, $\dot{q}$ and $\ddot{q}$ is the blend duration at each path point.

Assume that there are k path points and three consecutive path points are j, l, and m. Following notations are defined.

- q^j – value of joint variable q corresponding to path point j,
- $|\ddot{q}^j|$ – magnitude of constant blend acceleration at path point j,
- T_{jl} – total travel time between path points j and l,
- t_{jl} – travel time for the linear segment between path points j and l,
- t_j – duration of the blend around path point j,
- $\dot{q}^{jl}$ – constant joint velocity (or slope) of the linear segment connecting path points j and l.

For the via points, the blend duration t_{jl} near via point l is computed from specified acceleration at the via point $\ddot{q}^l$ and the constant linear velocities in two segments $\dot{q}^{jl}$ and $\dot{q}^{lm}$.

$$t_l = \frac{\dot{q}^{jl} - \dot{q}^{lm}}{\ddot{q}^l} \tag{7.25}$$

where $\ddot{q}^l = \text{sign}(\dot{q}^{jl} - \dot{q}^{lm})|\ddot{q}^l|$

The linear velocity in segment jl is given by

$$\dot{q}^{jl} = \frac{q^l - q^j}{T_{jl}} \tag{7.26}$$

The duration of the linear segment t_{jl}, assuming parabolic blend to be symmetric around the path points, is

$$t_{jl} = T_{jl} - \frac{1}{2}t_j - \frac{1}{2}t_l \tag{7.27}$$

Once again, there are many possible solutions, depending on the value of accelerations used in each blend. To compute the blend duration at the initial point $j = 1$, the fact that the joint velocity at the end of blend is the same as the velocity of the linear segment 12 (t_{12}), is used. This gives

$$\ddot{q}^1 t_1 = \frac{q^2 - q^1}{T_{12} - \frac{1}{2}t_1} \tag{7.28}$$

where $\ddot{q}^1 = \text{sign}(\dot{q}^2 - \dot{q}^1)|\ddot{q}^1|$

From Eq. (7.28) the blend time t_1 at the initial point and using t_1, the linear segment velocity and duration, $\dot{q}^{12}$ and t_{12} are computed as

$$t_1 = T_{12} - \sqrt{\frac{T_{12}^2\ddot{q}^1 - 2(q^2 - q^1)}{\ddot{q}^1}} \tag{7.29}$$

$$t_{12} = T_{12} - \frac{1}{2}t_2 - t_1 \tag{7.30}$$

and

$$\dot{q}^{12} = \frac{q^2 - q^1}{T_{12} - \frac{1}{2}t_1} \tag{7.31}$$

Finally, the blend duration at the goal point is determined. Similar to the starting point, the velocity continuity constraint, in the middle segments connecting path points (k–1) and k, gives

$$\ddot{q}^k t_k = \frac{q^k - q^{k-1}}{T_{(k-1)k} - \frac{1}{2}t_k} \tag{7.32}$$

where $\ddot{q}^k = \text{sign}(\dot{q}^{k-1} - \dot{q}^k)|\ddot{q}^k|$

The solution for blend duration, linear velocity, and its duration are

$$t_k = T_{(k-1)k} - \sqrt{\frac{T_{(k-1)k}^2 \ddot{q}^k - 2(q^k - q^{k-1})}{\ddot{q}^k}} \quad (7.33)$$

$$t_{(k-1)k} = T_{(k-1)k} - \frac{1}{2}t_{k-1} - t_k \quad (7.34)$$

$$\dot{q}^{(k-1)k} = \frac{q^k - q^{k-1}}{T_{(k-1)k} - \frac{1}{2}t_k} \quad (7.35)$$

Thus, all the requisite parameters to determine the time history of joint position, velocity, and acceleration are obtained using Eqs (7.25) through (7.35). The acceleration at each via point must be sufficiently large, as discussed in single segment case [see Eq. (7.21)]. To make computations simpler, same value of blend acceleration may be used for each via point.

7.3 CARTESIAN SPACE TECHNIQUES

In the previous section, some approaches to joint-space trajectory planning have been discussed. Another approach to trajectory planning is to use Cartesian coordinates and Cartesian space. In Cartesian space, the position and orientation of a rigid body can be clearly defined.

The problem of planning trajectories that enable the manipulator's end-effector to track a given path in Cartesian space is investigated in this section. The user specifies the desired end-effector path, the travelling time, and the tool orientations along the path.

To tackle the aforesaid problem, a parametric description for the desired end-effector path is required. This description specifies the spatial attributes of the requisite motion with respect to the base frame. First, the mathematics of parametric description of the path in Cartesian space is briefly reviewed. Then, some common techniques for Cartesian space trajectory planning are discussed.

7.3.1 Parametric Description of a Path

In parametric representation, the path (or curve) is represented as a function of single parameter, often this parameter is time. The arc length c of the curve is a function of time and can be used as parameter for the parametric representation of curves. Many other parameters can be used to describe a given path. However, in this discussion only the path coordinate c is used as the path parameter.

Figure 7.8 shows a Cartesian space path, $\mathcal{P}$ to be tracked by the end-effector of a given n-DOF manipulator, and the base frame $\{0\}$. If the path lies in plane it is a plane path, otherwise it is a 'twisted path'. Most tool-point trajectories are twisted paths.

Let P_s and P_g be the end-points of an open path $\mathcal{P}$ and $F(c)$ be a continuous vector function defined in the interval $[c_s, c_g]$. The path coordinate c of a generic point P on path $\mathcal{P}$ is the length of arc of $\mathcal{P}$ with extremes P and P_s. Usually, the starting point on the path P_s is the path coordinate origin and at origin of path, $c = 0$. This means that for any point P on the path, the path coordinate specifies the arc length of the path between P_s and P.

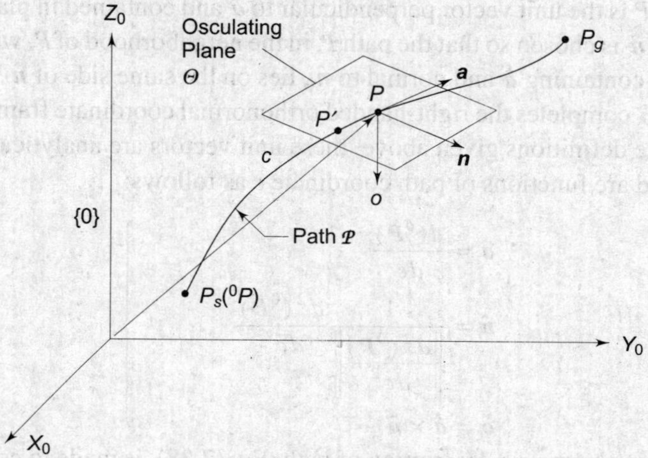

Fig. 7.8 *Path $\mathcal{P}$ relative to frame {0}*

It then follows that the path coordinate c can be used to express any point P on the path with respect to frame {0} as

$$ {}^{n}\boldsymbol{P} = \boldsymbol{F}(c) \tag{7.36}$$

Note that in Eq. (7.36), ${}^{0}\boldsymbol{P}$ is a 3×1 position vector of point P with respect to frame {0}. The vector function $F(c)$, thus, analytically expresses any point on the path $\mathcal{P}$ with respect to frame {0}. Hence, the path $\mathcal{P}$ can also be written as

$$ {}^{0}\mathcal{P} = \boldsymbol{F}(c) \tag{7.37}$$

Three unit vectors can be used to express point P, which form a right-handed orthonormal coordinate system with three orthonormal planes as shown in Fig. 7.9.

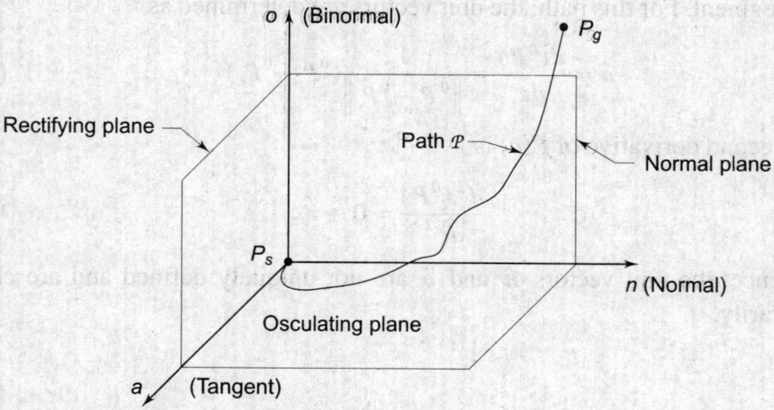

Fig. 7.9 *The orthonormal coordinate system for path $\mathcal{P}$*

First unit vector is the tangent unit vector $\hat{a}$, pointing in the increasing direction of c. The unit vector $\hat{a}$ lies in the osculating plane Θ at point P. This plane is the limit position of the plane containing the unit vector $\hat{a}$ and a point $P' \in \mathcal{P}$, when P' tends to P along the path, as shown in Fig. 7.8. The normal unit vector $\hat{n}$ at P is the unit vector perpendicular to $\hat{a}$ and contained in plane Θ. The direction of $\hat{n}$ is chosen so that the path $\mathcal{P}$, in the neighborhood of P, with respect to the plane containing $\hat{a}$ and normal to $\hat{n}$, lies on the same side of $\hat{n}$. Binormal unit vector $\hat{o}$ completes the right-handed orthonormal coordinate frame $\{n\ o\ a\}$.

As per the definitions given above, these unit vectors are analytically related to path $\mathcal{P}$ and are functions of path coordinate c as follows:

$$\left.\begin{aligned} \hat{\boldsymbol{a}} &= \frac{d({}^0\boldsymbol{P})}{dc} \\ \hat{\boldsymbol{n}} &= \frac{1}{\left\| \dfrac{d^2({}^0\boldsymbol{P})}{dc^2} \right\|} \frac{d^2({}^0\boldsymbol{P})}{dc^2} \\ \hat{\boldsymbol{o}} &= \hat{\boldsymbol{a}} \times \hat{\boldsymbol{n}} \end{aligned}\right\} \tag{7.38}$$

Use of this parametric description of path, Eq. (7.38), is made to compute the trajectory in Cartesian space. This is illustrated for two frequently encountered Cartesian space paths in manipulation tasks: straight line and circular motion, in the next sections.

7.3.2 A Straight-Line Path

Consider a straight-line path with start and goal points ${}^0\boldsymbol{P}_s$ and ${}^0\boldsymbol{P}_g$. Any point on the path can be expressed in terms of path parameter c, as

$$\boldsymbol{F}(c) = {}^0\boldsymbol{P} = {}^0\boldsymbol{P}_s + \frac{c}{\left\| {}^0\boldsymbol{P}_g - {}^0\boldsymbol{P}_s \right\|}({}^0\boldsymbol{P}_g - {}^0\boldsymbol{P}_s) \tag{7.39}$$

Note that $\boldsymbol{P}(0) = \boldsymbol{P}_s$ and $P\left\| {}^0\boldsymbol{P}_g - {}^0\boldsymbol{P}_s \right\| = \boldsymbol{P}_g$, represents the overall length of line segment. For this path, the unit vectors are determined as

$$\hat{\boldsymbol{a}} = \frac{d({}^0\boldsymbol{P})}{dc} = \frac{1}{\left\| {}^0\boldsymbol{P}_g - {}^0\boldsymbol{P}_s \right\|}({}^0\boldsymbol{P}_g - {}^0\boldsymbol{P}_s) \tag{7.40}$$

The second derivative of $\boldsymbol{F}(c)$ is

$$\frac{d^2({}^0\boldsymbol{P})}{dc^2} = 0 \tag{7.41}$$

Hence, the unit vectors $\hat{\boldsymbol{n}}$ and $\hat{\boldsymbol{o}}$ are not uniquely defined and are chosen arbitrarily.

7.3.3 A Circular Path

The circular path $\mathcal{P}(c)$, shown in Fig. 7.10, is specified using the following parameters

(a) Radius of circular path is ρ in plane $x'y'$.
(b) Unit vector of the circle axis, $\hat{\boldsymbol{r}}$, at the center of circle, directed according to right-hand rule,
(c) A point D on the axis of circle passing through the center of circle. Its position vector in frame {0} is ${}^0\boldsymbol{D}$.
(d) Position vector ${}^0\boldsymbol{P}_j$ in frame {0} of point on the circle (this serves as the path coordinate origin).

All the above parameters are expressed in terms of base frame {0}.

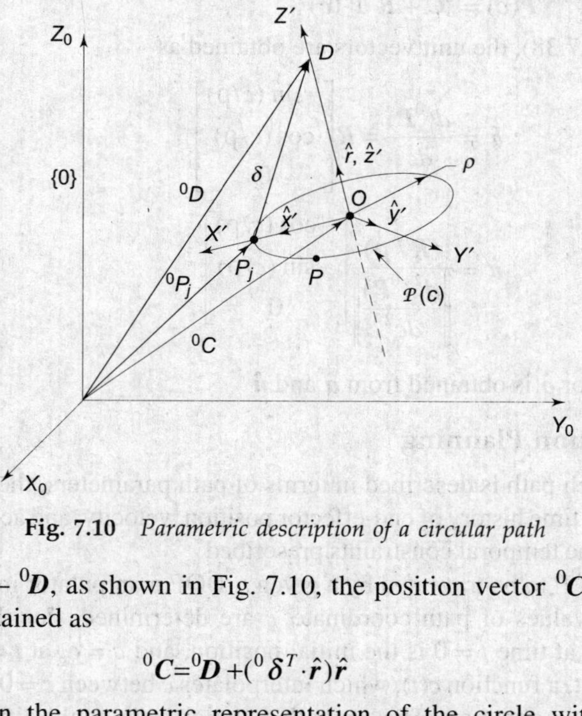

Fig. 7.10 *Parametric description of a circular path*

If ${}^0\delta = {}^0\boldsymbol{P}_j - {}^0\boldsymbol{D}$, as shown in Fig. 7.10, the position vector ${}^0\boldsymbol{C}$ of the circle centre O is obtained as

$$ {}^0\boldsymbol{C} = {}^0\boldsymbol{D} + ({}^0\boldsymbol{\delta}^T \cdot \hat{\boldsymbol{r}})\hat{\boldsymbol{r}} \tag{7.42} $$

First, obtain the parametric representation of the circle with respect to frame$\{x' \quad y' \quad z'\}$ defined as follows: $\hat{\boldsymbol{x}}'$ is the unit vector pointing from circle center O (which also serves as the origin O') towards $\boldsymbol{P}_j$, $\hat{z}'$ is the unit vector along circle axis $\hat{\boldsymbol{r}}$, while $\hat{\boldsymbol{y}}'$ completes the right-handed orthonormal coordinates frame.

Any point P on the circle $\mathcal{P}(c)$ can easily be described in terms of frame$\{x' \quad y' \quad z'\}$ as

$$ \boldsymbol{P}' = \begin{bmatrix} \rho\cos(c/\rho) \\ \rho\sin(c/\rho) \\ 0 \end{bmatrix} \tag{7.43} $$

where $\rho = \left\| {}^0\boldsymbol{P}_j - {}^0\boldsymbol{C} \right\|$ gives the circle radius. Thus, the description of the circular path $\mathcal{P}'$ with respect to farme $\{x' \, y' \, z'\}$, is given by

$$\mathcal{P}'(c) = \begin{bmatrix} \rho \cos(c/\rho) \\ \rho \sin(c/\rho) \\ 0 \end{bmatrix} \tag{7.44}$$

Finally, the parametric description of circle with respect to frame $\{0\}$ is obtained by mapping, $\mathcal{P}'$ to base frame as

$${}^0\mathcal{P}(c) = {}^0C + \boldsymbol{R}\,\mathcal{P}'(c) \tag{7.45}$$

where $\boldsymbol{R}$ is the rotation matrix of frame $\{x' \, y' \, z'\}$ with respect to base frame $\{0\}$. Thus, the position vector of any point on the circle is given by

$${}^0\boldsymbol{P}(c) = {}^0\boldsymbol{C} + \boldsymbol{R}\,\mathcal{P}'(c) \tag{7.46}$$

From Eq. (7.38), the unit vectors are obtained as

$$\hat{\boldsymbol{a}} = \frac{d({}^0\boldsymbol{P})}{dc} = \boldsymbol{R}\begin{bmatrix} -\sin(c/\rho) \\ \cos(c/\rho) \\ 0 \end{bmatrix} \tag{7.47}$$

$$\hat{\boldsymbol{n}} = \frac{(\boldsymbol{R}/\rho)}{\left\| \dfrac{d^2({}^0\boldsymbol{P})}{dc^2} \right\|}\begin{bmatrix} -\cos(c/\rho) \\ -\sin(c/\rho) \\ 0 \end{bmatrix} \tag{7.48}$$

The unit vector $\hat{\boldsymbol{o}}$ is obtained from $\hat{\boldsymbol{a}}$ and $\hat{\boldsymbol{n}}$.

7.3.4 Position Planning

Once the given path is described in terms of path parameters, the next aim is to determine the time history of end-effector position, velocity, and acceleration consistent with the temporal constraints prescribed.

For a path ${}^0\boldsymbol{P}$ to be traversed by a given n-DOF manipulator in a time t_g, first the extreme values of path coordinate c are determined. For the given path, assume $c = 0$ at time $t = 0$ is the initial position, and $c = c_g$ at $t = t_g$ is the goal position. Next, a function $c(t)$, which interpolates c between $c = 0$ and $c = c_g$ in a smooth way, is obtained. This smooth function $c(t)$ could either be a cubic polynomial or a linear segment with parabolic blends and is obtained as discussed in the case of joint-space schemes for point-to-point motion without via points.

Now that a time law is imposed on the path coordinate c, it is possible to obtain the time history of ${}^0\boldsymbol{P}$ by substituting the values attained by $c(t)$, at each instant of time, in the Eq. (7.38).

The linear velocity of the end-effector is determined as

$${}^0\dot{\boldsymbol{P}} = \dot{c}\frac{d({}^0\boldsymbol{P})}{dc} = \hat{\boldsymbol{a}}\,\dot{c} \tag{7.49}$$

where $\hat{\boldsymbol{a}} = \dfrac{d({}^0\boldsymbol{P})}{dc}$ as per Eq. (7.38).

Linear acceleration of the tool is obtained by differentiating Eq. (7.49) with respect to time,

$$ {}^0\ddot{\boldsymbol{P}} = \dot{c}^2 \frac{d^2({}^0\boldsymbol{P})}{dc^2} + \ddot{c}\frac{d({}^0\boldsymbol{P})}{dc} \tag{7.50} $$

Thus, for a Cartesian straight-line motion, ${}^0\boldsymbol{P}$ is obtained from Eqs (7.40), (7.49) and (7.50) as

$$ {}^0\ddot{\boldsymbol{P}} = \frac{\dot{s}}{\left\| {}^0\boldsymbol{P}_g - {}^0\boldsymbol{P}_s \right\|}({}^0\dot{\boldsymbol{P}}_g - {}^0\dot{\boldsymbol{P}}_s) = \hat{\boldsymbol{a}}\,\dot{c} \tag{7.51} $$

$$ {}^0\ddot{\boldsymbol{P}} = \frac{\dot{s}}{\left\| {}^0\boldsymbol{P}_g - {}^0\boldsymbol{P}_s \right\|}({}^0\ddot{\boldsymbol{P}}_g - {}^0\ddot{\boldsymbol{P}}_s) = \hat{\boldsymbol{a}}\,\ddot{c} \tag{7.52} $$

Similarly, for circular motion, from Eqs (7.47), (7.48), (7.49), and (7.50), the velocity and acceleration of point P on the circle are as

$$ {}^0\ddot{\boldsymbol{P}} = \boldsymbol{R}\dot{c}\begin{bmatrix} -\sin(c/\rho) \\ \cos(c/\rho) \\ 0 \end{bmatrix} \tag{7.53} $$

$$ {}^0\ddot{\boldsymbol{P}} = \boldsymbol{R}\begin{bmatrix} -\dfrac{\dot{c}^2}{\rho}\cos(c/\rho) - \ddot{c}\sin(c/\rho) \\ -\dfrac{\dot{c}^2}{\rho}\sin(c/\rho) + \ddot{c}\cos(c/\rho) \\ 0 \end{bmatrix} \tag{7.54} $$

7.3.5 Orientation Planning

The next step in Cartesian space schemes is to obtain the time sequence of orientations attained by the end-effector. But, these orientations cannot be specified by rotation matrices because interpolation of two-rotation matrices does not always yield a valid rotation matrix (because orthonormality of columns is not maintained).

Hence, in planning orientations, either the Fixed or Euler angle representation outlined in Chapter 2, is used. Supposing, *zyx*-Euler angle representation is used to express the end-effector orientation with respect to the base, the angles ϕ, θ, and φ are defined as the Euler angles. It will be useful to define 3×1 orientation vector:

$$ \alpha = [\phi \quad \theta \quad \varphi]^T \tag{7.55} $$

Because α is similar to ${}^0\boldsymbol{P}$, discussed in position planning, any path on the end-effector's orientation along with temporal constraints can be imposed. The steps outlined as follows generate a smooth trajectory joining

$\alpha_s = [\phi_s \quad \theta_s \quad \varphi_s]^T$ to $\alpha_g = [\phi_g \quad \theta_g \quad \varphi_g]^T$ in time $t = t_g$.

Step 1. Obtain the parametric description of the given path joining as to α_s to α_g.

Step 2. Impose a time law on path coordinate, c, as done in position planning in Cartesian space.

Step 3. Obtain the time sequence of values attained by $\alpha(t)$ based on $c(t)$.

Step 4. As explained earlier for position planning, obtain the time history of $\dot{\alpha}(t)$ and $\ddot{\alpha}(t)$.

For example, a linear segment joint α_s and α_g has the following expressions for $\dot{\alpha}(t)$ and $\ddot{\alpha}(t)$.

$$\dot{\alpha}(t) = \frac{\dot{c}}{\left\|\alpha_g - \alpha_s\right\|}(\alpha_g - \alpha_s) \tag{7.56}$$

$$\ddot{\alpha}(t) = \frac{\ddot{c}}{\left\|\alpha_g - \alpha_s\right\|}(\alpha_g - \alpha_s) \tag{7.57}$$

7.4 JOINT-SPACE VERSUS CARTESIAN SPACE TRAJECTORY PLANNING

The joint-space trajectory planning is simple to implement and joint coordinates fully specify the position and orientation of end-effector. But when it comes to specifying a task or defining an obstacle in workspace, Cartesian coordinates offer a better choice. It is easy to specify the position and orientation of a rigid body in space in Cartesian coordinates. The obstacles and the path followed by the end-effector (or tool point) can be expressed in terms of simple transformations and easy to visualize in Cartesian space.

The main drawback with joint-space trajectory planning is that it does not account for the existence of obstacles in the workspace of a manipulator. The other drawback is that due to the nonlinear nature of the manipulator's kinematics model, it is difficult to predict the resulting end-effector motion that will be produced by a particular trajectory executed in the joint space. The above-mentioned drawbacks could be overcome by employing Cartesian space trajectory planning.

One of the main drawbacks with Cartesian space techniques is its computational complexity. Because the manipulator control system controls the joint variables, it is essential to resort to inverse kinematics/dynamics and determine the values of joint variables at every instant of time. This requires much more computations, which reduce the trajectory-sampling rate in real-time operation. As a result, the trajectory-tracking accuracy decreases. This problem is not there in joint- space schemes, where planning is done directly in joint space.

The other limitation with Cartesian space schemes is that the planned paths may have singularities or may pass close to manipulator singularities. In such a case, when singular configuration of manipulator is reached, one or more joint velocities may increase infinitely. One approach to avoid excessive joint rates is to limit joint rates within manageable limits, while computing the trajectory and to increase the travel time in such a region. However, doing so would violate the temporal attributes of the specified motion and this will make the overall motion slower.

It may also happen that the intermediate points of a planned path are unreachable although the start and goal points lie within the workspace. This happens primarily due to presence of voids within the workspace created by the limits over the movements of individual joints. The solution is to identify such voids and avoid them while planning Cartesian paths.

Cartesian space planning requires more complex algorithms and more computational ability, so that in runtime the joints are servoed to follow the desired Cartesian paths. To sum it up, Cartesian space planning offers a better and simpler representation of the task but at a higher computational cost.

SOLVED EXAMPLES

Example 7.1 *Cubic spline trajectory*

The second joint of a SCARA manipulator is required to move from $\theta_2 = 30^\circ$ to 150° in 5 seconds. Find the cubic polynomial to generate the smooth trajectory for the joint. What is the maximum velocity and acceleration for this trajectory?

Solution The given parameters of the trajectory are:

$$\begin{aligned} &\theta^s = 30^\circ \qquad \theta^g = 105^\circ \\ &v^s = 0 \qquad v^g = 0 \\ &\qquad\qquad t^g = 5\,s \end{aligned} \tag{7.58}$$

The four coefficients of the cubic polynomial are computed from the above using Eq. (7.8) as:

$$\begin{aligned} a_0 &= \theta^s = 30.0 \\ a_1 &= 0 \\ a_2 &= \frac{3}{t_g^2}(\theta^g - \theta^s) = 6.0 \\ a_3 &= -\frac{2}{t_g^3}(\theta^g - \theta^s) = -1.2 \end{aligned} \tag{7.59}$$

Hence the polynomials interpolating the position, velocity and acceleration of the joint are

$$\begin{aligned} \theta(t) &= 30.0 + 6.0t^2 - 1.2t^3 \\ \dot{\theta}(t) &= 12.0t - 3.6t^2 \\ \ddot{\theta}(t) &= 12.0 - 7.2t \end{aligned} \tag{7.60}$$

As expected, the velocity profile is parabolic and acceleration profile is linear. The time sequence of position, velocity, and acceleration can be easily generated at any specified sampling rate. The plot of these functions will be identical to

Fig. 7.4. The maximum values of velocity and acceleration are 7.5 deg/s and 12 deg/s^2, respectively.

Example 7.2 ***Point-to-point motion with via points with specified velocities***

Determine the time-history of the position, velocity and acceleration of the end-effector of a 5-DOF manipulator, which moves from start to goal point via two intermediate points. The desired end-effector motion is specified in Table 7.1. Assume a cubic spline trajectory in each segment and the continuity of velocity at the via points is enforced.

Table 7.1 *Desired end-effector motion*

End-effector	*Path point (j)*			
	1	2	3	4
Position q^j (rad)	0	$3\pi/2$	$-\pi/2$	2π
Velocity $\dot{q}^j$(rad/s)	0	$-\pi/2$	π	0
Traversal time t_j (s)	0	3	5	2

Solution The given trajectory has two via points and, therefore, will have three segments and three cubic polynomials will be required to define the time-history of motion of end-effector. Let the three cubic polynomials be

$$\begin{aligned} P_1(t) &= a_{10} + a_{11}t + a_{12}t^2 + a_{13}t^3 \\ P_2(t) &= a_{20} + a_{21}t + a_{22}t^2 + a_{23}t^3 \\ P_3(t) &= a_{30} + a_{31}t + a_{32}t^2 + a_{33}t^3 \end{aligned} \tag{7.61}$$

where a_{ij} are the unknown coefficients of the three cubic polynomials.

The constraints imposed on these three polynomials are given by Eq. (7.12). Applying the constraints and substituting the boundary values from Table 7.1, for first segment cubic gives the following equations:

$$\begin{aligned} P_1(0) &= a_{10} = 0 \\ P_1(3) &= a_{10} + 3a_{11} + 9a_{12} + 27a_{13} = 3\pi/2 \\ \dot{P}_1(0) &= a_{11} = 0 \\ \dot{P}_1(3) &= a_{11} + 6a_{12} + 27a_{13} = \pi/2 \end{aligned} \tag{7.62}$$

The four coefficient for the first segment polynomial P_1 are computed from Eq. (7.62) as

$$\begin{aligned} a_{10} &= 0 \\ a_{11} &= 0 \\ a_{12} &= 2.09 \\ a_{13} &= -0.52 \end{aligned} \tag{7.63}$$

and the cubic for first segment is

$$P_1(t) = 2.09\,t^2 - 0.52\,t^3 \tag{7.64}$$

Similarly, for second cubic the constraints from Table 7.1 give

$$\begin{aligned} P_2(3) &= a_{20} + 3a_{21} + 9a_{22} + 27a_{23} = 3\pi/2 \\ P_2(8) &= a_{20} + 8a_{21} + 64a_{22} + 512a_{23} = -\pi/2 \\ \dot{P}_2(3) &= a_{21} + 6a_{22} + 27a_{23} = -\pi/2 \\ \dot{P}_2(8) &= a_{21} + 16a_{22} + 192a_{23} = \pi \end{aligned} \tag{7.65}$$

Equations (7.65) are solved to compute the from coefficients of polynomial P_2. The solution is

$$\begin{aligned} a_{20} &= -1.77 \\ a_{21} &= 7.36 \\ a_{22} &= -2.22 \\ a_{23} &= 0.16 \end{aligned} \tag{7.66}$$

and the cubic for the second segment is

$$P_2(t) = -1.77 + 7.36t - 2.22t^2 + 0.16t^3 \tag{7.67}$$

For the third cubic polynomial constraints from Table 7.1 give

$$\begin{aligned} q_3(8) &= a_{30} + 8a_{31} + 64a_{32} + 512a_{33} = -\pi/2 \\ q_3(10) &= a_{30} + 10a_{31} + 100a_{32} + 1000a_{33} = 2\pi \\ \dot{q}_3(8) &= a_{31} + 16a_{32} + 192a_{33} = \pi \\ \dot{q}_3(10) &= a_{31} + 20a_{32} + 300a_{33} = 0 \end{aligned} \tag{7.68}$$

The four constraints for the third cubic are computed as:

$$\begin{aligned} a_{30} &= 752.41 \\ a_{31} &= -267.04 \\ a_{32} &= 31.02 \\ a_{33} &= -1.18 \end{aligned} \tag{7.69}$$

and the cubic for the third and last segment is:

$$P_3(t) = 752.41 - 267.04t + 31.04t^2 - 1.18t^3 \tag{7.70}$$

The cubic polynomials P_1, P_2 and P_3 for the three segments of the point-to-point trajectory are, thus, give by Eqs. (7.64), (7.67) and (7.70). These cubic polynomials describing the position of end-effector are plotted in Fig. 7.11(a) and their first and second derivatives, which represent velocity and acceleration of end-effector, respectively, are plotted in Fig. 7.11(b) and Fig. 7.11(c).

From these figures, note that continuity of motion is maintained for position and velocity but there is discontinuity of acceleration at the via points.

Example 7.3 Point-to-point motion with via points with computed velocities

Consider the determination of point-to-point trajectory with two via points with the data in Example 7.2 where the velocities at via points are computed instead of being specified.

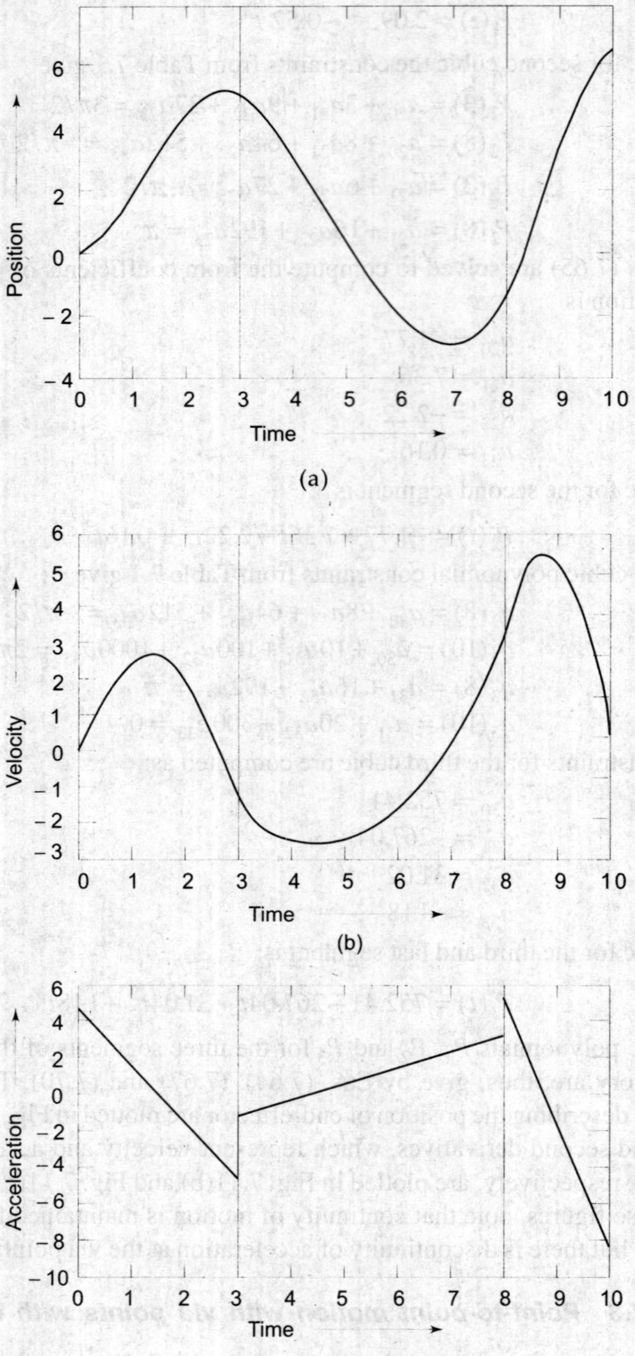

Fig. 7.11 *Time-history of position, velocity and acceleration for point-to-point motion with via points*

Solution As in Example 7.2, the starting and final velocities are zero, that is

$$\dot{q}^1 = \dot{q}^4 = 0 \tag{7.71}$$

The average velocities for the segments 1, 2 and 3 are computed using Eq. (7.14) and the data from Table 7.1. Thus

$$\bar{\dot{q}}^1 = \frac{q^2 - q^1}{T_1} = \frac{\frac{3\pi}{2} - 0}{3} = \frac{\pi}{2}$$

$$\bar{\dot{q}}^2 = \frac{q^3 - q^2}{T_2} = \frac{-\frac{\pi}{2} - \frac{3\pi}{2}}{2} = -\pi \tag{7.72}$$

$$\bar{\dot{q}}^3 = \frac{q^4 - q^3}{T_3} = \frac{2\pi + \frac{\pi}{2}}{3} = \frac{5\pi}{2}$$

According to Eq. (7.14), the via point velocities are computed from

$$\dot{q}^j = \begin{cases} 0 & \text{if } \operatorname{sign}(\bar{\dot{q}}^{j-1}) \neq \operatorname{sign}(\bar{\dot{q}}^j) \\ \frac{1}{2}(\bar{\dot{q}}^{j-1} + \bar{\dot{q}}^j) & \text{if } \operatorname{sign}(\bar{\dot{q}}^{j-1}) = \operatorname{sign}(\bar{\dot{q}}^j) \end{cases} \tag{7.73}$$

At via point 2

$$\dot{q}^2 = 0 \quad \text{because from Eq. (7.72)} \ \operatorname{sign}(\bar{\dot{q}}^1) \neq \operatorname{sign}(\bar{\dot{q}}^2) \tag{7.74}$$

and at via point 3

$$\dot{q}^3 = 0 \quad \text{since } \operatorname{sign}(\bar{\dot{q}}^2) \neq \operatorname{sign}(\bar{\dot{q}}^3) \tag{7.75}$$

Note that in this case the given path points positions result in zero velocity at both via points. The three cubic polynomials for the three segments of the trajectory are computed using the position and time constraints from Table 7.1 and velocity constraints from Eqs (7.71), (7.74) and (7.75). The three polynomials are as in Eq. (7.61).

Applying the given and computed constraints for the first polynomial, gives four equations as:

$$\begin{aligned} P_1(0) &= a_{10} = 0 \\ P_1(3) &= a_{10} + 3a_{11} + 9a_{12} + 27a_{13} = 3\pi/2 \\ \dot{P}_1(0) &= a_{11} = 0 \\ \dot{P}_1(3) &= a_{11} + 6a_{12} + 27a_{13} = 0 \end{aligned} \tag{7.76}$$

which give the four coefficients of $P_1(t)$ as:

$$\begin{aligned} a_{10} &= a_{11} = 0 \\ a_{12} &= 1.57 \\ a_{13} &= -0.35 \end{aligned} \tag{7.77}$$

Applying the constraints for the second joint yields the four equations as (see Eq. (7.65)):

$$\begin{aligned} P_2(3) &= a_{20} + 3a_{21} + 9a_{22} + 27a_{23} = 3\pi/2 \\ P_2(8) &= a_{20} + 8a_{21} + 64a_{22} + 512a_{23} = -\pi/2 \\ \dot{P}_2(3) &= a_{21} + 6a_{22} + 27a_{23} = 0 \\ \dot{P}_2(8) &= a_{21} + 16a_{22} + 192a_{23} = 0 \end{aligned} \tag{7.78}$$

which are solved to give four coefficients of the second segment cubic $P_2(t)$ as:

$$\begin{aligned} a_{20} &= -4.79 \\ a_{21} &= 7.24 \\ a_{22} &= -1.66 \\ a_{23} &= 0.10 \end{aligned} \tag{7.79}$$

Similarly, the solution for the four coefficients of the third polynomial $P_3(t)$ is

$$\begin{aligned} a_{30} &= 1.38 \times 10^3 \\ a_{31} &= -471.2 \\ a_{32} &= 53.01 \\ a_{33} &= -1.96 \end{aligned} \tag{7.80}$$

Substituting the computed values of the coefficients from Eqs (7.77), (7.79) and (7.80) into three cubic polynomials (Eq. (7.61)) gives the three cubics for the three segments as:

$$\begin{aligned} P_1(t) &= 1.57t^2 - 0.35t^3 \\ P_2(t) &= -4.79 + 7.24t - 1.66t^2 + 0.10t^3 \\ P_3(t) &= 1.38 \times 10^3 - 471.2t + 53.01t^2 - 1.96t^3 \end{aligned} \tag{7.81}$$

These cubic for the three segments are plotted and the time-history of position, velocity and acceleration are obtained as illustrated in Fig. 7.12. From these plots it is easy to observe that the continuity of position and velocity is maintained as in Example 7.2 and the acceleration is still discontinuous. Comparing these profiles with profiles in Fig. 7.11 reveals the effect of change in boundary conditions on the motion of the end-effector.

Example 7.4 ***Trajectory with a trapezoidal velocity profile***

The motion of a joint of 3-DOF arm is constrained by an actuator that can produce a maximum acceleration of 0.35 rad/s^2 and maximum velocity of 15 rad/s. If trapezoidal velocity profile is assumed, determine the trajectory if the joint moves by π rad in 10 s.

Solution A trapezoidal velocity profile implies a constant acceleration in the start phase, a cruise velocity and a constant deceleration in the arrival phase. The trajectory will be as shown in Fig. 7.3.

Assuming that the trajectory is symmetric and the traversal is with maximum possible acceleration, the parameters for the trajectory are:

$$\begin{aligned} q^s &= 0 & q^g &= \pi \\ \dot{q}^s &= 0 & \dot{q}^g &= 0 \\ t_s &= 0 & t_g &= 10 \text{ s} \\ v^c &= 15 \text{ rad/s} & \ddot{q}^c &= 0.35 \text{ rad/s}^2 \end{aligned} \tag{7.82}$$

From Eq. (7.16), at the mid-point of the trapezoidal trajectory

$$\begin{aligned} q^m &= \frac{q^s + q^g}{2} = \frac{\pi}{2} \\ t_m &= \frac{t^g}{2} = 5 \text{ s} \end{aligned} \tag{7.83}$$

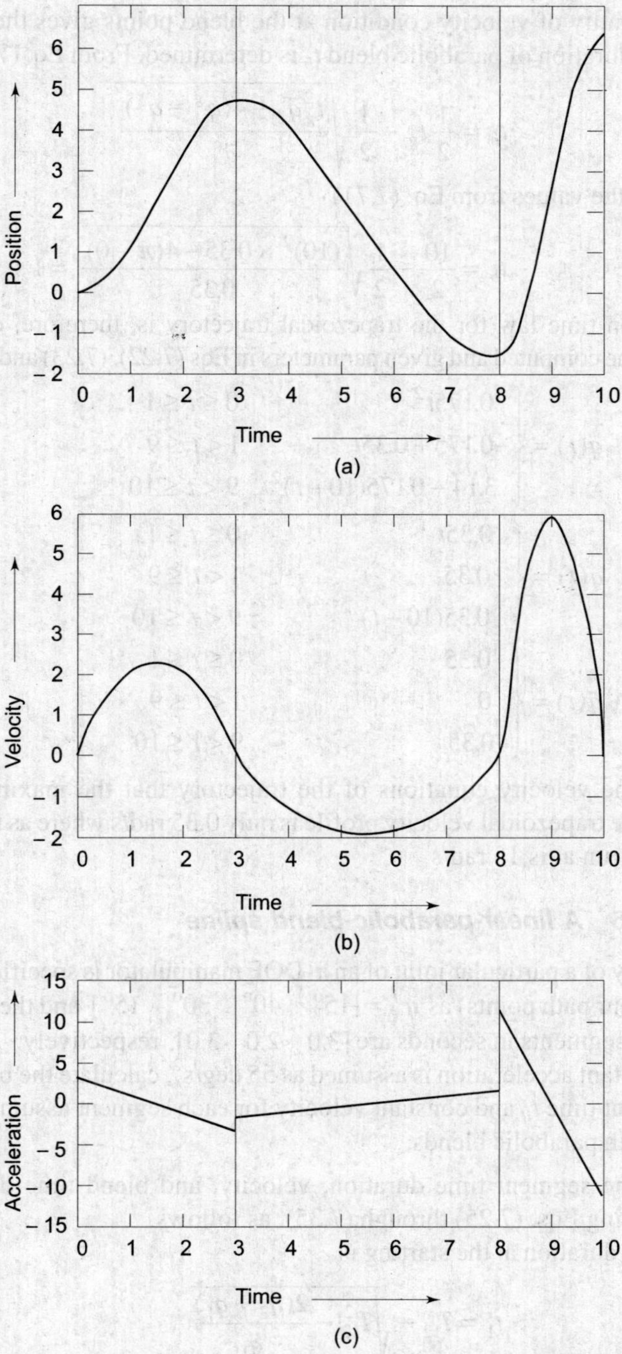

Fig. 7.12 *Time-history of position, velocity and accelerations for point-to-point motion with computed velocity at via points*

The continuity of velocity condition at the blend points gives the constraint from which duration of parabolic blend t_b is determined. From Eq. (7.20)

$$t_b = \frac{1}{2}t_g - \frac{1}{2}\sqrt{\frac{t_g^2\ddot{q}^c - 4(q^g - q^s)}{\ddot{q}^c}} \tag{7.84}$$

Substituting the values from Eq. (7.71)

$$t_b = \frac{10}{2} - \frac{1}{2}\sqrt{\frac{(10)^2 \times 0.35 - 4(\pi - 0)}{0.35}} = 1 \text{ s} \tag{7.85}$$

The motion time law for the trapezoidal trajectory is, therefore, obtained by substituting the computed and given parameters in Eqs (7.22), (7.23) and (7.24) as

$$q(t) = \begin{cases} 0.175t^2 & 0 \le t \le 1 \\ -0.175 + 0.35t & 1 < t \le 9 \\ 3.14 - 0.175(10-t)^2 & 9 < t \le 10 \end{cases}$$

$$\dot{q}(t) = \begin{cases} 0.35t & 0 \le t \le 1 \\ 0.35 & 1 < t \le 9 \\ 0.35(10-t) & 9 < t \le 10 \end{cases} \tag{7.86}$$

$$\ddot{q}(t) = \begin{cases} 0.35 & 0 \le t \le 1 \\ 0 & 1 < t \le 9 \\ -0.35 & 9 < t \le 10 \end{cases}$$

Note from the velocity equations of the trajectory that the maximum cruise velocity of the trapezoidal velocity profile is only 0.35 rad/s where as the actuator is capable to run at is 15 rad/s.

Example 7.5 *A linear-parabolic-blend spline*

The trajectory of a particular joint of an ***n***-DOF manipulator is specified with two via points (four path points) as $\boldsymbol{q}^j = [15^\circ \quad 40^\circ \quad 30^\circ \quad 15^\circ]$ and the travel time for the three segments in seconds are [3.0 2.0 3.0], respectively.

If the constant acceleration is assumed as 55 deg/s^2, calculate the blend time t_b linear segment time t_{jl} and constant velocity for each segment assuming a linear trajectory with parabolic blends.

Solution The segment time duration, velocity, and blend time duration are calculated using Eqs. (7.25) through (7.35), as follows.

The blend duration at the starting is

$$t_1 = T_{12} - \sqrt{T_{12}^2 - \frac{2(q_2 - q_1)}{\ddot{q}_1}}$$

or

$$t_1 = 3.0 - \sqrt{3.02 - \frac{2(40-15)}{55}} = 0.156 \text{ s} \tag{7.87}$$

The constant velocities for the linear segments are:

$$\dot{q}^{12} = \frac{40-15}{2-0.5\times 0.156} = 13.0 \text{ deg / s}$$

$$\dot{q}^{23} = \frac{30-40}{2.0} = -5.0 \text{ deg / s} \tag{7.88}$$

The second blend duration is computed from:

$$t_2 = \frac{\dot{q}^{23} - \dot{q}^{12}}{\ddot{q}^2}$$

where $\ddot{q}^2 = -55$ (for deceleration). Thus,

$$t_2 = \frac{-5.0-13.0}{-55} = 0.33 \text{ s}$$

and
$$t_{12} = T_{12} - t_1 - \frac{t_2}{2} = 3.0 - 0.156 - \frac{0.33}{2} = 2.679 \text{ s} \tag{7.89}$$

Also
$$t_4 = 3.0 - \sqrt{9 + \frac{2(15-30)}{55.0}} = 0.092 \text{ s} \tag{7.90}$$

and
$$\dot{q}^{34} = \frac{q^4 - q^3}{T_{34} - \dfrac{t_4}{2}} = \frac{15-30}{3.0 - \dfrac{0.092}{2}} = -5.078 \text{ deg / s} \tag{7.91}$$

Similarly, for the third segment:

$$t_3 = \frac{\dot{q}^{34} - \dot{q}^{23}}{\ddot{q}^3} = \frac{-5.078+5.0}{55} = 0.00142 \text{ s}$$

$$t_{23} = 2 - \frac{1}{2}(0.33) - \frac{1}{2}(0.00142) = 1.834 \text{ s} \tag{7.92}$$

$$t_{34} = 3 - \frac{1}{2}(0.00142) = 2.99 \text{ s}$$

The parameters for the trajectory are tabulated in Tables 7.2 and 7.3 with blend duration of each parabolic segment at each path point and linear velocity and duration of linear velocity for each segment, respectively.

Table 7.2 *Blend duration for parabolic blends at each path point*

Path Point	t_b *(s)*
1	0.156
2	0.330
3	0.001
4	0.092

Table 7.3 *Linear velocity and its duration for each segment*

Segment	$\dot{q}$ *(deg/s)*	t *(s)*
1–2	13.00	2.68
2–3	–5.00	1.83
3–4	–5.08	2.99

Example 7.6 ***Trajectory planning for a 3-DOF RRP arm***

For the 3-DOF manipulator arm discussed in Example 4.4, Fig. 4.7 design a linear trajectory with parabolic blends. The initial & final position of the end-effector is expressed by homogeneous transformation matrices $\boldsymbol{T}_s$ & $\boldsymbol{T}_g$ given below and time taken for traversal is 5 seconds.

$$\boldsymbol{T}_s = \begin{bmatrix} 0.354 & 0.866 & -0.354 & -0.106 \\ -0.612 & 0.500 & 0.612 & 0.184 \\ 0.707 & 0 & 0.707 & 0.212 \\ 0 & 0 & 0 & 1 \end{bmatrix} \tag{7.93}$$

$$\boldsymbol{T}_g = \begin{bmatrix} 0.583 & -0.766 & 0.272 & 0.027 \\ 0.694 & 0.643 & 0.324 & 0.032 \\ -0.423 & 0 & 0.906 & 0.091 \\ 0 & 0 & 0 & 1 \end{bmatrix} \tag{7.94}$$

Solution The forward kinematic model for the 3-DOF RRP arm was obtained in Example 4.4 and is given by Eq. (4.67) as:

$${}^0\boldsymbol{T}_3 = \begin{bmatrix} C_1C_2 & -S_1 & -C_1S_2 & -d_3C_1S_2 \\ S_1C_2 & C_1 & -S_1S_2 & -d_3S_1S_2 \\ S_2 & 0 & C_2 & d_3C_2 \\ 0 & 0 & 0 & 1 \end{bmatrix} = \begin{bmatrix} r_{11} & r_{12} & r_{13} & r_{14} \\ r_{21} & r_{22} & r_{23} & r_{24} \\ r_{31} & r_{32} & r_{33} & r_{34} \\ 0 & 0 & 0 & 1 \end{bmatrix} \tag{7.95}$$

The inverse kinematics of the arm was also carried out in Example 4.4 and solutions for the three joint variables were obtained as (see Eqs (4.71), (4.73) and (4.74)):

$$\begin{aligned} \theta_1 &= A\tan 2(-r_{24}, -r_{14}) \\ \theta_2 &= A\tan 2\left(\pm\sqrt{r_{14}^2 + r_{24}^2}, r_{34}\right) \\ d_3 &= +\sqrt{r_{14}^2 + r_{24}^2 + r_{34}^2} \end{aligned} \tag{7.96}$$

For the initial position of the end-effector specified by T_s, Eq. (7.93), the feasible solution obtained in Example 4.4 is (see Eq. (4.75)):

$$\begin{aligned} \theta_1^s &= -60^\circ \\ \theta_2^s &= 45^\circ \\ d_3^s &= 0.30 \text{ m} \end{aligned} \tag{7.97}$$

The joint variables corresponding to final position of the end-effector as specified by $\boldsymbol{T}_g$ are:

$$\begin{aligned} \theta_1^g &= 50^\circ \\ \theta_2^g &= -25^\circ \\ d_3^g &= 0.10 \text{ m} \end{aligned} \tag{7.98}$$

Now, to plan linear trajectory with parabolic blends for the three links, the default acceleration for each link should satisfy the constraint in Eq. (7.21).

$$\left|\ddot{q}^c\right| \geq \frac{4\left|q^g - q^s\right|}{t_g^2} \tag{7.99}$$

That is, for link 1

$$\ddot{q}_1{}^c \geq \frac{4 \times (50 - (-60))}{5^2} = 17.6 \quad \text{or choose} \quad \ddot{q}_1{}^c = 20.0 \ \text{deg/s}^2 \tag{7.100}$$

Similarly, for link 2 and link 3

$$\begin{aligned} \ddot{q}_2{}^c &\geq -11.2 \quad \text{or choose} \quad \ddot{q}_2{}^c = -15.0 \ \text{deg/s}^2 \\ \ddot{q}_3{}^c &\geq -0.032 \quad \text{or choose} \quad \ddot{q}_3{}^c = -0.05 \ \text{m/s}^2 \end{aligned} \tag{7.101}$$

The trajectory (time-history) for each link is computed using the Eqs (7.20), (7.22), (7.23) and (7.24).

For first link

$$t_{b1} = \frac{1}{2} \times 5 - \frac{1}{2}\sqrt{\frac{25 \times 20 - 4(50 - (-60))}{20}} = 1.63 \ \text{s} \tag{7.102}$$

Therefore,

$$q_1(t) = \begin{cases} -60 + 10t^2 & 0 \leq t \leq 1.63 \\ -86.57 + 32.6t & 1.63 < t \leq 3.37 \\ 50 - 10(5-t)^2 & 3.37 < t \leq 5.0 \end{cases}$$

$$\dot{q}_1(t) = \begin{cases} 20t & 0 \leq t \leq 1.63 \\ 32.6 & 1.63 < t \leq 3.37 \\ 20(5-t) & 3.37 < t \leq 5.0 \end{cases} \tag{7.103}$$

$$\ddot{q}_1(t) = \begin{cases} 20 & 0 \leq t \leq 1.63 \\ 0 & 1.63 < t \leq 3.37 \\ -20 & 3.37 < t \leq 5.0 \end{cases}$$

For second link, with $t_g = 5$ s and $\ddot{q}_2{}^c = -15 \ \text{deg/s}^2$

$$t_{b2} = 2.5 - \frac{1}{2}\sqrt{\frac{25 \times (-15) - 4(-25 - 45)}{-15}} = 1.24 \ \text{s} \tag{7.104}$$

$$q_2(t) = \begin{cases} 45 + 7.5t^2 & 0 \leq t \leq 1.24 \\ 56.53 - 18.6t & 1.24 < t \leq 3.76 \\ -25 + 7.5(5-t)^2 & 3.76 < t \leq 5.0 \end{cases}$$

$$\dot{q}_2(t) = \begin{cases} -15t & 0 \leq t \leq 1.24 \\ -18.6 & 1.24 < t \leq 3.76 \\ -15(5-t) & 3.76 < t \leq 5.0 \end{cases} \tag{7.105}$$

$$\ddot{q}_2(t) = \begin{cases} -15 & 0 \leq t \leq 1.24 \\ 0 & 1.24 < t \leq 3.76 \\ 15 & 3.76 < t \leq 5.0 \end{cases}$$

For third link, with $t_g = 5$ s and $\ddot{q}_3{}^c = -0.05$ m / s

$$t_{b3} = 2.5 - \frac{1}{2}\sqrt{\frac{25 \times (-0.05) - 4(0.10 - 0.30)}{-0.05}} = 1 \text{ s} \tag{7.106}$$

$$q_3(t) = \begin{cases} 0.3 - 0.025t^2 & 0 \le t \le 1 \\ 0.325 - 0.05t & 1 < t \le 4 \\ 0.10 + 0.025(5-t)^2 & 4 < t \le 5 \end{cases}$$

$$\dot{q}_3(t) = \begin{cases} -0.05t & 0 \le t \le 1 \\ 0.05 & 1 < t \le 4 \\ -0.05(5-t) & 4 < t \le 5 \end{cases} \tag{7.107}$$

$$\ddot{q}_3(t) = \begin{cases} -0.05 & 0 \le t \le 1 \\ 0 & 1 < t \le 4 \\ 0.05 & 4 < t \le 5 \end{cases}$$

EXERCISES

7.1. A one-degree of freedom manipulator with rotary joint is to move from 113° to 210° in 7 seconds. Find the coefficients of the cubic polynomial to interpolate a smooth trajectory. Plot the position, velocity and acceleration variation as a function of time.

7.2. A single cubic trajectory given by $q(t) = 30 + t^2 - 6t^3$ is used for a period of 3 seconds. Determine starting and goal position, velocity, and accelerations of the end-effector.

7.3. The joint in Example 7.1 has initial and final velocity of 1.0 deg/s and 1.2 deg/s at $\theta^s = 30°$ and $\theta^g = 105°$, respectively. Determine the smooth trajectory polynomial for the position of the joint.

7.4. The trajectory for a joint's motion between start and goal position in a pick-n-place operation is determined by dividing the motion in two segments. The interpolating polynomial for each segment is cubic and the acceleration is continuous at the via point. Determine the coefficients for the two cubics.

7.5. The motion of a joint from start to goal position is specified in terms of position, velocity and acceleration at the beginning and end of a path segment. Determine the coefficients of the fifth-degree (quintic) polynomial for interpolating the smooth trajectory in the segment.

7.6. The trajectory for point-to-point motion between two points is divided into three segments and a 4-3-4 trajectory plan is used. That is, the first segments is a fourth degree polynomial specifying the trajectory from start position to lift-off point; followed by a cubic polynomial in the mid trajectory segment up to the set-down point and the last segments trajectory is similar to the first segment, a fourth-degree polynomial, from set-down point to goal position. Determine the polynomial equations for the three segments. Assume appropriate constraints.

7.7. Determine the equations of polynomials for the three segments of a point-to-point trajectory between two points if a 3-5-3 trajectory plan is used.

7.8. The trajectory between two points is divided into five segments and five-cubic polynomial interpolation is to be used. The boundary conditions that the polynomials must satisfy are
(i) position constraints at start, lift-off, set-down and goal positions, and
(ii) continuity of velocity and acceleration at all the path points.
Determine the polynomial for each segment.

7.9. A rotary joint moves from –15° to +45° in 3 seconds. Determine the polynomial for a smooth trajectory, if the initial and final velocity and accelerations are zero.

7.10. A linear trajectory with parabolic blends is used instead of a cubic polynomial in Exercise 7.1. Determine the parameters for the trajectory. Assume constant acceleration of 30 deg/s^2.

7.11. Calculate the parameters for a two-segment linear spline with parabolic blends with acceleration in blend region to be 30 deg/s^2 and duration of each segment as 3 seconds. The three path points for the joint are 15°, –5° and 30°. Plot the trajectory.

7.12. Determine the trajectory for a pick-n-place cycle, which has to pass through three via points using piecewise-linear interpolation with parabolic blends for each segment. The path points are $\left[0^\circ \quad 10^\circ \quad 45^\circ \quad 30^\circ \quad 5^\circ\right]$ and the travel times for the segments are $[0.5 \quad 1.5 \quad 2.0 \quad 1.0]$ seconds respectively. Assume that the constant acceleration is constant at 25 deg/s^2 in each segment.

7.13. Find the trajectory for a Cartesian space straight line path from $[1.0 \quad 1.0 \quad 1.0]^T$ to $[-1.0 \quad 0 \quad 1.0]^T$.

7.14. Compute the joint trajectory for $q^s = 0$ and $q^g = 4$ with null initial and final velocities and accelerations. Assume $t_g = 1$.

7.15. The velocity profile of a trajectory is trapezoidal with constant acceleration segment of 0.5 second duration and constant velocity of 10 deg/sec. Determine the parameters of the smooth trajectory for interpolating the time sequence of position with this type of trajectory.

7.16. The end-effector of two link planar manipulator with each link 1 m long, discussed in Section 6.2, is to move from initial position $(x_s, y_s) = (0.9, 1.6)$ to final position $(x_g, y_g) = (-1.6, 0.5)$. Determine the coefficients of cubic polynomials for each joint to generate a smooth trajectory.

7.17. If linear spline with parabolic blend is used instead of a cubic polynomial in Exercise 7.16, determine the parameters of the polynomials to interpolate both joint trajectories.

7.18. For a three-degree of freedom, cylindrical manipulator, design a linear trajectory with parabolic blends. The initial position of the end-effector is expressed by the homogeneous matrix $\boldsymbol{T}_s$ and the goal position by $\boldsymbol{T}_g$ as

$$T_s = \begin{bmatrix} 0 & 1 & 0 & 200 \\ 1 & 0 & 0 & 100 \\ 0 & 0 & -1 & -100 \\ 0 & 0 & 0 & 1 \end{bmatrix} \quad (7.108)$$

$$T_g = \begin{bmatrix} 1 & 0 & 0 & -50 \\ 0 & -1 & 0 & 200 \\ 0 & 0 & 0 & 400 \\ 0 & 0 & 0 & 1 \end{bmatrix} \quad (7.109)$$

7.19. For a two-segments continuous-acceleration trajectory, the path points are $\theta^s = 10°$, $\theta^g = 30°$ and $\theta^m = 22°$. The duration of two segments are 1.2 second and 1.0 second, respectively. Taking the velocity at via point as 15 deg/s, plot the position, velocity and acceleration of the interpolated trajectory as a function of time.

7.20. Third joint of a 5-DOF manipulator is $\theta_3 = 60°$ at start of work cycle. It is to move by angle of 45° in 3 seconds. Find the cubic trajectory for this joint to achieve the specified motion. The joint motion starts from rest and comes to rest at the destination point. Plot position, velocity and acceleration of the joint as a function of time.

7.21. Two joints of a SCARA manipulator are to move by 30° and 45° in 2 seconds and 3 seconds, respectively. Assuming the trajectory to be cubic for both the joints, determine the coefficients a_{ij} for the two cubic polynomials:

$$\theta_1(t) = a_{10} + a_{11}t + a_{12}t^2 + a_{13}t^3$$

$$\theta_2(t) = a_{20} + a_{21}t + a_{22}t^2 + a_{23}t^3$$

Will the trajectories for the two joints be different if the joints are considered at some location other than 0° at the start point?

SELECTED BIBLIOGRAPHY

1. S. Chand and K. Doty, "Online Polynomial Trajectories for Robot Manipulators," *The International Journal of Robotics Research*, **4**, 1985.
2. M. Erdmann, "Using Backprojections for the Fine Motion Planning with Uncertainty," *The International Journal of Robotics Research*, **5**(1), 1986.
3. Miroslaw Galicki, "Real-Time Trajectory Generation for Redundant Manipulators with path constraints," The *International Journal of Robotic Research*, **20**(8), 676–693, 2001.
4. C.S. Lin and P.R. Chang, "Joint Trajectory of Mechanical Manipulators for Cartesian Path Approximation," *IEEE Tr on Systems, Man, and Cybernetics*, **13**, 1983.

5. C.S. Lin, P.R. Chang and J.Y.S. Luh, "Formulation and Optimization of Cubic Polynomial Joint Trajectories for Industrial Robots," *IEEE Tr on Automatic Control*, ***28***, 1066–1073, 1983.
6. T. Lozano-Perez, "A Simple Motion Planning Algorithm for General Robot Manipulators," *IEEE Journal of Robotics and Automation*, ***3***(3), 224–238, Jun 1987.
7. T. Lozano-Perez, M.T. Mason and R.H. Taylor, "Automatic Synthesis of Fine-Motion Strategies for Robots," *The International Journal of Robotics Research*, ***3***(1), 3–24, 1984.
8. J.Y.S. Luh and C.S. Lin, "Optimum Path Planning for Mechanical Manipulators," *Tr of ASME Journal of Dynamic Systems, Measurement, and Control*, ***102***, 142–151, 1981.
9. A.A. Maciejewski and C.A. Klein, "Obstacle Avoidance for Kinematically Redundant Manipulators in Dynamically Varying Experiments," *The International Journal of Robotics Research*, ***4***(3), 109–117, 1985.
10. D. Nenchev, "Tracking manipulator trajectories with ordinary singularities: A null space-based approach," *International Journal of Robotic Research*, ***14***, 399–404, 1995.
11. F. Pfeiffer and R. Johanni, "A Concept for manipulator trajectory planning," *IEEE Tr on Robotics and Automation*, ***3***, 115–123, 1987.
12. Z. Shiller and H.H. Lu, "Computation of Path Constrained Time Optimal Motions with Dynamics Singularities," *Tr of ASME Journal of Dynamic Systems, Measurement and Control*, ***114***(2), 34–40, 1992.
13. M. Walker and D. Orin, "Efficient Dynamic Computer Simulation of Robotic Mechanisms," *Tr of ASME Journal of Dynamic Systems, Measurement, and Control*, ***104***, 1982.
14. A.K.C. Wong, S.W. Lu and M. Rioux, "Recognition and Shape Synthesis of 3D Object Based on Attributes Hypergraphs," *IEEE Tr on Pattern Analysis and Machine Intelligence*, 279–290, 1989.

8

Control of Manipulators

A robot is required to carry out the specified task by moving its end-effector accurately and repeatedly. In the previous chapter, techniques to determine the set of joint-location time histories for a desired end-effector motion through space have been developed. In this chapter, the analysis and design of a manipulator control system that accepts the joint-location time history as input and causes the manipulator to track the commanded trajectory, is discussed. It is assumed that the reader has the basic knowledge of control systems.

The control requires the knowledge of the mathematical model and some sort of intelligence to act on the model. Whereas the required mathematical model is obtained form basic physical laws governing robot dynamics and associated devices, intelligence requires sensory capabilities and means for acting and reacting to sensed variables. A robot performs the specified tasks in its environment, which can be divided into two classes: *contact type* tasks and *noncontact type* tasks. The noncontact type tasks involve manipulation of the end-effector in space to do the desired work, while in the contact type tasks the end-effector interacts with the environment. In general, a perfect robot is one, which is not only able to control its movement but also controls the force that it applies on the environment. On the basis of the type of task performed, manipulators are also classified as *contact type manipulator* and *noncontact type manipulator*. The first half of this chapter is devoted to the control of noncontact type manipulators and the control of contact type manipulators involving force/torque control is discussed in the second half.

Individual joints of a manipulator are powered and driven by actuators that apply a force or a torque to cause motion of the links. The actuator commands that move the manipulator and achieve the specified end-effector motion, are provided by the manipulator control system. These commands are based on the "control set points" generated from the set of joint-torque time histories obtained by the trajectory planner. The actual joint and/or end-effector positions and their derivatives can be fed back to the control system to get accurate motion. The schematic diagram of a manipulator control system is shown in Fig. 8.1. The dotted lines for the feedback indicate that the control system may or may not employ feedback of the actual joint locations and velocities. The parameters q, $\dot{q}$, $\ddot{q}$, τ and so on, in the figure, have the usual meaning discussed in previous chapters.

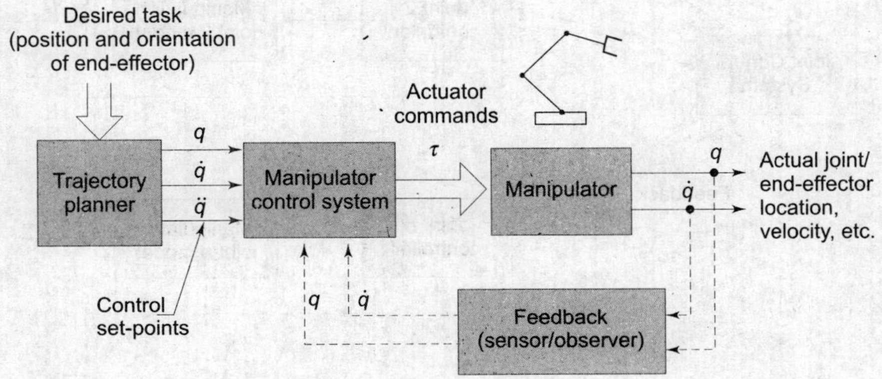

Fig. 8.1 *Block diagram of a manipulator control system*

The dynamic control of manipulator motions and/or interaction forces requires knowledge of forces or torques that must be exerted on the manipulator joints to move the links and the end-effector from the present location to the desired location, with or without the constraints of a particular planned end-effector trajectory, and/or planned end-effector force/torque. In both situations of the desired end-effector location and the desired end-effector force, the control of individual joint's location is important. Hence, it is an obvious requirement that each joint is controlled by a position servo.

If the manipulator is to move very slowly or move one joint at a time, then the control is simple because coupled dynamic forces are negligible. Each joint can be controlled independently by using a simple control system, which produces a joint actuator force/torque proportional to the required change in the joint variable. This is the proportional control algorithm. If the motions are fast, which are must for effective robot applications, all joints must move simultaneously. In this situation, the coupled dynamic forces are significant and it is known that dynamics is nonlinear and complex. Joints cannot move independently and complex control algorithm will be required. The typical robot control architecture for an n-DOF manipulator consists of a master control system to control and synchronize n joints. This master control is responsible for sending "set point" commands to each of the joint controllers. The n joint controllers use the set point information to command the joint actuator to move the joint. The joint controller may employ a feedback of current joint position. It may also periodically give feedback to the master controller. The schematic of typical control architecture for a manipulator control system is illustrated in Fig. 8.2.

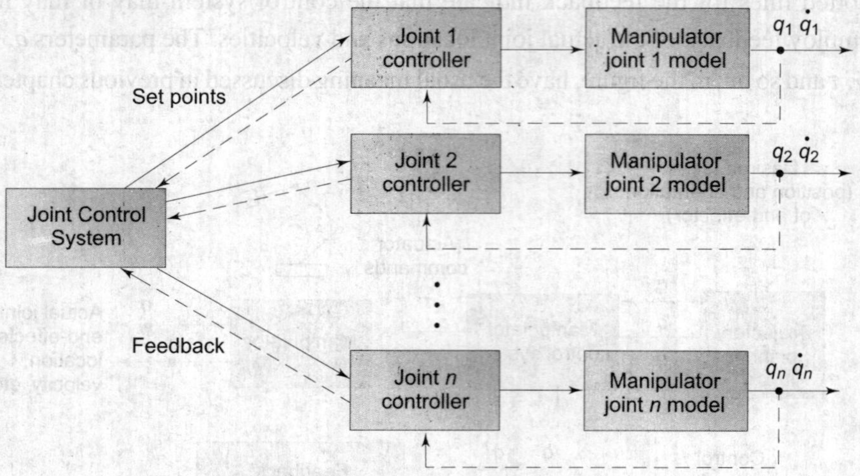

Fig. 8.2 *Robot control architecture for an n-DOF manipulator*

8.1 OPEN- AND CLOSE-LOOP CONTROL

In the case of *open-loop control* (no feedback), the actuator torque can be directly computed from the dynamic model of the manipulator. As seen in Chapter 6, Eq. (6.45), the dynamic behaviour of joint *i* of the manipulator is given by

$$\tau_i = \sum_j M_{ij}\ddot{q}_j + \sum_j \sum_k h_{ijk}\dot{q}_j\dot{q}_k + G_i \tag{8.1}$$

In this model, accurate computation of the parameters M_{ij}, h_{ijk}, and G_i is difficult. Moreover, this model does not include effects of friction, backlash and so on, at the joints and other external disturbances or noise. These effects are difficult to model and are highly nonlinear. Because of all these problems, the open-loop control has limited trajectory-tracking capabilities and is used in very few nonprecision applications.

The limitations of parametric inaccuracies and unmodelled nonlinear effects are overcome by using a *closed-loop control* scheme. Here, at every instant of time, the actual joint positions and velocities are measured directly by sensors such as encoders and tachometers mounted at the joints. These are used to compute the error between the desired and actual positions and velocities. A similar measurement and feedback is used for error in the desired end-effector position and/or force/torque. A control law then computes the joint torques required as a function of these servo errors using the dynamic model.

The interactive effects of inertial, Coriolis, centrifugal, and gravitational forces vary continuously with time. The computation of these from the dynamic model is complex and time consuming. This makes the actual real-time control extremely difficult.

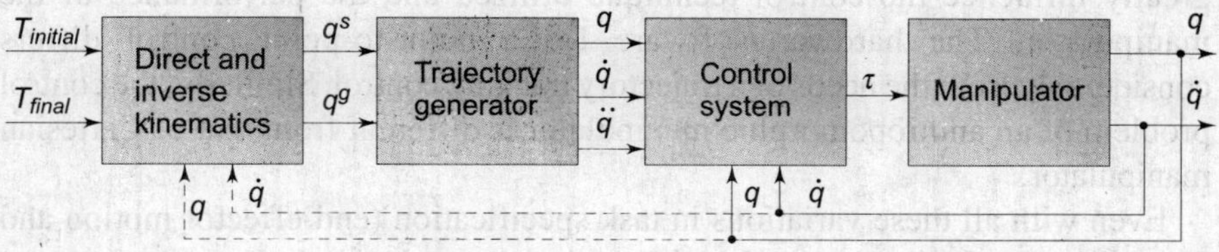

Fig. 8.3 *Operations for a point-to-point motion control of a manipulator*

The schematic description of the point-to-point motion control of an n-DOF manipulator arm is shown in Fig. 8.3. For this operation, the initial and final locations of the end-effector in Cartesian space, specified by the transformation matrices T_{intial} and T_{final}, serve as the input. The inverse kinematics model computes the desired end-effector locations in joint space. Then, a trajectory generator computes the joint-position time histories, based on the joint-space algorithms discussed in Chapter 7. Depending on the servo error computed from the base reference values and the sensor measurements, the control system commands the individual actuators to achieve the desired motion.

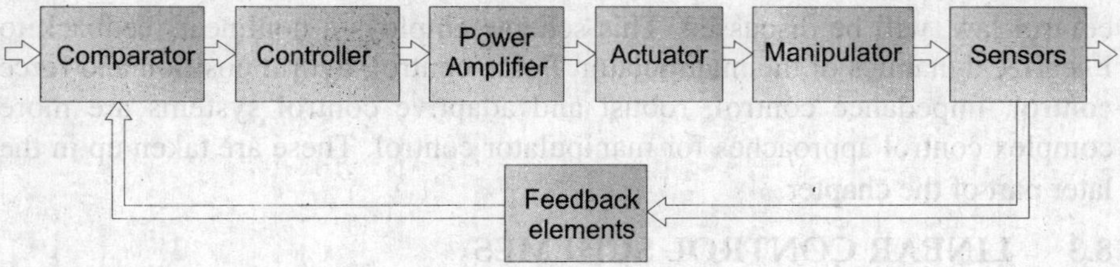

Fig. 8.4 *General block diagram for manipulator control system*

The block diagram in Fig. 8.4 illustrates the components of the control system to control the manipulator joints, consisting of actuators and the mechanical kinematics chain. A typical control system has a comparator (a small computer); a digital to analog converter (DAC); a controller to implement the control strategy; an amplifier to boost and transform the small command signals so that it can actuate the joint actuator; a sensor to detect the position and speed of joint and an analog to digital converter (ADC) in the feedback path.

8.2 THE MANIPULATOR CONTROL PROBLEM

The task to be performed by a manipulator may require either the execution of specific motions of the end-effector in free space and/or the application of specific contact forces at the end-effector, if the end-effector is constrained by the environment. For a given manipulator, many techniques can be used to control it to perform in the desired manner. The performance and range of applications of a manipulator is significantly influenced by the control scheme followed, as well as the way it is implemented. The mechanical design and configuration design also greatly influence the control technique utilized and the performance of the manipulator. The hardware/software for a point-to-point control differs considerably from the needs of a trajectory tracking control. Similarly, the control problem of an anthropomorphic manipulator is different from that of Cartesian manipulator.

Even with all these variations in task specification (end-effector motion and forces) the manipulator control problem can be stated as:

Find the joint actuator torques $\tau_i(t)$, required to produce a planned trajectory, that is, location, velocity, and acceleration [$\boldsymbol{q}(t)$, $\dot{q}(t)$, $\ddot{q}(t)$] and/or planned force/torque $\tau(t)$ at the end-effector, for the entire work cycle such that planned task is performed as specified.

Depending on the control law employed to compute the joint torques, the control systems are classified as either *linear* or *nonlinear*. First, the design of linear controllers based on an approximate linear model of the manipulator will be examined. The effects of time varying inertia, interaction torques and nonlinear torques, of the dynamic model are ignored to simplify the control system. Servo control strategies such as proportional derivative (*PD*) and proportional integral derivative (*PID*) are discussed.

Finally, the method of computed torque control, which is based on a nonlinear control law, will be discussed. This scheme employs a nonlinear feedback to linearize dynamics of the manipulator. Force control, hybrid position and force control, impedance control, robust and adaptive control systems are more complex control approaches for manipulator control. These are taken up in the later part of the chapter.

8.3 LINEAR CONTROL SCHEMES

There are a number of ways by which a controller can react to the error signal and supply the output for actuators. Some of the simple control strategies are

(i) The on/off or two-step control.
(ii) The proportional (*P*) control, which produces a control signal proportional to error.
(iii) The derivative (*D*) control, which produces a control signal proportional to the rate of change of error. Derivative control can be considered as an anticipatory control that measures the existing rate of change of error, anticipates incoming larger error, and applies correction before the larger error is arrived. Derivative control is never used alone.
(iv) The integral (*I*) control, which produces a control signal that is proportional to integral of the error with time. The integral controller can be considered as "looking-back", summing all the errors and then responding.
(v) Combination modes: Combinations of above strategies such as proportional plus derivative (*PD*), proportional plus integral (*PI*), proportional plus integral plus derivative (*PID*) are often used to achieve the desired performance.

In manipulator control, there are many situations where better control can be achieved with other strategies such as partitioned *PD* control, adaptive control, computed torque control and so on. In the following sections, three control strategies, their implementation, and performance are discussed for the position control problem.

The use of linear control schemes is only valid for those systems whose behavior can be mathematically modelled by linear differential equations. However, the manipulator dynamic model, as given by Eq. (8.1), is highly nonlinear. Hence, the linear control technique for manipulator control is essentially an approximate method and for this a manipulator joint should be modelled as a linear second-order system.

The approximate linear model of an individual joint is obtained by ignoring the configuration-dependent nature of link inertias, interaction inertias, and gravity torques. It will be seen that the linear controller, based on this approximate model, serves as a good approximation and results in a satisfactory trajectory-tracking performance for the highly geared (nondirect drive) joint, a joint having a gear box with large gear (speed) ratio. Such controllers are used in vast majority of industrial robots. This linear approximation will also serve as a basis for the study of more complex nonlinear control schemes.

It is clear from Fig. 8.2, that the manipulator control problem is a *multi-input, multi-output* (*MIMO*) problem, involving joint and end-effector locations, velocities, accelerations, and force vectors. To simplify the problem, each joint is considered to be independent and separately controlled. This single-joint model is assumed to have a single input (set point) and a single output (location, velocity etc.). Hence, the n-DOF manipulator is modelled as n-independent linear second-order systems and is controlled by n-independent *single-input, single-output* (*SISO*) control systems. Before developing and analyzing the linear second-order *SISO* model of the joint, the general second-order linear system characteristics are briefly brushed up first. The readers familiar with linear control systems may choose to skip the next section. At the same time, for a complete and detailed study, books on control systems may be consulted.

8.4 CHARACTERISTICS OF SECOND-ORDER LINEAR SYSTEMS

This section is a brush up of elementary concepts of control theory and second-order linear system behaviour. For a second-order system, the relationship between the *input* x and the *output* y is described by a differential equation of the form

$$a_2 \frac{d^2 y}{dt^2} + a_1 \frac{dy}{dt} + a_0\, y = b_0' x \tag{8.2}$$

where a_2, a_1, a_0, and b_0' are constants. The Laplace transform of this equation, with all initial conditions zero, is

$$a_2 s^2 Y(s) + a_1 s Y(s) + a_0 Y(s) = b_0' \, X(s) \tag{8.3}$$

The open-loop *transfer function* of this system is

$$G(s) = \frac{Y(s)}{X(s)} \frac{b_0'}{a_2 s^2 + a_1 s + a_0} \tag{8.4}$$

where, $Y(s)$ is the Laplace transform of output and $X(s)$ is the Laplace transform of input. An alternative and more familiar way of writing differential equation of second-order system is

$$\frac{d^2y}{dt^2} + 2\zeta\omega_n + \frac{dy}{dt} + \omega_n^2\, y = b_0\omega_n^2\, x; \text{ with } b_0 = b_0'/a_0 \tag{8.5}$$

where, ζ is the *damping ratio* and ω_n is the undamped *natural frequency* at which the system oscillates. In the Laplace domain, the open-loop transfer function $G(s)$ of this system is given by

$$G(s) = \frac{Y(s)}{X(s)} = \frac{b_0\omega_n^2}{s^2 + 2\zeta\,\omega_n s + \omega_n^2} \tag{8.6}$$

The above general form of mathematical model describes the system dynamics for many physical systems such as mass-damper-spring mechanical systems, *LCR* electrical circuits, fluid systems, thermal systems and so on.

The control of a system consists of varying the input so that the closed-loop dynamics of the system has the desired properties; say the output y follows the desired trajectory y_d. *Feedback* control explicitly uses the error signal, $e = y - y_d$, the difference between the measured output y and desired output y_d, as a part of control law, so as to reduce the error and system sensitivity to the parameters in dynamic model. The block diagram of this system with *unity negative feedback* is drawn in Fig. 8.5.

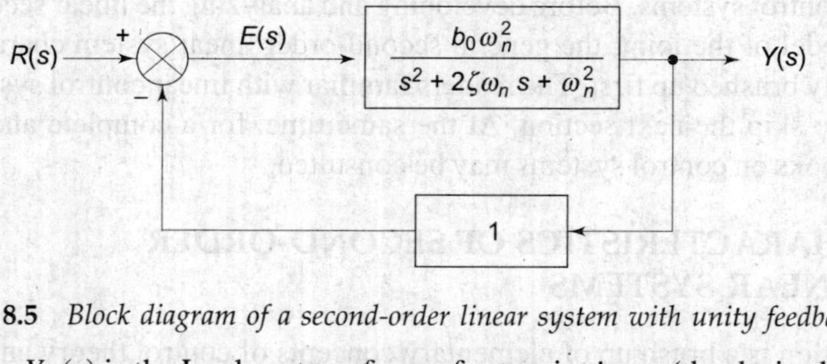

Fig. 8.5 *Block diagram of a second-order linear system with unity feedback*

If $R(s) = Y_d(s)$ is the Laplace transform of set-point *input* (the desired output), the closed-loop transfer function of the control system is

$$\frac{Y(s)}{R(s)} = \frac{G(s)}{1+G(s)} = \frac{b_0\omega_n^2}{s^2 + 2\zeta\,\omega_n s + (1+b_0)\omega_n^2} \tag{8.7}$$

The second-order linear *characteristic equation* of the closed-loop system, from Eq.(8.7), is written in simplified (standard) form as

$$s^2 + 2\zeta\omega_n s + \omega_n^2 = 0 \tag{8.8}$$

where ζ and ω_n in Eq. (8.8) can be found from the denominator of Eq. (8.7) as

$$\omega_n\sqrt{(1+b_0)} \to \omega_n, \quad \frac{\zeta}{\sqrt{(1+b_0)}} \to \zeta$$

The *roots* of the characteristics equation, r_1, r_2 which characterize the system's time response to a given input, are

$$r_1, r_2 = -\zeta \omega_n \pm \omega_n \sqrt{\zeta^2 - 1} \tag{8.9}$$

These roots determine the nature of time response of the system. If roots r_1 and r_2 are real, then the response of the system is sluggish and nonoscillatory. If r_1 and r_2 are complex conjugate, then the response of the system is oscillatory. The system response analysis helps in fixing the controller parameters that ensure closest possible trajectory tracking performance of a joint.

To study the nature of response of a second-order system, unit-step input is considered. For a unit-step input, $R(s) = 1/s$, the output of the system, from Eq.(8.6), is given by

$$Y(s) = G(s)\,R(s) = \frac{b_0 \omega_n^2}{s(s^2 + 2\zeta\,\omega_n s + \omega_n^2)} \tag{8.10}$$

This can be rearranged as

$$Y(s) = \frac{b_0 \omega_n^2}{s(s + r_1)(s + r_2)} \tag{8.11}$$

where, r_1 and r_2 are the roots of the characteristic equation, and are given by Eq.(8.9).

Depending on the value of the damping ratio ζ, these roots can be real and unequal or real and equal or complex conjugate. These three classes of roots characterize the three types of time response of the system. These three classes of roots arise with value of damping ratio greater than unity, equal to unity, and less than unity, respectively and are discussed below.

Case (i) Damping ratio greater than unity

A damping ratio greater than unity ($\zeta > 1$) gives the roots r_1 and r_2 as real and unequal. The roots r_1 and r_2 are

$$\begin{aligned} r_1 &= -\zeta\omega_n + \omega_n\sqrt{\zeta^2 - 1} \\ r_2 &= -\zeta\omega_n - \omega_n\sqrt{\zeta^2 - 1} \end{aligned} \tag{8.12}$$

The time response is obtained by taking the inverse Laplace transform of Eq. (8.12), which gives

$$y(t) = 1 - \frac{e^{-\zeta\omega_n t}}{2\sqrt{(\zeta^2 - 1)}} \Big[\left(\sqrt{(\zeta^2 - 1)} - \zeta\right) e^{-\omega_n \sqrt{(\zeta^2 - 1)}\,t} + \left(\sqrt{(\zeta^2 - 1)} + \zeta\right) e^{\omega_n \sqrt{(\zeta^2 - 1)}\,t} \Big] \tag{8.13}$$

In this case, the system has a sluggish and nonoscillatory behavior (time response), and system is *overdamped*.

Case (ii) Damping ratio equal to unity

For a damping ratio equal to unity ($\zeta = 1$), both r_1 and r_2 are real and equal with $r_1 = r_2 = -\omega_n$ and the time response of the system is

$$y(t) = \left[1 - e^{-\omega_n t} - \omega_n t e^{-\omega_n t}\right] \tag{8.14}$$

Under this condition, the system exhibits the fastest possible nonoscillatory, overshoot-free response and is *critically damped.*

Case (iii) Damping ratio less than unity

The roots r_1 and r_2 are complex conjugates of each other when the damping ratio is less than unity ($\zeta < 1$). The time response, $y(t)$ is

$$y(t) = 1 - \frac{e^{-\zeta\omega_n t}}{\sqrt{1-\zeta^2}} \sin\left(\omega_n\sqrt{(1-\zeta^2)}\ t + \tan^{-1}\frac{\sqrt{(1-\zeta^2)}}{\zeta}\right) \tag{8.15}$$

In this case, the system has an oscillatory behavior and is '*underdamped*'.

From the above analysis, it is observed that the linear second-order system is stable for all values of damping ratio, although the behaviour may be oscillatory or nonoscillatory.

The system (time) responses for unit-step input, Eqs (8.13), (8.14), and (8.15), are plotted in Fig. 8.6 for these typical values of damping coefficient (ratio) ζ.

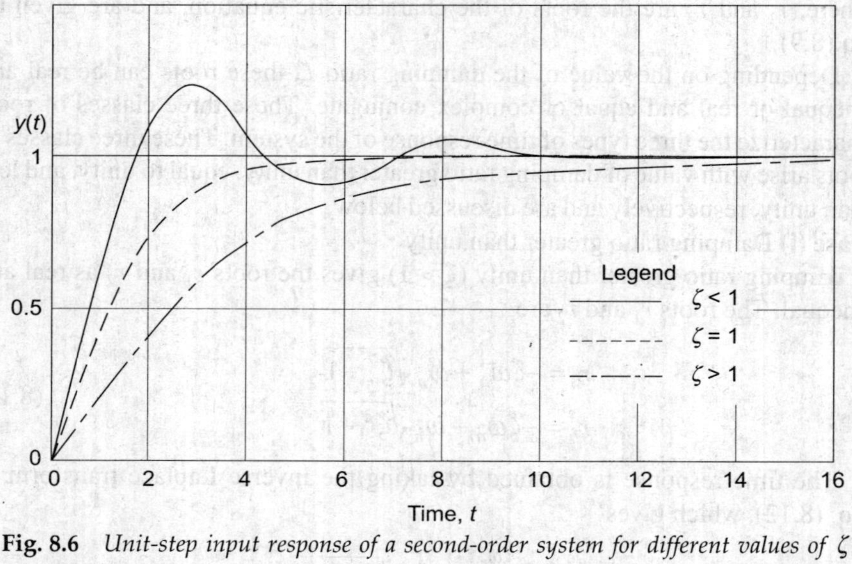

Fig. 8.6 *Unit-step input response of a second-order system for different values of* ζ

A robotic control system cannot be allowed to have an oscillatory response for obvious reasons. For instance, in a pick-n-place operation, an oscillating end-effector may strike against the object before picking it to manipulate. Hence, highest possible speed of response and yet nonoscillatory response, dictates that the controller design parameters shall be chosen to have the damping ratio $\zeta = 1$ or at least close to it but less than unity.

8.5 LINEAR SECOND-ORDER SISO MODEL OF A MANIPULATOR JOINT

A simplified linear model of one rotary joint of a manipulator is developed in this section for applying the linear control schemes. A common drive system for many industrial robots consists of an electric motor as the actuator and a reduction gearbox to provide the torque to the joint.

The schematic of an actuator-gear-link assembly of a rotary joint is shown in Fig. 8.7. The gear ratio 'η' is defined as

$$\eta = \frac{\text{number of teeth on the gear of link shaft}}{\text{number of teeth on the gear of actuator shaft}} = \frac{\eta_2}{\eta_1} \tag{8.16}$$

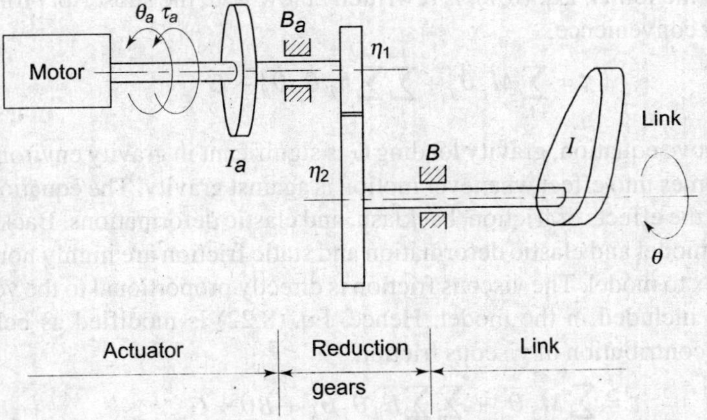

Fig. 8.7 *Schematic of actuator-gear-link mechanism for a manipulator joint*

Note that the gear ratio η is greater than 1 for a reduction gear. The gear ratio causes a reduction of the actuator speed $\dot{\theta}_a$ to $\dot{\theta}$, the speed of the link and increase of the actuator torque τ_a to τ, the torque applied to the link (the load) as given by

$$\tau = \eta \tau_a$$
$$\dot{\theta} = \frac{1}{\eta}\dot{\theta}_a \tag{8.17}$$
$$\theta = \frac{1}{\eta}\theta_a$$

where θ_a is the angular displacement of the actuator shaft and θ is the joint (link) angular displacement.

The link torque τ, when reflected to the actuator side, is scaled down by a factor of η, the gear ratio. If I_a is the inertia of rotor of the actuator and B_a is the viscous friction coefficient at actuator bearings, for the rotational mass-damper system, the torque required at the actuator shaft is given by

$$\tau_a = I_a \ddot{\theta}_a + B_a \dot{\theta}_a + \frac{\tau}{\eta} \tag{8.18}$$

or $$\tau_a = I_a\ddot{\theta}_a + B_a\dot{\theta}_a + \frac{1}{\eta}(I\ddot{\theta} + B\dot{\theta}) \tag{8.19}$$

where I is the inertia of the load and B is the viscous friction coefficient for load. Using Eq. (8.17), Eq. (8.19) is written in items of actuator variables as

$$\tau_a = \left(I_a + \frac{I}{\eta^2}\right)\ddot{\theta}_a + \left(B_a + \frac{B}{\eta^2}\right)\dot{\theta}_a \tag{8.20}$$

or in terms of link variables as

$$\tau = \left(I + \eta^2 I_a\right)\ddot{\theta} + \left(B + \eta^2 B_a\right)\dot{\theta} \tag{8.21}$$

The link torque τ is obtained from the dynamic model of the joint. The dynamic equation of the joint i, Eq. (8.1), is rewritten below with the subscript i for joint i dropped for convenience.

$$\tau = \sum_j M_j\ddot{\theta}_j + \sum_j\sum_k h_{jk}\dot{\theta}_j\dot{\theta}_k + G \tag{8.22}$$

In the above equation, gravity loading G is significant in gravity environments only and comes into effect whenever motion is against gravity. The equation does not include the effects of friction, backlash, and elastic deformations. Backlash is difficult to model and elastic deformation and static friction are highly nonlinear and complex to model. The viscous friction is directly proportional to the velocity and can be included in the model. Hence, Eq. (8.22) is modified as below to include the contribution of viscous friction.

$$\tau = \sum_j M_j\ddot{\theta}_j + \sum_j\sum_k h_{jk}\dot{\theta}_j\dot{\theta}_k + B\dot{\theta} + G \tag{8.23}$$

where B is the viscous friction coefficient at the link bearings.

The terms in Eq. (8.23) are split into linear decoupled and nonlinear interacting terms. The two linear decoupled terms are: $M\ddot{\theta}$, from the first summation term in Eq. (8.23) for link i (remember that subscript i has been dropped) and the third term $B\dot{\theta}$. The remaining terms in Eq. (8.23) are nonlinear interacting terms and are grouped together and denoted as 'd', that is

$$d = \sum_{\substack{j=1 \\ j\neq i}}^{n} M_j\ddot{\theta}_j + \sum_j\sum_k h_{jk}\dot{\theta}_j\dot{\theta}_k + G \tag{8.24}$$

Thus, Eq. (8.23) is written as

$$\tau = M\ddot{\theta} + B\dot{\theta} + d \tag{8.25}$$

Note that the Eq. (8.25) is written for joint i with subscript i dropped.

Substituting τ from Eq. (8.25) into Eq. (8.18) and rearranging, the actuator torque is obtained as

$$\tau_a = (I_a + \frac{M}{\eta^2})\ddot{\theta}_a + (B_a + \frac{B}{\eta^2})\dot{\theta}_a + \frac{d}{\eta} \tag{8.26}$$

In the case of highly geared manipulators ($\eta >> 1$), it is quite reasonable to ignore the contribution of the nonlinear interacting term d and still guarantee a good trajectory-tracking performance by the controller. Neglecting the nonlinear terms, the actuator torque reduces to the simple form

$$\tau_a = I_{eff}\,\ddot{\theta}_a + B_{eff}\,\dot{\theta}_a \tag{8.27}$$

where I_{eff} is the effective inertia and B_{eff} is the effective damping with

$$\begin{aligned} I_{eff} &= I_a + M/\eta^2 \\ B_{eff} &= B_a + B/\eta^2 \end{aligned} \tag{8.28}$$

Because the configuration dependent term M is reduced by a factor η^2, with $\eta >> 1$, I_{eff} is almost constant for highly geared manipulators.

It may be useful to summarize the assumptions made in obtaining the model given by Eq. (8.27):

(a) The links are rigid bodies.
(b) Link inertia is constant.
(c) Nonlinear interacting terms such as the interactive and centrifugal torques and gravity torque are ignored.
(d) Nonlinear effects such as coulomb friction and external disturbance are ignored.
(e) Backlash is ignored.

Thus, Eq. (8.27) is a *SISO* model of a rotary joint of a manipulator as a second-order linear system with actuator torque τ_a as the input and angular displacement θ_a as the output. To construct a controller for an n-DOF manipulator, n-independent *SISO* linear control systems may be designed.

In the above discussion, a rotary joint model has been considered. The model of rotational joint holds analogously for prismatic joint also. Before proceeding to examine the design of controllers based on this approximate joint model, the characteristics of actuators employed to drive the robotic links are discussed because the actuator also influences the controller design.

8.6 JOINT ACTUATORS

The actuators used in robotic systems to drive the links are broadly classified as:

(i) Hydraulic actuators,
(ii) Pneumatic actuators, and
(iii) Electric actuators

Hydraulic actuators can produce large force/torque to drive the manipulator joints without the use of reduction gearing and are easily applied for robotic position control. But hydraulic systems are cumbersome and messy and require a great deal of equipment such as pumps, actuators, hoses, and servo valves. In applications where position and/or torque must be accurately controlled, hydraulic actuators prove disadvantageous due to friction of seals, leakage, viscosity of oil, and its complex temperature dependence.

Pneumatic actuators possess all the disadvantages of hydraulic actuators except that these are relatively cleaner. Pneumatic actuators are difficult to control accurately due to high friction of seals and compressibility of air (or other pneumatic medium).

The most popular choice for actuators for robotic systems is the electric motor. Although these do not match the power-to-weight ratio of hydraulic or pneumatic actuators, these are easy to control and interface. This makes them attractive for small-to-medium sized manipulators.

The electric actuators generally require reduction gears of high ratios. The high-gear ratio linearizes the system dynamics and reduces the coupling effects. This is an added advantage of the electric actuators but at the cost of increased joint friction, elasticity, and backlash. On the other hand, use of hydraulic or pneumatic actuators to directly drive the joint minimizes the drawbacks due to friction, elasticity, and backlash. Electric actuators are subclassified into four categories.

(a) AC motor,
(b) DC motor,
(c) Stepper motor, and
(d) Other devices such as solenoid.

Amongst electric motors, DC motors are most straightforward to interface and control and, hence, are the most commonly used. AC motors and stepper motors are less utilized because AC motors are difficult to control and stepper motors have low torque capability. Moreover, stepper motors cannot be used for Cartesian space trajectory tracking operations (like plotting). The model of a typical DC motor is discussed in the following section keeping in view its applications in developing control strategies. Models for other types of motors and other electrical actuators can be easily found in other books.

8.6.1 Model of a DC Motor

The schematic diagram of an armature-controlled permanent magnet DC motor is shown in Fig. 8.8. Motors with field excitation provided by permanent magnets are preferred for robotic applications. Such motors are labelled permanent magnet DC (PMDC) motors.

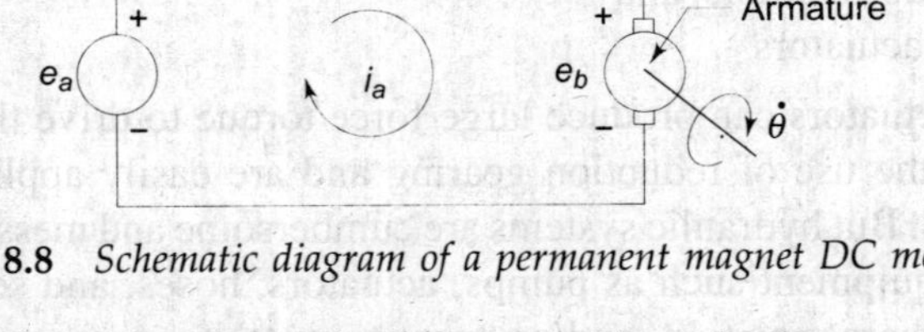

Fig. 8.8 *Schematic diagram of a permanent magnet DC motor*

With reference to the schematic diagram in Fig. 8.8, the control signal to the DC Motor is applied at the armature terminals in the form of armature voltage e_a. The motor torque τ_a is related to the armature current i_a as

$$\tau_a = K_T\, i_a \tag{8.29}$$

where K_T is the motor torque constant. Observe, that the armature current directly controls the torque generated by the actuator. The back emf e_b induced in the armature winding is given by

$$e_b = K_b \dot{\theta}_a \tag{8.30}$$

where K_b is the back emf constant. Applying the Kirchoff's voltage law for the circuit of Fig. 8.8, gives

$$e_a = e_b + i_a R_a \tag{8.31}$$

where e_a is the emf across armature and R_a is the armature resistance. Substituting e_b from Eq. (8.30) gives

$$e_a = K_b \dot{\theta}_a + i_a R_a \tag{8.32}$$

Thus, the armature voltage is adjusted, depending on the commands generated by the manipulator control system, which specify the torque required from the actuator. The continuous adjustment of the armature voltage e_a is carried out by the motor driver circuitry so that a desired current i_a (corresponding to required torque τ_a) flows through the armature windings. The schematic of a motor driver circuitry is shown in Fig. 8.9.

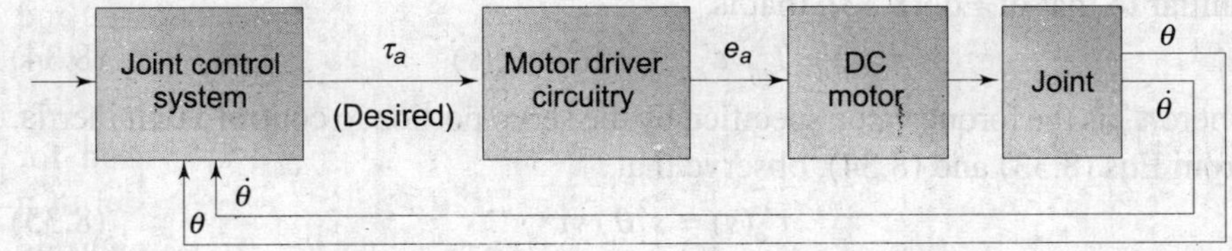

Fig. 8.9 *Control of a manipulator joint driven by a DC motor*

8.7 PARTITIONED PD CONTROL SCHEME

With the development of linear second-order *SISO* model of a manipulator joint and modelling of the actuator, it is now possible to study various linear control strategies for independent joint control. The first scheme discussed is a modified *PD* control.

A linear controller based on a *Partitioned Proportional Derivative* (*PPD*) control strategy is slightly different from a simple *PD* controller. It is useful for systems that are more complex. The controller is partitioned into two parts: a model-based portion and a servo portion, such that the joint parameters (I_{eff} and B_{eff} in this case) appear only in the model-based portion. The approximate model of the manipulator joint developed in previous section is used to implement this controller. The block diagram of this controller is shown in Fig. 8.10. Note that the portion to the right of the partition line is the physical system, the model-based portion and to the left of the partition line is the servo-based control system, usually implemented in a computer or microcontroller.

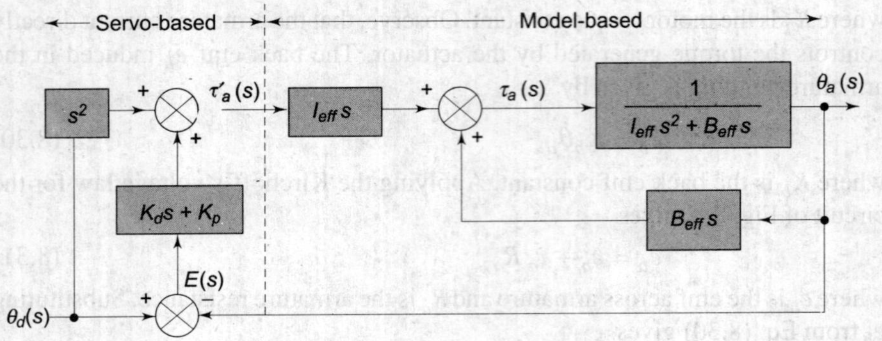

Fig. 8.10 *A partitioned* PD *controller for trajectory control of a rotary joint*

The model-based portion of the control law computes the command torque based on the system parameters I_{eff} and B_{eff}. The Laplace of the approximate model of the joint in Eq. (8.27) is

$$\tau_a(s) = (I_{eff}\, s^2 + B_{eff}\, s)\theta_a(s) \tag{8.33}$$

The *model-based control law* defines the actuator torque using a structure similar to that of Eq. (8.33), that is

$$\tau_a(s) = I_{eff}\, \tau'_a(s) + B_{eff}\, s\theta_a(s) \tag{8.34}$$

where τ'_a is the torque value specified by the servo portion to control a unit inertia. From Eqs (8.33) and (8.34), observe that

$$\tau'_a(s) = s^2\theta_a(s) \tag{8.35}$$

This is the equation of motion for a system with *unit inertia*. Thus, the model-based law effectively reduces the system to that of unit-inertia system. In other words, the physical system appears to be a unit-inertia system to the servo portion.

The servo portion of the controller is based on the error in actuator position and its derivative. These servo errors are computed as

$$E(s) = \theta_d(s) - \theta_a(s) \tag{8.36}$$

and
$$sE(s) = s\theta_d(s) - s\theta_a(s) \tag{8.37}$$

where $\theta_d(s)$ is the desired joint displacement of the actuator shaft.

The servo portion of the control law makes use of the feedback to modify the system behavior. Because the model-based portion of the control law has made the system appear as a unit inertia, the servo portion control law design becomes very simple. The controller is simply required to control a system of unit inertia with no friction and no stiffness.

Assuming that the sensors are capable of detecting position and velocity, a proportional derivative (*PD*) control scheme can be used to control the system of unit mass/inertia. The control system using the *PD* control scheme with position and velocity feedback for the control of unit inertia is shown in Fig. 8.11. This is the simplified model of the *PPD* controller shown in Fig. 8.10.

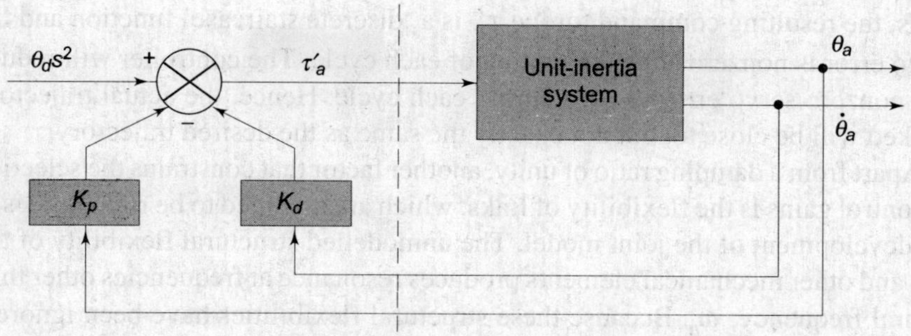

Fig. 8.11 *Simplified* PPD *controller for control of unit-inertia system*

The control law for the servo portion is specified with two controller gain parameters, the proportional constant K_p, and the derivative constants K_d as

$$\tau'_a(s) = s^2\theta_d(s) + K_d sE(s) + K_pE(s) \tag{8.38}$$

Combining Eq. (8.38)with Eq. (8.35), the control law becomes

$$s^2\theta_a(s) = s^2\theta_d(s) + K_d sE(s) + K_pE(s) \tag{8.39}$$

The error dynamics of the *PPD* controller, using Eq. (8.37), is given by

$$s^2E(s) + K_d sE(s) + K_pE(s) = 0 \tag{8.40}$$

Equation (8.40) is identical to the characteristics equation of the second-order linear system, Eq. (8.8). Thus, the partitioned *PD* controller reduces the system to that of a linear second-order system. Comparing Eq. (8.40) with Eq. (8.8) gives the natural frequency of the system as:

$$\omega_n = \sqrt{K_p} \tag{8.41}$$

and damping ratio as

$$\zeta = \frac{K_d}{2\sqrt{K_p}} \tag{8.42}$$

From Eq.(8.40), it is clear that any desired second-order performance of the control system is possible by the choice of proper *control gains*, K_p and K_d.

As discussed in Section 8.1, the controller for the manipulator should be critically damped, that is $\zeta = 1$. Hence, from Eqs (8.41) and (8.42), for critically damped performance

$$K_d = 2\sqrt{K_p} = 2\omega_n \tag{8.43}$$

The Eq. (8.43) specifies the methodology for setting the control gains K_p and K_d, for the critically damped controller performance.

A direct consequence of control law, Eq. (8.40), is that the *servo error* at any instant of time is zero provided there is no initial error and the computation time for the computer is zero, that is, the actuator torque is computed as a continuous function of time.

In reality, the time taken to compute the servo error, the *PD* law control gains and to command a new value of torque, is nonzero and is known as the *cycle time*.

Thus, the resulting command torque τ'_a is a 'discrete staircase' function and the servo error is nonzero at the beginning of each cycle. The controller will reduce this nonzero servo error to zero during each cycle. Hence, the actual trajectory tracked will be close to, but not exactly the same as the desired trajectory.

Apart from a damping ratio of unity, another factor that constrains the selection of control gains is the flexibility of links, which are assumed to be rigid bodies in the development of the joint model. The unmodelled structural flexibility of the link and other mechanical elements produces resonance at frequencies other than natural frequency, ω_n. Because these structural flexibilities have been ignored, the controller must be designed so as not to excite these unmodelled resonances.

The lowest unmodelled resonance, which corresponds to the maximum value of the effective inertia seen by the actuator, I_{max} has a resonance frequency

$$\omega_{res} = \omega_0 \sqrt{\frac{I_0}{I_{\max}}} \tag{8.44}$$

where ω_0 is the structural frequency when the effective inertia is I_0.

To prevent exciting these structural oscillations and also to ensure structural stability, the controller natural frequency ω_n must be limited to $0.5\,\omega_{res}$. That is

$$\omega_n \leq 0.5\,\omega_{res}$$

or

$$\omega_n \leq 0.5\,\omega_0 \sqrt{\frac{I_0}{I_{\max}}} \tag{8.45}$$

From Eqs (8.41) and (8.45),

$$K_p \leq (0.5\,\omega_0)^2 \frac{I_0}{I_{\max}} \tag{8.46}$$

Thus, the control gains K_p and K_d for the *PD* controller are selected in accordance with Eqs (8.43) and (8.46).

8.7.1 Effect of External Disturbance

The performance of the partitioned *PD* controller in the presence of an external disturbance or *noise* is examined now. Figure 8.12 shows the partitioned *PD* (*PPD*) controller with an additional input of disturbance torque τ_{dist}. The error dynamics in the Laplace domain, given by Eq. (8.40), is modified as

$$s^2E(s) + K_d sE(s) + K_p E(s) = \tau_{dist}(s) \tag{8.47}$$

The system with disturbance input will be stable if and only if the disturbance is bounded. In other words, for all bounded disturbances, the controller will keep the system stable. Assume that the disturbance is as a step function of magnitude *K* and is given by

$$\tau_{dist}(s) = \frac{K}{s} \tag{8.48}$$

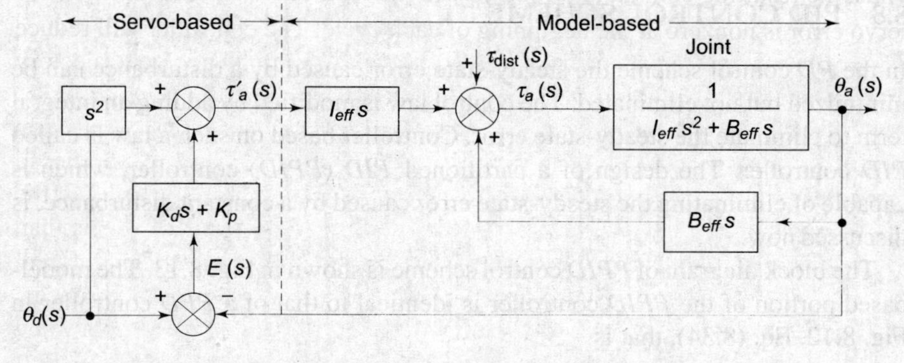

Fig. 8.12 PPD *controller with external disturbance*

From Eqs (8.47) and (8.48), the error dynamics becomes

$$(s^2 + K_d s + K_p)E(s) = \frac{K}{s}$$

or

$$E(s) = \frac{K}{s(s^2 + K_d s + K_p)} \tag{8.49}$$

The steady-state error is defined as

$$e_{ss} = \lim_{s \to 0} |\, s\, E(s)| \tag{8.50}$$

Substituting Eq. (8.49) in Eq. (8.50), and applying the limit gives the steady state error in actuator shaft position as

$$e_{ss} = \frac{K}{K_p} \tag{8.51}$$

Thus, the steady-state error is inversely proportional to the proportional gain K_p of the *PD* controller. Because one of the purposes of a controller is to provide for disturbance rejection, that is, to maintain good performance, K_p is chosen at the maximum possible value that does not excite the structural oscillations. This maximum value of K_p, which minimizes the steady-state error, is obtained from Eq. (8.46) as

$$K_{p_{\max}} = (0.5\omega_0)^2 \frac{I_0}{I_{\max}} \tag{8.52}$$

The reader must understand that the above discussion is for the simplest case of a constant disturbance and is intended to show good disturbance rejection performance of a *PD* controller.

Recall that in the development of the linear *SISO* model in Section 8.5, the nonlinear terms were grouped together and neglected. The effect of the nonlinear term *d* in Eq. (8.25) can now be considered as a disturbance.

The presence of steady-state error means that the end-effector will not reach the target point and will always be slightly away from it.

8.8 PID CONTROL SCHEME

In the *PD* control scheme the steady-state error caused by a disturbance can be minimized but not eliminated. The control law is modified by adding an integral term to eliminate the steady-state error. Controller based on such a law is called *PID* controller. The design of a partitioned *PID* (*PPID*) controller, which is capable of eliminating the steady-state error caused by a constant disturbance, is discussed now.

The block diagram of *PPID* control scheme is shown in Fig. 8.13. The model-based portion of the *PPID* controller is identical to that of a *PPD* controller in Fig. 8.12, Eq. (8.34), that is

$$\tau_a(s) = I_{eff}\tau'_a(s) + B_{eff}s\theta_a(s) \tag{8.53}$$

The servo portion includes, in addition to the position and derivative components, an integral of the servo error *E*(*s*). The *PID* servo law is, thus, obtained from Eq. (8.38) by adding the integral term as

$$\tau'_a(s) = s^2\theta_d(s) + K_d s\theta_a(s) + K_p\theta_a(s) + \frac{K_i\theta_a(s)}{s} \tag{8.54}$$

where K_i is the integral constant. From Eqs (8.35), (8.53), and (8.54), the error dynamics for the *PPID* controller is written as

$$s^2E(s) + K_d sE(s) + K_pE(s) + \frac{K_i}{s}E(s) = 0 \tag{8.55}$$

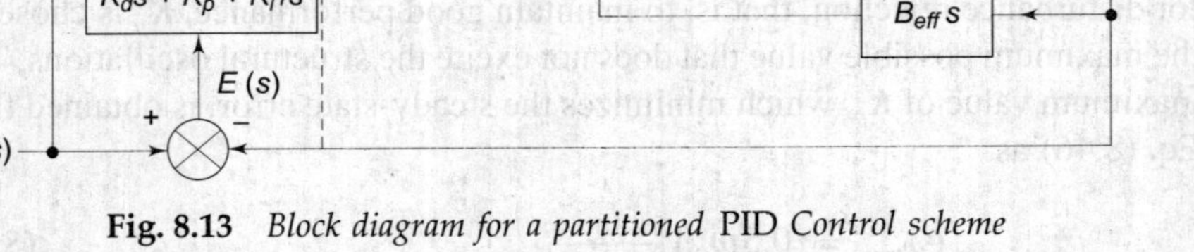

Fig. 8.13 *Block diagram for a partitioned* PID *Control scheme*

In the presence of a constant disturbance as a step function of constant magnitude, $\tau_{dist}(s) = K/s$, the error dynamics is given by

$$s^2E(s) + K_d sE(s) + K_pE(s) + \frac{K_i}{s}E(s) = \frac{K}{s} \tag{8.56}$$

or

$$E(s) = \frac{K}{s^3 + K_d s^2 + K_p s + K_i} \tag{8.57}$$

The steady-state error for the *PID* controller in the presence of a disturbance is

$$e_{ss} = \lim_{s \to 0} s\,E(s) = \lim_{s \to 0} \frac{sK}{s^3 + K_d s^2 + K_p s + K_i} \tag{8.58}$$

On simplifying, it gives

$$e_{ss} = 0 \tag{8.59}$$

Hence, the *PID* controller causes the manipulator to reach the desired target position even in the presence of a constant disturbance.

It is observed that the *PPID* controller is a third-order system, while *PPD* controller is of second order. In a third-order system, stability criterion must be applied at the final design stage and the servo gains are modified to ensure a good stability margin.

8.9 COMPUTED TORQUE CONTROL

In the previous sections, each joint of a manipulator was viewed as an individual system to be controlled and linear controllers based on *PD* and *PID* control schemes were examined for the simplified *SISO* model. More realistic problem is the control of an n-DOF manipulator as a single system. A control scheme that makes direct use of the complete dynamic model of the manipulator to cancel the effect of gravity, Coriolis and centrifugal force, friction, and the manipulator inertia tensor, is discussed in this section.

The dynamic model of the n-DOF manipulator based on rigid body dynamics gives the $n \times 1$ vector of joint torques τ, Eq. (D.16), Appendix D, as

$$\tau = \boldsymbol{M}(\boldsymbol{q})\ddot{\boldsymbol{q}} + \boldsymbol{H}(\boldsymbol{q}, \dot{\boldsymbol{q}}) + \boldsymbol{G}(\boldsymbol{q}) \tag{8.60}$$

where $\boldsymbol{M}(\boldsymbol{q})$ is the $n \times n$ inertia matrix, $\boldsymbol{H}(\boldsymbol{q}, \dot{\boldsymbol{q}})$ is the $n \times 1$ vector of centrifugal and Coriolis torques, $\boldsymbol{G}(\boldsymbol{q})$ is the $n \times 1$ vector of gravity torques and $\boldsymbol{q}(t)$, $\dot{\boldsymbol{q}}(t)$, $\ddot{\boldsymbol{q}}(t)$ denote the $n \times 1$ vectors of joint positions, velocities and accelerations, respectively. Recall that $\boldsymbol{q}$ is the generalized joint variable.

The overall contribution of friction and other non-rigid-body effects can be included in dynamic model as a function, $\boldsymbol{F}(\boldsymbol{q}, \dot{\boldsymbol{q}})$ of joint positions and velocities. Thus, the Multi-Input-Multi-Output (*MIMO*), nonlinear dynamic model of an n-DOF manipulator becomes

$$\tau = \boldsymbol{M}(\boldsymbol{q})\ddot{\boldsymbol{q}} + \boldsymbol{H}(\boldsymbol{q}, \dot{\boldsymbol{q}}) + \boldsymbol{F}(\boldsymbol{q}, \dot{\boldsymbol{q}}) + \boldsymbol{G}(\boldsymbol{q}) \tag{8.61}$$

The design of a nonlinear control system based on the complicated model given by Eq. (8.61), is considered now. Because this controller is based on a more accurate (though not the exact) dynamic model of a manipulator, it provides better trajectory-tracking performance than linear controllers do. The controller discussed here employs the *computed torque control law* to modify the system and effectively *decouple* and *linearize* it.

Similar to the *PPD* and *PPID* laws, the computed torque control scheme also comprises two portions—a model-based and a servo portion. The model-based portion defines the $n \times 1$ vector of control torques τ using a structure identical to that of Eq. (8.61) as

$$\tau = M(q)t' + H(q, \dot{q}) + F(q, \dot{q}) + G(q) \tag{8.62}$$

where τ' is the $n \times 1$ torque vector specified by the servo portion. Comparing Eqs (8.61) and (8.62), it is observed that

$$\tau' = \ddot{q} \tag{8.63}$$

Thus, the model-based portion effectively linearizes as well as decouples the system dynamics by employing a nonlinear feedback of the actual positions and velocities of joints. The schematic representation of this nonlinear control scheme is shown in Fig. 8.14.

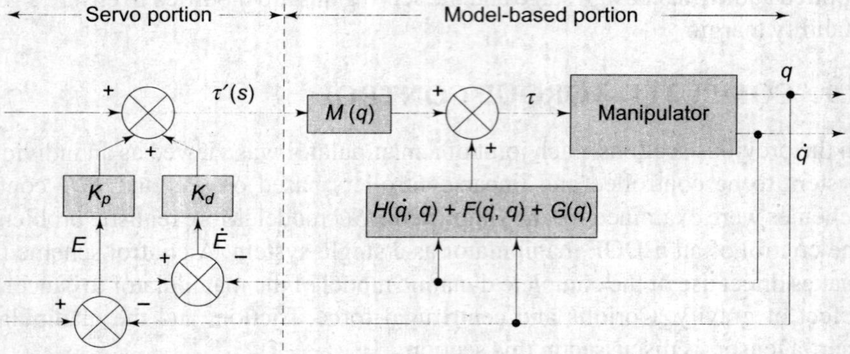

Fig. 8.14 *A nonlinear controller based on the computed torque control law*

The control law of the servo portion is based on the $n \times 1$ vectors E and $\dot{E}$ of servo errors in joint positions and velocities, respectively. These errors are defined as

$$E(t) = q_d(t) - q(t) \tag{8.64}$$

and
$$\dot{E}(t) = \dot{q}_d(t) - \dot{q}(t) \tag{8.65}$$

where q and q_d denote the $n \times 1$ vectors of actual and desired joint positions, respectively.

The servo portion of the computed torque control scheme is, therefore, defined as

$$\tau' = \ddot{q}_d + K_d\dot{E} + K_pE \tag{8.66}$$

where K_p, K_d are the $n \times n$ matrices of position and velocity gains, respectively. Usually, K_p and K_d are chosen as diagonal matrices with constant gains. This serves to decouple the error dynamics of the individual joints. The model of error behavior or error dynamics is obtained from Eqs (8.63) and (8.66) as

$$\ddot{q} = \ddot{q}_d + K_d\dot{E} + K_pE$$

or with $\ddot{E} = \ddot{q}_d - \ddot{q}$

$$\ddot{E} + K_d\dot{E} + K_pE = 0 \tag{8.67}$$

This shows that the error dynamics of a closed-loop system is specified by a second-order linear error equation.

This vector equation is decoupled if $\boldsymbol{K}_p$ and $\boldsymbol{K}_d$ are diagonal matrices with constant gains. Hence, the error equation could be written on a joint-by-joint basis. For joint i the error equation is

$$\ddot{e}_i + K_{di}\,\dot{e}_i + K_{pi}\,e_i = 0 \tag{8.68}$$

with K_{pi} and K_{di} are the position and velocity gains, respectively, for joint i.

For critically damped performance of joint i, the relationship between K_{di} and K_{pi} is, as obtained in Eq. (8.43). That is

$$K_{di} = 2\sqrt{K_{pi}} \tag{8.69}$$

Observe that the computed torque controller employs nonlinear feedback to linearize error dynamics and provides a better trajectory tracking performance than linear controllers, since it is based on a more accurate dynamic model. But, the biggest limitation of this approach is that it is computationally more intensive and, hence, highly expensive when compared to linear controllers. Inaccuracies in the parameters of the manipulator in the dynamic model are other factors, which limit the manipulator's performance. Discrete time control, robust control, impedance control, force control and so on, are the methods employed to overcome these disadvantages and tackle more sophisticated control schemes. Some of these are discussed in next section.

8.10 FORCE CONTROL OF ROBOTIC MANIPULATORS

The control systems for non-contact type of tasks have been dealt with in previous sections, where the manipulator moves in space and its position is controlled, using various approaches. A contact task can be used to perform an assembly, do drilling, wiping a surface and so on. In these applications, the interacting force(s)/torque(s) must be regulated such that force(s)/torque(s) remain within specific range(s). A simple position-controlled manipulator is not suitable for these applications. Take the example of a high stiffness end-effector in contact with a fragile material such as glass. Say, it is required to pick a glass full of water. If the environment is not known with exactness, or there is an uncertainty in stiffness of the contact model, then even for small variation in these parameters, a large force may be applied on the environment. This may damage the environment (glass tumbler in this case) and no useful work may be done. Another example of limitation of position control is that it is impossible to scratch paint from a glass pane with a high stiffness end-effector, using position regulation. If this is attempted, either the glass pane will break or the manipulator would wave the scraping tool with no contact with the glass pane.

An alternative is to sense and control the force/torque that is exerted on the environment instead of position control of manipulator. The ability to measure and control contact force(s) generated at the end-effector offers a possible alternative for extending the effective precision of a manipulator

The force control is entirely dependent on the force sensing. Force sensing is possible by locating sensors at the actuator, or at the tip of the end-effector, or from the environment itself, or at the manipulator wrist. The last one is most commonly used. The determination of actual force/torque may require geometric transformations, depending on the sensor location.

8.11 DESCRIPTION OF FORCE-CONTROL TASKS

In this framework, first force-control approach is discussed for a situation in which the manipulator is partially constrained to move due to contact with a surface. A simplified model of interaction between the end-effector and the environment is used to understand the partially constrained tasks. Every contact task may be divided into subtasks. A subtask is defined by a particular contact situation, which may have specific constraints in some directions and freedom in others.

The interaction task is described using the *natural constraints* and/or *artificial constraints*. These are

- The natural constraints are the constraints imposed by the environment and are defined by the geometry of the task to be performed, either as position constraint or as force constraint on the end-effector of the manipulator.
- The strategy for executing a task imposes artificial constraints. The manipulator can control only those variables, which are not subject to natural constraints. That is, the reference values of the variables controlled by the manipulator are termed as *artificial constraints*.

These constraints are for each degree of freedom of the end-effector and the two sets are complimentary to each other. The following example illustrates these two classes of constraints.

Consider the situation where a robot is required to (i) scrap the paint from a glass pane and (ii) wipe it clean. For scrapping paint from the glass pane the end-effector is not free to move through the glass surface and a *position constraint* exists naturally along the x-direction of the frame $\{c\}$ attached to the tip of scrapper, as shown in Fig. 8.15.

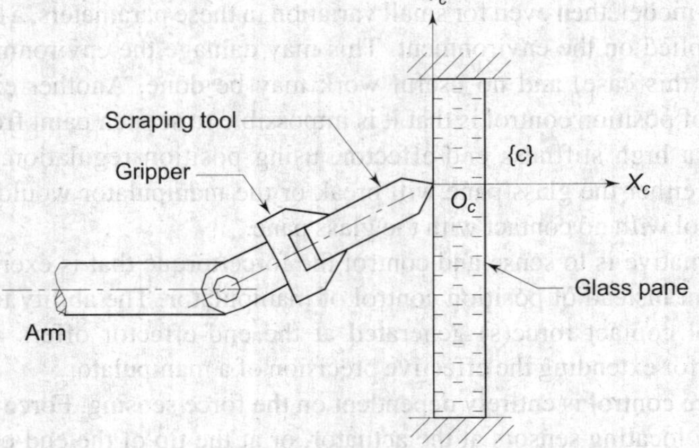

Fig. 8.15 *Scraping paint from a glass pane*

If the surface is frictionless, as in the case of wiping of glass pane with soap, the end-effector cannot apply an arbitrary force tangent to the surface and a *natural force constraint* exists. This is, the tangential force along *y*- and *z*-directions must be zero. Three additional natural constraints arise from the fact that no reaction torque is available at O_c the origin of frame $\{c\}$, and hence the torque about the *x*-, *y*- and *z*-axes must be zero.

For scraping of paint from the window pane, a specific motion of end-effector is required and a force has to be applied by the end-effector in the direction in which the end-effector hand is not free to move, that is, perpendicular to glass surface. The motion is specified by the user as the trajectory for sweeping the entire surface of the glass. The *artificial position constraint*s on the trajectory are specified in terms of desired position and orientation trajectory of the end-effector or its velocity. Similarly, if excessive force is applied it can break or damage the surface and if small force is applied, paint will not be scrapped. Hence, this force must be controlled by the manipulator and it is the *artificial constraint* associated with the task. Notice that this force is not a variable that is subjected to natural constraints.

In general, for each subtask configuration a generalized surface can be defined with position constraint along the normal to the surface and force constraint along the tangent. If a position constraint is naturally defined normal to the surface, then an artificial force constraint exists normal to the surface and is user defined. Similarly, in the direction in which natural force constraints exist, a user defined artificial position constraint exists and is to be specified. It is worth noting that the position constraint is a vector comprising of position as well as orientation of the end-effector and the force constraint is a vector of forces in three principal directions and their respective torque components.

In other words, position and force along each degree of freedom are determined by either a natural or an artificial constraint. Thus, the number of natural constraints and the number of artificial constraints are together equal to the number of degrees of freedom of the constrained (task) space, which is six in general. The natural constraint may be expressed in terms of artificial constraints or vice-versa, in certain situations. For instance, in the above example of paint scrapping the friction between the scrapping tool and the glass surface relates the tangential force along *y*- and *z*-axes with the normal force along *x*-axis.

Position constraints are usually indicated by giving values for components of velocity of the end-effector, $\boldsymbol{V}$, which is a vector, comprising of linear as well as angular velocities because in many cases it is simpler to specify a position constraint as a "velocity equals zero" constraint.

As a second example, consider the *peg-in-hole* assembly operation as illustrated in Fig. 8.16. The figure shows an assembly sequence used to put a round peg into a round hole. The peg is brought down onto the surface to the side of the hole and then slid along the surface until it reaches above the hole. It is then inserted until the peg reaches the bottom of the hole, at which time the assembly is complete. A critical and difficult to meet requirement in this assembly is that the axes of the peg are always in alignment with the axes of the hole. This assembly task can be divided into four subtask domains. First, the arm is free in the space, the second, the peg just touches the surface and slides towards the hole, the third, the peg is just inserted into the hole and starts moving down the hole, and lastly, it touches the bottom of the hole.

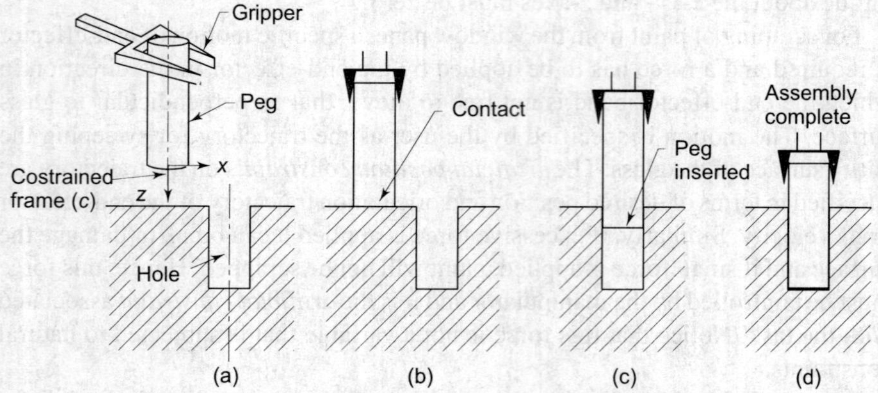

Fig. 8.16 *Peg-in-hole assembly operation tasks*

These four subtasks are illustrated in Fig. 8.16. For each of the subtasks, natural and artificial constraints keep on changing as the assembly is performed. The sequence of the planned artificial constraints causes the task to proceed in the desired manner. This requires substantial task planning and the controller has to switch from one control strategy to another. This is explained with the above example of peg-in-hole assembly. A frame $\{c\}$ is attached to the end of peg in which the constraints are defined for the assembly. For the first subtask, the arm is free to move in space and the natural constraints become that of force because the end-effector will not be able to apply arbitrary force in free space in any direction. Therefore, the natural constraints for the first subtask, with respect to the frame $\{c\}$ attached to the tip of the peg are

$$\boldsymbol{F} = 0 \tag{8.70}$$

Then the position trajectory becomes the artificial constraint in which the task is of moving the peg only in z-direction with user-defined velocity v_z. Hence, the artificial constraints are

$$\boldsymbol{V} = [0 \;\; 0 \;\; \mathrm{v}_z \;\; 0 \;\; 0 \;\; 0]^T \tag{8.71}$$

As soon as the peg touches the surface, the force-sensing system senses a force in z-direction. When this force reaches a threshold value, the control of manipulator is switched to second stage of the strategy. As seen in Fig. 8.16, in the second subtask, the peg is not free to move in z-direction or to rotate about x- or y-axes and it is not free to apply force in any of the three directions. The strategy for sliding along the surface in x-direction with velocity v_{slide}, while maintaining surface contact, with a contact force $f_{contact}$, is described by the artificial constraints. These natural and artificial position and force constraints are tabulated in Table 8.1.

Table 8.1 *Natural and artificial constraints for second subtask*

Natural Constraints	*Artificial Constraints*
$v_z = 0$	$v_x = v_{slide}$
$\omega_x = 0$	$v_y = 0$
$\omega_y = 0$	$\omega_z = 0$
$f_x = 0$	$f_z = f_{contact}$
$f_y = 0$	$\tau_x = 0$
$\tau_z = 0$	$\tau_y = 0$

A velocity in z-direction (crossing the threshold of zero) detects the third subtask, as the peg slightly falls into the hole, as soon as it reaches completely above the hole. This signals a change of natural constraints and artificial constraints. These new constraints are tabulated in Table 8.2 with v_{insert} as the velocity with which peg is inserted into hole.

Table 8.2 *Natural and artificial constraints for third subtask*

Natural Constraints	*Artificial Constraints*
$v_x = 0$	$v_z = v_{insert}$
$v_y = 0$	$\omega_z = 0$
$\omega_x = 0$	$f_x = 0$
$\omega_y = 0$	$f_y = 0$
$f_z = 0$	$\tau_x = 0$
$\tau_z = 0$	$\tau_y = 0$

A force in z-direction, when the assembly is complete, detects the final stage. A point to note is that sensed variable is one, which is not being controlled. For example, in case of the third subtask, the velocity in z-direction is monitored (sensed), while the force in z-direction is controlled.

The above example is a simple one. In real situations, determining task strategies may be quite complex. The planning including uncertainty, which is always present in real systems, in practical situations and automation of this is still a research topic.

8.12 FORCE-CONTROL STRATEGIES

The duality between position and force control requires special control strategy. From the various force control methods, two broad categories emerge for force control: *pure force control* and *impedance control*. In pure force control, the contact force is directly tracked to the desired force, as closely as possible. This scheme can be applied only when the manipulator is in contact with environment. In contrast, the impedance control methodology controls position to regulate the force. In this, the dynamic position-force relation, known as *target impedance*, is controlled. If the end-effector has been compliant enough, like human hand picking a glass tumbler filled with water or using a sponge for washing the window pane, the job could even have been done with position control itself.

There are two basic methodologies to achieve this compliant motion namely: *hybrid position/force control* and *impedance control*.

The first approach is based on the observation that when the end-effector is in contact with the environment, the workspace of the end-effector can naturally be decomposed into a *position subspace* and a *force subspace*. These two subspaces correspond to the directions in which the end-effector is, respectively, free to move and constrained by the environment. The compliance is achieved by explicitly controlling force in force subspace and position in position subspace. But, this requires significant task planning and control law switching in implementation. Additionally, this approach is prone to robustness problem, during transition between unconstrained and constrained motion. A *robust control* means: design of a *controller* with low sensitivity to parameter variations, disturbances, unmodelled dynamics, and other sources of uncertainties.

Alternatively, the impedance control of compliant motion proposes that the control objective be the regulation of the "mechanical impedance" of the end-effector. Thus, the objective of impedance controller is to maintain a desired dynamic relationship between the end-effector position and contact forces. This scheme can be robustified using *PID* controllers. Other robustification schemes can be functional analytical method or Lyapunov-based variable structure approach.

The hybrid position/force control and impedance force control, with a thrust to direct adaptation in impedance scheme, are discussed in the following sections.

8.13 HYBRID POSITION/FORCE CONTROL

The interaction task between the manipulator and environment is defined in terms of the natural constraints and artificial constraints with reference to a constraint frame. The control of artificial constraint variables is required to accomplish the task, as explained in Section 8.11. Because the definition of artificial constraints involves both position and force variables, the control structure is termed as hybrid position/force control. This approach divides the force control problem into three subproblems:

(a) Position control in the direction in which a natural force constraint exists.
(b) Force control in the direction in which a natural position constraint exists.
(c) A control switching law to implement arbitrary mixing of the above modes to perform the task.

The hybrid position/force control is studied by applying it to a simple single degree of freedom system. Later, it will be extended to control an *n*-degree of freedom manipulator.

Consider a single degree of freedom system with mass m in contact with the environment. The contact with environment is modelled as a spring of stiffness k_e. The system to be controlled is, therefore, a mass attached to the spring or a single degree of freedom spring-mass system as illustrated in Fig. 8.17. Unmodeled or disturbance force (due to friction, backlash, etc.) applied to system can be considered as f_{dist}. The variable to be controlled is the force exerted on the environment, f_e and f is the control force applied to mass by the actuator.

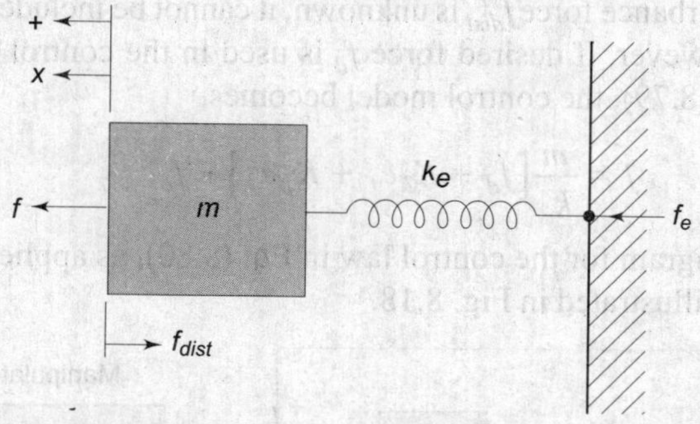

Fig. 8.17 *Single degree of freedom mass-spring system*

The force balance gives the differential equation of the spring-mass system for a displacement x of mass/spring as

$$f = m\ddot{x} + k_e x + f_{dist} \tag{8.72}$$

The reaction force (f_e) exerted by the environment, which is to be controlled, is the force from spring

$$f_e = k_e x \tag{8.73}$$

The spring constant k_e is typically very high, of the order of 10^6 N/m. The control problem for the spring-mass system is to maintain f_e to a desired value f_d. The force error, therefore, is defined as

$$e_f = f_d - f_e \tag{8.74}$$

The control system dynamics is obtained by differentiating Eq. (8.74) and substituting Eq. (8.73) in the resulting equation. The result is

$$\ddot{e}_f = \ddot{f}_d - \ddot{f}_e = \ddot{f}_d - k_e\ddot{x} \tag{8.75}$$

Substituting $\ddot{x}$ from system model in Eq. (8.72) into Eq. (8.75), gives

$$\ddot{e}_f = \ddot{f}_d - \frac{k_e}{m}(f - f_e - f_{dist}) \tag{8.76}$$

The above expression is based on the assumption that environment is fixed and exactly known. Assuming that a *PD* control scheme is used in the controller so that e_f satisfies a desired closed-loop dynamics. The *PD* control law equation is

$$\ddot{e}_f = -K_d\dot{e}_f - K_p e_f \tag{8.77}$$

where, K_p and K_d are proportional and derivative gains.

For a unity feedback, the closed-loop control law in Laplace domain becomes:

$$s^2 e_f(s) + sK_d e_f(s) + K_p e_f(s) = 0 \tag{8.78}$$

The control action, that is the actuator force is obtained by equating right-hand side of Eq. (8.76) and Eq. (8.77) as

$$f = \frac{m}{k_e}\left[\ddot{f}_d + K_d\dot{e}_f + K_p e_f\right] + f_e + f_{dist} \tag{8.79}$$

Because disturbance force f_{dist} is unknown, it cannot be included directly in the control law. However, if desired force f_d is used in the control law in place of $(f_e + f_{dist})$ in Eq.(8.79), the control model becomes,

$$f = \frac{m}{k_e}\left[\ddot{f}_d + K_d\dot{e}_f + K_p e_f\right] + f_d \tag{8.80}$$

The block diagram for the control law in Eq. (8.80), as applied to the spring-mass system, is illustrated in Fig. 8.18.

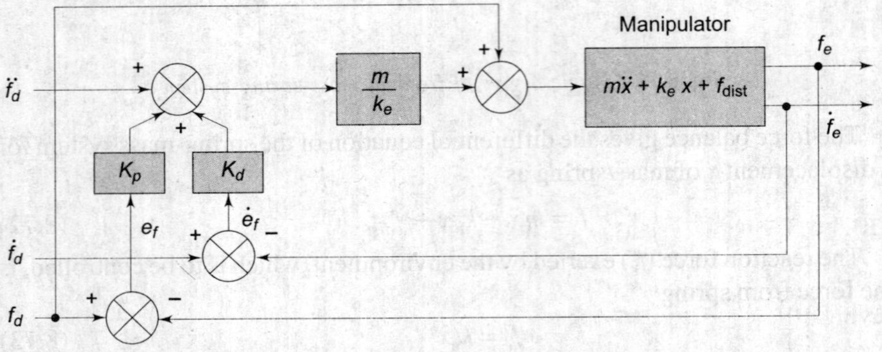

Fig. 8.18 *Force control for spring-mass system*

In practical applications, the requirements may be quite different from ideal situation in Fig. 8.18. For example, the task may require application of constant forces, which means that the inputs of the control system $\dot{f}_d$ and $\ddot{f}_d$ are permanently set to zero. Sensed force (f_e) is a noisy process and, hence, its derivative $\dot{f}_e$ would also be a noisy signal. Alternative approach is to estimate $\dot{f}_e$ from other variables that can be accurately sensed. For example, differentiation of Eq. (8.73) gives

$$\dot{f}_e = k_e\,\dot{x} \tag{8.81}$$

As joint velocity ($\dot{x}$) can be sensed quite accurately, with constant k_e, $\dot{f}_e$ can be obtained from Eq. (8.81) accurately. This is a better estimation of $\dot{f}_e$ free from noise for use in control system. The new control law is, therefore, written as:

$$f = m\left[\frac{K_p}{k_e}e_f + K_d\dot{x}\right] + f_d \tag{8.82}$$

The control system block diagram for the new control law is shown in Fig. 8.19. This control law has the advantage that it does not require the input of the desired force derivatives. The controller gains are chosen such that the system becomes less sensitive to variations in k_e.

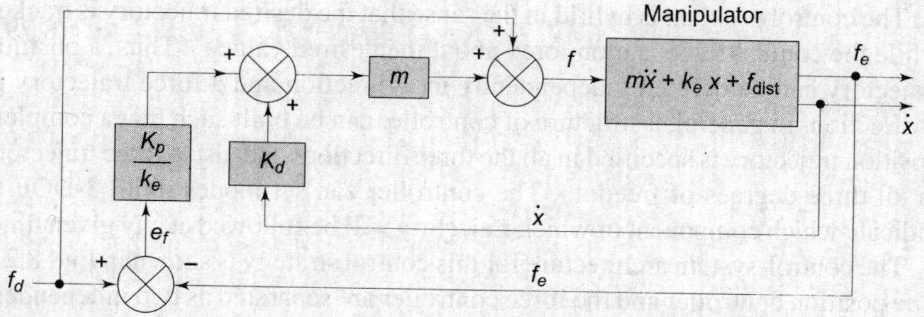

Fig. 8.19 *Practical force control system for a spring-mass system*

Hybrid position/force control can use *PID* controller to improve steady-state performance. In addition, systems are made robust with respect to variations in k_e, which is associated with the elastic deformation of the environment and the manipulator. The environment stiffness k_e is not known with exactness and quite often changes with space and time.

8.13.1 Control Architecture for Hybrid Position/Force Control Scheme

Having developed the hybrid position/force control as a force control servo, it can be easily applied to multidegree of freedom systems. Consider a Cartesian configuration manipulator with 2-DOF, having two prismatic joints as shown in Fig. 8.20. For simplicity, it is assumed that each link has mass m and slides on frictionless linear bearings. Also, assume that the joint motions are lined up exactly with the constraint frame $\{c\}$ as the manipulator follows the environment surface, as in Fig. 8.20.

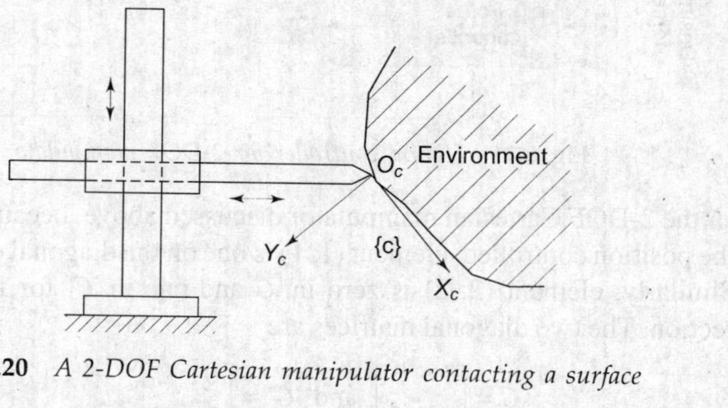

Fig. 8.20 *A 2-DOF Cartesian manipulator contacting a surface*

The trajectory to follow is defined by the environment surface while maintaining a desired contact force f_d. The contact between the manipulator end-point and the surface is assumed frictionless. This means that the environment interaction force f_e is always normal to the surface. The normal to the surface is oriented in y-direction of frame $\{c\}$; hence, force control is required in that direction only and position control in x-direction.

The control problem is hybrid in the sense that the desired trajectory is tracked while the contact force is monitored about the desired value f_d. Thus, a position trajectory can be defined independently in x-direction and a force trajectory in y-direction. In general, a structure of controller can be built such that a complete position trajectory is specified in all the three directions and also a force trajectory in all three degrees of freedom. The controller can set modes in all 3-DOF to indicate which component of which trajectory will be followed at any given time.

The control-system architecture for this control strategy is shown in Fig. 8.21. The position controller and the force controller are separated as two independent systems. These interact with each other through the artificial position and force constraints. The control mode—position or force—to control each joint of the manipulator is decided on whether position control is in effect or force control is in effect.

Two $n \times n$ diagonal matrices C and C' are used to describe the switching between position and force control modes. Recall that the position control and force control due to artificial constraints are mutually exclusive. This fact and the artificial constraints fix the contents of matrix C and C'. The diagonal elements of the matrices are ones and zeros, such that, where a one is present in C, indicating position control in effect, a zero is present in C'. Where a zero is present in C, a one is present in C' and force control is in effect.

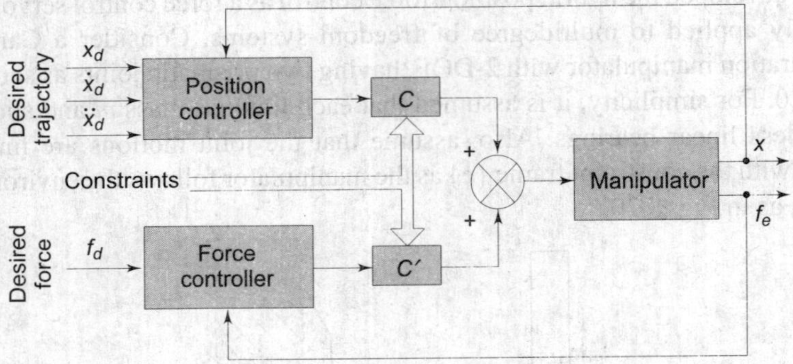

Fig. 8.21 *Hybrid controller for 2-DOF manipulator*

For the 2-DOF Cartesian manipulator discussed above, because x component is to be position controlled, element (1, 1) is one on the diagonal of C and zero on C'. Similarly, element (2, 2) is zero in C and one in C' for force control in y-direction. The two diagonal matrices are

$$C = \begin{bmatrix} 1 & 0 \\ 0 & 0 \end{bmatrix} \text{ and } C' = \begin{bmatrix} 0 & 0 \\ 0 & 1 \end{bmatrix} \tag{8.83}$$

Note that C' matrix has the zero and ones on diagonal inverted from C.

8.14 IMPEDANCE FORCE/TORQUE CONTROL

The impedance control specifies the dynamic relationship between force and position rather than the position and force constraints and switching law between them as in hybrid force control discussed above. The dynamic relationship between the force and position describes the impedance or the mechanical stiffness of the system.

Impedance control provides a unified framework for considering both constrained and unconstrained motion control problems and possesses certain advantage for many applications. For example, this approach can often be implemented with little off-line task planning, it provides robustness to uncertainties and disturbances and can accommodate stable transition between unconstrained and constrained motion. In impedance control approach, the magnitude of contact force depends on the reference position trajectory for the end-effector as well as on the environmental stiffness. There is no builtin mechanism in impedance control to ensure that excessive contact forces are not generated as a result of poor choice of the reference position trajectory or inexact estimates of the environmental stiffness. The class of applications that require accurate regulation of the contact force or demand that the contact force lie within an acceptable range, require trajectory adaptation. In the ideal case, where the parameters of the environment (i.e. location and stiffness) are known exactly, a reference position trajectory can be synthesized a priori to produce the desired contact force. However, in practical cases, where the environmental parameters are not known exactly, the impedance control scheme tends to exhibit poor force tracking characteristic. Hence, a mechanism is required to be developed for providing force-tracking capabilities within the impedance control framework.

This can be done by two simple force-control schemes for automatically generating the reference position trajectory to provide force-tracking capabilities on-line. The first scheme is based on generating the reference position on-line as a function of force-tracking error. This is called *direct adaptation*. The second scheme estimates environmental parameters on-line, using indirect adaptive approach and the required position is computed based on these estimates.

8.14.1 Force Tracking Characteristic of Impedance Control

Consider an n-DOF manipulator operating in n-dimensional task space with $n \leq 6$. When the tool-tip comes in contact with a hard surface, as shown in Fig. 8.22, it cannot penetrate the surface and a natural position constraint is

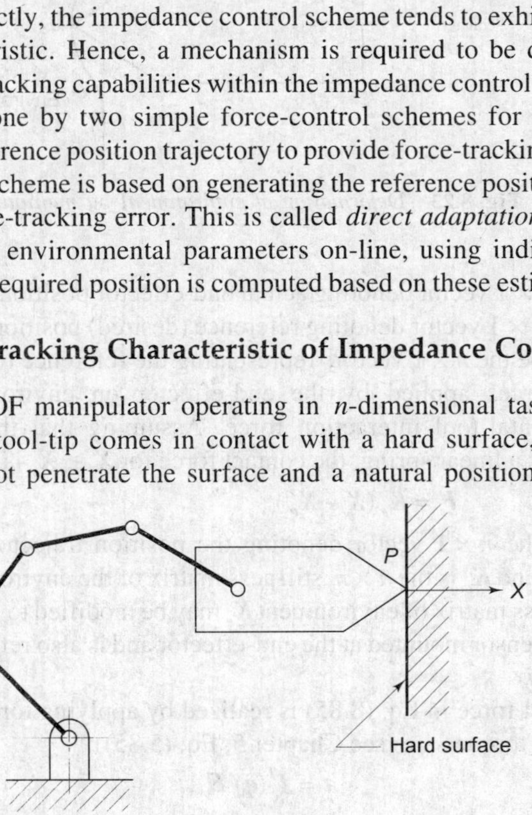

Fig. 8.22 *Interaction of tool-tip with environment*

defined. It no longer makes sense to attempt to control position in the direction perpendicular to surface, that is, the x-direction in Fig. 8.22. However, tool-tip force along x direction needs control to maintain contact with the surface, which is the artificial constraint. Other artificial and natural constraints can be identified as in Section 8.11.

The contact force between the tool-tip and the surface produces a deformation of the environment (the surface) because no real surface is infinitely hard. The environment deforms in the vicinity of the point of contact, as shown in Fig. 8.23. This elastic deformation or the elastic behaviour of environment can be modeled as a generalized spring of stiffness $\boldsymbol{K}_e$. The spring constant is a positive-definite $n \times n$ diagonal stiffness matrix called *environment stiffness matrix* and is defined as

$$\boldsymbol{K}_e = \text{diag}\{k_1 \quad k_2 \quad \cdots \quad k_n\} \tag{8.84}$$

where the scalar k_i denotes the environment stiffness along dimension i.

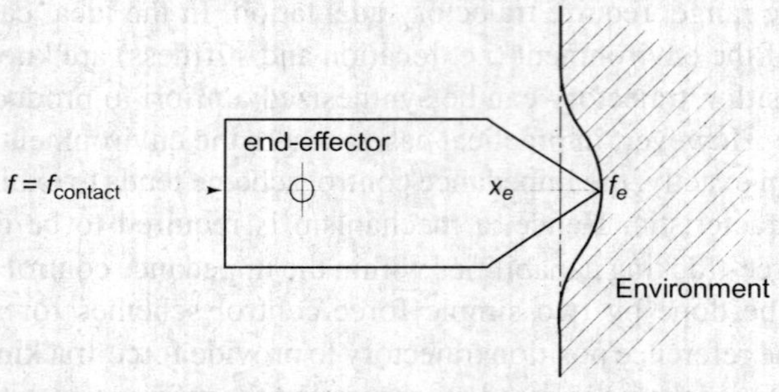

Fig. 8.23 *Deformation of environment by manipulator*

Let

$\boldsymbol{X}$ be the $n \times 1$ vector denoting actual end-effector position trajectory,

$\boldsymbol{X}_d$ be the $n \times 1$ vector denoting reference (desired) position trajectory, and

$\boldsymbol{F}_d$ and $\boldsymbol{F}$ be the $n \times 1$ vectors representing the reference (desired) and actual contact forces applied by the end-effector on environment. $\boldsymbol{F}$ is the environmental-tool interaction force. Assuming that the environment is modeled as a linear spring, the contact force for $\boldsymbol{X} \geq \boldsymbol{X}_e$ is

$$\boldsymbol{F} = \boldsymbol{K}_e(\boldsymbol{X} - \boldsymbol{X}_e) \tag{8.85}$$

where $\boldsymbol{X}_e$ is the $n \times 1$ vector denoting the position trajectory of undeformed environment and $\boldsymbol{K}_e$ is the $n \times n$ stiffness matrix of the environment

The stiffness matrix of environment $\boldsymbol{K}_e$ may be modified to include stiffness of force/torque sensor mounted at the end-effector and is also referred as *equivalent stiffness matrix*.

The contact force in Eq. (8.85) is realized by applying torques at joints. The required joint torques are (see Chapter 5, Eq. (5.85)):

$$\tau = \boldsymbol{J}^T(\boldsymbol{q})\,\boldsymbol{F} \tag{8.86}$$

where $J(q)$ is the $6 \times n$ manipulator Jacobian matrix and q is $n \times 1$ joint displacement vector. The Jacobian also relates the infinitesimal joint displacements δq to the infinitesimal end-effector displacement $(X - X_e)$,

$$X - X_e = J(q)\ \delta q \tag{8.87}$$

Combining Eqs. (8.85)–(8.87) gives

$$\tau = (J^T(q)\ K_e\ J(q))\delta q \tag{8.88}$$

The term in brackets, in Eq. (8.88) is called the *joint stiffness matrix*, K_q, that is

$$K_q = J^T(q)\ K_e\ J(q) \tag{8.89}$$

Note that K_q is not diagonal and may degenerate at manipulator singularities. Equation (8.88) relates the joint displacement q and the joint torques τ with the environment stiffness K_e, that is, the variables directly relevant to the control system. The *impedance control* is developed on these relationships. Let E represent the *displacement error* as the difference between the desired end-effector location and the actual location, that is

$$E = X_d - X \tag{8.90}$$

It is assumed that E is same as the infinitesimal deflection of end-effector from its reference location. A *PD*-plus-gravity control law to control the dynamic interaction between the manipulator and its environment, is

$$\tau = J^T(q)\left[K_p E + K_d \dot{E}\right] + G(q) \tag{8.91}$$

where K_p and K_d are the proportional and derivative gain constant matrices; $G(q)$ is the gravitational torque matrix; $J(q)$ is the manipulator Jacobian and E is the displacement error vector. The matrices K_p and K_d in Eq. (8.91), if assumed to represent "stiffness" and "damping" of the manipulator as seen from the environment, $(K_p E + K_d \dot{E})$ becomes the equation for spring-damper system and gives a force which is transformed into a joint torque vector through the use of $J^T(q)$.

This simple model is representative of many practical situations in which the stiffness term dominates the interaction dynamics. It is to be noted that all Cartesian position and force variables are expressed relative to common fixed world coordinate frame such that one of its axes is normal to the environment. The block diagram in Fig. 8.24 illustrates this impedance control scheme.

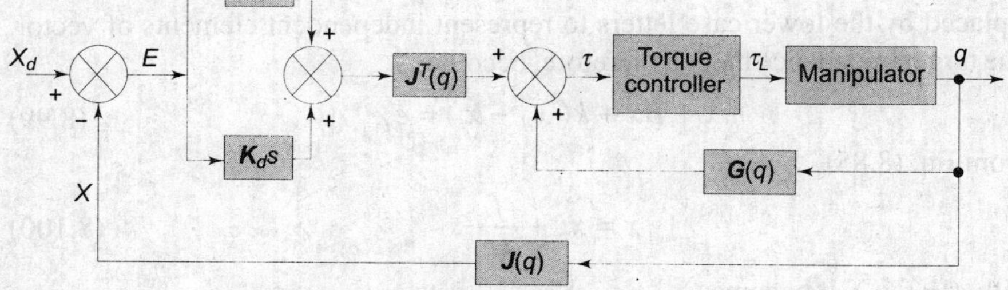

Fig. 8.24 *Impedance control of a manipulator*

As stated earlier, the objective of impedance control is to establish a desired user specified dynamic relationship, referred to as *target impedance*, between the end-effector position $\boldsymbol{X}$ and contact force $\boldsymbol{F}$. Typically, the target impedance is chosen as a linear second-order system so that the dynamic relation mimics a damper-spring-mass system. The model for target impedance is, thus

$$\boldsymbol{M}\ddot{\boldsymbol{X}}(t) + \boldsymbol{B}\dot{\boldsymbol{X}}(t) + \boldsymbol{K}\boldsymbol{X}(t) = \boldsymbol{F}(t) \tag{8.92}$$

To accomplish force tracking, the $n \times 1$ desired force (set point) vector $\boldsymbol{F}_d$ is specified by the user and assumed to be constant during the contact task and the target impedance is driven by the force error $\boldsymbol{E}_f = (\boldsymbol{F}_d - \boldsymbol{F})$, instead of $\boldsymbol{F}$. Thus, in terms the force error $\boldsymbol{E}_f$ and position error $\boldsymbol{E}$, Eq. (8.92) is written as:

$$\boldsymbol{M}\ddot{\boldsymbol{E}}(t) + \boldsymbol{B}\dot{\boldsymbol{E}}(t) + \boldsymbol{K}\boldsymbol{E}(t) = \boldsymbol{E}_f(t) \tag{8.93}$$

$$\boldsymbol{M}\big[\ddot{\boldsymbol{X}}_d(t) - \ddot{\boldsymbol{X}}(t)\big] + \boldsymbol{B}\big[\dot{\boldsymbol{X}}_d(t) - \dot{\boldsymbol{X}}(t)\big] + \boldsymbol{K}\big[\boldsymbol{X}_d(t) - \boldsymbol{X}(t)\big] = \boldsymbol{E}_f(t) \tag{8.94}$$

where $\boldsymbol{M}$, $\boldsymbol{B}$, and $\boldsymbol{K}$ are $n \times n$ constant diagonal mass, damping, and stiffness matrices of target impedance, respectively, and are specified by the user.

During the free space motion, there is no contact with environment, hence

$$\boldsymbol{F} = \boldsymbol{F}_d = 0 \quad \text{or} \quad \boldsymbol{E}_f = 0 \tag{8.95}$$

Hence, the target impedance model given in Eq. (8.94) for the free space motion becomes

$$\boldsymbol{M}\big[\ddot{\boldsymbol{X}}_d(t) - \ddot{\boldsymbol{X}}(t)\big] + \boldsymbol{B}\big[\dot{\boldsymbol{X}}_d(t) - \dot{\boldsymbol{X}}(t)\big] + \boldsymbol{K}\big[\boldsymbol{X}_d(t) - \boldsymbol{X}(t)\big] = 0 \tag{8.96}$$

For a stable system $(\boldsymbol{X}_d - \boldsymbol{X}) \to 0$ as $t \to \infty$, hence, the end-effector position tracks the reference position trajectory $\boldsymbol{X}_d$.

When the end-effector comes into contact with the environment, the dynamic interaction between the manipulator and environment is dictated by the model given by Eq. (8.94). Now to produce a constant force $\boldsymbol{F}$, desired displacement $\boldsymbol{X}_d$ is specified to penetrate into the environment by a constant amount, which means

$$\ddot{\boldsymbol{X}}_d = \dot{\boldsymbol{X}}_d = 0 \tag{8.97}$$

with this, Eq. (8.94) reduces to

$$\boldsymbol{M}\ddot{\boldsymbol{X}}(t) + \boldsymbol{B}\dot{\boldsymbol{X}}(t) + \boldsymbol{K}\big[(\boldsymbol{X}_d(t) - \boldsymbol{X}(t)\big] = \boldsymbol{E}_f(t) \tag{8.98}$$

In Cartesian space control, each Cartesian variable can be considered independently. Thus, the vectors in the target impedance equation, Eq. (8.98), are replaced by the lower case letters to represent independent elements of vector. The target impedance model, therefore, becomes

$$m\ddot{x} + B\dot{x} + k(x_d - x) = e_f \tag{8.99}$$

From Eq. (8.85)

$$x = x_e + \frac{f}{k_e} \tag{8.100}$$

with $e_f = f_d - f$, it becomes

$$x = x_e + (f_d - e_f)/k_e \tag{8.101}$$

Substituting for x and its derivatives in Eq. (8.99) and rearranging gives

$$m\ddot{e}_f + B\dot{e}_f + (k + k_e)e_f = m\ddot{f}_d + B\dot{f}_d - kf_d + k\,k_e(x_d - x_e) \qquad (8.102)$$

The steady-state force error is obtained by equating all time derivatives in Eq. (8.102) to zero. This gives

$$e_{ss} = \frac{k}{(k + k_e)}[-f_d + k_e(x_d - x_e)] \qquad (8.103)$$

which can be written in a concise form as

$$e_{ss} = k_{eq}\left[-\frac{f_d}{k_e} + x_d - x_e\right] \qquad (8.104)$$

where k_{eq} is the equivalent stiffness given by

$$k_{eq} = \frac{k\,k_e}{k + k_e} \qquad (8.105)$$

From Eq. (8.103), it is easily seen that if $x_d = x_e + f_d/k_e$, then $e_{ss} = 0$. In other words, if the precise location (x_e) of the environment and the exact value of the environment stiffness (k_e) are known a priori, the desired position trajectory x_d can be synthesized accordingly to exert the desired contact force. But, in practice x_e and k_e are never known exactly, as a result desired force f_d cannot be exerted.

In addition, apart from uncertainties in environmental parameters (x_e and k_e), there will be residual position errors because of imperfections of robot position control system. If this residual position error is denoted by ε, it is going to contribute an additional force error as $k_e(m\ddot{\varepsilon} + B\dot{\varepsilon} + k\varepsilon)$ and the Eq. (8.102) will become

$$m\ddot{e}_f + B\dot{e}_f + (k + k_e)e_f = m\ddot{f}_d + B\dot{f}_d + kf_d + k\,k_e\,x_e + k_e[m\ddot{\varepsilon} + B\dot{\varepsilon} + k(\varepsilon - x_d)] \qquad (8.106)$$

If the position control system is stable, in the steady state, $\ddot{\varepsilon} = \dot{\varepsilon} = 0$. Equation (8.103) will be modified to

$$e_{ss} = k_{eq}\left[\frac{f_d}{k_e} + x_e + \varepsilon - x_d\right] \qquad (8.107)$$

It is seen that the residual position error ε contributes to the force error e_f and causes an increase in e_{ss}, as is expected. This scheme is called *pure impedance scheme* and can be implemented by the block diagram shown in Fig. 8.25.

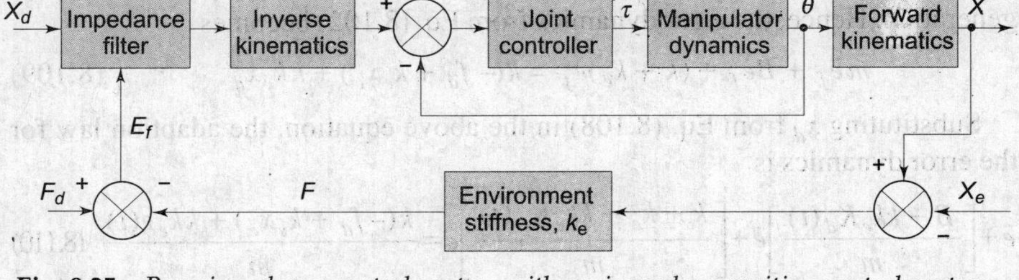

Fig. 8.25 *Pure impedance control system with an inner-loop position-control system*

8.14.2 Adaptive Control

An adaptive control system can be built for the impedance control system discussed above. The control system is made adaptive to actually remove the steady-state force errors. An adaptive control system based on the direct adaptation scheme is shown in Fig. 8.26. The direct adaptation scheme enables the application of a desired contact force f_d, without requiring a priori knowledge of environment parameters, x_e and k_e.

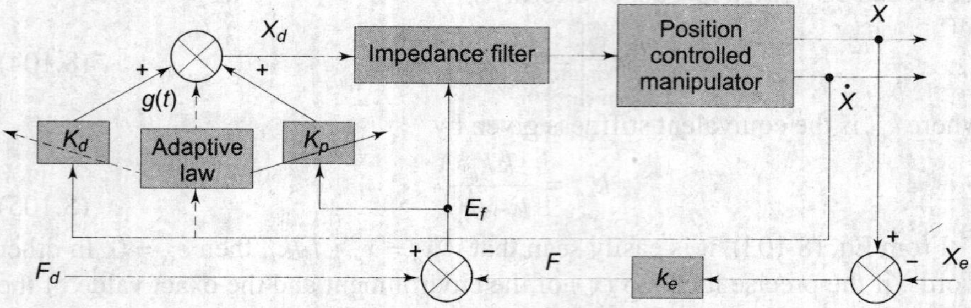

Fig. 8.26 *Direct adaptive impedance control scheme*

Underlying concept of direct adaptive control scheme is to generate the reference position x_d as a function of the force tracking error. A possible adaptive control law is

$$x_d(t) = g(t) + K_p(t)e_f(t) + K_d(t)\dot{e}_f(t) \tag{8.108}$$

where $K_p(t)$ and $K_d(t)$ are the adaptive proportional and derivative feedback gains acting on the force error $e_f(t)$ and the error rate $\dot{e}_f(t)$, respectively, and $g(t)$ is the auxiliary signal to be generated by the adaptive scheme.

Note that the force error rate $\dot{e}_f(t)$ is ordinarily not available because f is a noisy signal. One approach is to filter the force/torque signal to remove the high-frequency noise and then differentiate the filtered signal. This approach works well when the robot moves slowly into contact with a soft environment. An alternative approach is to replace $\dot{e}_f(t)$ by $-k_e\dot{x}$ as suggested by Eq. (8.101), and noting that f_d is constant. In this case, since k_e is an unknown positive constant, it can be absorbed in adaptation gains and in the weighing factor. This approach is suitable when the sampling rate is high and the end-effector velocity measurement is accurate.

The model of target impedance, Eq. (8.92), is used for the adaptive control law generation. Hence, the error dynamics from Eq. (8.102) becomes

$$m\ddot{e}_f + B\dot{e}_f + (k + k_e)e_f = k(-f_d + k_e x_e) + kk_e x_d \tag{8.109}$$

Substituting x_d from Eq. (8.108) in the above equation, the adaption law for the error dynamics is

$$\ddot{e} + \left[\frac{B + kk_e K_d(t)}{m}\right]\dot{e} + \left[\frac{k + k_e + kk_e K_p(t)}{m}\right]e = \frac{k(-f_d + k_e x_e) + kk_e g(t)}{m} \tag{8.110}$$

This equation represents the "adjustable system" in the *model-reference adaptive control* (MRAC) framework. If desired behaviour of the force tracking is represented by

$$\ddot{e}_m + 2\zeta\omega_n \dot{e}_m + \omega_n^2 e_m = 0 \tag{8.111}$$

where ζ and ω_n are the user-defined damping ratio and undamped natural frequency of the force-error dynamics, respectively, then an adaptation law for $g(t)$, $K_p(t)$, and $K_d(t)$, which ensure that the response of the error dynamics of Eq. (8.110) tends to that of reference model of Eq. (8.111), can be developed by a proportional integral (*PI*) adaptation law.

SOLVED EXAMPLES

Example 8.1 *Linear second order system model*

Consider the rotational system with a rotational mass at the end of a shaft as shown in Fig. 8.27. The input, a torque τ, is applied to the disk with a moment of inertia I about the axis of shaft, which is fixed at the other end. The torsional stiffness of the shaft is k and the bearing friction constant is B. What is the condition for the system to be critically damped?

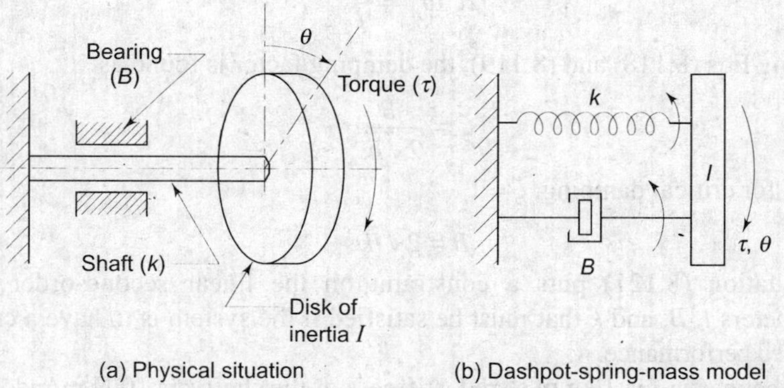

(a) Physical situation (b) Dashpot-spring-mass model

Fig. 8.27 *Rotating mass at the end of a shaft*

Solution First, the differential equation (mathematical model) of the system is obtained. Consider the dashpot-spring-mass model of the system shown in Fig. 8.27(b). The torque balance gives the equation

$$I\frac{d^2\theta}{dt^2} = \tau - B\frac{d\theta}{dt} - k\theta \tag{8.112}$$

or

$$I\frac{d^2\theta}{dt^2} + B\frac{d\theta}{dt} + k\theta = \tau \tag{8.113}$$

This is the second-order linear differential equation model of the torsional system. The Laplace transform of Eq. (8.113) gives

$$I s^2\theta(s) + B s\theta(s) + k\,\theta(s) = \tau(s) \tag{8.114}$$

where $\tau(s)$ is Laplace transform of input and $\theta(s)$ is the Laplace transform of output. Then, the transfer function of the system is

$$G(s) = \frac{\theta(s)}{\tau(s)} = \frac{1}{Is^2 + Bs + k} \tag{8.115}$$

and the characteristics equation for the torsional system is

$$Is^2 + Bs + k = 0$$

or

$$s^2 + \frac{B}{I}s + \frac{k}{I} = 0 \tag{8.116}$$

The condition for critical damping requires the damping factor ζ equal to 1. The standard form of second-order differential equation, Eq. (8.8), is

$$s^2 + 2\zeta\,\omega_n s + \omega_n^2 = 0 \tag{8.117}$$

Comparing the Eq. s (8.116) and Eq. (8.117) gives the system parameters as

$$\omega_n = \sqrt{\frac{k}{I}} \tag{8.118}$$

and

$$2\zeta\,\omega_n = \frac{B}{I} \tag{8.119}$$

Solving Eqs (8.118) and (8.119), the damping factor is found as

$$\zeta = \frac{B}{2\sqrt{Ik}} \tag{8.120}$$

Thus, for critical damping, $\zeta = 1$

$$B = 2\sqrt{Ik} \tag{8.121}$$

Equation (8.121) puts a constraint on the linear second-order system parameters *I, B,* and *k* that must be satisfied if the system is to have a critically damped performance.

If the inertia *I* is 1 kg m s^2/rad, stiffness of the shaft *k* is 10 N m/rad, bearing friction (damping coefficient) *B* is 2 N m s/rad and a constant input torque τ of 2 Nm is applied, then the response of the system can be found as follows:

Substituting the values of above parameters with $\tau(s) = \frac{2}{s}$ for step input in Eq.(8.114), gives

$$s^2\,\theta(s) + 2s\,\theta(s) + 10\,\theta(s) = \frac{2}{s} \tag{8.122}$$

or

$$\theta(s) = \frac{2}{s(s^2 + 2s + 10)} \tag{8.123}$$

From Eqs. (8.118) and (8.119)

$$\omega_n^{\ 2} = 10 \quad \text{or} \quad \omega_n = 3.16 \text{ rad/s} \tag{8.124}$$

and $$\zeta = 0.316 \tag{8.125}$$

Therefore, the system is underdamped and the time domain response is obtained from Eq. (8.15)

$$\theta(t) = 0.2 - 0.2108\, e^{-t} \sin(3t - 71.57) \tag{8.126}$$

Example 8.2 *Model-based control*

Consider a manipulator with a linear, second-order dynamic model as:

$$\tau = I\ddot{\theta}_o + B\dot{\theta}_o \tag{8.127}$$

where I is total inertia and B is effective friction. The actuator gain is K_m in N m s/rad. The manipulator is required to follow the desired trajectory defined by $[\ddot{\theta}_d \quad \dot{\theta}_d \quad \theta_d]$. A model-based control system as shown in Fig. 8.28 is deployed for the dynamic control of the manipulator.

Determine the error dynamics of the system and the condition for critically damped response of the system.

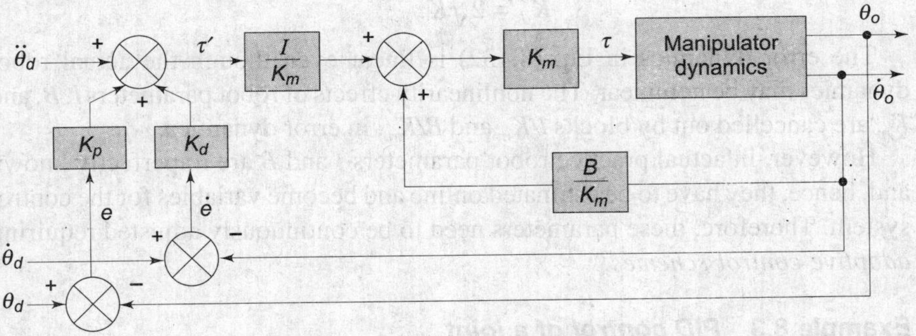

Fig. 8.28 *Model-based trajectory following robot control*

Solution The given dynamic model of the manipulator is

$$\tau = I\ddot{\theta}_o + B\dot{\theta}_o$$

From the control system block diagram in Fig. 8.28 the torque τ is

$$\tau = K_m\left[\frac{I}{K_m}\tau' + \frac{B}{K_m}\dot{\theta}_o\right] \tag{8.128}$$

and $$\tau' = \ddot{\theta}_d + K_d\dot{e} + K_p e \tag{8.129}$$

where error e and its derivative are defined as

$$\begin{aligned} e &= \theta_d - \theta_o \\ \dot{e} &= \dot{\theta}_d - \dot{\theta}_o \end{aligned} \tag{8.130}$$

Combining Eqs (8.127) and (8.128) gives

$$I(\ddot{\theta}_d + K_d\dot{e} + K_p e) + B\dot{\theta}_o = I\ddot{\theta}_o + B\dot{\theta}_o \tag{8.131}$$

or $$\ddot{\theta}_d - \ddot{\theta}_o + K_d\dot{e} + K_p e = 0 \tag{8.132}$$

Defining the second derivative of error as $\ddot{e} = \ddot{\theta}_d - \ddot{\theta}_o$, the system's dynamic performance in error domain is given by

$$\ddot{e} + K_d\dot{e} + K_p e = 0 \tag{8.133}$$

The error dynamics in Eq. (8.132) is independent of parameters *I, B,* and K_m because of cancellation of these terms. But in the model-based control strategy employed in the control system in Fig. 8.28, signals proportional to these are provided, requiring perfect knowledge of parameters *I, B,* and K_m.

As per Eq. (8.132), when excited by initial conditions, the error and its derivatives, the system dynamics seeks zero steady-state error, that is, $e_{ss} = 0$. This means that the system follows the trajectory demanded through command inputs θ_d, $\dot{\theta}_d$, and $\ddot{\theta}_d$.

The gains K_p and K_d are adjusted for critically damped response, for which they are related as (see Eq. (8.43)),

$$K_d = 2\sqrt{K_p}$$

The error dynamics in Eq. (8.132) is linear even though the actual robot dynamics may be nonlinear. The nonlinearity effects of robot parameters *I, B,* and K_m are cancelled out by blocks I/K_m and B/K_m, in error dynamics.

However, in actual practice robot parameters *I* and *B* are imperfectly known and, hence, they have to be estimated online and become variables for the control system. Therefore, these parameters need to be continuously adjusted requiring *adaptive control schemes*.

Example 8.3 *PID control of a joint*

Consider a servomotor driven joint-link with no viscous damping as shown in Fig. 8.29 initially at $\theta = 0$ rad. The *PID* controller block diagram for the joint controller is as shown in Fig. 8.30 with the following specific parameter values:

$$\begin{aligned} I_{eff} &= 1 \text{ kg m}^2 \\ K_T &= 10 \text{ N m/A} \\ R_a &= 5\ \Omega \\ A &= 20 \text{ V/V} \\ K_b &= 0.1 \text{ V/rads/s} \approx 10.5 \text{ V/100 rpm} \\ K_t &= 20.94 \text{ V/1000 rpm} = 0.2 \text{ V/rad/s} \\ m &= 1 \text{ kg} \end{aligned} \tag{8.134}$$

Determine the transient performance of the joint if its final position is $\pi/4$ radians, that is, $\theta_d(s) = \dfrac{0.785}{s}$.

Solution Assuming no disturbance and no proportional, integral, or derivative control, the transfer function of the joint controller is obtained as

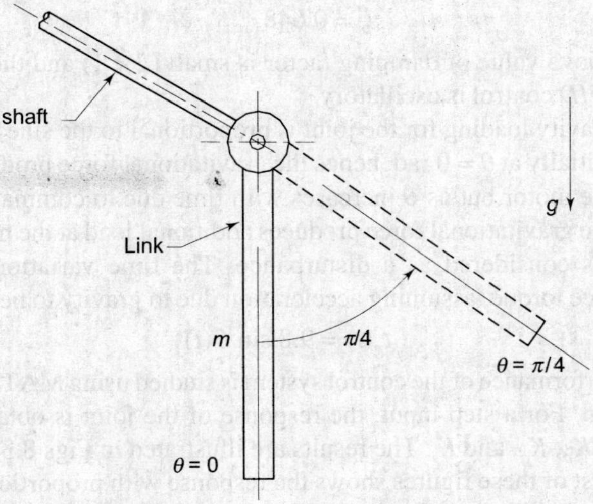

Fig. 8.29 *Servomotor driven joint-link*

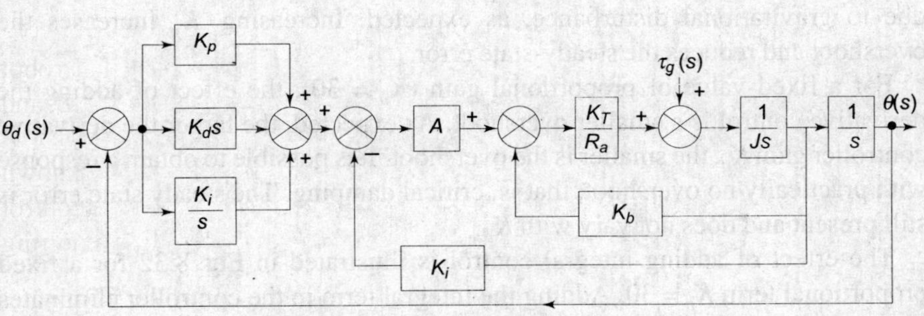

Fig. 8.30 *PID Controller for a joint*

$$\frac{Q(s)}{Q_d(s)} = \frac{\dfrac{AK_T}{R_a}}{s\left(Js + \dfrac{K_T}{R_a}(K_b + AK_t) + \dfrac{AK_T}{R_a}\right)} \tag{8.135}$$

Substituting the value of given parameters

$$\frac{Q(s)}{Q_d(S)} = \frac{40}{s^2 + 8.2\ s + 40} \tag{8.136}$$

This clearly is a linear second-order system. Therefore, the characteristic equation for the joint control system, without gravity loading and no *PID* gain constants, is

$$s^2 + 8.2\ s + 40 = 0 \tag{8.137}$$

Comparing Eq. (8.137) with Eq. (8.8), gives

$$\omega_n = 6.32 \text{ rad/s} \tag{8.138}$$

and $$\zeta = 0.648 \tag{8.139}$$

The above value of damping factor is small ($\zeta < 1$) and the system response without *PID* control is oscillatory.

The gravity loading for the joint is proportional to the sine of θ. Note that the joint is initially at $\theta = 0$ rad, hence the gravitational force produces no additional load at the motor but as θ increases with time due to command signal to reach $\theta = \pi/4$ the gravitational force produces additional load at the motor. This gravity loading is considered as a disturbance. The time variation of gravitational disturbance torque, assuming acceleration due to gravity to be 9.8 kg m/s^2, is

$$\tau_g(t) = 9.8 \sin \theta(t) \tag{8.140}$$

The performance of the control system is studied using MATLAB for computer simulation. For a step input, the response of the joint is obtained for different values of K_p, K_d, and K_i. The results are illustrated in Figs 8.31 and 8.32.

The first of these figures shows the response with proportional controller and the effect of adding a derivative control. It is observed that with only proportional control (K_d, K_i, = 0) there is an overshoot as well as steady state error is present due to gravitational disturbance, as expected. Increasing K_p increases the overshoot and reduces the steady-state error.

For a fixed value of proportional gain (K_p = 30), the effect of adding the derivative control is a smaller overshoot. As expected, the larger the derivative controller gain K_d, the smaller is the overshoot. It is possible to obtain a response with practically no overshoot, that is, critical damping. The steady state error is still present and does not vary with K_d

The effect of adding integral control is illustrated in Fig. 8.32 for a fixed proportional term K_p = 30. Adding the integral term in the controller eliminates the steady-state error.

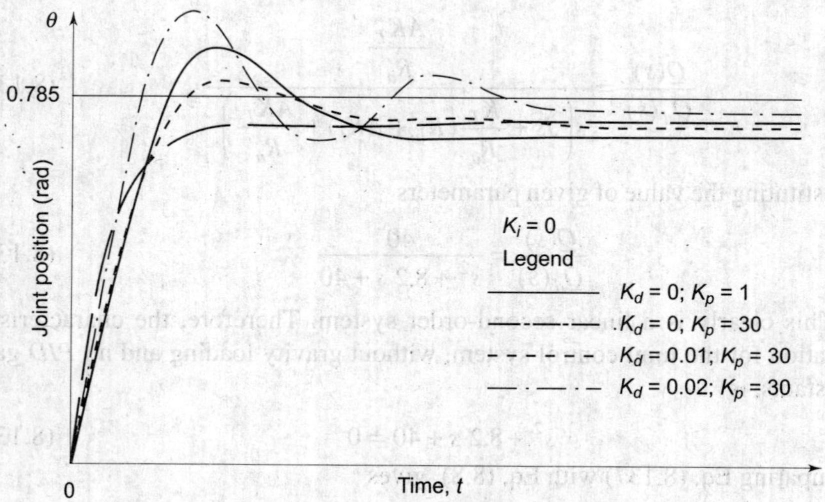

Fig. 8.31 *Step response of system in Fig. 8.30 for* PD *control*

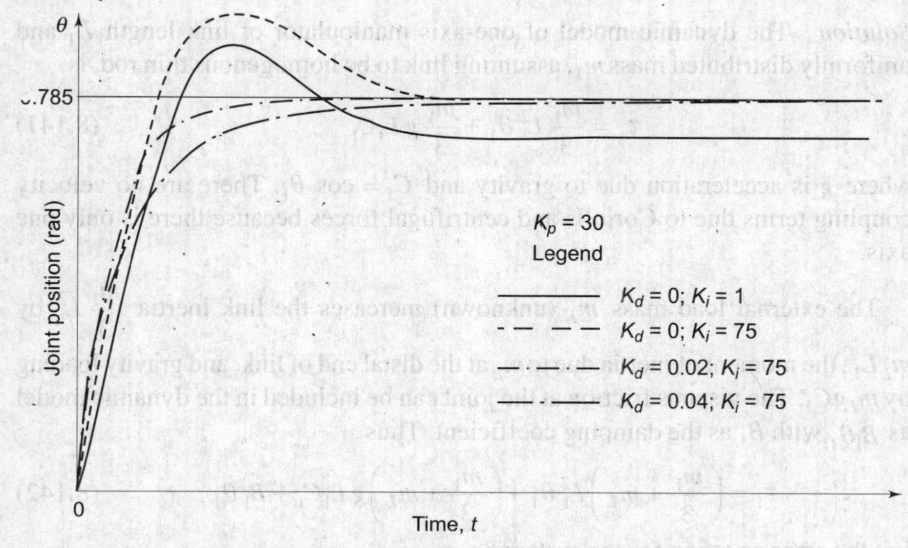

Fig. 8.32 *Step response of system with* PID *control*

It is possible to get the desired performance from the system by choosing the appropriate values of parameters of *PID* controller. A stepwise thumb-rule for selecting the parameters can be

1. Adjust K_p until the system step response is either critically or slightly underdamped with $K_i = K_d = 0$.
2. Next, adjust K_i to reduce the steady-state error to acceptable value (or eliminate completely). Normally, K_i is greater than K_p.
3. Finally, increase K_d until the step response is again critically damped or slightly overdamped.

Example 8.4 *The one link manipulator*

Consider the controller design for a single-link manipulator shown in Fig. 8.33. The load carried by the link is modeled as a point mass m_L at the distal end of the link.

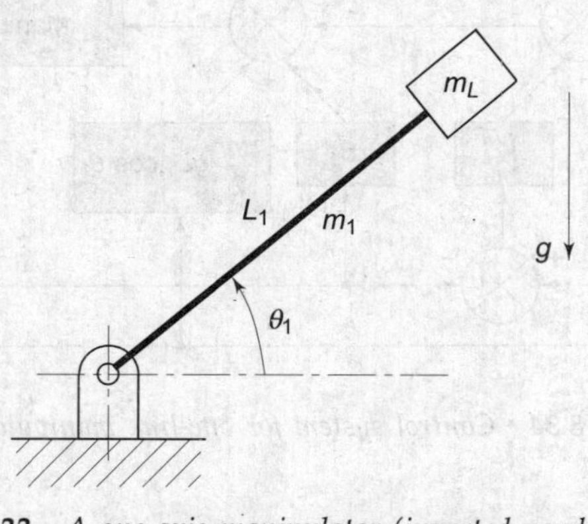

Fig. 8.33 *A one-axis manipulator (inverted pendulum)*

Solution The dynamic model of one-axis manipulator of link length L_1 and uniformly distributed mass m_1, assuming link to be homogenous thin rod, is

$$\tau_1 = \frac{m_1}{3} L_1^2 \ddot{\theta}_1 + \frac{m_1}{3} g L_1 C_1 \tag{8.141}$$

where g is acceleration due to gravity and $C_1 = \cos\,\theta_1$. There are no velocity coupling terms due to Coriolis and centrifugal forces because there is only one axis.

The external load mass m_L (unknown) increases the link inertia $\frac{m_1}{3} L_1^2$ by $m_L L_1^2$, the moment of inertia due to m_L at the distal end of link, and gravity loading by $m_L g C_1$. The viscous friction at the joint can be included in the dynamic model as $B_1 \dot{\theta}_1$ with B_1 as the damping coefficient. Thus

$$\tau_1 = \left(\frac{m_1}{3} + m_L\right) L_1^2 \ddot{\theta}_1 + \left(\frac{m_1}{3} + m_L\right) g L_1 C_1 + B_1 \dot{\theta}_1 \tag{8.142}$$

Let the error equation for the system be

$$I\ddot{e} + K_d \dot{e} + K_p e = 0 \tag{8.143}$$

where $$I = \left(\frac{m_1}{3} + m_L\right) L_1^2 \tag{8.144}$$

Substituting, $e = \theta_d - \theta_1$, Eq. (8.143) becomes

$$I\ddot{\theta}_d + K_d\left(\dot{\theta}_d - \dot{\theta}_1\right) + K_p(\theta_d - \theta_1) = I\ddot{\theta}_1 \tag{8.145}$$

Hence, the torque to be supplied by the actuator motor is

$$\tau = I\ddot{\theta}_d + K_d\left(\dot{\theta}_d - \dot{\theta}_1\right) + K_p(\theta_d - \theta_1) + B_1 \dot{\theta}_1 + \frac{m}{2} g L_1 C_1 \tag{8.146}$$

If the manipulator has to follow a desired trajectory defined by $\left[\ddot{\theta}_d \;\; \dot{\theta}_d \;\; \theta_d\right]$, these become the input to the control system. The block diagram shown in Fig. 8.34 is a realization of such a control system.

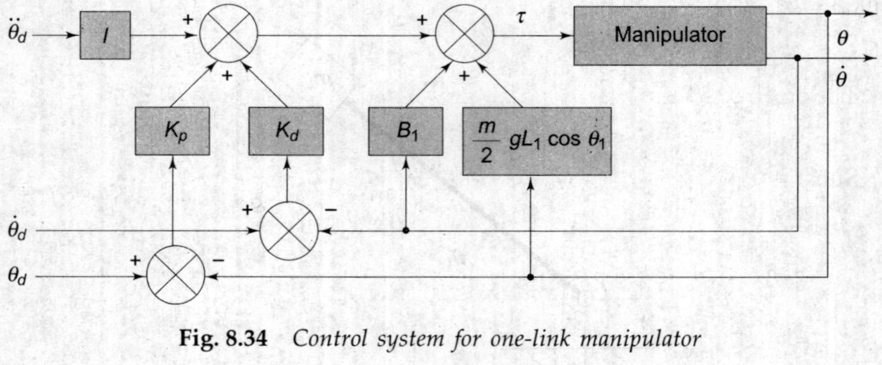

Fig. 8.34 *Control system for one-link manipulator*

Example 8.5 ***Impedance control***

Consider the 2-DOF, RR manipulator, as shown in Fig. 8.35. Drive the impedance control law for the manipulator assuming no gravity loading.

Solution The Jacobian for RR manipulator was obtained in Example 5.4 for links of unit length ($L_1 = L_2 = 1$). The significant Jacobian for RR manipulator with link lengths L_1 and L_2 is (see Exercise 5.5)

$$\boldsymbol{J}(q) = \begin{bmatrix} -L_1S_1 - L_2S_{12} & -L_2S_{12} \\ L_1C_1 + L_2C_{12} & L_2C_{12} \end{bmatrix} \tag{8.147}$$

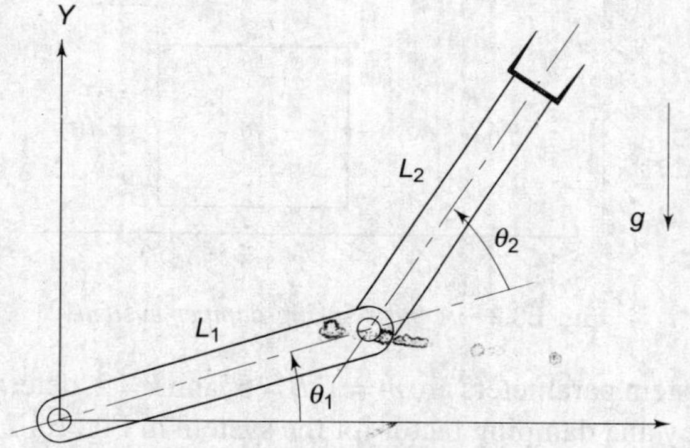

Fig. 8.35 *The 2-DOF RR manipulator*

where $C_1 = \cos\theta_1$, $S_1 = \sin\theta_1$, $C_{12} = \cos(\theta_1 + \theta_2)$, and $S_{12} = \sin(\theta_1 + \theta_2)$. This Jacobian indicates singularities at $\theta_2 = 0$ (fully stretched) and $\theta_2 = \pi$ (folded). The impedance control law is given as, Eq. (8.91)

$$\tau = \boldsymbol{J}^T(\boldsymbol{q})[\boldsymbol{K}_p\boldsymbol{E} + \boldsymbol{K}_d\dot{\boldsymbol{E}}] + \boldsymbol{G}(\boldsymbol{q}) \tag{8.148}$$

Assume that the controller stiffness matrix $\boldsymbol{K}_p$ and the controller-damping matrix $\boldsymbol{K}_d$ are diagonal and are given as

$$\begin{aligned} \boldsymbol{K}_p &= diag[K_{p1} \quad K_{p2}] \\ \boldsymbol{K}_d &= diag[K_{d1} \quad K_{d2}] \end{aligned} \tag{8.149}$$

The error vector is

$$\boldsymbol{E} = [e_1 \ e_2]^T \tag{8.150}$$

Substituting values, the control law, Eq. (8.148) gives

$$\tau_1 = -(L_1S_1 + L_2S_{12})\left(K_{p1}e_1 + K_{d1}\dot{e}_1\right) + (L_1C_1 + L_2C_{12})\left(K_{p2}e_2 + K_{d2}\dot{e}_2\right) + g_1 \tag{8.151}$$

$$\tau_2 = -L_2S_{12}\left(K_{p1}e_1 + K_{d1}\dot{e}_1\right) + L_2C_{12}\left(K_{p2}e_2 + K_{d2}\dot{e}_2\right) + g_2 \tag{8.152}$$

The controller gains $\boldsymbol{K}_p$ and $\boldsymbol{K}_d$ are selected on the basis of the manipulator task defined by natural and artificial constraints while keeping track of damping characteristics.

EXERCISES

8.1 A mass less shaft carries an inertia of 2 kg m^2. If the shaft stiffness is 250 Nm/rad, determine the resonance (natural) frequency of the shaft.

8.2 Consider the mass-spring-damper system shown in Fig. E8.2. Obtain the open-loop transfer function for the system. Determine the characteristic equation of the system if a unit feedback is used in the closed-loop system. From this determine the natural frequency and the damping factor for the system.

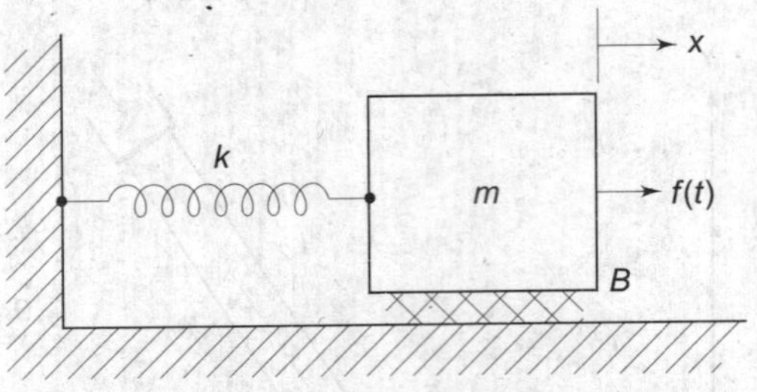

Fig. E8.2 *A mass-spring-damper system*

8.3 If the system parameters are $m = 1, B = 5$, and $k = 2$, determine the natural frequency and damping factor for the system in Fig. E8.2.

8.4 Determine the motion of the system in Fig. E8.2 of the block initially at rest is released from a position $x = 3$. Take the system parameters from Exercise 8.3.

8.5 A constant force of 2 N is applied to the mass in Fig. E8.2. Find the displacement (response) of mass if damping coefficient is 2 N s/m, the spring stiffness 10 N/m and the mass is 1 kg. Assume all the initial conditions are zero.

8.6 Find the time response $y(t)$ for a unit step input if the Laplace domain transfer function of a linear second-order system is

$$G(s) = \frac{6}{s^2 + 14s + 49} \tag{8.153}$$

What is the type of damping?

8.7 The transfer system of system is

$$G(s) = \frac{0.2}{0.1s^2 + 0.6s + 1} \tag{8.154}$$

Determine the natural frequency, damping ratio, and the time response of the system for a unit step input.

8.8 A joint drive system consists of a DC servomotor with total inertia of 0.02 kg m^2 and bearing friction of 0.5 N/s and a gearbox with gear ratio of 32. The link inertia is 5 kg m^2 and the link bearing friction is 2 N/s. Determine the effective inertia and effective damping for the joint. Assume all shafts are rigid and mass less.

8.9 The link inertia of third joint of a 5-DOF manipulator varies between 4 and 9 kg m^2. The joint is driven by a DC motor with armature inertia of 0.05 kg m^2 through a gearbox with a gear ratio of 64. Determine the minimum and maximum effective inertia for the joint.

8.10 Consider the mass-spring-damper system shown in Fig. E8.2. A force f is applied to the mass to push it in the positive x-direction. Design a control system for regulating the position of mass, using a
(a) proportional derivative (*PD*) control strategy.
(b) partitioned proportional derivative (*PPD*) control strategy.
Give the control system block diagram and the mathematical model for the control system.

8.11 If the parameters of the system in Exercise 8.10 are: $m = 1$, $B = 5$, and $k = 2$ (same as Exercise 8.3) determine the controller gains K_p and K_d for the two controller laws in Exercise 8.10 that would give a critically damped performance.

8.12 For the mass-spring-damper system shown in Fig. E8.2, design a *PPID* control system. Give the schematic block diagram of the controller and its mathematical model.

8.13 For *PPID* controller in Exercise 8.11 determine the controller gains. K_p, K_d, and K_i for the critically damped performance of the system and zero steady-state error, take the system parameters as given in Exercise 8.3.

8.14 A *lumped model* is many times used to have estimate of lowest resonance frequency of beams and shafts. From the energy analysis it can be shown that a uniform beam of mass m can be replaced by a mass less beam of same length with a lumped point mass of 0.23 m at the end of beam. Likewise, a distributed inertia of I can be replaced by a lumped inertia of 0.33 I at the end of the shaft.
A link of 6.52 kg mass has an end-point lateral stiffness of 2400 N/m. Determine the resonance frequency of the link assuming the drive system to be rigid.

8.15 A shaft of stiffness 2000 N m/rad has inertia of 2 kg m^2. Determine the resonance frequency of the shaft assuming a lumped model if the joint uses a direct drive.

8.16 The shaft in Exercise 8.15 is driven by the motor with a gearbox having a gear ratio of $\eta = 110$. If the motor shaft has inertia of 0.1 kg m^2, what is the resonance frequency due to the flexibility of the shaft?

8.17 The dynamic equations for a 2-DOF manipulator are

$$\tau_1 = m_1 L_1^2 \ddot{\theta}_1 + m_2 L_1 L_2 \dot{\theta}_1 \dot{\theta}_2 + B_1 \dot{\theta}_1$$

$$\tau_2 = m_2 L_2^2 (\ddot{\theta}_1 + \ddot{\theta}_2) + B_2 \dot{\theta}_2 + m_2 g L_1 C_1 \tag{8.155}$$

Design a controller for the system.

8.18 For the two-link manipulator discussed in Example 6.1, design a computed torque controller. Assume the friction and other effects to be negligible.

8.19 A prismatic object is moved around on a surface plate with one of its face in contact with surface plate. Assuming the surface to be frictionless and rigid, identify the variables subjected to artificial constraints and natural constraints. What artificial constraint(s) exists for variables not subjected to natural constraint(s)?

8.20 What are the natural and artificial constraints for the task of opening a mineral water bottle with a manipulator? Show the constraint frame for the task.

8.21 The manipulator task for a manipulator is to use its end-effector to wind a mechanical clock. The winding task is equivalent to turning a crank as shown in the Fig. E8.21. Choose the appropriate constraint frame and identify all the natural and artificial constraints.

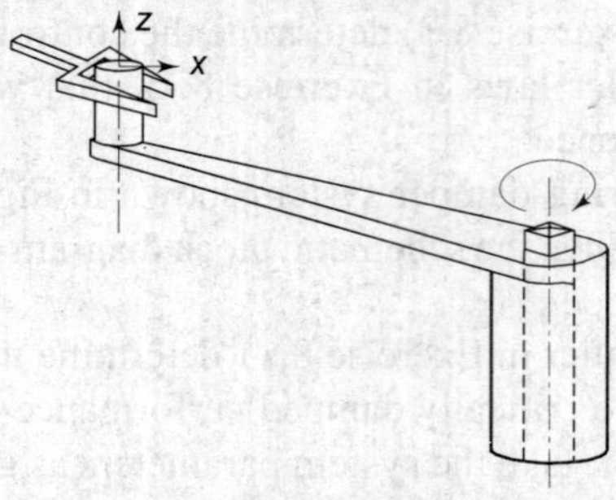

Fig. E8.21 *Manipulator turning a crank*

8.22 For the two link planar manipulator in Fig. 8.35, design the hybrid position force controller to follow a surface defined as

$$x = \cos(t);\ y = \sin(t) \qquad (8.156)$$

while maintaining a constant contact force f_d with the friction surface. Draw the block diagram of the controller.

8.23 For the task of driving a screw of pitch p at a desired angular velocity ω_d using a screwdriver, determine the natural and artificial constraints. The schematic of the task is shown in Fig. E8.23.

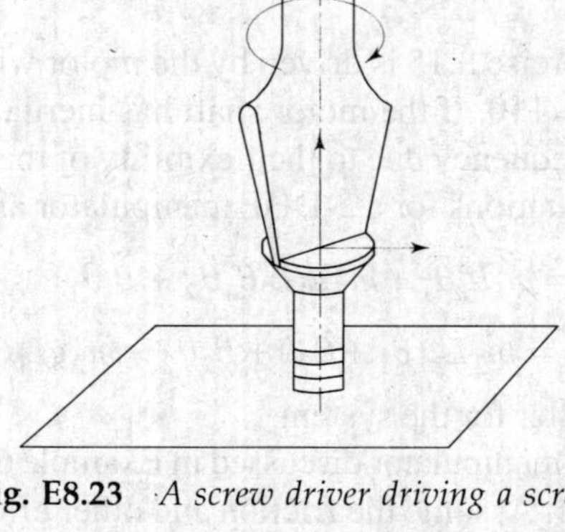

Fig. E8.23 *A screw driver driving a screw*

8.24 Determine the partitioned *PD* control law for the three axes SCARA manipulator shown in Fig. E8.24.

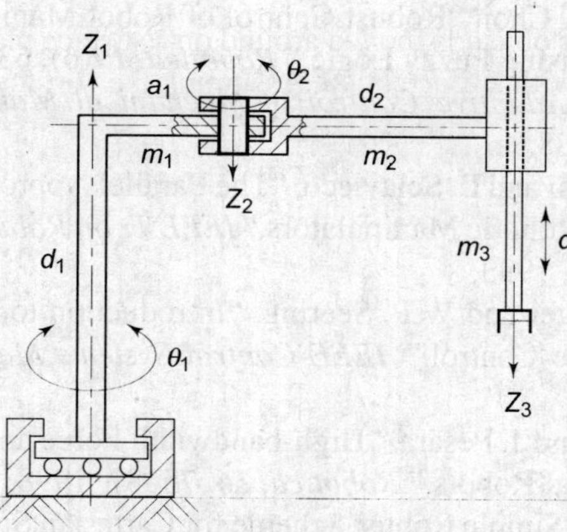

Fig. E8.24 *A 3-DOF SCARA manipulator*

8.25 For the three axes SCARA manipulator shown in Fig. E8.24 with the gravity loading terms as $\boldsymbol{G}(\boldsymbol{q}) = [0\ 0\ -mL_3g]^T$, determine the controller law for impedance control. If the manipulator task is to draw an arc in the *x-y* plane, comment on the choice of the controller gains.

8.26 What are the major differences between the open-loop and closed-loop servo systems?

8.27 Study the DC servomotor and permanent magnet stepper motor. Tabulate the major differences between DC servomotor and stepper motor in terms of their suitability as actuators for robotics manipulators.

8.28 Find details of hydraulic actuators and draw the schematic of a hydraulic joint actuation system. Describe the various components of the system.

8.29 For a robot controller it is proposed to implement partitioned proportional integral (*PPI*) control strategy. Develop the block diagram and mathematical model for *PPI* Controller.

8.30 Describe the architecture of the hybrid position/force control and compare it with impedance control.

SELECTED BIBLIOGRAPHY

1. C. Abdallah, D. Dawson and P. Dorato, M. Jamshidi, "Survey of Robust Control for Rigid Robots," *IEEE-Control Systems Magazine*, ***11***(2), 24–30, 1991.
2. C.H. An, C.G. Atkeson and J.M. Hollerbach, *Model-based Control of a Robot Manipulator*, MIT Press, Cambridge, Mass, 1988.
3. Khaled R. Atia and M.P. Cartmell, "A New Methodology for Designing PD Controllers," *Robotica*, ***19***(3), 267–274, 2001.
4. R.J. Anderson and M.W. Spong, "Hybrid Impedance Control of Robotic Manipulators," *IEEE Journal of Robotics and Automation*, **4**, 549–556, 1988.

5. Hyeung-Sik Choi, "Robust Control of Robot Manipulators with Torque Saturation using Fuzzy Logic," *Robotica*, ***19***(6), 631–639, 2001.
6. J.J. Craig, *Adaptive Control of Mechanical Manipulators*, Addison-Wesley, 1988.
7. S. Chiaverini and L. Sciavicco, "The Parallel Approach to Force/Position Control of Robotic Manipulators," *IEEE Tr on Robotics and Automation*, ***4***, 361–373, 1993.
8. S.D. Eppinger and W.P. Seering, "Introduction to Dynamic Models for Robot Force Control," *IEEE-Control Systems Magazine*, *7*(2), 48–52, 1987.
9. E. Freund and J. Pesara, "High-bandwidth Force and Impedance Control for Industrial Robots," *Robotica*, ***16***, 75–87, 1998.
10. T.C. Hsia, "Simple Robust Scheme for Cartesian Space Control of Robot Manipulators", *International Journal of Robotics and Automation*, 167–174, 1994.
11. S. Jung, T.C. Hsia and R.G. Bonitz, "Force Tracking Impedance Control for Robot Manipulators with an Unknown Environment: Theory, Simulation and Experiment," *Int. Journal of Robotic Research*, ***20***(9), 765–774, 2001.
12. O. Khatib, "A Unified Approach to Motion and Force Control of Robot Manipulators: The Operational Space Formulation," *IEEE Journal of Robotics and Automation*, ***3***(1), 43–53, 1987.
13. Hariharan Krishnan, "Design of Force/Position Control Laws for Constrained Robots, Including Effects of Joint Flexibility and Actuator Dynamics," *Robotica*, ***17***(1), 41–48, 1999.
14. C.S.G. Lee and B.H. Lee, "Resolved Motion Adaptive Control for Mechanical Manipulators," *Tr. ASME, J of Dynamic Systems, Measurements and Control*, ***106***(2), 134–142, 1984.
15. J.Y.S. Luh, W.D. Fisher and R.P.C. Paul, "Joint Torque Control by a Direct Feedback for Industrial Robots," *IEEE Tr on Automatic Control*, ***28***(2), 153–161, Feb 1983.
16. M.T. Mason, "Compliance and Force Control for Computer Controlled Manipulators," *IEEE Tr on Systems, Man, and Cybernetics*, ***6***, 418–432, 1981.
17. R. Ortega and M.W. Spong, "Adaptive Motion Control of Rigid Robots: A Tutorial," *Automatica*, ***25***, 877–888, 1989.
18. M.S. de Queiroz, S. Donepudi, T. Burg and D.M. Dawson, "Model-based Control of Rigid-link Flexible-joint Robots: An Experimental Evaluation," *Robotica*, ***16***, 11–21, 1998.
19. N.N. Sharma and R.K. Mittal, "Performance Analysis of Rule Based Fuzzy Controlled Robotic Manipulator using ANOVA" *Proc. of National Seminar on Current Applications of Computers in Design Engineering*, Jodhpur, India, 23-30, March 3–4, 2001.
20. J.-J.E. Slotine and W. Li, "Adaptive Manipulator Control: A Case Study," *IEEE Tr on Automatic Control*, ***33***, 995–1003, 1988.

21. M.W. Spong and M. Vidyasagar, *Robot Dynamics and Control*, John Wiley & Sons, New York, 1989.
22. M.W. Spong, "On the Robust Control of Robot Manipulators," *IEEE Tr on Automatic Control*, ***37***, 1782–1786, 1992.
23. D.E. Whitney, "Historical Perspective and State of the Art in Robot Force Control," *The International Journal of Robotics Research*, ***6***(1), 3–14, 1987.
24. T. Yoshikawa, "Dynamic Hybrid Position/Force Control of Robot Manipulators – Description of Hand Constraints and Calculation of Joint Driving Force," *IEEE Journal of Robotics and Automation*, ***3***, 386–392, 1987.
25. J.S.-C. Yuan, "Closed-Loop Manipulator Control Using Quaternion Feedback," *IEEE Journal of Robotics and Automation*, ***4***, 434–440, 1988.

9

Robotic Sensors and Vision

A robot with a closed-loop control system is able to carry out the specified work cycle with the help of information about joints and end-effector from the sensors. The joint and end-effector position, velocity, and possibly acceleration are fed back by means of optical encoders, tachometers, or other sensors, placed on the manipulator joints or joint actuators to ensure that the manipulator moves in the desired manner. Such a robot can accurately move but is unable to sense and respond to any change in its work environment.

For example, in a pick-n-place operation the robot is required to pick parts from a specific location, say "pick point", on the moving conveyor and place them into different bins based on some specified criterion, as shown in Fig. 9.1. The manipulator will move the gripper to pick point and gripper will close to grip the object at the pick point irrespective of the presence or orientation of the part and

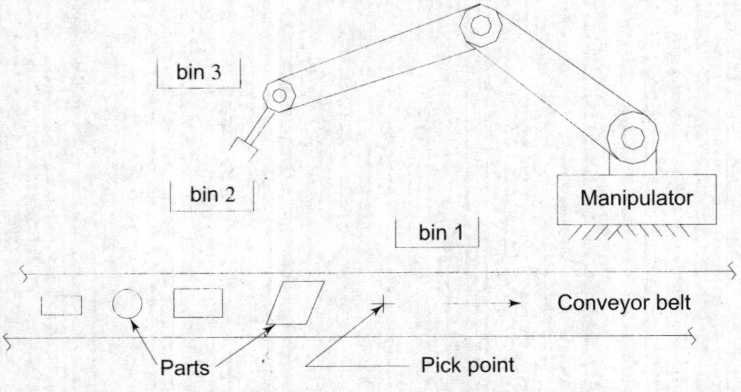

Fig. 9.1 *A robot performing a simple task of pick-n-place: picking parts from a moving conveyor and placing them into bins 1, 2, or 3*

will move to the next position even if no part is picked up. That is, it will execute the work cycle it is programmed to without any concern for 'correctness' of the work.

Similarly, the manipulator will continue to move through the specified path even if some obstacle comes in its path. It will collide with the obstacle rather than intelligently move around the obstacle and avoid it.

Therefore, for a robotic manipulator to operate effectively and intelligently and to enable it to work in unstructured environment, it must be equipped with sensors, which give information about itself and its environment. The computer-cum-controller operating the robot must receive information from the sensors and process it in real-time for decision making. Robots capable of such decision making are called *intelligent robots*, the environment sensing is called *intelligent sensing*, and the sensors used for such operations are called *intelligent sensors*. The intelligent sensors need associated hardware and software to process the signals from the robot's data *acquisition system*. The performance, capability, and flexibility of a robot are greatly dependent on its data-acquisition system. A robot without intelligent sensors is handicapped and can only do very specific tasks, while a robot that is able to "see" and "feel" is easier to train to perform complex tasks.

9.1 THE MEANING OF SENSING

First, what does it mean to sense something? Can you name the five human senses? What does each sense tell us about the surroundings? Senses often seem to overlap. In other words, two different senses can sometimes detect the same things. For example, to find out if you are near a wall, you can open your eyes and look, stick out your hand to feel, or even yell to hear whether your voice is reflected. Human sensing is considered as the best sensing and man-made sensors are still far inferior to human and other natural sensors in many respects.

9.1.1 The Human Sensing

What you learn from your senses depends on what your brain does with the information it receives from your sense organs, such as your eyes, ears, and skin. For example, a big piece of "steel" may look heavy to you, but that is because you already know that it is steel and steel is heavy. You would have to pick it up (or try to) and feel it to really know, or if you cannot guess the material it is made of. Your brain is trained to know what information from each sense organ means and how to interpret it.

Your brain also knows how to interpret the different kinds of signals one organ can send to the brain. For example, a sudden sensation on your finger might be interpreted as "It is really sharp!" or "That's hot!" or even "This is really heavy!" How is all of that possible from just the tip of your finger? First, the nerves (sensors) in your finger are extremely sensitive (and there are many of them), so they can pick up very small details. Second, your brain does a lot of work of processing and organizing the information that the nerves in your finger send to it.

9.1.2 The Problem of Robot Sensing

Much of what your senses "tell" you depends on what you already know about the kinds of things you might be sensing. What a particular sensation makes you think and conclude or do is a result of complex interaction between your sense organs and your brain. Robots "know" nothing, which is why sensing with physical sensors to solve these problems with sensor's output and computers is very difficult.

What seems like a very simple sensing task becomes quite complex when applied to the robot. For example, for most people, it is pretty easy to tell if there is an apple on the table in front of them. You just open your eyes and look, immediately you know, "Yes, there it is", or "No, there are no apples here." You can also do it with your eyes closed by just reaching out and feeling or smelling for one. For a robot, it is not that simple.

Assume you have a robot with a light sensor that resembles a colour video camera. You would like to "teach" (program) your robot so that it can determine whether there is an apple in front of it. How will it be done? You may try like this—put something in front of the robot, and "tell" it is not an apple or it is an apple. How many things do you need to show? What about the apple size and shape variations?

Well, a good first step would be to let the robot understand what an apple is. How would you explain the apple to your robot?

The problem of robot sensing is not only difficult as illustrated above, it is greatly limited by the sensor capabilities. It is extremely difficult to build sensor systems that are as sensitive as human senses.

Properly designing or choosing and using sensors is an essential part of robotics, and sensors of all types are used in many different kinds of robots for different tasks. For example, some robots use only simple position sensors, while some may have a combination of special cameras and advanced software to gather information about their surroundings, and orient themselves based on the behaviour of other robots.

9.2 SENSORS IN ROBOTICS

Sensors are used in robotic systems for a variety of functions. The task planning and control algorithms, discussed in previous chapters, require, both on-line measurements of the parameters characterizing the internal state of the manipulator as well as its work environment. The most common and minimal use of sensors is to provide information about the status of links and joints of the manipulator and about the working environment of the robot. In addition, sensors may be deployed to provide data for inspection and quality control, safety monitoring, and to detect and resolve interlocks in the workcell.

The major functions of sensors in robots can be grouped into five basic categories.

1. Status sensors,
2. Environment sensors,
3. Quality control sensors,
4. Safety sensors, and
5. Workcell control sensors.

The role of the sensors in robotics under each of these categories is described below.

9.2.1 Status Sensors

The primary use of sensors on a robot is to sense position, velocity, acceleration, and/or torque/force at each joint of the manipulator for position and motion control. These sensors form an essential part of the basic or internal closed-loop control systems and are called *internal sensors* or *state* (or *status*) *sensors*. Internal sensors give feedback on the status of the manipulator itself. The degree of accuracy that can be achieved by a manipulator depends on the resolution and accuracy of internal sensors. Internal sensors must also be cost effective because they are required for each axis of the manipulator.

9.2.2 Environment Sensors

The second major function for sensors is to extract features of the objects in the workcell or surrounding environment of the robot. This knowledge is utilized by the computer controller to modify or adapt to a given situation. For example, if the robot has to process several types of different parts, each requiring a different sequence of actions by the robot, such as adjusting the gripper orientation or applying the exact gripping force, it must determine the required parameters for each part. Sensors for these functions are generally placed in the environment of the robot or are external to the manipulator and are called *external* or *environment sensors*. There are exceptions where an external sensor may be mounted on the manipulator such as a wrist force-torque sensor. Another example is a camera mounted on the wrist to "see" the part in workcell before gripping it.

External sensors would be used to perform one or more of the following functions:

- detect presence of workpiece.
- determination of position and/or orientation of workpiece and of objects to be articulated, or other objects present in the workcell.
- workpiece identification.
- determination of workpiece properties such as size, shape and so on.
- detect, identify obstacles in the environment, and provide information about their size, shape, location, speed and so on.
- provide information about the manipulator-environment interaction forces and torques.
- provide information about environmental variables such as temperature, humidity, and so on.

- determination of the position and orientation of the end-effector, joints, and links of the manipulator.

Sometimes, the accuracy requirements in a given application are more stringent than the inherent accuracy and repeatability of the manipulator. For example, the task of assembling two parts that have very small allowance and require close alignment that is smaller than the resolution of the manipulator. In such situations, the feedback from external sensors, such as vision, can be used to improve the accuracy of positioning of the robot.

All the information from the external sensors would have to be processed by the computer in real time to guide the manipulator in the execution of its programmed work cycle. An important sensing method is the vision system that might be employed to determine such characteristics as part location and orientation, size and shape, and many other properties. Robotic vision is discussed in detail in the later part of this chapter.

9.2.3 Quality Control Sensors

Inspection and quality control is an important function for robotic sensors. Traditionally, quality control is performed on statistical sampling basis, using manual inspection techniques. Because sensors can be used to determine a variety of part quality characteristics, the use of sensors permits 100 percent inspection. The external sensors used for environmental feedback can also be used for detecting faults and failures in the finished product. The inspection process can be made a part of the programmed work cycle and the sequence of operations performed by the manipulator may be linked to the result of inspection process. Computer vision, ultrasonic, sonic, and other sensors can be used for inspection. Vision system can provide a large variety of information about the work piece in addition to its position and orientation.

9.2.4 Safety Sensors

An important function of the sensors in robotics is safety and hazard monitoring. The safety of the workers and other equipment in the work environment, and that of the manipulator itself, are important concerns. For example, if there is power failure, all the links of the manipulator may fall to zero-gravity position instantaneously. This may injure the human beings in the vicinity or may damage the equipment or the manipulator itself. A solution is to sense the power failure and apply breaks to prevent the uncontrolled falling of links due to gravity.

9.2.5 Workcell Control Sensors

Another major use of sensor technology in robotics is to implement interlocks in workcell. An interlock in the work cycle is a situation that requires sequencing of tasks such that completion of a task must be ensured before proceeding to the next task. For example, a part must arrive on the conveyor before the gripper of the manipulator can pick it. This kind of verification of interlocked tasks can be done by incorporating the signals from a variety of sensors in the robot program. Sensors may also be used to detect and resolve the interlocks in the workcell.

All the five categories of sensor functions described above require sensors to be an integral part of robot control system to accomplish specific control function. An intelligent robot requires an intelligent data acquisition system and multilevel hierarchical closed-loop control system. Figure 9.2 shows the multilevel control architecture of a computer-based intelligent robotic manipulator.

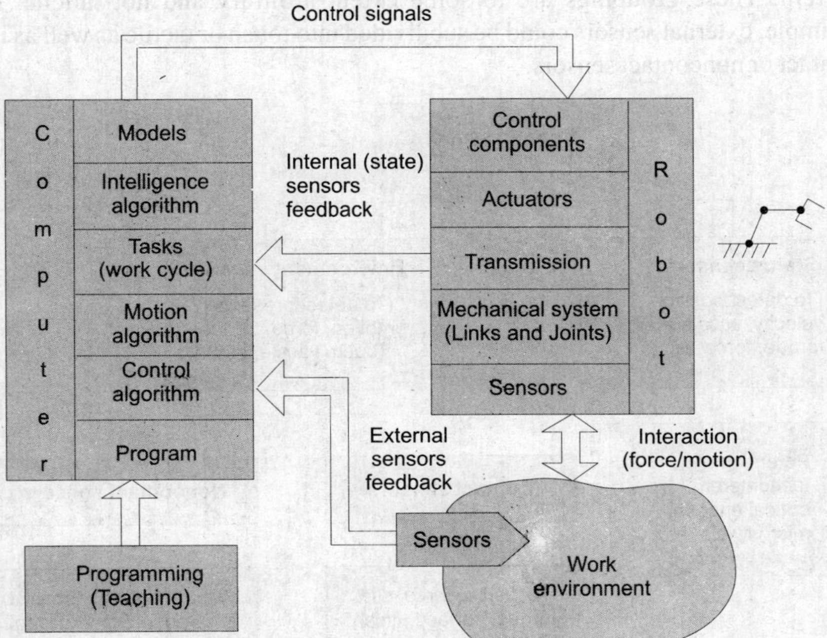

Fig. 9.2 *Architecture of a computer-based intelligent robotic manipulator*

Fusion of the available sensory data with task planning characterizes the robot as an intelligent robot, which is capable of perception of action. The control algorithms can guarantee the coordinated motion of the mechanical structure in correspondence to the task planning.

The signals from all the sensors on the manipulator and in the environment are fed to the computer-controller, which analyses them as per system models, task programs, control algorithms, motion algorithms, and intelligence algorithms and generates the control signals. These control signals are applied to the actuators to command and control the manipulator. The interaction of the manipulator with the work environment is sensed by the sensors and again fed back and the cycle goes on.

A broad classification of robotic sensors is discussed briefly in the next section.

9.2.6 Classification of Robotic Sensors

Robotic sensors can be classified in a number of broad groups in view of their applicability and use in robotics like property sensors, functional sensors, acoustic sensors, optic sensors and so on.

The schematic in Fig. 9.3 gives a possible classification hierarchy of the robotic sensors. The tasks and specific sensors in each group are also mentioned therein. These groupings are to some extent arbitrary and not unique. For example, external sensors could be subdivided into touch or tactile as well as into contact or noncontact sensors.

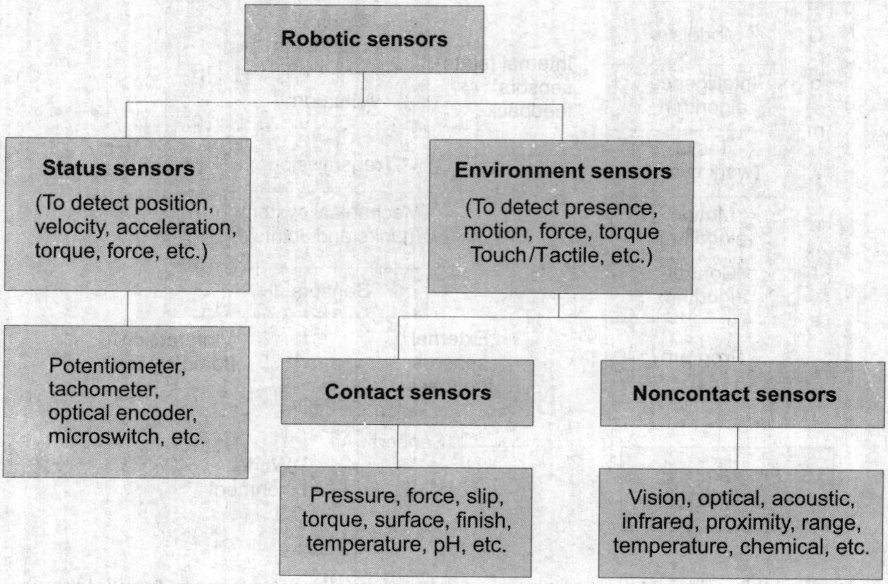

Fig. 9.3 *A possible classification hierarchy of robotic sensors*

Contact or touch sensors are one of the most common sensors in robotics. These are generally used to detect a change in position, velocity, acceleration, force, or torque at the manipulator joints and/or the end-effector. There are two main types, bumper and tactile. Bumper type detect whether they are touching anything, the information is either "yes" or "no". They cannot give information about how hard is the contact or what they are touching. Tactile sensors are more complex and provide information on how hard the sensor is touched, or what is the direction and rate of relative movement.

The technologies used for sensors in this group are electric, electromagnetic, electronic, and optic. Some common robotic contact type sensors are potentiometers, tachometers, optical encoders, accelerometers, linear variable differential transformers (LVDT), and force-torque sensors.

The second important group is of the noncontact sensors, which are used to get parametric information about the environment, task, objects in the workspace and to detect presence, distance, or features of the workpiece. The sensors in this group include proximity sensors, range sensors, temperature sensors, vision systems and so on. Various technologies such as optic, acoustic, infrared, electromagnetic and so on are used for sensors in this group. Proximity sensors are used in robotics to detect the presence of an object within a specified distance or find the accurate distance of the object. These are useful in object grasping or collision avoidance.

The vision system is probably the most powerful sensor as it can detect various features of the workpiece such as size, shape, geometry, distance, colour, texture, surface finish, and other attributes of the workpiece. It can also detect characteristics of the environment such as position and orientation of parts and obstacles, changes in these with time, presence of smoke, humidity, temperature gradient, and so on.

Some common kinds of sensors used in robotics are discussed here. The complete description of various sensor technologies and sensors and their detailed analysis can be found elsewhere.

9.3 KINDS OF SENSORS USED IN ROBOTICS

Many types of sensors are used robotics. In this section, some common sensors suitable for robotic applications are discussed. These include touch and tactile sensors in the contact type and proximity, range and physical property sensors in the noncontact type.

Position sensing is the most fundamental need for robots to control position and orientation or motion of joints and links of the robot.

The measure of the position and rotation of the manipulator, that is, position of its end-effector or its joints is needed to get an accurate assessment of manipulator motions and positions. In many applications, there is a need to accurately know the position of the end-effector (tool or gripper) or to find the coordinate locations of objects in workspace, or to track motion to provide real-time adjustments. For all these, rotation sensors are very important in robotics to keep track of joint speeds and motions of end-effectors. There are very few devices to directly give absolute position for linear motions. Often linear motion is converted to rotary motion using rack and pinon drive and rotary sensors such as encoders are calibrated to give absolute linear position.

In industrial robotic manipulators currently manufactured, the revolute joints far outnumber the prismatic joints. Thus, sensing of rotation or angular position is the fundamental requirement, as stated above. Reliable position sensors are widely implemented in robots in the form of potentiometers, tachometers, encoders, resolvers, and accelerometers.

Proximity and touch sensors are generally used in intelligent grasping of work piece, while force-torque sensors provide feedback for object manipulation and control of its interaction with the environment. For example, these sensors are useful to prevent slipping or damage of the object once grasped or to apply the right torque to tighten a nut. In contrast, the gross guidance information for a manipulator is provided by range and vision sensing. Vision sensing can also provide other useful information in addition to gross guidance information. The vision sensing fundamentals are discussed in later part of this chapter.

9.3.1 Acoustic Sensors

Acoustic or sonic sensors are commonly used for a wide variety of noncontact presence, proximity, distance measuring, or navigation applications. Sonic sensors can be simple or very complex. They typically transmit a short burst of ultrasonic sound towards the target, which reflects the sound back to the sensor. The system then counts the time for the echo to return to the sensor and calculates the distance of the target using the speed of sound in the medium. The wide variety of sensors currently available differ from one another acoustically. They operate at different frequencies and have different radiation patterns.

Ultrasonic sound is vibrations at a frequency above the range of human hearing, usually >20 kHz. Most ultrasonic sensors use a single transducer to both transmit the sound pulse and receive the reflected echo, typically operating at frequencies between 40 kHz and 250 kHz.

Silicon-based ultrasonic sensors and acoustic wave sensors are relatively new and extremely versatile microelectromechanical system (MEMS) sensors that are just beginning to realize their potential in robotics. They are competitively priced, inherently rugged, very sensitive, intrinsically reliable, and consistent in performance. These can be used for a large variety of applications, such as presence, proximity, velocity, acceleration, force torque, humidity, temperature, and so on.

Acoustic wave sensors send out acoustic waves, as the acoustic wave propagates through or on the surface of the material, changes in the characteristics of the propagation path affect the velocity and/or amplitude of the wave. These changes can be monitored by measuring the frequency or phase characteristics of the reflected wave. These changes can then be correlated to the corresponding physical quantity being measured. Array of ultrasonic sensors or acoustic wave sensors are useful for imaging and scene analysis. For a limited class of applications acoustical imaging may be complementary or even competitive to optical imaging.

9.3.2 Optic Sensors

Optic or light-based sensors are noncontact sensors and wide range of optic sensors based on different techniques are available for use in robotics. Optic sensors other than camera, used in vision systems, are discussed here.

Simple light sensors work with visible light and work on the principle of detecting the change in the intensity of light sent and received. The signal can be binary 0 (black or no light) or 1 (white or bright), or magnitude of light, say, between 0 and 100.

Optic sensors in nonvisible light include infrared, ultraviolet, and laser-beam sensors. An infrared sensor based tracking sensor system allows a full 3-D tracking within a space of few meters, with resolution of a fraction of a millimeter. Another eye safe laser based scanning sensor is used for navigation on automated guided vehicles, robots, and other material handling equipment.

9.3.3 Pneumatic Sensors

Applications of pneumatic sensors on robots are oldest and still convenient or suitable in some cases. Pneumatic sensors may be contact-less devices or contact sensors. They can be used as proximity sensors, touch sensors or pressure or force sensors.

Pneumatic proximity sensors are useful in workpiece grasping as they can provide accurate distance between fingers and workpiece, orientation of workpiece, finger closure signals and so on. Pressure-or vacuum-based sensors are very sensitive and are suitable for handling delicate and fragile objects. Counter-pressure based sensors are used in detecting or identifying the features such as edges, holes, shape, etc. and detecting the movement.

Some other conventional sensors that are used on robot are proximity and micro-switches, synchros, resolvers, differential transformers, eddy current sensors, hall-effect sensors, optical interrupters, tachometers and accelerometers.

9.3.4 Force/Torque Sensors

The most significant problem with force/torque sensor (FTS) is mounting it on the joint. For force/torque sensor to be sensitive, it must be flexible (not rigid). This makes the overall joint much more flexible contrary to rigidity required for precision and accuracy and causes a loss of controllability of the joint.

It is possible to estimate the joint torques and inertias without the use of force/torque sensors. Estimation algorithms based on actuator power input and inertia estimation are available. These methods are complex and very sensitive to propagated errors.

Apart from the joint forces/torques, the interaction between robot (its end-effector) and environment generates forces/torques that must be controlled to preserve the integrity of task being performed. For measuring these interaction forces, sensors are mounted between the wrist and the end-effector and are known as *Wrist Force Torque Sensors* (WFTS).

A strain-gauge based WFTS is shown in Fig. 9.4. The sensor has two annular rings, outer ring and inner ring, which are connected to wrist end and the end-effector end, respectively. These rings are connected to each other by a cross and on the arms of this cross, eight pairs of strain gauges are mounted as shown. By suitably selecting and connecting the gauges in the Wheatstone bridge the sensor detects three force components along the three principal axes and three moments along the three axes.

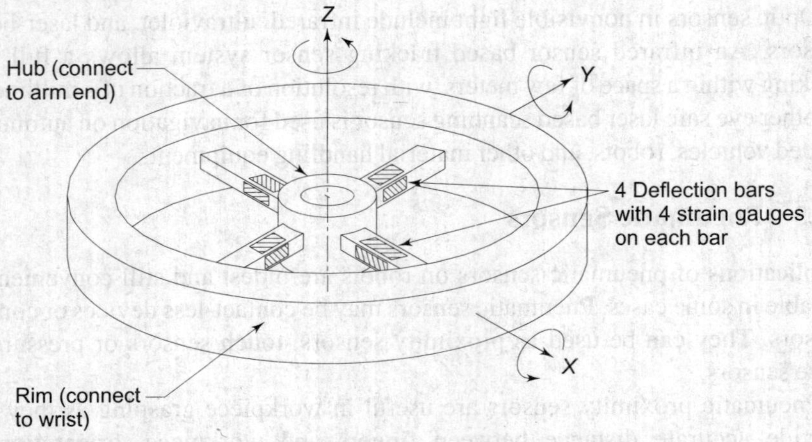

Fig. 9.4 *A six-component strain-gauge based wrist force-torque sensor*

Many other alternate designs are available for WFTS based not only on strain gauges but also on sonic and optic sensors. It is important to note that WFTS should not affect the positioning accuracy of the manipulator and, therefore, it must have high stiffness. Some other requirements of WFTS are: it must be small, light in weight, compact in design, sensitive, linear, and with low internal friction.

9.3.5 Optical Encoders

For position sensing in robotics, optical encoders are frequently used sensors because of their simple construction, low cost, ease of application and versatility. Optical encoder converts linear or angular displacement into digital code or pulse signals. Optical encoders are of two types:

Absolute encoders: They provide actual position relative to a fixed reference (zero) position. Their output is a digitally coded signal with distinct digital code indicative of each particular least significant increment of resolution.

Incremental encoders: These sense the position from previous position. Their output is a pulse for each increment of resolution but these make no distinction between increments.

The optical encoder is made up of three basic component: a light source, a rotary (or translatory) disc and a sensor, as illustrated in Fig. 9.5. The light source can be an incandescent lamp or an infrared light-emitting diode (LED). The disc has alternate opaque and transparent sectors, which are etched by means of a photographic process on a plastic disc or slots are cut on a metal disc. The sensor is a photodiode or a phototransistor, which senses the light (or no light) passing through sectors or slots of the disc. Finally, a conditioning electronics detects the light and dark signals and converts these signals into usable form of pulses or digital code.

The two types of optical encoders, incremental and absolute are described in detail in the following paragraphs.

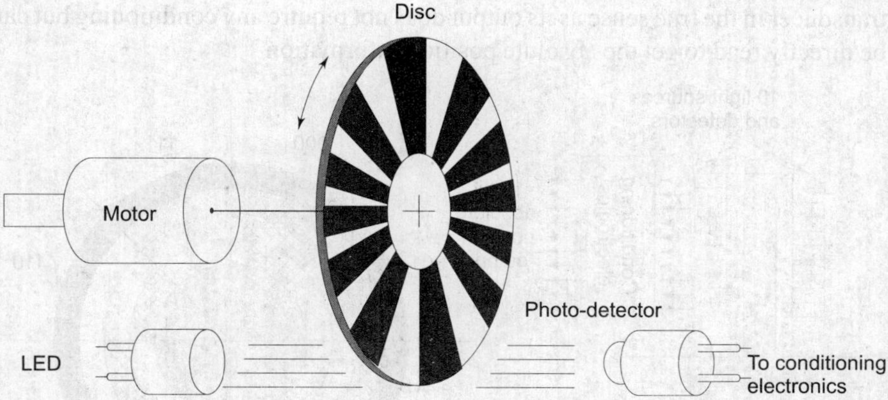

Fig. 9.5 *Components of an optical shaft encoder*

Incremental Optical Encoder The incremental optical encoder disc is connected to the motor shaft and rotates with motor. The rotation of the disc permits the light to reach the sensor, whenever the transparent slot comes in front of LED, thereby generating a series of pulses as output of sensor. These pulses are fed to a counter, which counts the number of pulses; the count being a measure of angle (or distance) through which the shaft has moved. By sampling the counter at regular intervals by means of clock pulses, the speed of rotation of the shaft can be obtained. The resolution of an incremental encoder is given by:

Basic resolution = (360°/m) where m = No. of sectors on disc, each sector is half transparent and half opaque. A typical incremental encoder having a disc with 2500 sectors (or lines) gives a basic resolution of 0.144°. Using another disk as a mask, using two or four optoelectronic channels or using multiplier circuits increases the accuracy and resolution of the incremental optical encoder. For example, if the pulses are increased by a factor of 4, by use of any of these approaches, the above incremental encoder will give a resolution of 0.036°.

Absolute Encoder Absolute optical encoder consists of a multiple track light source, a multi-track receiver and a multi-track rotary disk. The basic construction of absolute encoder in shown in Fig. 9.6(a) for the measurement of angular displacement. The absolute encoder disc has concentric circles of slots also known as tracks or channels and radial sectors to generate the pulses. As many LED-photodiode pairs as the number of tracks are needed. These may be suitably spread round the tracks to avoid signal interference.

The transparent and opaque slots in the tracks are arranged in such a way that the sequential output from the encoder is a number in binary code. The transparency is regarded as '1' and opaqueness as '0'. The tracks and slots of a 3-bit absolute encoder disc are illustrated in Fig. 9.6(b).

The number of bits in the binary code will be equal to the number of tracks on the disc. Thus, for an absolute encoder with a disc of 10 tracks, there will be 10 bits and the number of positions that can be detected is 2^{10} or 1024. This encoder will have a resolution of 360°/1024 = 0.35°. An absolute encoder is a digital transducer in the true sense as its output does not require any conditioning but can be directly read to get the absolute position information.

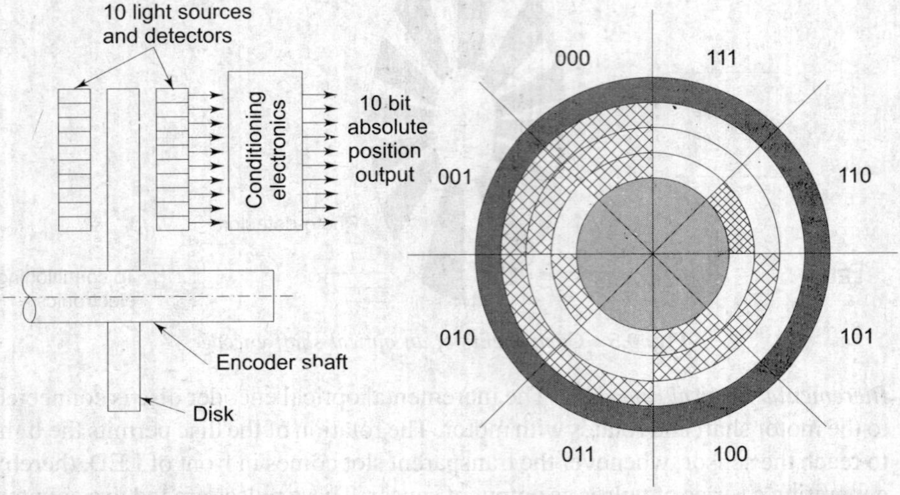

(a) Arrangement of light sources and detectors for 10-bit output

(b) A 3-bit 3 track absolute encoder disc

Fig. 9.6 *Absolute optical encoder and its disc.*

The chief disadvantage of optical encoders is that the output is volatile when power is off. There are ways to overcome this.

9.3.6 Choosing the Right Sensor

To decide what type of sensor or which sensor would be best for a particular situation, following guidelines may be used.

- Carefully analyze the nature of information needed.
- Find what information will be provided by the sensor.
- Determine how the sensor signals will be used.
- For the variable to be sensed determine: its nominal value, range of values, accuracy required, required reliability and speed of measurement etc.
- Estimate the environmental variations under which the sensing is to be done.
- Examine the nature of task dependent on the sensed signals, for example, critical tasks require reliable sensing.

Based on the above guidelines all possible sensors that can do the specified task are identified taking into account such other factors as sensitivity, linearity, in the desired range, maintainability, life, power consumption, ruggedness, availability and cost. Remember that there is always more than one way to sense the desired quantity and more than one type of sensor can do the job, hence, out of the several sensors available to get the desired information, one that is 'optimum' should be deployed.

9.4 ROBOTIC VISION

The most powerful sensor, which can equip a robot with large variety of sensory information, is robotic vision. Robotic, or synonymously, "computer" or "machine" vision systems are among the most complex sensory systems in use. Robotic vision may be defined as the process of acquiring and extracting information from images of 3-D world.

Robotic vision is primarily targetted at manipulation and interpretation of image and use of this information in robot operation control. Robotic vision requires two aspects to be addressed. One, provision for visual input, and two, processing required to productively utilize the visual information in a computer-based system. The architecture of vision systems, the components for visual input, image representation, and image storage are discussed first. The processing of the acquired image to extract the information from it is discussed later.

9.5 INDUSTRIAL APPLICATIONS OF VISION-CONTROLLED ROBOTIC SYSTEMS

Vision systems can provide information about the position, orientation, identity, and condition of each part in the surroundings. This valuable knowledge can be used to automate the manipulation of objects, plan robot motions to avoid collision with obstacles, or decide how to grasp an object. Although the volume of data produced by a vision system is largest, the amount of information present in an image cannot be matched by any other sensory system. The effective use of robotic vision system makes assembly, quality control, parts handling, and classification tasks more robust.

Using a single camera, it is possible to track multiple objects in visually cluttered environments. A vision-controlled industrial robot can be deployed for a number of different applications.

9.5.1 Presence

The simplest application is to find the presence or absence of a part at a specific location, say on a conveyor belt or in a bin. Many other simple sensors such as proximity or touch sensors can be used for this, but the visual detection of presence combined with other applications discussed next gives much more accurate and versatile information.

9.5.2 Object Location

The parts or obstacles in the visual fields can be located accurately and their position and orientation can be determined with precision. Accurate coordinate assessment is used for tracking motion of manipulator, end-effector, objects, or obstacles. Sometimes, for this purpose special identification markings are made on the parts.

9.5.3 Pick and Place

The manipulator can be guided to pick parts from a specific location after its presence has been detected or from any imaged location in the workcell and place it at the desired location. The gripper can be oriented according to the part's orientation to hold the part properly. To pick a part from a moving conveyor or to put on it is inherently easier than to lift it from a storage bin or place it there.

9.5.4 Object Identification

The objects captured in an image can be identified and distinguished from each other. This may be a part of sorting process, involving determination of location and pick-and-place tasks to effect physical movement of parts.

9.5.5 Visual Inspection

This is a very powerful and potential application area for automated quality control using robots, while they are performing other tasks. Visual inspection is based on extracting specific quantitative measurements of desired parameters from an image.

The robot involved in visual inspection can do sorting of 'good' and 'bad' parts or control the manufacturing process based on measured values of parameters.

9.5.6 Visual Guidance

The image of the scene can be used for accurate specification of relative positions of the manipulator and the part in the scene as well as their relative movements. An example of application of visual guidance is the assembly operation, which requires accurate position and orientation control of two parts to be fitted together. To assemble the parts, one part is held stationary, while the other is maneuvered by the robot, whose movements are controlled by the visual feedback. Another example can be of guiding the motion of the manipulator through stationary or mobile obstacles in the environment of the robot.

Potential industrial applications for vision-controlled robots are numerous and with the help of vision systems robots can be made to perform difficult and complex tasks. From the above applications, it should be clear that the task that can be performed depends on what information is obtained from the image.

The processing hardware and the concepts involved in robotic vision are considerably complex as compared to other sensory systems. The subject of robotic vision is subjective in nature and generalized analytic solutions are yet to be found. Even after years of research, robotic vision remains a challenge. It can be said that the robotic vision is still in the early stages of development but it holds the future as the most powerful robot sensory technology.

9.6 PROCESS OF IMAGING

The basic process of imaging, getting an image for computer processing, from the light source to an algebraic image array is depicted by the schematic in Fig. 9.7. The light source illuminates the object and the camera captures the reflected light. The image formed in the camera is converted into analog signal (voltage) with the help of suitable transducers. Finally, the analog voltages are digitized and converted into an algebraic array. This array is the "image" to be processed and interpreted by the computer according to predefined algorithms.

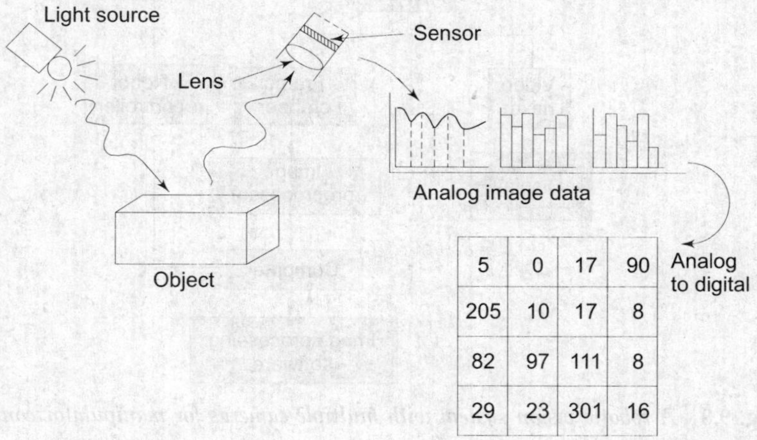

Fig. 9.7 *Capturing an image and its digitization for further computer processing—a schematic*

9.7 ARCHITECTURE OF ROBOTIC VISION SYSTEMS

Robot vision system is divided into subsystems based on the way the vision systems are generally implemented. A robotic vision system consists of one or more cameras acting as sensors. The vision interface unit connects the cameras to the video display and the hardware to convert the image into a suitable digital form. This hardware includes a video-buffering unit called a *frame grabber* and an *image preprocessor*.

A computer is utilized to process the image, using the image processing software for extraction of information about the scene, interpretation of the image, recognition of the objects and so on for the desired application.

A schematic diagram of a typical robotic vision system architecture is shown in Fig. 9.8. On the basis of the processing output, the computer generates command signals, which are sent to the manipulator through the robot controller, for the manipulator operation control. Also video signals are sent to video display for the operator.

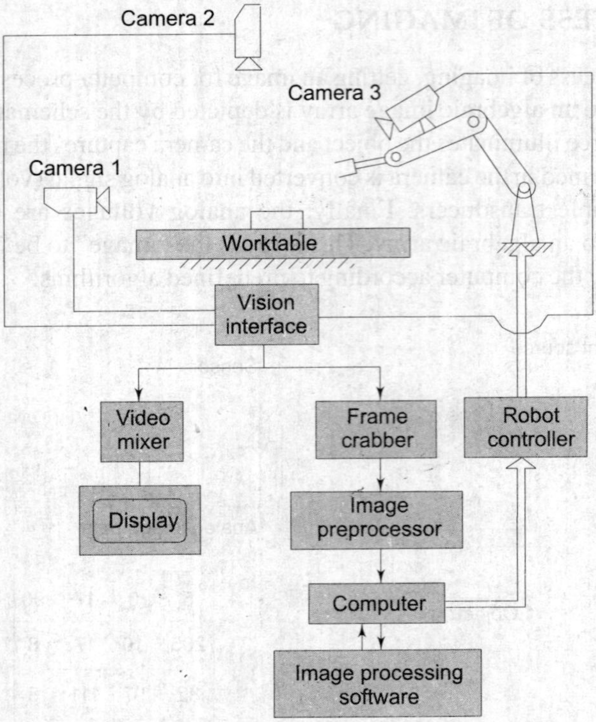

Fig. 9.8 *A robotic vision system with multiple cameras for manipulator control*

9.7.1 Stationary and Moving Camera

An elementary robotic vision system consists of only one stationary camera mounted over-seeing the workspace, (say, camera 2 in Fig. 9.8). This produces a single perspective 3-D view of the scene of activities. Multiple stationary cameras give multiple perspectives of the scene. A camera can also be mounted on the manipulator, for example, camera 3 in Fig. 9.8. Such a mobile camera generates an unlimited number of perspectives. This camera can be moved by the manipulator around the object to get specific views of the object and much better visual information. This is called *active sensing* as the manipulator itself positions the sensor on the basis of previous experience or closest to the task being performed. The disadvantages of mobile camera use are, one, the additional mass of the camera is undesirable on more than one account and two, the camera is subjected to continuous motion and rough handling. The disadvantages of using a stationary camera are (i) stationary camera requires a complete preknowledge of the 3-D scene, and (ii) some parts of the scene may be hidden from the camera even with use of multiple stationary cameras.

Stationary camera and mobile camera, both, have computational and control problems. The fundamentals of robotic vision are introduced here by restricting to the case of a single overhead stationary camera for getting a 2-D image with constrained lighting. The setup for this idealized system is shown in Fig. 9.9.

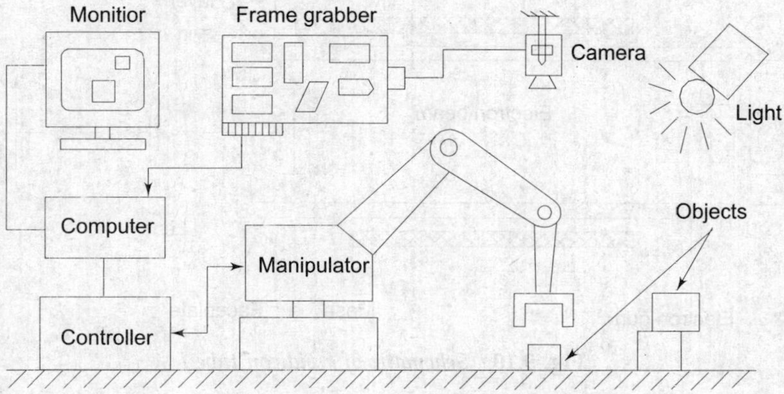

Fig. 9.9 *Vision system setup with single overhead stationary camera*

9.8 IMAGE ACQUISITION

The first link in the vision chain is the camera. It plays the role of robotic "eye" or the sensor. This is the imaging component or the noncontact or remote sensor. The visual information is converted into electrical signals in the camera and when *sampled* spatially and *quantized*, these signals give a *digital image* in real-time, by a process called *digitizing*.

The robotic vision cameras are essentially optoelectronic transducers, which convert optical input signal to electrical output signal. They fall in the domain of TV cameras. There is a variety of camera technologies available for imaging. Some of these are: black-and-white *vidicon* tube; solid-state cameras based on *charge-coupled devices* (CCD), charge injection devices (CID), and *silicon bipolar sensor* cameras. Two of these technologies are described here.

9.8.1 Vidicon Tube

The basic structure of the vidicon camera tube is shown in Fig. 9.10. The optical image is formed on the glass faceplate coated with a thin photosensitive layer composed of a large number of tiny photoresistive elements. The resistance of the element decreases with increasing illumination. Once the image forms on the faceplate, a charge is accumulated, which is function of the intensity of the impinging light over a specified time, from which an electrical video signal is derived.

The charge built up is 'read' by scanning the photosensitive layer by a focused electron beam produced by the electron gun at the rear of the tube. The scanning is controlled by a deflection coil mounted along the length of tube. The electron beam is made to scan the entire surface, typically, 30 times per second, line by line, consisting of over 500 scan lines for the whole image, as shown in Fig. 9.10. Each complete scan is called a *frame*.

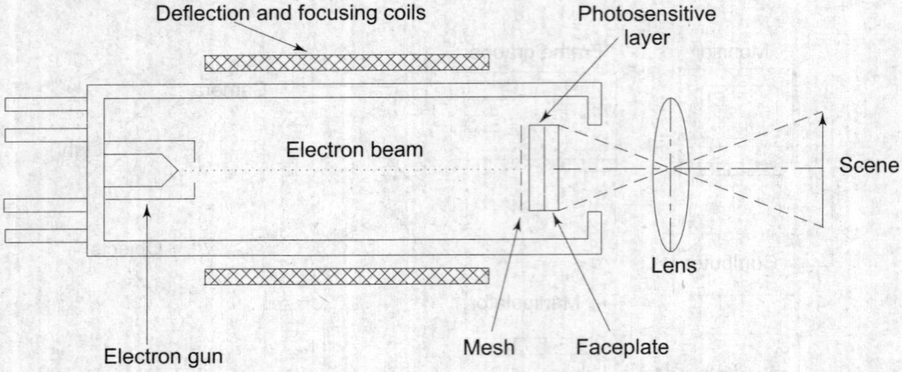

Fig. 9.10 *Schematic of a vidicon tube*

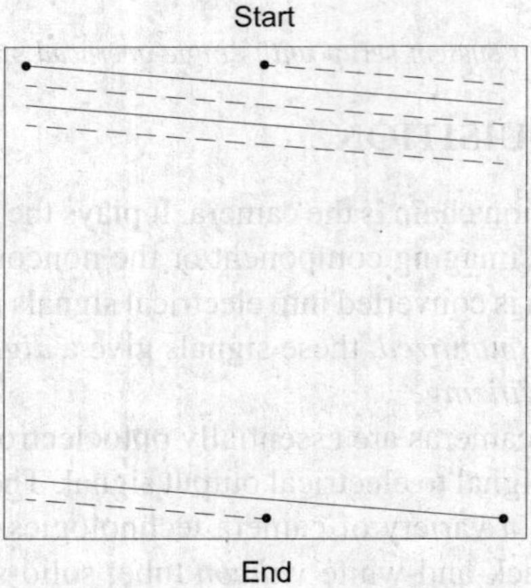

Fig. 9.11 *Electron beam scanning pattern for one frame of image*

Scanning at higher line rates per frame, scanning even and odd lines simultaneously at 60 times per second (shown with dashed and continuous lines in Fig. 9.11) are some of the techniques utilized to reduce flicker, etc.

9.8.2 Charge-Coupled Device (CCD)

The charge-coupled device falls in the category of solid-state semiconductor devices. A monolithic array of closely spaced metal oxide semiconductor forms the photosensitive layer. A simplified construction of CCD is illustrated in Fig. 9.12.

The light is absorbed on the photoconductive substrate and charge accumulates around the isolated "wells" under control of electrodes, as shown in the Fig. 9.12. Each isolated well represents a pixel. Charges are accumulated for the time it takes to complete a single image scan. The charge built up is proportional to the intensity of image. Once the charge is accumulated, it is transferred by the electrodes, line by line, to the registers.

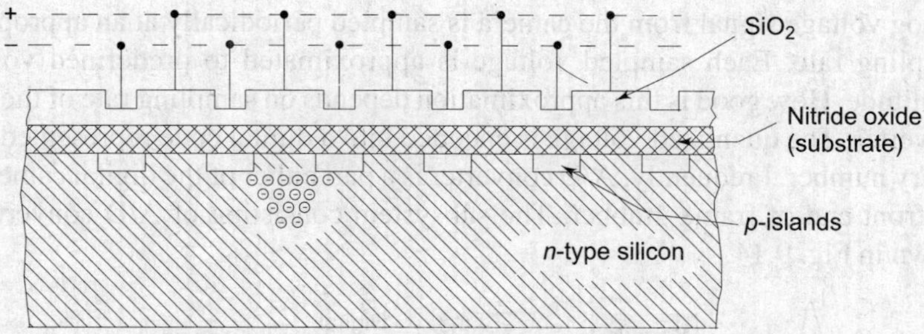

Fig. 9.12 *Charge-coupled device (CCD) sensor structure*

9.9 DESCRIPTION OF OTHER COMPONENTS OF VISION SYSTEM

A complete vision system consists of hardware and software for performing the functions of sensing and processing the image (the scene) and utilizing the results obtained to command the robot. A block diagram of the components of a vision system, arranged schematically, is shown in Fig. 9.13 and their functions are explained below.

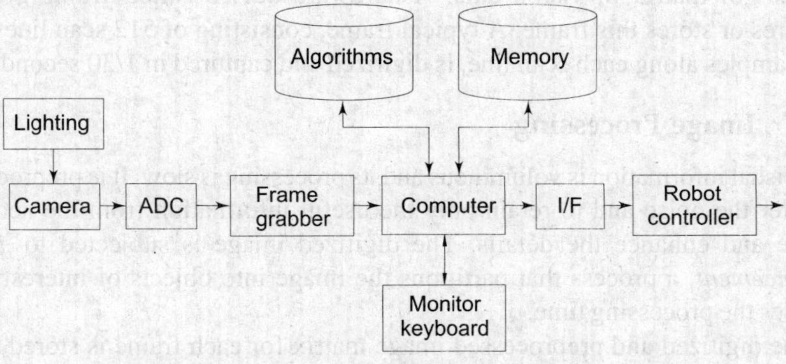

Fig. 9.13 *Schematic showing components of a vision system*

9.9.1 Illumination

The image presented to the camera is light reflected from the environment. This varies in wavelength and intensity throughout the image and is directly dependent on the illumination of the scene or lighting, that is, type, luminous power and placement of light sources. A poor lighting produces low-contrast images, shadows, and noise in a 2-D vision system, the desired contrast can often be accomplished by using a controlled lighting system. A 3-D vision system may require more sophisticated lighting system. In a single camera system, triangulation is used to detect shape and depth. Some systems use two cameras to get 2-D images to achieve a stereoscopic view of the scene.

9.9.2 Analog-to-Digital Conversion and Frame Grabber

Analog-to-digital (A/D) conversion is required to convert analog picture signal from the camera into digital form that is suitable for computer processing. The analog voltage signal from the camera is sampled periodically at an appropriate sampling rate. Each sampled voltage is approximated to predefined voltage amplitude. How good is this approximation depends on sampling rate of the A/D converter. The quantized voltage is encoded into a digital code represented by a binary number. Frequently, A/D converter is a part either of the digital camera or the front end of frame grabber. The subsystem consisting of A/D converter is shown in Fig. 9.14.

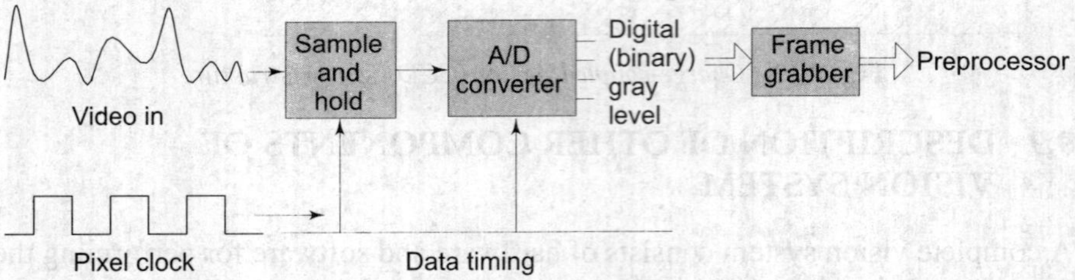

Fig. 9.14 *Configuration of analog video to digital conversion subsystem*

The complete scan of one image, a frame, after analog to digital conversion, consists of matrix of vision data. A hardware device called *frame grabber* captures or stores this frame. A typical frame, consisting of 512 scan lines with 512 samples along each scan line, is digitized and captured in 1/30 second.

9.9.3 Image Processing

The visual information is voluminous and its processing is slow. It is preprocessed to filter the noise and to retain only the useful information from the acquired image and enhance the details. The digitized image is subjected to *image enhancement*, a process that partitions the image into objects of interest. This reduces the processing time.

The digitized and preprocessed image matrix for each frame is stored in the memory and then subjected to processing as per the image processing algorithms for computation of features, *recognition*, and *interpretation* of image. The computation and extraction of features such as size, shape and so on is called image *description*; the process of identifying the objects in the image is called recognition and interpretation is the process of assigning meaning to the environment containing an ensemble of recognized objects.

Typical complex tasks that can be performed by a vision system are

- determination of geometrical features of parts like size, shape, etc.
- determination of workpiece orientation and location,
- determination of speed and direction of motion of parts and obstacles,
- determination of difference between colour, texture etc. of objects,
- recognition of parts presented to the camera in varying positions,
- recognition of parts touching or piled on each other,
- recognition of parts that have high degree of variability,
- recognition of environment despite poor lighting or image degradations from other reasons,

- distinguishing between good and faulty parts,
- distinguishing between very similar parts,

The preprocessed image frame is stored in computer memory for further processing. The methods of image representation are discussed in next section.

9.10 IMAGE REPRESENTATION

A fundamental requirement of vision system is to generate an appropriate method of representing image data. The method should provide for convenient and efficient storage and processing by computer and it should encapsulate all the information, which defines important characteristics of the scene in the image.

The 2-D image can be described by an image density function, I. The value of the function $I(x, y)$, at spatial coordinates (x, y) in the x-y plane is the intensity or brightness of light at the point. The reflected light falling on the photosensitive target is a measure of image density function.

9.10.1 Digitization

For computer-based processing, a digital form of the image is required. A suitable approximation of the intensity function $I(x, y)$ is made for the convenience of representation and processing. This approximation is known as the digitization and is carried out in two stages; spatial digitization and amplitude digitization. These are explained below.

(a) Spatial Digitization The continuous image density function I is sampled at regular discrete points in the 2-D plane. This is referred to as image sampling where the picture area $I(\Delta x \times \Delta y)$ centered at each discrete point is identified as a picture element or pixel.

For spatial digitization, the spatial coordinates (x, y) are sampled at discrete intervals. If there are m samples along x coordinate and n samples along y coordinate, this would give a total of $m \times n$ pixels for the image. Often the samples m and n are powers of 2 for obvious reasons. The digitized image is characterized by the *spatial resolution* of $m \times n$ pixels. A spatially digitized image, generally, consists of a square array of $n \times n$ equally distributed pixels. The spatial digitization is illustrated in Fig. 9.15.

The scanning of image formed on the target in camera progresses from upper left corner to the lower right corner, a convention taken from the normal raster scan of television. The origin is in upper left corner, x-axis is from left to right, and y-axis is from top to bottom. A typical scanning standard is 525 scan lines, which naturally digitizes y-axis. The natural sampling of y-axis is extended to x-axis so that the image intensity is defined at discrete points in space producing a 525 × 525 pixel digitized image. The image of one face of a dice (a cube), as shown in Fig. 9.16, illustrates the variation of the intensity function $I(x, y)$. The dice is made of white plastic and has black dots on each face, and is placed on a gray background.

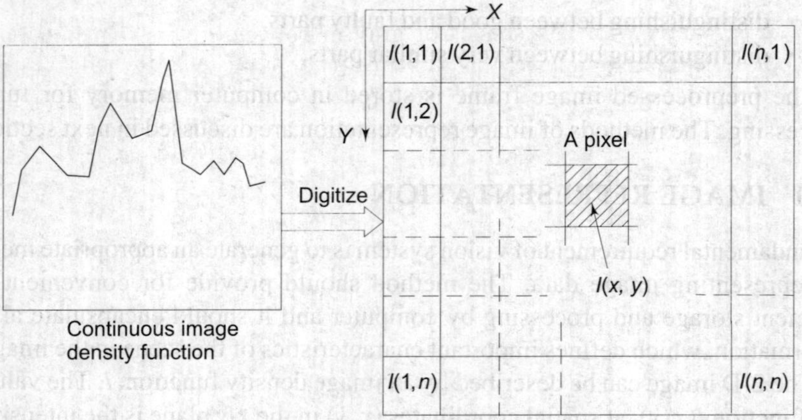

Fig. 9.15 *Spatial digitization giving n × n pixel grid*

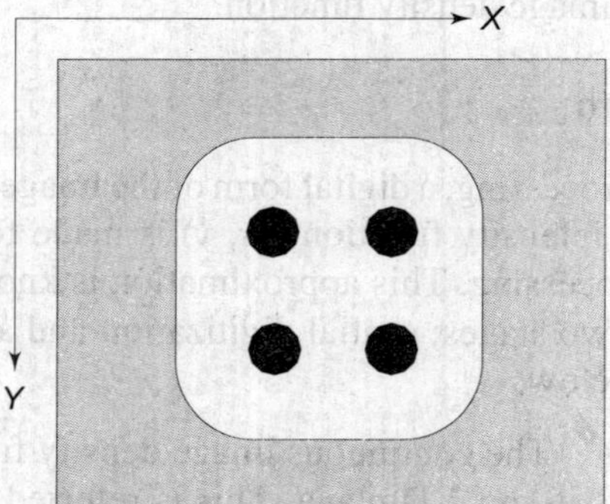

Fig. 9.16 *The gray scale picture of one face of dice illustrating the variation of intensity function I(x, y)*

(b) Amplitude Digitization The intensity value at each pixel in the spatially digitized image can have any magnitude (amplitude) between the zero and maximum intensity. The local intensity magnitude at a pixel is, too, encoded or quantized in amplitude digitization or intensity quantization. This is done by approximating the value to one of a fixed number of unique levels. The full range of quantized image intensity levels available in a particular image is called *gray scale* and the image intensity level at spatial coordinates (x, y) is called *gray level*. The number of gray levels within a gray scale is chosen to be a power of two.

If the gray scale of an image is quantized into 2^d gray levels (possible intensity values or shades of gray for a pixel), the intensity level $I(x, y)$ of any pixel in the digitized image array will be

$$0 \leq I(x, y) \leq 2^d - 1 \tag{9.1}$$

where d is an integer. Working with integer powers of 2 has a number of advantages for both hardware and software.

The 2^d gray levels will require d bits of storage for each pixel. For example, an 8-bit storage can give $2^8 = 256$ gray levels with, say, 0 corresponding to black and 255 to white, with 254 shades of gray in between. Note that human eye perception is limited to approximately 64 (=2^6) gray levels only. The amplitude digitization process is illustrated in Fig. 9.17. The analog video signal voltage after spatial digitization is plotted as "mountain peaks" over the pixel array in Fig. 9.17(a). Assuming an amplitude digitization with four gray levels ($d = 2$), the intensity values I are converted to respective gray levels using Fig. 9.17(b) and the digitized gray level mountains are shown in Fig. 9.17(c). This amplitude digitized image can more conveniently be represented and stored as a 3 × 3 matrix shown in Fig. 9.17(d).

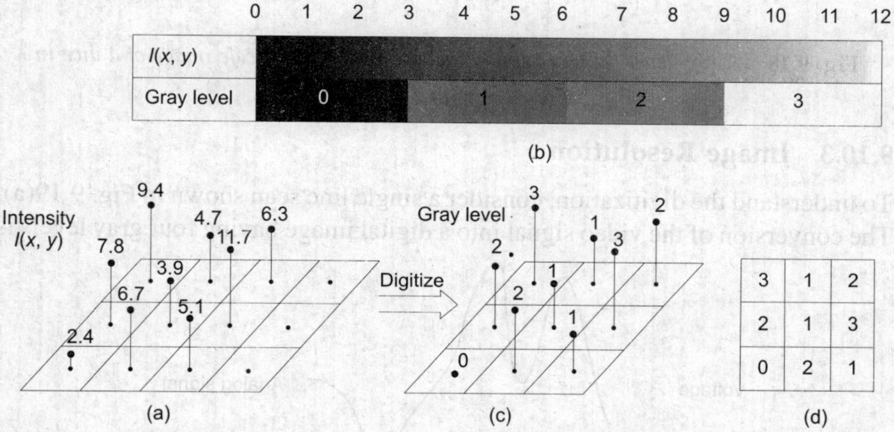

Fig. 9.17 *The amplitude digitization of a spatially digitized image*

To understand the process of digitization and effect of digitization parameters on the digital image, first, the simplest form of digital image is discussed.

9.10.2 A Binary Image

An amplitude digitization with just two possible levels of gray, that is $d = 1$, produces an image with array of pixels, which are either black or white. Such image is known as a *binary image* or a *black-and-white image*. The two-valued pixels require only one-bit of storage space each, say 0 for black and 1 for white.

A gray-scale image can be easily converted into a binary image. It requires only fixing of a thresholding image intensity level I^T to quantize the image pixels into one of the two values. Thus, a binary image representation $I'(x, y)$ can be obtained from a gray-scale image $I(x, y)$ by applying the rule:

$$\begin{aligned} I'(x, y) &= 1 \quad \text{if } I(x, y) \geq I^T \\ I'(x, y) &= 0 \quad \text{if } I(x, y) < I^T \end{aligned} \tag{9.2}$$

With one bit of storage requirement with each pixel, a binary image is simple to store and manipulate. One possible binary image approximation of the face of the dice in Fig. 9.16 is illustrated in Fig. 9.18. Note the loss of information about the boundaries of the dice.

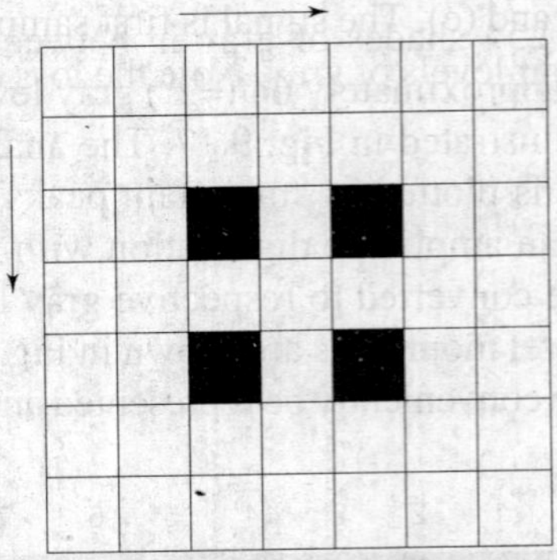

Fig. 9.18 *A two-level (binary) approximation of the gray-scale image of a dice in Fig. 9.16 with 7 × 7 grid*

9.10.3 Image Resolution

To understand the digitization, consider a single line scan shown in Fig. 9.19(a). The conversion of the video signal into a digital image having four gray levels is

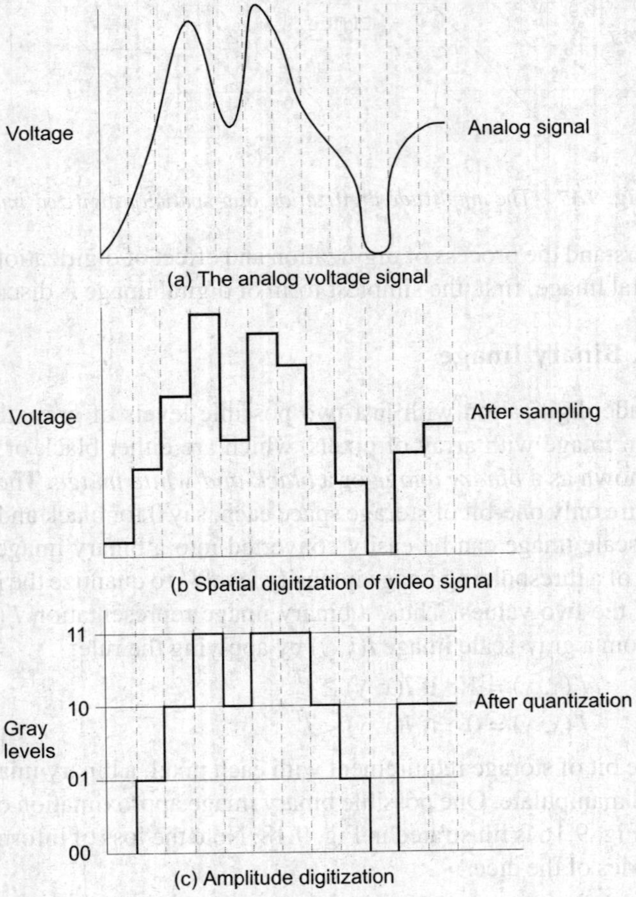

Fig. 9.19 *Two levels of digitization of image to get a digital image*

illustrated in Fig. 9.19(b) and (c). The signal is first sampled at discrete intervals and then threshold into four levels of gray. Note the loss of visual information in going from (a) to (c).

In the process of digitization of a gray-scale image, two important parameters are the number of pixels (spatial resolution) and the number of gray-scale levels (amplitude digitization). These two parameters determine the extent of detail the image can convey. The effect of varying spatial resolution is illustrated in Fig. 9.20 and amplitude digitization in Fig. 9.21.

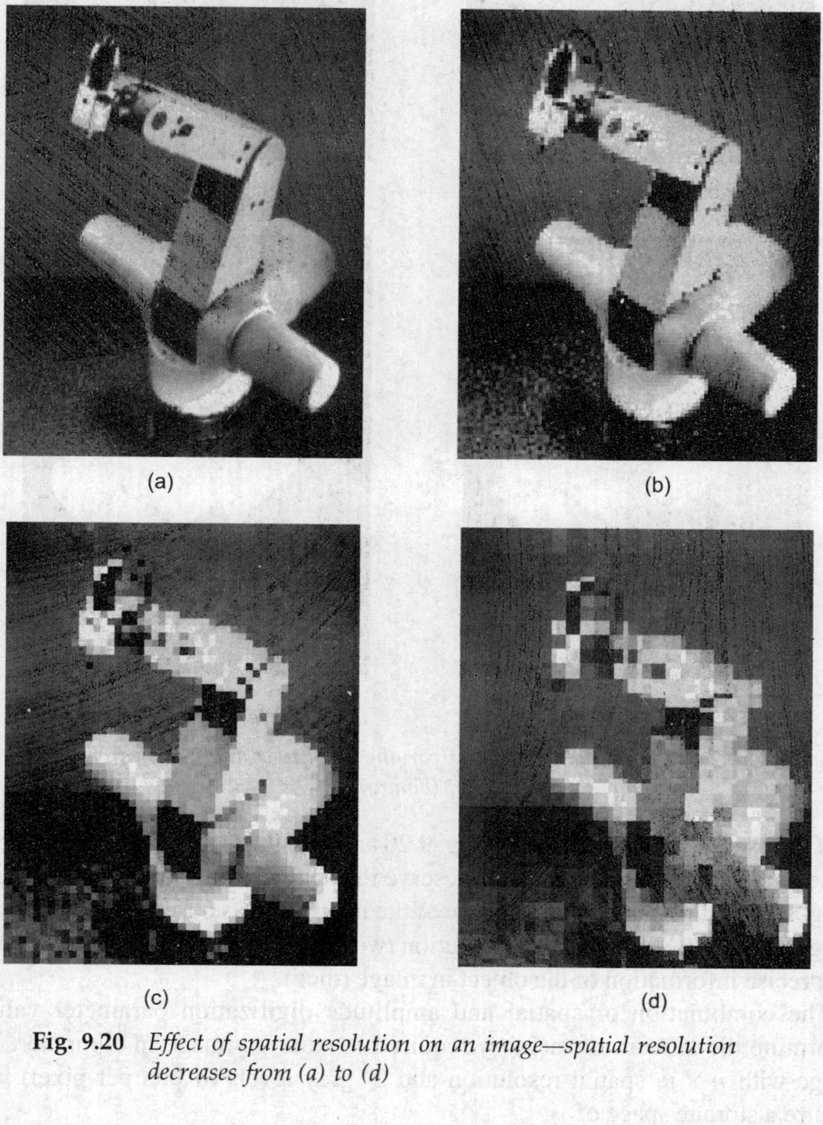

Fig. 9.20 *Effect of spatial resolution on an image—spatial resolution decreases from (a) to (d)*

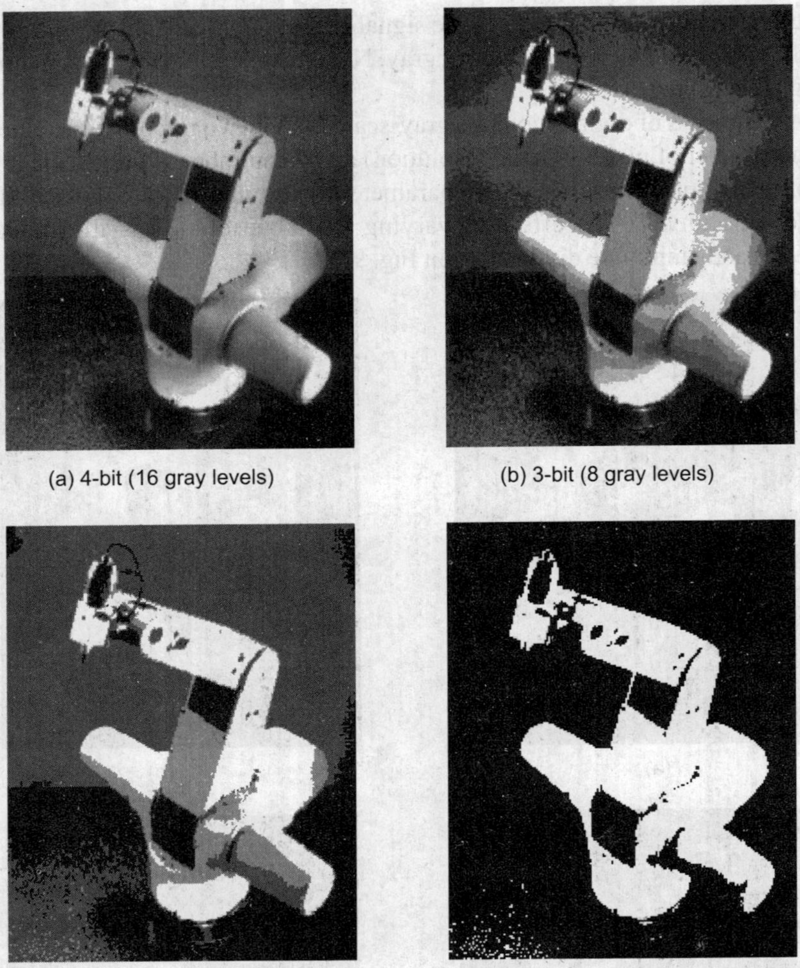

(a) 4-bit (16 gray levels)

(b) 3-bit (8 gray levels)

(c) 2-bit (4 gray levels)

(d) 1-bit (2 gray levels)

Fig. 9.21 *Illustration of effect of varying amplitude digitization from 8-bit in Fig. 9.20(a) to 1-bit (binary)*

It is observed from the figures (Fig. 9.20 and Fig. 9.21) that a proper choice of these parameters must be made to preserve the desired information in the digital image, depending on the intended use of the image. It was observed in the binary image in Fig. 9.18 that the low-resolution two-level digitization did not preserve the precise information of the object in image (dice).

The combination of spatial and amplitude digitization parameter values determine the amount of memory required to store an image in computer. An image with $n \times m$ spatial resolution and 2^d gray levels (d bits per pixel) will require a storage space of

$$\text{Number of bits} = m \times n \times d \tag{9.3}$$

For example, a single image digitized to 1024×1024 pixels with 8-bits (one byte) storage for each pixel, giving 256 gray levels, will require one megabyte (8 megabit) of memory storage.

The intensity amplitude quantization is usually from 1 to 8-bits with one bit for binary (two gray levels) or black-and-white image to eight bits for 256 gray-levels values from 0 (black) to 255 (white).

In a 2-D image, x- and y-axes are typically digitized from 64 to 1024 pixels, giving spatial resolution of 64×64 to 1024×1024 pixels, requiring a total of 4,096 to 1,048,576 pixels, respectively. These spatial resolutions will require a minimum of 4,096 bits for two gray levels to 8,388,608 bits for 256 gray levels of amplitude digitations, respectively. This means a single image may require 512 to 1,048,576, 8-bit bytes of storage, depending on the image resolution chosen. Typically, a 512×512 pixels array with 16 gray levels is considered a good resolution.

The digitized 16×16 pixels image for the dice in Fig. 9.16 with 4-gray levels is shown in Fig. 9.22. The 8×8 dice face is assumed to be bright (white) which is placed on a gray background and the number dots on the dice are assumed black.

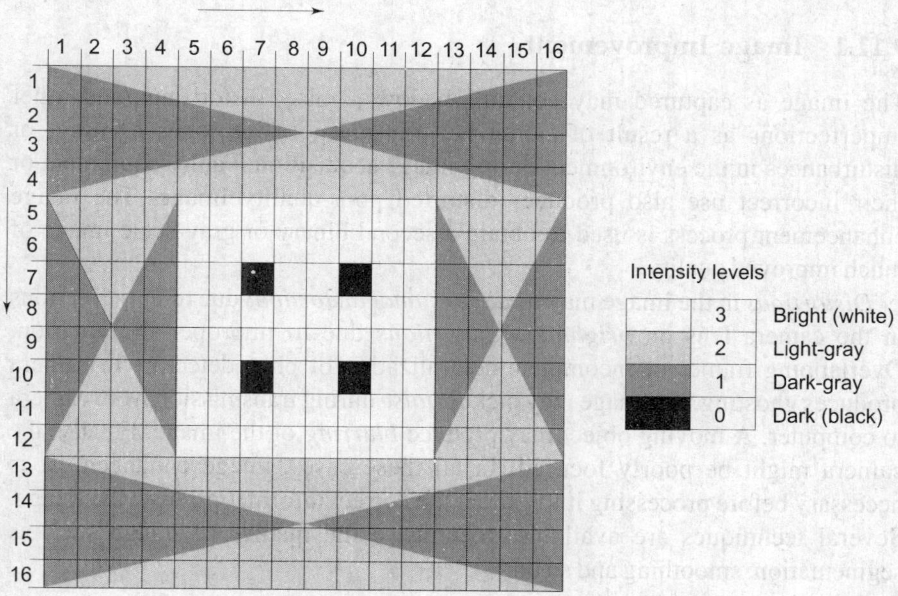

Fig. 9.22 *Digitized 16 × 16 pixel image of dice with 4-gray levels*

The image size (number of bits) given by Eq. (9.3) represents only a lower bound on the storage requirements. For example, if the computer hardware has 8-bit memory words, even for one-bit (two gray levels, binary image), for each pixel one byte will be used and the actual storage size will be eight times larger. The storage may be further increased on account of trade-off between storage and a computationally convenient organization of data into computer memory for faster image processing. Hence, the image size determined from spatial resolution and gray scale is an approximation to match the hardware capabilities.

9.11 IMAGE PROCESSING

The digitized image is stored in the computer memory for processing. There are two processing objectives: (i) to extract the spatial parameters of the objects and (ii) to recognize the objects in the image. On a 2-D visual image, integral operations can be applied to extract many useful features about the image. Some of these are:

1. Largeness (size or area).
2. Center of area or gravity (location).
3. Inclination (orientation).
4. Axis of least moment of inertia.
5. Other features (shape, geometry, radius etc.)
6. Similarity or dissimilarity (recognition).

Before an image is processed for drawing conclusions, it is usually preprocessed to remove the noise and other imperfections. Improving the quality to facilitate extraction of the desired information is the first part of image processing.

9.11.1 Image Improvement

The image as captured may contain shadows, noise, distortions, and other imperfections as a result of sampling, transmission, improper lighting, or disturbances in the environment during image acquisition. Faulty equipment or their incorrect use also produces distorted poor-quality image. The image enhancement process is used to obtain a second binary or gray-scale image of much improved quality.

Distortions in the image may be *dimensional distortions* due to imperfections in the camera lens or *brightness distortions* due to improper illumination. Overlapping frames or incomplete neutralization of photodetectors in camera produces ghosting. An image may pickup *noise* during transmission from camera to computer. A moving object may produce *blurring* of the image. Finally, the camera might be poorly focused. In all these cases, image enhancement is necessary before processing it to extract the correct information from the image. Several techniques are available to improve the quality of image such as segmentation, smoothing and so on.

9.11.2 Segmentation

A scene may contain one or more objects, which must be distinguished from the background. *Segmentation* is the process of identifying groups of related pixels for locating connected regions or areas of image having similar characteristics. This helps in identifying edges or distinct entities representing possible objects and separating them from the background. Segmentation process divides the image into constituent parts. There are many techniques to segment an image. Three important techniques are discussed here.

(a) Gray-Level Histogram The simplest method to segment a gray-scale image into background and foreground (object) area is to convert each pixel to a binary value, representing either background or object. In a digitized image, it is possible to measure the frequency with which each of the quantized levels (grayscale) occurs.

The plot of this frequency for pixels of each gray level *g* is known as gray-level histogram. For a particular image, with a light object on a dark background, the histogram may look like one shown in Fig. 9.23.

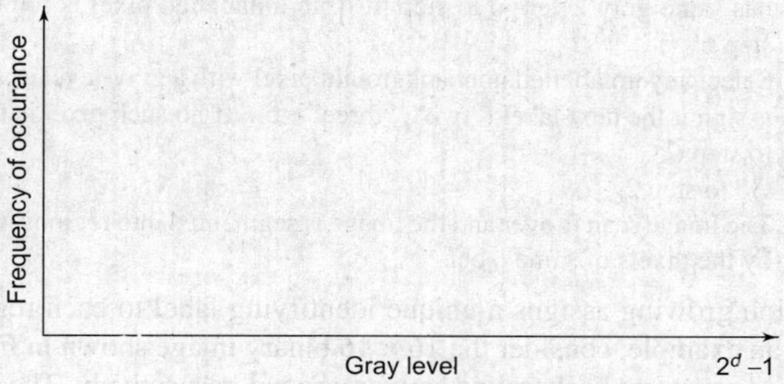

Fig. 9.23 *Gray-scale histogram for 2^d gray levels*

Each image generates a unique gray-level histogram but it is not possible to work backwards and get the image from the histogram. The histogram in Fig. 9.23 has two peaks. Clearly, the first peak is the background and the second peak represents the object or vice versa. The few scattered stray pixels represent the noise. This method may not work if there are several objects in the image or objects have holes. Sophisticated enhancement techniques may be applied to the histogram. For example, *equalization* may be applied to redistribute gray levels of pixels having low probability of use, which increases the contrast. Other information such as lack of contrast, image saturation and so on, can be easily obtained by examining the minimum and maximum gray levels found in the histogram. The conclusions drawn from the histogram can be employed in robotic vision control systems to control the lighting, camera position and so on, automatically.

(b) Region Growing This segmentation technique is based on grouping of pixels having similar attributes into regions. This is also called *pixel aggregation*. The gray-scale raw image is scanned and the region is grown by appending or connecting together the neighbouring pixel that have same property say gray-level, texture, colour and so on. Often the gray-level, that is, the intensity, is the image attribute of interest. Each region is labelled with a unique integer number. A region growing and labelling algorithm is given below assuming gray-level as the property and any pixel belonging to background (gray level = 0) is skipped.

Algorithm 9.1 ➢ Region Growing and Labeling Algorithm

Step 1 Select any pixel as "seed" (a non-background pixel), note it's gray-level and assign it a label 1.

Step 2 Evaluate each unlabelled nonbackground pixel in neighbourhood of the seed pixel. Assign same label to all the neighbouring pixels having same gray level.

Step 3 Select one of the neighbouring pixels with the same label and call it the seed, go to step 2. If none of the neighbours of all the pixels in the region has same gray level go to step 4. If no unlabelled pixel is found, go to step 6.

Step 4 Select any unlabelled nonbackground pixel with a gray-level as seed and assign it the next label ("two", "three" etc.). If no such pixel is found go to step 6.

Step 5 Go to step 2.

Step 6 The image scan is over and the image is segmented into regions identified by the pixels of same label.

The region growing assigns a unique identifying label to each region in the image. As an example, consider the 16 × 16 binary image shown in Fig. 9.24(a) where 0 and 1 represent background and foreground, respectively. The segmented image using the region-growing algorithm is shown in Fig. 9.24(b). Clearly, three different regions, labelled as 1, 2, and 3, are identified. In a nonbinary gray-level image ($d > 1$), regions of different gray-levels may adjoin each other. The adjoining regions may either be merged to get larger regions, if they differ marginally in the intensity values, or left as isolated pixel group. The recognition system can proceed with the segmented information.

0	0	0	0	0	0	0	0	0	0	0
0	0	0	0	0	0	0	0	0	0	0
0	1	1	1	1	1	0	0	0	0	0
0	1	1	1	0	0	0	0	0	0	0
0	1	1	1	0	0	0	0	1	1	0
0	0	0	0	0	0	0	1	1	1	1
0	1	1	0	0	0	0	0	1	1	0
0	0	1	1	1	1	0	0	0	0	0
0	0	0	0	1	1	1	1	0	0	0
0	0	0	0	0	0	0	0	0	0	0
0	0	0	0	0	0	0	0	0	0	0

(a) A binary image

0	0	0	0	0	0	0	0	0	0	0
0	0	0	0	0	0	0	0	0	0	0
0	1	1	1	1	1	0	0	0	0	0
0	1	1	1	0	0	0	0	0	0	0
0	1	1	1	0	0	0	0	2	2	0
0	0	0	0	0	0	0	2	2	2	2
0	3	3	0	0	0	0	0	2	2	0
0	0	3	3	3	3	0	0	0	0	0
0	0	0	0	3	3	3	3	0	0	0
0	0	0	0	0	0	0	0	0	0	0
0	0	0	0	0	0	0	0	0	0	0

(b) Segmented image after applying region growing

Fig. 9.24 *Region growing for segmentation to identify objects in image*

(c) Edge Detection Another effective technique for segmentation is edge detection. The outline boundary of an object within an image is equivalent to identifying the edges of the object that separate the object from its background.

Algorithms to identify whether an image pixel lies on the edges of an object or not, are known as edge detection algorithms. A common method is based on the intensity change or intensity discontinuity that occurs in adjacent pixels at the boundary or edge of an object. The idea underlying most of the edge detection algorithms is the computation of local gradient of image intensity. The gradient is zero in all regions of constant intensity and is nonzero wherever intensity changes. A high local intensity gradient, indicating a sudden intensity transition, is symptomatic of high probability of presence of a boundary or edge. Thus, the magnitude of the first derivative of intensity function can be used to detect edges in the image.

For binary images, the edge detection can be done by a simple edge following procedure. In a binary image, the change in intensity from 0 to 1 indicates the presence of edge. The image is scanned in a "left-right, top-down scan" until an object pixel (gray level = 1) is found. From there on, the edge is followed by examining the neighbours of the pixels.

To implement this edge following procedure for binary images, a formal concept of defining neighbours of a pixel p is required. There are several methods of defining neighbourhood. For instance, the two horizontal and two vertical neighbours of a pixels p at (x, y) in Fig. 9.25 are given by coordinates

$$(x-1, y), (x+1, y), (x, y-1), (x, y+1)$$

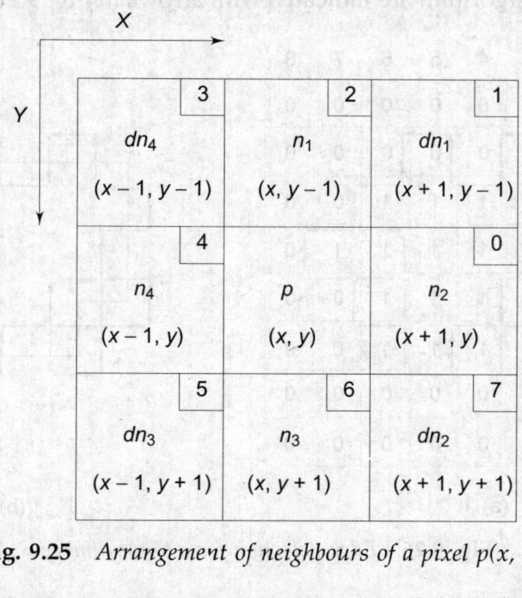

Fig. 9.25 *Arrangement of neighbours of a pixel p(x, y)*

This set of pixels is called 4-*neighbours* and the four members of the 4-neighbours set are labelled as n_1, n_2, n_3, and n_4. In addition, there are four diagonal neighbours of p, labelled as dn_1, dn_2, dn_3, and dn_4, and their coordinates are:

$$(x-1, y+1), (x+1, y+1), (x+1, y-1), (x-1, y-1)$$

These, together with the 4-neighbours constitute the 8-neighbours of p. These neighbours are illustrated in Fig. 9.25. The numbers in the top right hand corner of each cell indicates the direction of transversal.

The edge-detection procedure for binary images with 4-neighbours is given in Algorithm 9.2.

Algorithm 9.2 ➢ Edge-detection Algorithm for Binary Images

The pixel intensities are 0 and 1 for background and object, respectively.

Step 1 Do a row-by-row (left to right and top to bottom) scan of the image, starting from origin until an object pixel (value = 1) is found. Record this location as 'start'.

Step 2 Turn 90° in counterclockwise direction and step into the neighbour.

Step 3 If the new pixel is same as 'start' go to step 7.

Step 4 If new pixel is an object pixel, go to step 2.

Step 5 If new pixel is a background pixel (value = 0), turn 90° clockwise and step into the neighbour.

Step 6 If the new pixel is same as 'start' go to step 7 else, go to step 4.

Step 7 Stop when the 'start' pixel in step 1 is reached a second time. Edge detection of one object is complete.

As an example, the Algorithm 9.2 is applied to the binary image in Fig. 9.26(a) and the resulting object boundary is shown in Fig. 9.26(b). The pixels traversed according to the algorithm are indicated with arrows in Fig. 9.26(a).

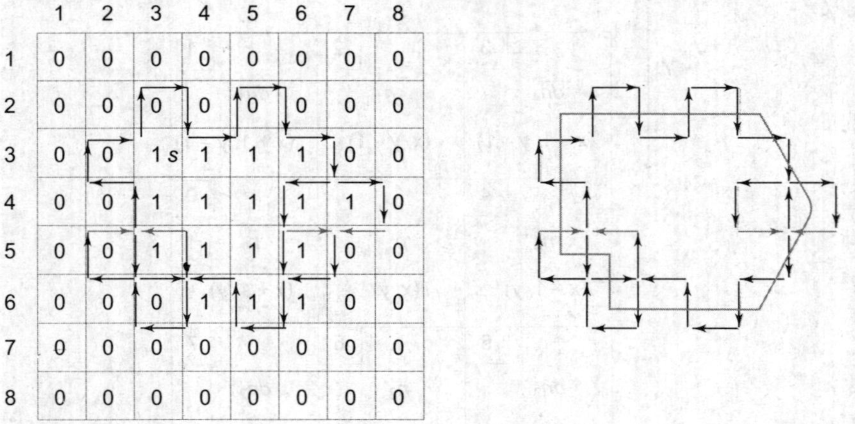

(a) (b)

Fig. 9.26 *Edge detection for a binary image*

In case of gray-level images, the local gradient of intensity function has to be computed and to decide whether it is an edge point or not, the familiar procedure of thresholding is applied. The local gradient function $G(x, y)$ is computed for each pixel and the $n \times n$ array of gradient values is obtained. On this array, the algorithm 9.2 is applied with minor modifications. Instead of gray level 0 and 1, the gradient function is used to detect edges. For the local gradient function $G(x, y)$, an 'edge threshold' value G^T is used to detect the edge points as:

$$G(x, y) \geq G^T \rightarrow \text{pixel } (x, y) \text{ is an edge point}$$
$$G(x, y) < G^T \rightarrow \text{pixel } (x, y) \text{ is not an edge point}$$

The choice of G^T affects the edge detection and an ideal choice is difficult. The edge detection for a gray level image using local gradient function is illustrated in Fig. 9.27. The local gradient function array with gradient scale between 0 and 10 is shown in Fig. 9.27(a). The edges detected with two different values of threshold values $G^T = 6$ and $G^T = 8$ are shown in Fig. 9.27(b) and (c).

0	1	1	2	3	5	4	1
1	10	9	7	8	7	5	3
0	7	5	4	4	3	8	0
4	10	3	4	5	0	9	4
3	10	1	1	10	7	4	1
4	9	5	2	8	4	2	2
2	7	8	7	6	3	0	2
1	1	4	3	3	2	5	4

(b) Array of local gradient function ($0 \leq G \leq 10$)

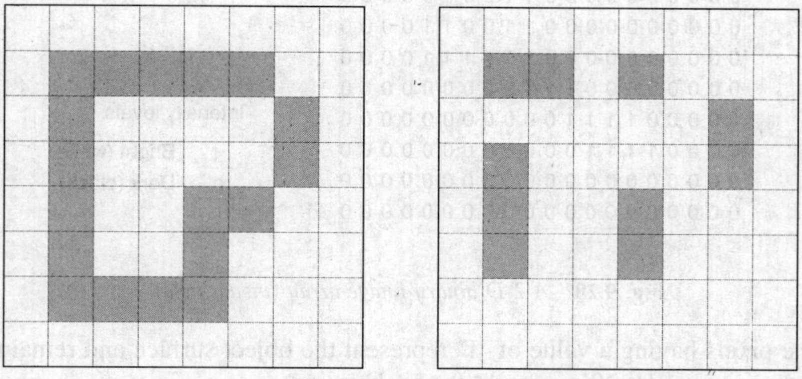

(b) $G^T = 6$ (c) $G^T = 8$

Fig. 9.27 *Edge detection in gray level image with two different threshold values of local gradient function G(x, y)*

9.11.3 Smoothing

Smoothing operations are iterative in nature and are used to improve quality of the image. The image intensity is modified using local techniques based on the assumption that the pixel value for a pixel in some sense is similar to that of its neighbours. There are several techniques for smoothing images.

One such technique is neighbourhood averaging. This simple technique generates a smooth image $I'(x, y)$, from image $I(x, y)$, such that the intensity at every point (x, y) is obtained by averaging the intensity values of pixels of $I(x, y)$ in the predefined neighbourhood of (x, y). The neighbourhood can be chosen as any suitably sized square or rectangular array of pixels around (x, y). The degree of smoothing is proportional to the size of neighbourhood used. A major disadvantage of smoothing is that it blurs the edges and other sharp details.

9.11.4 Object Descriptors

An object can be recognized by its characteristics, like geometric features (shape, size, area, perimeter etc.) and other features, for example, number of line segments, corners, arcs, or holes. To respectively identify a given object in different images such characteristics are computed from the image. The extracted geometric features of an image are known as object descriptors. Obviously, any single feature, say perimeter, cannot be a good unique descriptor to identify an object. Therefore, several descriptors are simultaneously obtained for a given object and then, the object may be identified with good statistical confidence level. The image processing for computation of the features from the image data is discussed now. Consider a 2-D binary image shown in Fig 9.28.

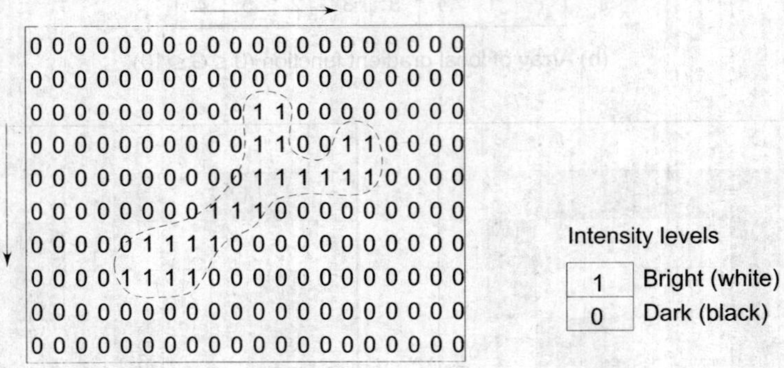

Fig. 9.28 *A 2-D binary image array (visual field)*

The pixels having a value of '1' represent the object surface and remaining pixels having a value '0' represent the background.

The image intensity function for a binary image of an object O is:

$$I(x, y) = \begin{cases} 1 & \text{for all } (x, y) \in O \text{ (object)} \\ 0 & \text{for all } (x, y) \notin O \text{ (background)} \end{cases} \tag{9.4}$$

This functional definition assumes uniform brightness of object everywhere and background is clearly distinguished. The binary image intensity function for image in Fig. 9.28 is shown in Fig. 9.29. In the view field of $m \times n$ pixels, the object image lies within a rectangular window with corners (a_1, b_1) and (a_2, b_2) as shown in Fig. 9.29.

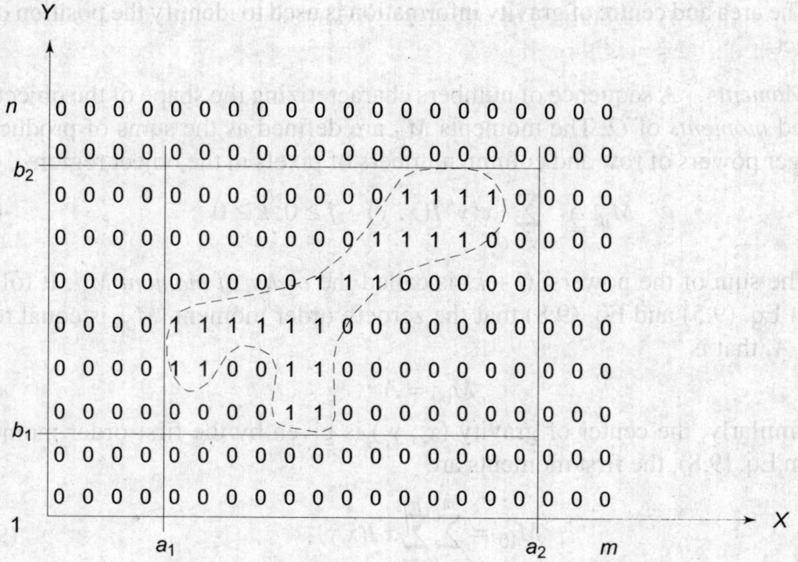

Fig. 9.29 *Rectangular window for image in Fig. 9.28*

The intensity function in Eq. (9.4) is the mathematical model of the visual field in the form of a $m \times n$ pixel matrix of 0's and 1's. This can be used to extract useful information about the image. Image area, center of gravity, orientation and many other attributes can be computed from the image intensity function. The calculation of these image characteristics is discussed next.

(a) Image Area Area, size, or largeness, A of the image is obtained by integrating the function $I(x, y)$ over the view field. Because the image function is a discrete function, integration can be replaced by summation. The area of object O is, thus, given by

$$A = \sum_{x=1}^{m} \sum_{y=1}^{n} I(x, y) \tag{9.5}$$

However, computations will be drastically reduced if the summation is carried out over the image portion of visual field, that is, the window between (a_1, b_1) and (a_2, b_2) as shown in Fig. 9.29. Summation over the entire view field of $m \times n$ pixels is considered here for simplicity. Note that the area of binary picture is the sum of all 'ones' in the $m \times n$ pixel matrix.

(b) Centre of Gravity The centre of gravity or centroid (x_c, y_c) of the image is

$$x_c = \frac{1}{A} \sum_{x=1}^{m} \sum_{y=1}^{n} x\, I(x, y) \tag{9.6}$$

and

$$y_c = \frac{1}{A} \sum_{x=1}^{m} \sum_{y=1}^{n} y\, I(x, y) \tag{9.7}$$

The area and centre of gravity information is used to identify the position of the object.

(c) Moments A sequence of numbers characterizing the shape of the object O is called *moments* of O. The moments M_{jk} are defined as the sums of products of integer powers of row and column numbers of pixels in the object region,

$$M_{jk} \underset{=}{\Delta} \sum_{x,\,y \in O} x^j y^k I(x, y) \quad j \geq 0, k \geq 0 \tag{9.8}$$

The sum of the powers, $(j + k)$ is called the *order of moment* M_{jk}. It follows from Eq. (9.5) and Eq. (9.8) that the zeroeth order moment, M_{00} is equal to the area A, that is

$$M_{00} = A \tag{9.9}$$

Similarly, the center of gravity (x_c, y_c) is given by the first-order moments. From Eq. (9.8), the first moments are

$$M_{10} = \sum_{x=1}^{m}\sum_{y=1}^{n} x\, I(x,y) \tag{9.10}$$

and

$$M_{01} = \sum_{x=1}^{m}\sum_{y=1}^{n} y\, I(x,y) \tag{9.11}$$

Because the size A is given by M_{00}, Eq. (9.9) the normalized first moments provide (x_c, y_c) as

$$x_c = \frac{M_{10}}{M_{00}} \tag{9.12}$$

$$y_c = \frac{M_{01}}{M_{00}} \tag{9.13}$$

Once the centroid of the image is known, moments can also be defined about the center of gravity (x_c, y_c). These are called central moments and are defined as:

$$\mu_{jk} = \sum_{x,y \in O} (x - x_c)^j (y - y_c)^k I(x, y) \quad j \geq 0,\ k \geq 0 \tag{9.14}$$

The central moments are equal to moments if centroid is located at the origin. The central moments are invariant to translation of image. Because the object is balanced about the centre of gravity, the first-order central moments are always zero, that is

$$\mu_{10} = \sum_{x=1}^{m}\sum_{y=1}^{n} (x - x_c)\, I(x, y) = 0 \tag{9.15}$$

$$\mu_{01} = \sum_{x=1}^{m}\sum_{y=1}^{n} (y - y_c)\, I(x, y) = 0 \tag{9.16}$$

The second-order central moments of a given image are computed as follows:

$$\mu_{20} = \sum_{x=1}^{m}\sum_{y=1}^{n}(x - x_c)^2 \ I(x, y) \tag{9.17}$$

$$\mu_{02} = \sum_{x=1}^{m}\sum_{y=1}^{n}(y - y_c)^2 \ I(x, y) \tag{9.18}$$

$$\mu_{11} = \sum_{x=1}^{m}\sum_{y=1}^{n}(x - x_c)(y - y_c) \ I(x, y) \tag{9.19}$$

The second-order central moments also give important information about the image. In particular, the moments μ_{02} and μ_{20} represent the moments of inertia of image corresponding to x- and y-axes through the centroid.

(d) Orientation The orientation or inclination of the image can also be computed from the second-order central moments. Consider an arbitrary direction represented by a line L passing through centroid (x_c, y_c) of object and making an angle β with the x-axis, as shown in Fig. 9.30. The moment of inertia of image about line L will depend on the angle β. The angle at which this moment is minimized is defined as the inclination of the binary image with respect to x-axis, that is, angle β. The angle β is given by

$$\beta \cong \frac{1}{2} \ A\tan 2 \ (2\mu_{11}, \ \mu_{20} - \mu_{02}) \tag{9.20}$$

The angle β defines the orientation of the object.

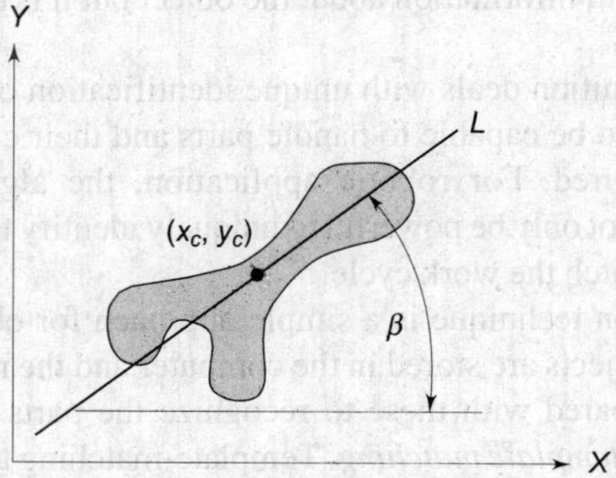

Fig. 9.30 *Orientation of object in Fig. 9.29*

(e) Other Attributes Many other attributes of image can be similarly obtained. Some of these are defined here.

$$\text{Eccentricity} = \frac{\text{Maximum chord length}}{\text{Minimum chord length}} \tag{9.21}$$

The maximum chord length is along the principal axis or major axis of object and minimum chord length is perpendicular to major axis.

The maximum chord length or major diameter D of object O is defined as

$$D = \max_{i,j} \sqrt{\left(x_j - x_i\right)^2 + \left(y_j - y_i\right)^2} \tag{9.22}$$

where $p_i = (x_i, y_i)$ and $p_j = (x_j, y_j)$ are pixels in the boundary of O.

$$\text{Thinness} = \frac{(\text{Perimeter})^2}{A} \tag{9.23}$$

$$\text{Roundness} = \frac{\left(x_c^2 + y_c^2\right)}{A} \tag{9.24}$$

$$\text{Aspect ratio} = \frac{\text{Length of rectangle enclosing object}}{\text{Width of rectangle enclosing object}} \tag{9.25}$$

For the image in Fig. 9.29,

$$\text{Aspect ratio} = \frac{a_2 - a_1}{b_2 - b_1} \tag{9.26}$$

This may be used to find the rectangle, which gives minimum aspect ratio.

These definitions are not unique and alternative ways of finding these and other features are available in literature.

9.11.5 Object Recognition

Identification of objects contained in the scene is an important task for image processing. Using the features extracted, as defined in the Section 9.11.4, it is possible to find useful information about the object but it is difficult to say what the object is.

The object recognition deals with unique identification of each object in the image. For a robot to be capable to handle parts and their classification, object recognition is required. For robotic application, the algorithms for object recognition should not only be powerful to uniquely identify the object but should be fast enough to match the work cycle.

Image comparison technique is a simple approach for object identification. Images of known objects are stored in the computer and the regions identified in the image are compared with these to recognize the parts in the image. This method is known as *template matching*. Template-matching technique is a subset of more general and vast *pattern recognition* techniques.

Once the image has been improved and segmented, it is compared against the template. In practice, it is rare for an image to match the template exactly. The recognition of the object is probabilistic. The closer the image matches with the template, more probable it is that the object is in fact the object represented by the template. Different templates can be stored for object recognition. A template can be the full image of the standard object, which can be swelled or shrunk to fit the image size, or it can be a set of features of the object in the image such as its area, diameter, aspect ratio and so on. Sufficient number of features must be stored to minimize errors in recognition process. The model template may also be obtained and stored during the vision system training.

SOLVED EXAMPLES

Example 9.1 *Image size*

An image is digitized with 256 samples per line and 256-line scan. Determine the number of bits required to represent the images as (a) binary image (b) gray-level image with 4-bits for coding gray levels.

Solution

(a) For black and white image
number of gray levels = $2^1 = 2$
number of bits = $256 \times 256 \times 2 = 131{,}072$ bits. (9.27)

(b) For gray-scale image
number of gray levels = $2^4 = 16$
number of bits = $256 \times 256 \times 16 = 1{,}048{,}576$ bits (9.28)

Example 9.2 *Run-length encoding*

Consider the $m \times n$ binary image shown in Fig. 9.31 where $m = n = 13$. Suppose the image is scanned from left to right and top to bottom in the order: $I(1, 1)$, $I(1, 2)$, …, $I(m, n)$. Determine the storage requirements of the image.

0	0	0	0	0	0	0	0	0	0	0	0	0
0	0	0	0	0	0	0	0	0	0	0	0	0
0	0	0	0	1	1	1	1	1	0	0	0	0
0	0	0	0	1	1	1	1	1	0	0	0	0
0	0	1	1	1	1	1	1	1	1	1	0	0
0	0	1	1	1	0	0	0	1	1	1	0	0
0	0	1	1	1	0	0	0	1	1	1	0	0
0	0	1	1	1	0	0	0	1	1	1	0	0
0	0	1	1	1	1	1	1	1	1	1	0	0
0	0	0	0	1	1	1	1	1	0	0	0	0
0	0	0	0	1	1	1	1	1	0	0	0	0
0	0	0	0	0	0	0	0	0	0	0	0	0
0	0	0	0	0	0	0	0	0	0	0	0	0

Fig. 9.31 *A 13 × 13 binary image of an object with a hole*

Solution The simplest storage of the image is with one byte per pixel. This would require 169 bytes for the 13 × 13 image in Fig. 9.31.

Images are often compressed or encoded to save on storage space. One possible compression technique is to pack as many pixels per byte as possible. An image with 4-gray levels requires two bits per pixel and, thus, four pixels can be packed per byte. Applying this to a binary image, 8 pixels can be packed per byte and the storage reduces by a factor of 8. For the image in Fig. 9.31, this packing would require 22 bytes.

Another type of compression can be achieved by using some encoding scheme. One such simple scheme for binary images is called *run-length encoding*. In this encoding, the image information is converted to *count of pixels* that have same intensity value. These numbers are stored in a one-dimensional array or vector. As long as the maximum count of pixels is less than 256 (2^8), one count will require one byte only. One byte is added to this list to store the starting pixel colour (0 or 1).

Applying this encoding to Fig. 9.31, the first pixel is a 0, and scanning from left to right and top to bottom, there are 30 zeros before a '1' is encountered giving first count of pixels as 30. The sequence of run lengths this far is [0, 30, ..]. This process of scan is continued and the total run-length code is:

$$\boldsymbol{I} = [0, 30, 5, 8, 5, 6, 9, 4, 3, 3, 3, 4, 3, 3, 3, 4, 3, 3, 3, 4, 9, 6, 5, 8, 5, 30]^T \quad (9.29)$$

Thus, the storage requirement for the encoded image is 26 bytes, including the byte for the color (value) of the starting pixel. Note that the maximum possible run length can be mn = 169 bytes, and minimum possible storage requirement can be only 2 bytes.

Storage of the unpacked binary image requires 169 bytes, and storage of the packed binary image requires only 22 bytes. If the part in Fig. 9.31 did not have a hole in the middle, then the run-length code would be even shorter, only 20 bytes, while the packed binary image would still occupy 22 bytes.

Example 9.3 *Region growing and labeling*

Identify the regions in the binary image given in Fig. 9.32 using the region growing and labeling algorithm.

1	1	1	0	0	0	0	0
1	1	0	0	0	0	0	0
1	1	0	0	1	1	1	1
0	0	0	0	1	1	1	1
0	0	0	1	1	1	1	1
0	0	1	1	1	0	0	0
0	0	1	1	1	0	0	0
0	0	1	1	1	0	0	0

Fig. 9.32 *A 8 × 8 binary image*

Solution This example illustrates how the region-labeling algorithm works, using the 8 × 8 binary image shown in Fig. 9.32. In algorithm 9.1, step 1 requires identification of "seed" pixel. The image *I*(*k*, *j*) is scanned from the left to right and top to bottom to identify an unlabeled foreground pixel. The first seed is found at address (1, 1) and is given a label '*f*', at which point step 2 of algorithm 9.1 is invoked. Note that the four neighbours of an unlabeled foreground pixel are searched in the counterclockwise order. The neighbours of pixel (1, 1) with same gray level are found as pixels: (1, 2) and (2, 1). These are assigned label '*f*'. Next step of algorithm 9.1 requires to iteratively examine the neighbouring pixels of all pixels with label '*f*' to grow the region. To keep track of unexamined neighbours with a particular assigned label a stack can be used. The empty stack is the condition for termination of step 3 of algorithm 9.1.

This is completion of first region growing and labeling and control returns to step 4 of algorithm 9.1. The partially labeled image array *I*(*k*, *j*) now has the labels and values shown in Fig. 9.33 with pixels of first identified region marked with '*f*'.

f	*f*	*f*	0	0	0	0	0
f	*f*	0	0	0	0	0	0
f	*f*	0	0	1	1	1	1
0	0	0	0	1	1	1	1
0	0	0	1	1	1	1	1
0	0	1	1	1	0	0	0
0	0	1	1	1	0	0	0
0	0	1	1	1	0	0	0

Fig. 9.33 *Partially labeled image after labeling of first region*

The search for unlabeled foreground pixels is resumed in step 4 of algorithm 9.1. The pixel (3, 5) is the next seed, which is labeled as '*s*' representing the second region. At this point, control returns to step 2 of algorithm 9.1 and steps 2 and 3 are continued until the growing and labeling of the second region is complete.

The search for another unlabeled foreground pixels resumes at step 4 of the algorithm. In this case, no new seed is found, because all foreground pixels have been labeled. The control goes to step 6, the termination of algorithm 9.1. The final labeled and segmented image with two regions is shown in Fig. 9.34. Background pixels ('0') are not marked in Fig. 9.34 for clarity.

f	*f*	*f*					
f	*f*						
f	*f*			*s*	*s*	*s*	*s*
				s	*s*	*s*	*s*
			s	*s*	*s*	*s*	*s*
		s	*s*	*s*			
		s	*s*	*s*			
		s	*s*	*s*			

Fig. 9.34 *Segmented image with two identifies and labeled regions*

Example 9.4 *Histogram and binary conversion*

An 8 × 8 gray level image with 16 gray level intensity values is given in Fig. 9.35.

(a) Construct the histogram of the image and obtain the threshold value.

(b) Convert the image into a binary (black-and-white) image using the threshold value determined in part (a).

6	4	0	1	2	4	3	5
3	13	13	14	13	11	13	3
4	8	10	13	14	10	11	4
3	4	7	13	14	4	2	2
5	5	3	14	13	5	0	6
5	10	13	15	14	14	12	3
5	12	14	13	11	10	9	2
2	4	5	5	5	4	8	4

Fig. 9.35 *An 8 × 8 gray level image with 16 gray levels*

Solution The histogram is drawn by getting the frequency count of the gray intensity levels. The frequency counts obtained from Fig. 9.35 are tabulated in Table 9.1.

Table 9.1 *Frequency counts of different gray levels for image in Fig. 9.35*

Gray level	*Frequency*	*Gray level*	*Frequency*
0	2	8	1
1	1	9	1
2	5	10	4
3	6	11	3
4	9	12	2
5	10	13	9
6	2	14	7
7	1	15	1

The histogram for the gray level image is given in Fig. 9.36. From the histogram it is clearly seen that a threshold of gray level = 8 can be used as image intensity value to convert the gray level image into a binary image. Thus, the binary image is obtained from the gray level image by applying the rule

$$I'(x, y) = \begin{cases} 1 & \text{if } I(x\ y) \geq 8 \\ 0 & \text{if } I(x\ y) < 8 \end{cases} \tag{9.30}$$

where $I(x, y)$ represent the binary image density.

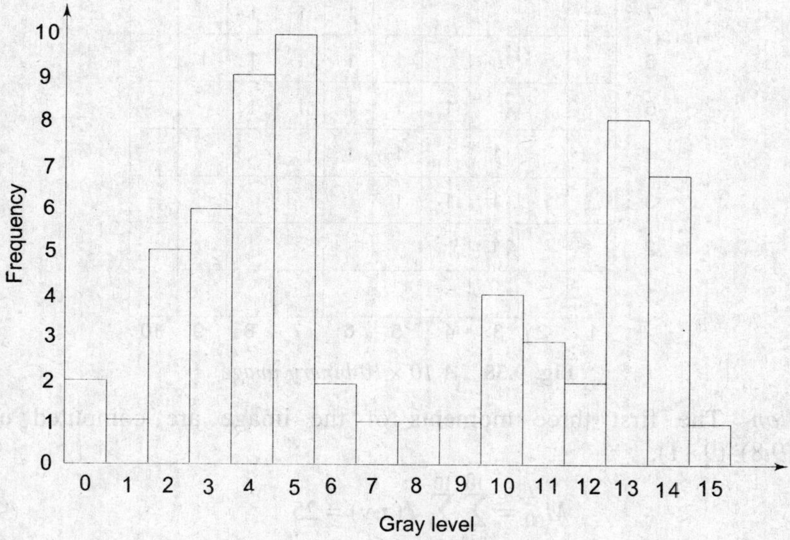

Fig. 9.36 *The histogram for the gray level image in Fig. 9.35*

Apply the rule in Eq. (9.30) to image in Fig. 9.35 binary image is obtained. The final 8 × 8 binary image is given in Fig. 9.37.

0	0	0	0	0	0	0	0
0	1	1	1	1	1	1	0
0	1	1	1	1	1	1	0
0	0	0	1	1	0	0	0
0	0	0	1	1	0	0	0
0	1	1	1	1	1	1	0
0	1	1	1	1	1	1	0
0	0	0	0	0	0	0	0

Fig. 9.37 *The binary (black-and-white) image of the object in Fig. 9.35*

From the binary image in Fig. 9.37 the object is easily identified as capital "*I*" whereas it was not visible in gray level image in Fig. 9.35.

Example 9.5 ***Object descriptors for a binary image***

Compute the area, centroid, aspect ratio, and orientation of the binary image shown in Fig 9.38.

10										
9							1			
8						1	1	1		
7						1	1	1		
6					1	1	1	1	1	
5				1	1	1	1	1		
4			1	1	1	1				
3			1	1	1					
2			1							
1										
	1	2	3	4	5	6	7	8	9	10

Fig. 9.38 *A 10 × 10 binary image*

Solution The first three moments of the image are computed using Eqs (9.8)–(9.11).

$$M_{00} = \sum_{x=1}^{10}\sum_{y=1}^{10} \boldsymbol{I}(x,y) = 25 \tag{9.31}$$

$$M_{10} = \sum_{x=1}^{10}\sum_{y=1}^{10} x\,\boldsymbol{I}(x,y) = 147 \tag{9.32}$$

and
$$M_{01} = 136 \tag{9.33}$$

From Eq. (9.9), the area is

$$A = M_{00} = 25 \tag{9.34}$$

The centroid (x_c, y_c) is computed using Eqs (9.12) and (9.13) as

$$x_c = \frac{M_{10}}{M_{00}} = 5.88 \cong 6 \text{ (say)} \tag{9.35}$$

$$y_c = \frac{M_{01}}{M_{00}} = 5.44 \cong 5 \tag{9.36}$$

Note that the area and centroid can also be computed using Eqs (9.5) – (9.7). The aspect ratio is computed using Eq. (9.26). From Fig. 9.38,

$$\text{Aspect ratio} = \frac{9-3}{9-2} = 0.86 \tag{9.37}$$

To compute the orientation central moments are required. The second-order moments are computed using Eqs (9.17) – (9.19) as

$$\mu_{20} = \sum_{x=1}^{10}\sum_{y=1}^{10}(x - x_c)^2 \; I(x, y) = 73 \tag{9.38}$$

Similarly,
$$\mu_{02} = 85 \tag{9.39}$$

and
$$\mu_{11} = 51 \tag{9.40}$$

The inclination β to x-axis is computed using Eq. (9.20).

$$\beta = \frac{1}{2} A\tan 2\;(2\;\mu_{11},\; \mu_{20} - \mu_{02}) = \;41.65^{\circ} \tag{9.41}$$

EXERCISES

9.1 Make a list of internal or status sensors used in robotic manipulators. Give brief description of their working and use.

9.2 Give situations where robot will require noncontact sensors. Identify suitable noncontact sensors for these applications and explain their working.

9.3 Briefly describe the working of some contact sensors used in robotics. Give advantages and disadvantages of each.

9.4 A robotic vision system is used to aid the robot to pick parts from a conveyor belt. Parts are only of one type and are placed randomly on the conveyor. What information must the vision system acquire from the image to make the robot work intelligently?

9.5 What is quantization error? How it can be corrected or minimized?

9.6 Make a list of limitations of two-dimensional vision systems.

9.7 Extract all possible features from the image in Fig. 9.24 and Fig. 9.29.

9.8 Determine first five moments and central moments for the image in Fig. 9.26 and Fig. 9.30

9.9 Determine the area, orientation, and centroid for the objects in the 11×11 pixel image in Fig. 9.22.

9.10 What tasks can be performed by a robotic vision system? Briefly explain.

9.11 Obtain a binary image from the gray-scale image in Fig. 9.25 by choosing appropriate thresholding level. For the binary image

(a) Identify the object(s) in the image-using region growing.

(b) Using edge detection, obtain the edges of the object(s).

(c) Determine all the object descriptors.

9.12 Compile a list of sensors that might be used in robotic systems. For each sensor, give an application.

9.13 Tabulate the advantages and disadvantages of different sensors, which can be used to obtain controlling information in robotic systems.

9.14 Discuss the industrial and nonindustrial applications of vision-controlled robots.

9.15 Show how image storage requirements depend on spatial and amplitude digitization parameters.

9.16 For a given picture, estimate how the storage (number of bits) increases when the number of rows and number of columns are doubled in the spatial digitization grid

9.17 Using the run-length coding scheme, determine the run-length code for the binary image in Fig. 9.36. What will be the storage requirement for the run-length code of the image?

9.18 Derive the expression for calculating the area on a binary image from the run-length code discussed in Exercise 9.17.

9.19 Determine the object descriptors for the binary image shown in Fig. E9.19.

0	0	0	0	0	0	0	0	0	0	0	0	0	0	0	0
0	0	0	0	0	0	0	0	0	0	0	0	0	0	0	0
0	0	1	1	1	1	1	1	1	1	0	0	0	0	0	0
0	0	1	1	1	1	1	1	1	1	0	0	0	0	0	0
0	0	1	1	0	0	0	0	1	1	0	0	0	0	0	0
0	0	1	1	0	0	0	0	1	1	0	0	0	0	0	0
0	0	1	1	0	0	0	0	1	1	0	0	0	0	0	0
0	0	1	1	0	0	0	0	1	1	0	0	0	0	0	0
0	0	1	1	1	1	1	1	1	1	0	0	0	0	0	0
0	0	1	1	1	1	1	1	1	1	0	0	0	0	0	0
0	0	0	0	0	0	0	1	1	1	1	0	0	0	0	0
0	0	0	0	0	0	0	0	1	1	1	1	0	0	0	0
0	0	0	0	0	0	0	0	0	1	1	1	1	0	0	0
0	0	0	0	0	0	0	0	0	0	1	1	1	1	1	0
0	0	0	0	0	0	0	0	0	0	0	1	1	1	1	0
0	0	0	0	0	0	0	0	0	0	0	0	0	0	0	0

Fig. E9.19 *Binary image of an object with one hole*

9.20 The local gradient function values of an 8×8 image are shown in Fig. E9.20. Determine the edges of the object in the image by choosing the appropriate threshold value of gradient function.

2	5	6	3	3	2	1	2
1	9	1	8	9	9	2	1
1	8	6	3	4	7	10	2
3	8	4	2	6	2	9	3
4	10	5	6	5	6	8	3
6	9	6	5	4	2	9	1
3	5	8	9	10	8	6	2
1	2	1	3	1	3	2	2

Fig. E9.20 *The local gradient function of an 8 × 8 an image*

9.21 The binary image of a scene is shown in Fig. E9.21. Determine the boundaries of the object(s) in the image using:
(a) Region growing
(b) Edge detection

0	0	0	0	0	0	0	0	0	0	0
0	0	0	0	0	0	1	0	0	0	0
0	0	0	0	0	1	1	1	0	0	0
0	0	0	0	1	1	1	1	1	0	0
0	0	0	0	1	1	1	1	1	1	0
0	0	0	0	1	1	1	1	1	0	0
0	0	0	0	1	1	1	1	0	0	0
0	0	0	1	1	0	0	0	0	0	0
0	0	1	1	0	0	0	0	0	0	0
0	1	1	0	0	0	0	0	0	0	0
0	0	0	0	0	0	0	0	0	0	0

Fig. E9.21 *The binary image of an object*

9.22 Calculate the object descriptors of for the binary image array in Fig. E9.21.

9.23 The spatially digitized output of an image acquisition system is the gray scale array shown in Fig. E9.23.

1	1	1	1	2	2	2	1
0	2	3	3	3	3	3	2
1	1	2	3	6	6	3	2
2	2	0	6	7	6	2	1
1	2	4	6	7	6	2	1
0	3	4	5	5	5	3	2
1	3	2	2	2	2	2	2
2	1	1	1	1	1	2	1
1	1	0	0	1	1	2	1

Fig. E9.23 *Spatially digitized gray scale image with 8-gray levels*

(a) Obtain a gray scale histogram for the image based on 3-bit (eight gray-level) per pixel.

(b) Convert the image to a binary image by choosing an optimal value of threshold intensity, and sketch the binary image array.

(c) Compute the image descriptors like area, aspect ration, centre of gravity, orientation, eccentricity, thinness and roundness for the binary image obtained in part (b).

9.24 Consider the binary image of an object given in Fig. E9.24. Determine the boundary of the object imaged using (a) edge detection (b) region growing algorithms discussed in the text. Assume that the object is white (1s) and is placed on a block background (0s).

0	0	0	0	0	0	0	0	0	0	0	0	0	0	0	0
0	0	0	0	1	1	1	1	1	1	1	1	0	0	0	0
0	0	0	0	1	1	1	1	1	1	1	1	0	0	0	0
0	0	0	0	1	1	0	0	0	0	1	1	0	0	0	0
0	0	0	0	1	1	0	0	0	0	1	1	0	0	0	0
0	0	0	0	1	1	0	0	0	0	1	1	0	0	0	0
0	0	0	0	1	1	1	1	1	1	1	1	0	0	0	0
0	0	0	0	1	1	1	1	1	1	1	1	0	0	0	0
0	0	0	0	1	1	1	1	1	0	0	0	0	0	0	0
0	0	0	0	1	1	0	1	1	1	0	0	0	0	0	0
0	0	0	0	1	1	0	0	1	1	1	0	0	0	0	0
0	0	0	0	1	1	0	0	0	1	1	0	0	0	0	0
0	0	0	0	1	1	0	0	0	0	1	1	1	0	0	0
0	0	0	0	1	1	0	0	0	0	0	1	1	1	0	0
0	0	0	0	1	1	0	0	0	0	0	1	1	1	0	0
0	0	0	0	0	0	0	0	0	0	0	0	0	0	0	0

Fig. E9.24 *The binary image of a white object on black background*

9.25 A gray scale digitized image with 6-bit amplitude digitization is represented by the 10×10 array of pixels shown in Fig. E9.25. Construct a gray level histogram for the image and choose an appropriate threshold value to convert the image into a black and white image. Obtain the black and white image grid.

11	17	16	15	17	21	17	16	16	4
15	12	10	59	61	60	61	60	56	56
26	11	12	10	60	57	60	61	60	62
48	42	19	10	15	15	13	60	60	61
60	56	12	11	11	18	4	12	16	57
60	62	63	26	12	13	11	7	10	12
61	59	62	59	19	13	9	6	14	10
60	60	59	54	61	58	13	16	12	12
63	62	58	53	60	63	62	56	13	8
28	17	15	10	19	27	61	51	10	6

Fig. E9.25 *A 10 × 10 gray scale image grid*

9.26 For the image in Fig. E9.24, determine (a) the area, (b) aspect ratio and (c) the centroid of the image.

SELECTED BIBLIOGRAPHY

1. N. Andreff, R. Horaud and B. Espiau, "Robot Hand-Eye Calibration Using Structure-from-Motion," *Int. J. of Robotics Research*, ***20***(3), 228–248, 2001.
2. T.O. Binford, "Survey of Model Based Image Analysis Systems," *Int. J. of Robotics Research*, ***1***(1), 18–64, 1982.
3. J.B. Burns, A.R. Hanson and E.M. Riscman, "Extracting Straight Lines," *IEEE Trans. Pattern Anal. and Mach. Intell.*, ***8***(4), 425–455, 1986.
4. J. Canny, " A Computational Approach to Edge Detection," *IEEE Trans. Pattern Anal. Mach. Intell.*, ***8***, 679–698, 1986.
5. Micheal C., Fairhurst, *Computer Vision for Robotic Systems*, Prentice Hall International (UK), 1988.
6. B.K.P. Horn, *Robot Vision*, The MIT Press, McGraw-Hill Book Co., New York, 1986.
7. F. Janabi-Sharifi and W.J. Wilson, "Automatic Selection of Image Features for Visual Servoing", *IEEE Trans. on Robotics and Automation*, ***13***(6), 890–903, 1997.

8. M.H. Lee, "Tactile Sensing: New Directions, New Challenges," *Int. J. of Robotic Research, **19**(7)*, 636–643, 2000.
9. P. Levi and L. Vajta, "Combined 2-D and 3-D Robot Vision System," *J of Robotic Systems*, ***11***(1), 187–194, 1994.
10. S.B. Morris, *Automated Manufacturing Systems: Actuators, Controls, Sensors and Robotics*, McGraw-Hill, New York, 1995.
11. J. Sklansky, "Image Segmentation and Feature Extraction," *IEEE Trans. Systems Man Cybernetics*, ***8***, 237–247, 1978.
12. C.C. Yang and M.M. Marefat, "Spatial quantization errors in active vision inspection," *Proc IEEE Int. Conf. Systems, Man and Cybernetics*, San Antonio, TX, October 2–5, 1994.

10

Robot Applications

Robotics has rapidly moved from theory to applications and from the research labs to industries over the last 20 or so years. In previous chapters, various aspects associated with design and analysis of robot manipulators, were discussed to develop a complete familiarization with the robotics technology. The mathematical models developed help in design of robots, calculating their motions, control them and plan or determine the trajectory and frame transformations required for performing specified tasks during the workcycle.

One of the key features of a robot is its versatility. As a designer, developer, planner or user of this technology one has to be familiar with the various issues involved in the applications of robots. As the technology is new, the prospective user or buyer of robot technology, who is accustomed to buy conventional items, will find the robot applications a complex subject.

This chapter is intended to introduce the industrial applications of robotics. Robotics is going to be a prominent component of manufacturing industry, which will affect human labour at all levels, from managers of production to shop floor unskilled workers. A programmable robot with a number of degrees of freedom and different configurations can perform specific and diverse tasks with the help of a variety of end-effectors. On the industrial scene it can be reprogrammed and adapted to changes in process or production line. Robots are also finding many applications outside of the industry, in research, hospitals, space, supermarkets, services sector, farmhouses, and even in homes as pets. The applications outside of a factory are much more complex, diverse and are based on human imagination.

When talking about robot applications in industry as well as other places, one needs to be concerned about the safety. After all, a robot, as it is today and going to be for decades to come, is a dumb machine, which is supposed to obey the commands.

Although robots are becoming common in the workplace, it is important to remember that robots are not "super workers". They have some real shortcomings and are to be understood as tools or machines people use.

Today's robots:

- Are not creative or innovative.
- Cannot react to unknown situations.
- Have no human feelings.
- Have no consciousness.
- Do not think independently.
- Cannot make complicated decisions.
- Do not learn from mistakes or otherwise.
- Do not adapt quickly to the changes in their environment.

The current-day applications of robots can be categorized into two broad areas: industrial applications and non-industrial applications. The industrial applications are discussed in depth and the non-industrial applications are listed at the end of chapter.

It is important to remember always that a robot is not conscious of what it is doing. It can only move its end-effector to well defined positions in well defined manner and perform the task there like grasping or ungrasping, drilling, paint spraying or scraping, welding, assembly, inspection or whatsoever.

10.1 INDUSTRIAL APPLICATIONS

Of the robots in the world today about 90% are found in industries. These robots are referred to as *industrial robots* and are regarded as "Steel Collar Workers". Of these more than 50% are deployed in automotive industries. Robots are useful in the industries in many ways. In today's economy, industry needs to be efficient to cope with the competition. Installing robots in the industry is often a step to be more competitive because robots can do certain tasks more efficiently than humans. Some of the tasks robots can do better are:

- Handling dangerous materials.
- Assembling products.
- Spray finishing.
- Polishing and cutting.
- Inspection.
- Repetitive, backbreaking and unrewarding tasks.
- Tasks involving danger to humans or dangerous tasks

Robots offer an excellent means of utilizing technology to make a given manufacturing operation more profitable and competitive. The main advantage offered for the industrial needs is the improved productivity and quality offered by the robots. However, the technology is relatively new on the industrial scene. Its use in the manufacturing processes is greatly limited for multiple reasons.

Robots applications in the industries today are primarily in four fields:

(i) Material handling,
(ii) Operations,
(iii) Assembly, and
(iv) Inspection.

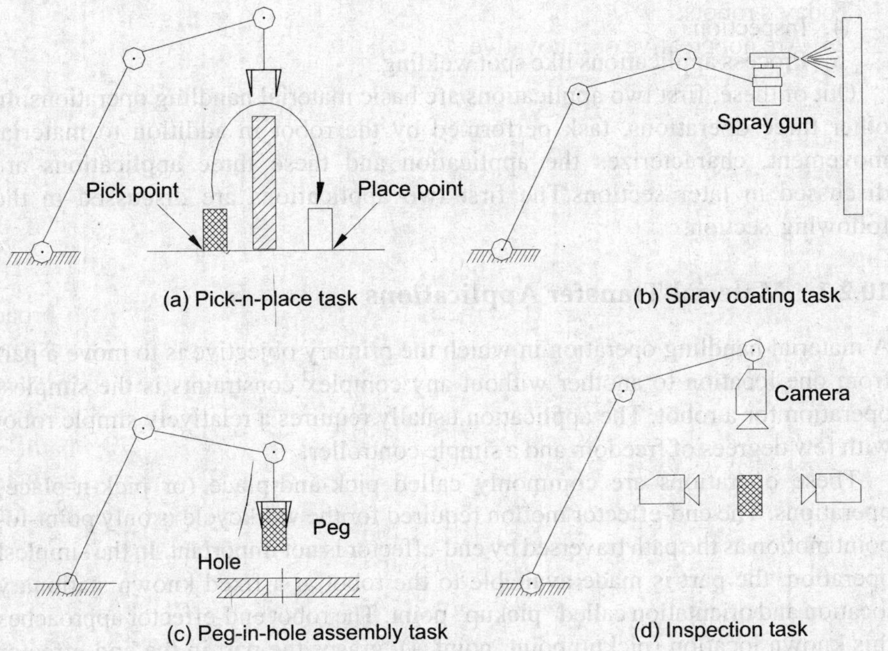

Fig. 10.1 *Four characteristic industrial applications of robots*

Typical examples of four applications are pictured in Fig. 10.1. The operations involving picking the material, workpiece or tool from one place and placing it at the desired place, are the material handling operations. These include material transfer and machine loading/unloading. Common material handling applications are in hazardous environment of foundry, die casting, plastic moulding, forging operations and handling dangerous materials. The end-effector for these applications is a suitable gripper to hold the material, that may be radioactive or red-hot material. During workcycle in operations category, a robot performs a process with the help of 'some' tools as its end-effector. Operations which current day robots perform are: arc welding, spot-welding, spray-painting etc. The work is usually brought and positioned for the robot to perform the task. Automotive industry requires all these applications and has been the leading user of the robots in this category.

10.2 MATERIAL HANDLING

The most basic robot application is one in which the robot is required to pick a part or other material from one location and place it at another location. Many tasks performed by a robot require this basic pick-and-place operation. Some examples are:

1. Material transfer applications.
2. Machine loading/unloading applications.
3. Assembly operation.

4. Inspection.
5. Process applications like spot welding.

Out of these, first two applications are basic material handling operations. In other three operations, task performed by the robot in addition to material movement, characterizes the application and these three applications are discussed in later sections.The first two applications are discussed in the following section.

10.2.1 Material Transfer Applications

A material handling operation in which the primary objective is to move a part from one location to another without any complex constraints is the simplest operation for a robot. The application usually requires a relatively simple robot with few degrees of freedom and a simple controller.

These operations are commonly called pick-and-place (or pick-n-place) operations. The end-effector motion required for the workcycle is only point-to-point motion as the path traversed by end-effector is not important. In the simplest operation, the part is made available to the robot at a fixed known stationary location and orientation called "pickup" point. The robot end-effector approaches this known location (pickup point, point *A*), grasps the part in the end-effector (usually a gripper), moves away from this point to safe distance (point *B*), moves close to the position where the part is to be placed (point *C*) and places the part at the desired location (delivery point, point *D*). This forms the typical workcycle (*A-B-C-D*) of the pick-n-place operation and is shown in Fig. 10.2.

In the pick-n-place operation it is essential that the part is made available to the robot at the stationary pickup point in the specific orientation and position by some mechanical feeding device or conveyor; the part placed at the delivery position (desired position) is moved away before the next part is delivered by the robot.

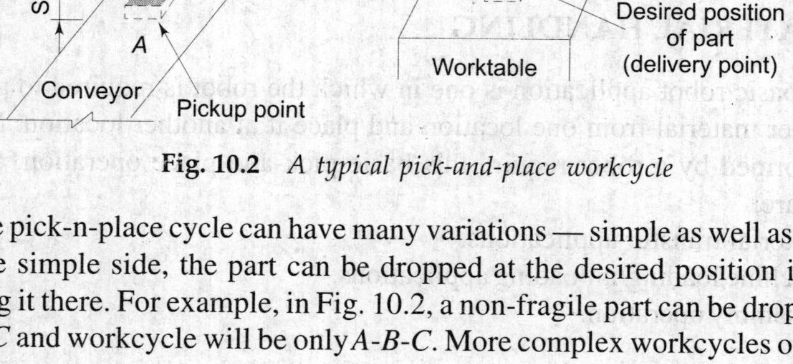

Fig. 10.2 *A typical pick-and-place workcycle*

The pick-n-place cycle can have many variations — simple as well as complex. On the simple side, the part can be dropped at the desired position instead of placing it there. For example, in Fig. 10.2, a non-fragile part can be dropped from point *C* and workcycle will be only *A-B-C*. More complex workcycles of material transfer operations include picking the part from a moving pickup point; changing motion pattern from cycle to cycle; changing delivery point location based on some attributes of the part and instead of just a two-point (pick-and-place) workcycle, or moving the part through multiple points in the workcycle to avoid obstacles. More complex pick-n-place applications are machine loading/unloading and palletizing. These are discussed next.

10.2.2 Machine Loading and Unloading Application

The loading of material or part and unloading it from a machine is the other material handing operation that is well suited for robotic applications. The robot's task is to pick the part from a specific location (the pickup point), and place it in desired position and orientation into the work holding device of the machine, which may be a chuck, vice etc. Once the part is loaded into the machine, some manufacturing operations are performed on the part by the machine and then the part is unloaded from the machine. For unloading, the robot picks the part from the work holding device of the machine and places it at the delivery point or loads it into another machine.

In the machine loading and unloading operation, the timing of the robot and the machine must be coordinated. The robot's cycle time must match the machine's cycle time. For this coordination, the robot controller must establish communication with the machine or monitor the machine operation with the help of suitable sensors and controllers.

The machine loading/unloading application is best characterized by a robot-centered workcell of the robot, the production machine, and some form of part delivery and removal systems. The workcell may have more than one production machine, to perform same or different operation on the part. A typical robot workcell with two machines is shown in Fig. 10.3. More complex workcells are possible with more machines and more robots to perform all required part handling of a multi-machine multi-operation manufacturing process.

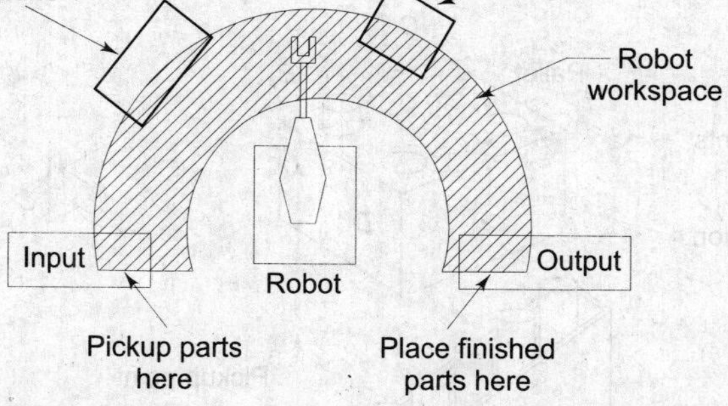

Fig. 10.3 *A robot centered workcell for machine loading/unloading application*

The machine loading and/or unloading is not restricted to loading/unloading of parts only. Robots can be used to load and/or unload tools, raw materials, etc. Robots are successfully used for machine loading and/or unloading application in a large variety of manufacturing operations. Some common manufacturing operations where robots are successfully used are:

- Metal working operations like forging, etc.
- Plastic moulding.
- Die casting.

These operations, as well as the environment of the workplace, are hazardous for the human operator. Hence, the use of robots for these and similar manufacturing operations is easily justified.

10.2.3 Palletizing Application

Many material handling operations require stacking of parts, that is, to palletize them. Like machine loading and unloading applications palletizing also involves two processes: Palletizing and depalletizing. Palletizing is the process of stacking or storing the material, parts or cartons on a pallet in a specified manner. Individual parts are picked from a fixed pickup point and moved for placing them on the pallet in palletizing operation and are picked up from the pallet, moved and placed at a fixed delivery point in depalletizing operation. The pickup point for palletizing could be the delivery point (of the finished product) of a manufacturing process while the delivery point of the depalletizing could be the pickup point for the raw material of an assembly line.

Palletizing operation is a pick-n-place operation with a major difference, that is, the placing point or delivery point is not fixed but changes with every part (every cycle) until the pallet is full. A typical pallet and palletizing operation is illustrated in Fig. 10.4. The part is picked up from point *A* (the pickup point), moved along the specified path (*B-C-D*) and is placed into the pallet in the 'next' location (the delivery point). The next location for the next part is easily computed from the part count, size of the pallet, size of part and location of corner of pallet. If the pallet contains bins or slots, one for each part (for example, cold drink cans crate is such a pallet), size of each bin and number of bins in the pallet specify the size of pallet.

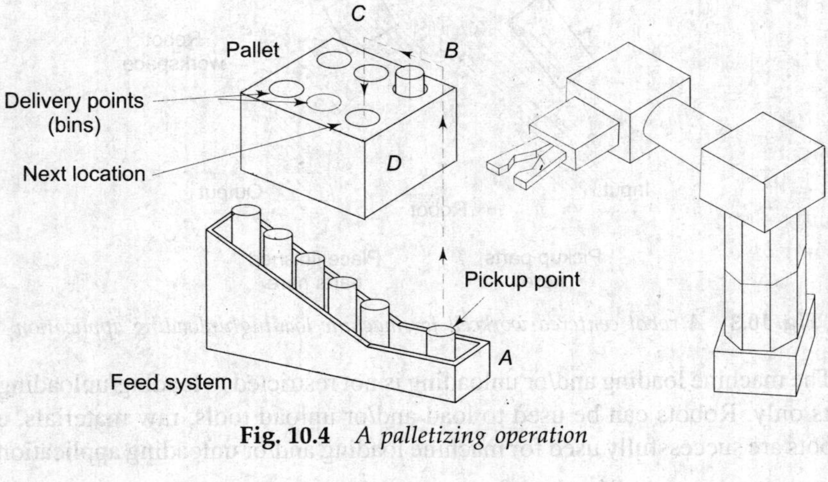

Fig. 10.4 *A palletizing operation*

Robots can be easily programmed to perform palletizing operation. The program computes the delivery point for each part while keeping the count of number of parts to detect when pallet is full.

The robot repeats the cycle of pick-n-place with a different placing point for each cycle. When the pallet is full it is removed manually or otherwise and an empty pallet is placed in its place. These tasks are coordinated with the robot controller with the help of suitable sensors and microswitches in the workcell.

In common palletizing operations the pallet is two-dimensional, that is, only one layer of parts is stacked, mostly in a horizontal plane. More complex palletizing operations have inclined or vertical pallets. A still more complex operation is one in which parts are stacked one over the other giving more than one layer of stacked parts. The computation of next delivery point location is in three dimensions in this case. Figure 10.5 is an example of stacking of cartons (or say breads or books) on a pallet.

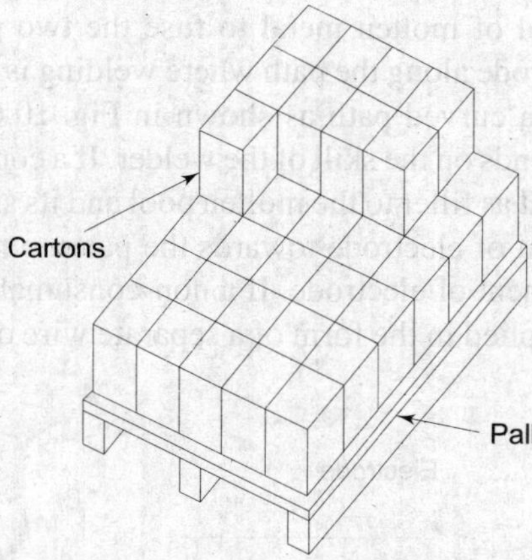

Fig. 10.5 *A multi-layer pallet example—stacking of cartons*

Depalletizing operation is reverse of palletizing operation. In depalletizing, the pickup point is different for each part while the delivery point is fixed.

10.3 PROCESSING APPLICATIONS

A large number of robot applications are more than just material handling applications. A class of these applications where the end-effector is a tool instead of a gripper are classified as processing applications. With the tool the robot manipulator performs some manufacturing process function. The type of the tool depends upon the operation to be performed by the robot. Sometimes, the tool is permanently attached to the end of the wrist while in other situations the tool may be grasped in the gripper. The later provides greater flexibility for the tool changing. Some of the processes where robots are efficiently utilized are:

- Welding.
- Spot-welding.

- Spray painting.
- Paint scraping.

Today, robots are used in the industries for many other processing applications like drilling and other machining operations, polishing, water jet cutting, etc. The arc welding application is discussed here in detail as an example.

10.3.1 Arc Welding

Welding application is most common process applications for industrial robots. Arc welding requires continuous long welding to make a welded joint between two parts. A human operator, known as the welder, using the arc welding equipment, generally performs the arc welding. The arc welding equipment consists of an electrode holder, an electrode, a low voltage high current electric supply and electric cables. The arc is initiated between the electrode and the metal pieces to be joined and this arc produces temperatures sufficient to melt the parts locally and form a pool of molten metal to fuse the two pieces together. The welder moves the electrode along the path where welding is to be done. It may be a straight-line path or a curved path as shown in Fig. 10.6. The accuracy and quality of the weld depends on the skill of the welder. If a consumable electrode is used it contributes metal as filler to the molten pool and its size reduces requiring (i) additional movement of electrode towards the parts to maintain the arc, and (ii) frequent replenishment of electrode. If a non-consumable electrode is used, filler metal may be supplied in the form of a separate wire or rod continuously.

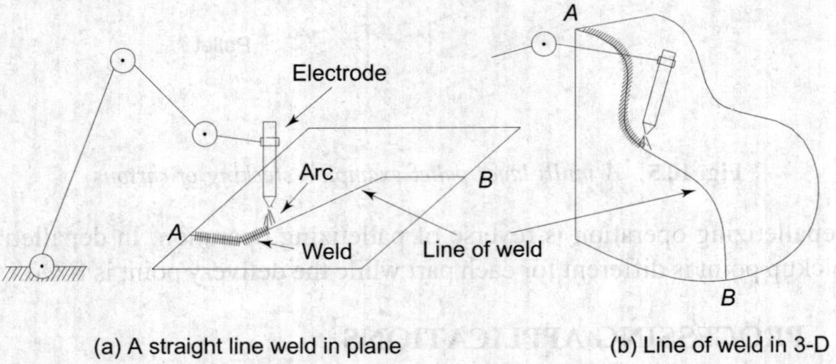

Fig 10.6 *Straight and curved weld trajectories for continuous arc welding*

The arc produced emits ultraviolet radiation that is injurious to human vision and can cause permanent vision damage. The arc welding has other hazards like high temperature, molten metal, flying sparks and toxic fumes. All these are a potential threat to the human operator. Because of all these factors and hazards it is logical to consider use of robots for arc welding.

10.3.2 Robot for Arc Welding Application

In utilizing a robot for continuous arc welding following issues are to be tackled:

- The arc welding process is usually in low quantity and changes often.

- The components to be welded have variations and may require the arc-welding path to change from part to part.
- The edges of parts to be welded are irregular.
- The path to be negotiated by the electrode is complex. For a robot negotiating a straight-line motion is as difficult as negotiating a curved path.
- In the case of consumable electrode arc welding, the robot is required to move the electrode towards the parts being welded to maintain the arc, as the electrode is consumed.
- The quality of weld depends on the speed of electrode movement. If speed is low, more metal will deposit and if it is more less metal will deposit and weld may be weak.

All these difficulties are compensated by the skill and judgment of welder.

10.3.3 Arc Welding Robot Requirements

The arc-welding robot has to overcome all the above difficulties though it cannot posses the skill and judgment of the welder. To perform satisfactory welding operations, the manipulator should be capable of moving its tool-point (the end of electrode) along a trajectory in three-dimensional space.

It must be capable of continuous path movement, as a point-to-point robot cannot do continuous arc welding. In addition, it should have a feed mechanism for consumable electrode or filler metal wire. The workcell controller should coordinate the robot motion; electrode or wire feed, the spark gap, the welding current and other activities in the workcell.

A human operator or another material handling robot may carry out the loading/unloading of parts. If another robot is used, the workcycle of both robots must be synchronized. If two workstations are used within the workspace of robots, the welding robot can perform the welding cycle while the material-handling robot performs the load/unload cycle.

The workspace of the welding robot should be large enough to accommodate size of the parts to be welded. A five degrees of freedom manipulator can weld parts in a plane while six degrees of freedom are required for welding complex contours. Robot programming for continuous arc welding requires a mechanism to feed the contour along which welding is to be performed as well as it may require interpolation algorithms for interpolating straight line or curved paths between two points.

The capabilities of arc welding robots increase manifold with the use of sensors that can track the welding path and the weld produced. These can help in overcoming most of the difficulties and compensate for irregularities and variations in the process parameters.

10.4 ASSEMBLY APPLICATIONS

Assembly is the final stage of manufacturing and it is manual labour intensive. It accounts for as much as 40 to 50 percent of human labour required to get the finished product. Given the high rate of occurrence of assembly operations in manufacturing industry, automated assembly systems have been traditionally applied to high-volume products requiring large investment in custom-designed equipment to perform the specific assembly. However, majority of products requiring assembly operations have low or medium production volumes and large investments in specialized assembly equipment cannot be economically justified. Robotics or other programmable and flexible automation systems are ideal for these applications.

Assembly means fitting two or more discrete parts together to form a new product or subassembly. The final complex assembled product is obtained by sequentially assembling components to the subassembly. The assembly operation involves considerable amount of handling, positioning and orienting of parts and applying controlled force to mate them together properly. Sometimes, the assembly may be of a temporary nature. It is made to perform a task. For example, assemblies of screwdriver with screw or spanner with nut, to tighten the later, are temporary assemblies.

The basic difference between the assembly operation and other applications discussed in the previous sections is that assembly requires a close interaction between the two parts being assembled and the assembled parts maintain their relationship with each other adding value to the product formed by assembly. Assembly operation can be mating of two parts, placing two parts together and fastening them with a third part like screw, nut-bolt, rivet, etc. or joining them (It may be noted that welding application discussed above assumes that the parts to be joined are correctly assembled by some means before welding). These operations require interaction between the parts being assembled and make the assembly operation more complex compared to other applications.

10.4.1 The Assembly Task

The elementary assembly operation involving assembly of mating a peg in a hole can be described as one of the most basic problem in robotics. For human beings putting a peg in a hole is an extremely simple and trivial task, which any normal human being can perform without any problem. However, surprisingly it is very difficult to have peg-in-hole assembly being carried out by a robot manipulator. The enormous difficulties that occur clearly show how complex apparently simple tasks as this really are. The reason is partly due to a physical limitation of the mechanical compliance of the wrist and arm of the manipulator and partly due to lack of mating strategy allowing the successful execution of the task irrespective of whatever is the initial position of peg and hole axes.

In the assembly task of peg-in-hole, if for some reasons the hole is not exactly in the location expected or the peg is grasped wrongly, a manipulator would be most of the time unable to perform the assembly successfully without any human help. The uncertainties embedded in the physical world make the task most difficult.

10.4.2 Peg-in-hole Assembly

The most common assembly task, as defined above, involves mating two parts or inserting one part (the peg) in to the other part (the hole). The peg-in-hole assembly can have different difficulty levels. Two types of peg-in-hole assembly tasks are shown in Fig. 10.7.

The assembly of a round peg in a round hole is illustrated in Fig. 10.7(a). For this operation, the orientation of peg about its own axis is not required for correct assembly; hence, a robot manipulator with a minimum of 5 degrees of freedom can perform the task. If the assembly involves a part that is not symmetric about its own axis, for example, assembly of a square peg in a square hole as shown in Fig. 10.7(b), the robot manipulator requires one additional degree of freedom.

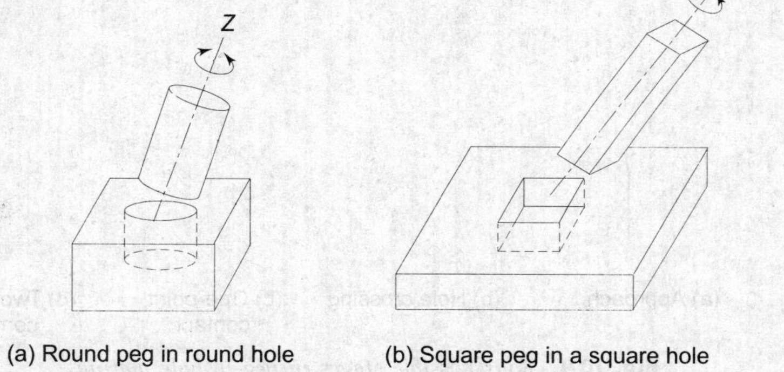

Fig. 10.7 *Two types of peg-in-hole assembly operations*

More complex peg-in-hole assembly involves assembly of multiple pegs into multiple holes simultaneously. An example of multiple peg-in-hole assembly is illustrated in Fig. 10.8. Assembling of electronic parts such as resistors, transistors or integrated circuits on the printed circuit board is a common situation in electronics industry. In multiple peg-in-hole assembly, all the pegs must be aligned with the corresponding holes for the assembly to take place.

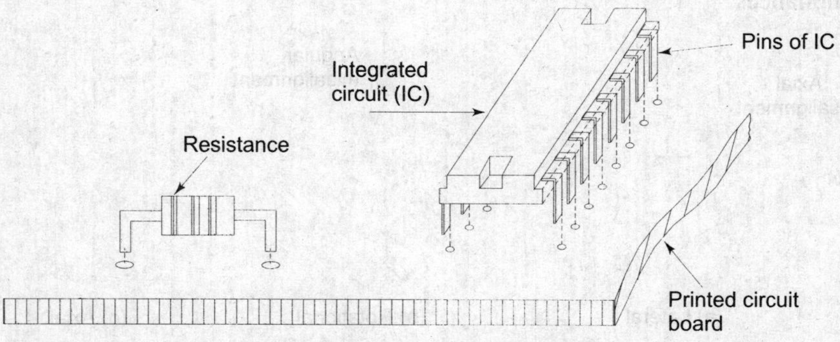

Fig. 10.8 *Examples of assembly involving multiple peg-in-hole operation*

10.4.3 Steps in Assembly

The assembly operation can be viewed as a four-step process. The four steps are:
(i) Approach,
(ii) Hole crossing,
(iii) One-point contact, and
(iv) Two-point contact.

These four stages are shown in Fig. 10.9. In the last stage, the peg may get stuck due to application of wrong forces and moments to the peg. In order to limit the occurrence to such situations and to increase peg-in-hole mating success rate, special hardware devices or sophisticated sensors, controllers and software exploiting mechanical *compliance* are utilized in the otherwise rigid robot manipulator.

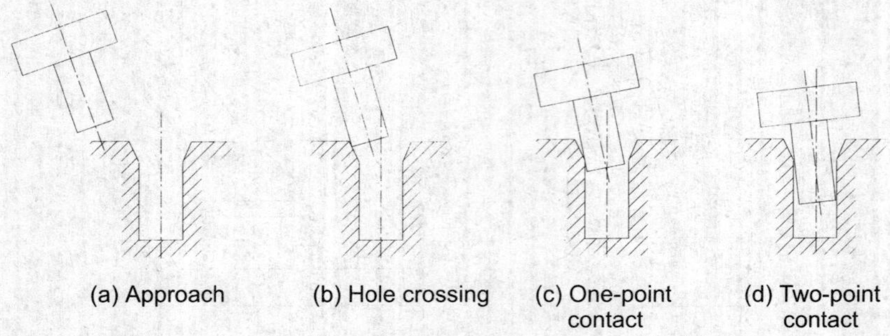

Fig. 10.9 *Four possible stages in peg-in-hole mating*

10.4.4 The Compliance

The compliances can be defined as allowed or initiated movement of the peg for the purpose of alignment with the hole. Three possible misalignments or position errors between the peg and hole are lateral, rotational and axial. These are illustrated in Fig. 10.10. To correct these misalignments, the peg must have three types of compliance: lateral compliance, rotational compliance and axial compliances.

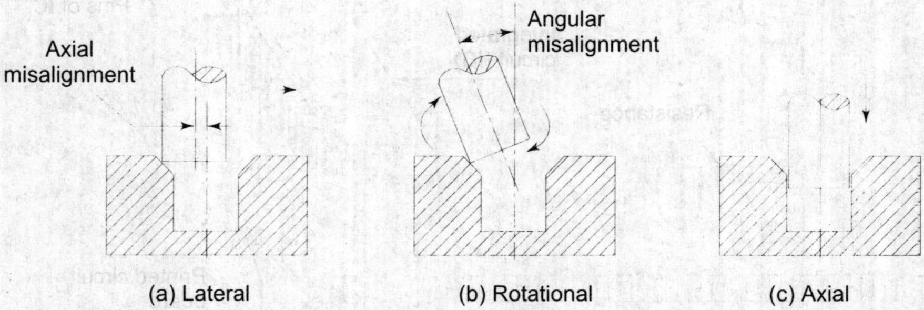

Fig. 10.10 *Three types of compliance required to ease out assembly*

Lateral Compliance The misalignment between the axis of peg and the axis of hole, assuming the axes to be parallel, can be corrected by allowing movement of peg in lateral direction. This is known as lateral compliance and is shown in Fig. 10.10(a).

Rotational Compliance If the axes of peg and hole are not parallel, Fig. 10.10(b), assembly of the two will not take place. Small angular misalignments are corrected by rotational compliance.

Axial Compliance Once the peg is above the hole, the axial compliance, Fig. 10.10(c), eases the assembly. The axial motion (compliance) in conjunction with lateral and rotational compliance pushes the peg into hole slightly and the assembly is initiated.

In addition to peg-in-hole assembly operation, a robot for performing several other tasks requires compliance. Some examples are:

(a) Turning doorknob to open the door.
(b) Playing a musical instrument.
(c) Scraping paint from a glass pane.
(d) Performing surgery.

10.4.5 Providing Compliance

The compliance is provided in a robot to correct for the lateral and angular errors during assembly so that the peg does not get stuck in the hole and assembly is successful. There are two ways compliance can be provided to the peg, see Fig. 10.11.

(i) Active compliance.
(ii) Passive compliance.

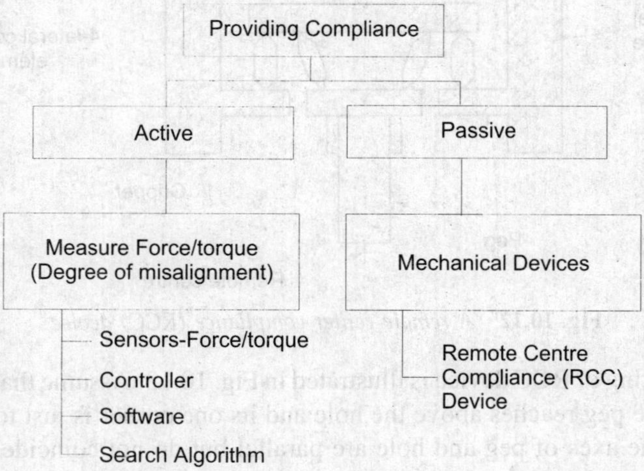

Fig. 10.11 *Two methods of providing compliance to the peg*

Active Compliance

The active compliance is based on exploiting sensory data to get successful assembly. It is provided by active force/torque sensing at the wrist of the manipulator (see section 9.4.4) to measure the interaction forces between the peg and hole. The degree of misalignment is computed from the information gathered about mating from the force/torque sensor. The robot controller is programmed to maneuver the joints of the robot manipulator according to a specified algorithm to minimize the misalignments and to make successful assembly. Alternative sensing by optical sensors and vision systems can also be used to provide active compliance. Using these sensors in addition to force/torque sensor increases the success rate of assembly.

Passive Compliance

The second method of providing compliance, to correct the misalignment errors during the assembly, is to use passive mechanical device known as *Remote Center Compliance* (RCC) device. The RCC device can provide all three types of compliances — lateral, rotational and axial and is attached between the wrist-end and the gripper. A typical RCC device consists for four lateral and four rotational compliance elements (springs) as pictured in Fig. 10.12. All the eight elements together provide the axial compliance.

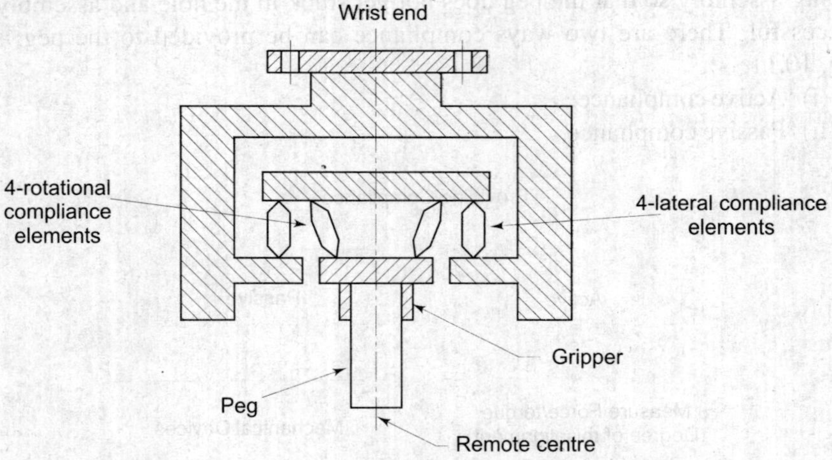

Fig. 10.12 *A remote center compliance (RCC) device*

The working of RCC device is illustrated in Fig. 10.13. Assume that during the assembly the peg reaches above the hole and its one corner is just touching the chamfer. The axes of peg and hole are parallel but do not coincide. When the robot tries to push in the peg, the forces at the contact point between peg and hole generate a lateral force. Under the lateral force the lateral compliance elements deform and allow the peg to move laterally without the movement of the manipulator. This stage is shown in Fig. 10.13(a). Consequently, the peg moves horizontally and slides into the hole.

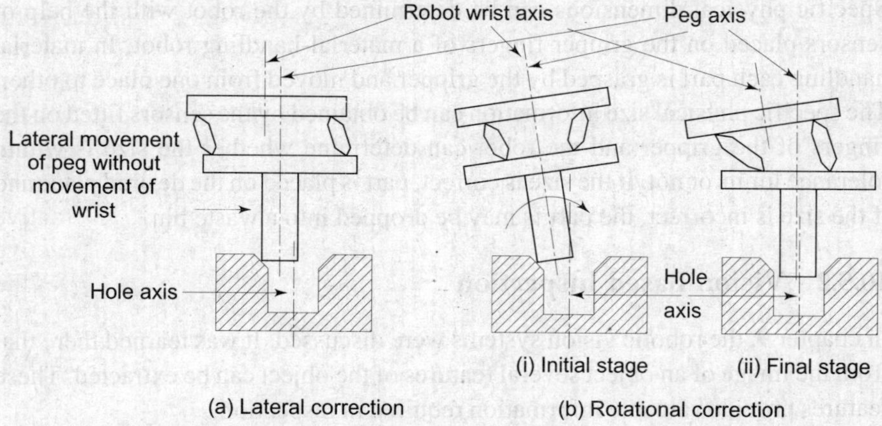

Fig. 10.13 *Use of RCC device to correct lateral and rotational errors*

When there is angular misalignment between the axes of peg and hole, their axes are not parallel. Assume that the manipulator is pushing the peg at an angle as illustrated in Fig. 10.13(b)(i). As the peg comes in contact with the corner of the hole, an angular moment is generated and angular compliance elements of RCC device deform. The deformation of angular elements results in making axis of peg parallel to axis of hole, facilitating the assembly while the axis of the manipulator is still misaligned. This stage is shown in Fig. 10.13(b)(ii). The vertical components of forces generated by the contact of peg with hole, deform (compress) all the eight compliance elements providing axial compliance. This axial compliance pushes the peg into the hole once the lateral and angular position errors are corrected, accomplishing the start of assembly.

10.5 INSPECTION APPLICATION

Applications of robots in inspection is a growing area. Quality control in industrial manufacturing operation is essential in the competitive world. A manual inspection of the finished product as well as the in-process parts is a tedious and monotonous task for a human operator. In large quantity production, 100 percent inspection is impossible and sampling and procedures of statistical quality control are resorted to. Robotics technology is expected to play a significant role in making 100 percent inspection possible.

Inspection function is required in every stage of manufacturing from raw materials to finished products. Robots can be used to inspect physical dimensions, surface finish, and other characteristics of the raw materials, intermediate stages of parts, finished parts, sub-assemblies or finished products. To perform the inspection tasks robot requires various sensors or vision systems.

10.5.1 Sensor Based Inspection

Specific physical dimensions can be determined by the robot with the help of sensors placed on the gripper fingers of a material-handling robot. In material handling each part is grasped by the gripper and moved from one place to other. The specific physical size information can be obtained by the sensors fitted on the fingers of the gripper and the robot can determine whether the size is within tolerance limits or not. If the size is correct, part is placed on the desired place and if the size is incorrect, the part is may be dropped into a waste bin.

10.5.2 Vision Based Inspection

In chapter 9, the robotic vision systems were discussed. It was learned there that from the image of an object several features of the object can be extracted. These features provide valuable information required in inspection.

A typical robotic vision system is capable of analyzing a two-dimensional scene. The two-dimensional view of the scene is obtained from a single camera. If more features of the object are required to be inspected, the camera is maneuvered into different positions by the robot manipulator, see Fig. 10.1(d). Alternately, the robot manipulator can be used to present the part to a stationary vision system (camera), in different orientations. Thus, robot's task is a sort of material handling task.

In the design of a robotic vision inspection system, several factors must be considered. These factors are:

- Proper lighting of the workspace.
- Camera with required resolution to extract dimensions and other features to the desired accuracy.
- The field of view should be large enough to accommodate the part.
- Robot should have sufficient degrees of freedom to manipulate the camera or the part, as the case be.

10.5.3 Testing

Testing is another associated quality control activity. Testing is done to determine attributes not determined by inspection, like functional performance, correctness of operations, strength, defects in material, load, fatigue and environmental effects, and other similar attributes that affect the performance of the product in the hands of the users. Robots can be used for performing various types of tests, which may be destructive or non-destructive. Usually special equipment and sensors are attached to robots for performing tests.

It is visualized that the future manufacturing will be characterized by a very high level of integration between robots, manufacturing processes, inspection, testing and the human worker, if at all present on the scene.

10.6 PRINCIPLES FOR ROBOT APPLICATION AND APPLICATION PLANNING

In the preceding part of this chapter various industrial applications of robots were described and various technical issues for different type of applications of robots were discussed. To apply robotics for industrial applications some common principle are presented here. There are four basic principles for robot applications in industries:

1. If the conditions at the workplace repeat without significant unstructured variability, then a simple robotics solution is possible.
2. If the workcycle changes often but within limits, robotics solutions are possible due to reprogrammability of the robots.
3. For all changes that are imposed due to a robot installation, the robot behavior and the workplace must be capable of change. It is important not to try to use robots to imitate human functions.
4. Recognize and exploit or create robot-compatible mechanical states or environments and eliminate or reduce the need for human skill and judgment.

When planning for a robot application to perform a task in industry several stages have to be taken into consideration.

1. Workplace analysis and evaluation: a thorough study of what is the task, including subtasks, process plans etc., their analysis and evaluation must be done to justify robotics application.
2. Recognition of requirements of alternative methods for automation: the layout of the workplace and the interlinking of robot and other machines in the workplace are major issues.
3. Selection of optimal method: at this stage simulation and graphical emulation are commonly used tools.
4. Search for solutions to implement the selected methods: the selection of the right robot (DOF, workspace, etc.) and the selection of peripheral devices such as gripper and sensors.
5. Economic analysis: is the final step in robot application planning and is discussed in the next section.

10.7 JUSTIFICATION OF ROBOTS

With the basic knowledge of industrial applications of robots in mind, it is important that each robot application be evaluated, both technically and economically. This section provides a checklist to aid in the technical evaluation along with a strategy for economic justification of the application of a robot. In addition, since the decision to use or not to use a robot also requires non-monetary factors, a "qualitative" justification section is included.

10.7.1 Technical Justification for Potential Applications

Though robots are capable of performing many tasks, the complexity involved with using a robot for a specific application may require considerable time and effort in the setup stages to justify deployment of the robot. Each application should be evaluated on a case-by-case basis, based on the complexity of the application, the run length and the costs involved.

To evaluate the use of a robot for a particular application a checklist is given in Table 10.1. The checklist is based on factors such as the requirements for the application, the physical environment of the application and the operational strategies that influence the application.

To use the checklist in Table 10.1, read the propositions, and then grade the answer to each question in terms of a five-point scale with 1 for 'strongly disagree' to 5 for 'strongly agree'. A score of 3 means 'neither agree nor disagree'. Based on the grade points awarded to the questions in the checklist compute an overall score. This overall score can be an indicator for making the decision.

Table 10.1 *Checklist for evaluating robot applications*

	Applications Requirements
A.1	Monotonous/repetitive operation
A.2	Medium complex operation (based on required positioning and orientation moves which must be done to complete operation)
A.3	No complex judgment or decision making required
A.4	Cycle time is greater than 3 seconds
A.5	Need to position and orient part or tool
A.6	Shape, size or features change from piece to piece
A.7	Require high positioning repeatability/accuracy
A.8	Large payloads
A.9	Need to move to multiple positions or through points which are path dependent
	Physical Environment
B.1	Workstation is well organized (orderly and repeatable)
B.2	Production volume >20,000 and <1,000,000 pieces/year
B.3	Hazardous environment
B.4	Motion of robot need not be coordinated with other machines
B.5	Machines in workstation can receive parts automatically
B.6	Machines in workstation need not communicate with robot

An overall score greater than 50 out of a maximum of 75 clearly favours the application of robot for the task. It is also possible to give weights to various propositions and a weighted overall score can be computed. Also some propositions may be very vital for the application and a score of 1 or 2 in them may be a deciding factor for the Yes/No answer. For example, if the answer to preposition "No complex judgment or decision making required" is 'strongly disagree' (a grade point score 1), the decision may be not to go for use of robot for the application. Similarly, if the user finds that many of the scores associated with the answers are 3 or less the use of robots for the application should be reconsidered. A cost analysis should precede the decision in favour of robot use.

10.7.2 Quantitative Justification

Cost is a single parameter commonly used for quantitative justification because it is possible to convert most of the quantifiable parameters into costs. The savings that accrue as a result of application of robots in manufacturing processes are the negative costs. When trying to cost justify the deployment of a robot, it is important to calculate the costs and savings over the lifetime of the robot. Typically, it is not possible to justify a robot to replace a human worker based purely on cost comparisons for one operation. Instead, cost justifying the robot requires evaluating the return on investment of the robot over the life of the robot and how quickly the robot would payback the investment. Note that much of the "savings" is in non-traditional areas such as quality improvement. As a guide, Table 10.2 contains many of the costs and savings involved with the use of a robot.

Table 10.2 *Significant costs and savings involved in deployment of robot*

Costs Involved
Robot and its peripherals
Tooling
Installation
Layout changes
Equipment modifications
Maintenance
Safety equipment
Training
Applications engineering
Potential Savings Areas
Labor
Increase in production
Reduced scrap, raw material
Manual tooling
Supervision
Worker training costs
Energy Savings
Supplies & Materials needed by Operator
Increase in Production Capacity
In-process and finished inventories

The cost justification process must take into consideration additional costs such as:

- Cost of maintaining the robot system (including the support equipment),
- Cost of redesigning the product so that it can be robotically worked on,

- Cost of redesigning the manufacturing system or workstation, and
- Cost of adding other support equipment that would be used by the robot to perform the application.

Standard methods of economic analysis like break-even analysis, present worth, return on investment etc. can be used for detailed economic analysis and arriving at an answer.

10.7.3 Qualitative Justification

To justify a robot for a manufacturing application many factors come into play. Several of these are not traditionally economically based, but instead apply to issues such as quality and safety. The justification for applications of robots should also consider these qualitative factors to arrive at a conclusion. Some of the qualitative factors that must be evaluated when deciding whether or not to use a robot for a particular application are the following:

Quality

Consistency of quality is one of the main reasons for deployment of robots in manufacturing process. It is difficult but possible to quantify quality and cost of quality. For example, in applications where the robot is applying adhesives, paint, or lubricants etc. robot can consistently apply the same amount of material on each part. The increase in quality can be measured in terms of lower scrap rates, less raw material use and more consistent parts.

Productivity

A robot may not be better than a human operator in performing the task once but the robot can maintain a constant speed and quality when performing the task repetitively. A robot never gets sick or needs to rest. It can work tirelessly 24 hours a day – 7 days a week or 365 days a year. It is called a 24/7 or 24/365 worker. This results in productivity increase that may be directly measurable through increased production quantities. Use of robots also reduces dependency on human labour and gives better delivery schedules, improved customer relations and goodwill. These benefits are difficult to quantify.

Hazards

Using a robot for certain application eliminates the need for a human operator to interact with dangerous material, hazardous environment or hazardous conditions such as radioactive materials, hot workpiece, heavy loads, excessive noise, or toxic gases. The cost of hazard for a human operator is difficult to quantify.

Flexibility

Deployment of robots increases the flexibility of the product mix, production schedule and production volume that can be processed in a workcell. The increased flexibility is a definite advantage but is difficult to quantify.

All the above factors are difficult to quantify into costs or savings, nevertheless these must be considered in the overall decision-making in favour or against the application of robots.

10.8 ROBOT SAFETY

A robotic system is an integration of robots, machines, computerized information channels and human beings, where no element can be considered perfect or immune from eventual failure and malfunction. Since the humans work rather close to the robots this increases the risk of mutual damage. All these factors necessitate the formulation of safety guidelines that indicate how the conditions of conflict can be minimized. Only when all the system components are functioning safely and reliably, the high productivity levels associated with robotic systems can be realized.

The humans can be divided into five groups that are at risk of direct injury from a robot:

1. *Workers:* A robot is deployed in the industry for specific application. There will be human workers in close vicinity of the robot for doing tasks invariably overshooting into the workspace of the robot. These workers are at the greatest risk of getting injured.
2. *Programmers:* A robot programmer using online programming method is in direct contact with the robot. This closeness with the robot's work envelope has its inherent danger of injury.
3. *Maintenance engineers:* A maintenance engineer is also in direct contact and is at risk from the same dangers as programmers. Additionally, because maintenance procedures often require that safety interlocks be disconnected, the inherent risk is greater for the maintenance engineer.
4. *Casual observers:* The casual observer is inquisitive. If the robot is not rigidly guarded, then a casual observer may move towards a robot that looks stationary and be injured when it continues or resumes its operation.
5. *Others outside the assumed danger zone:* Components manipulated by the robot can slip or fly out of the grippers and strike persons well outside the assumed danger zone of the robot if the surrounding area is not properly secured.

In a practical sense, safety procedures and devices allow the authorized entry of humans into a robot's working envelope with minimum risk of injury. Hardware devices and sensors would monitor all anticipated reasonable access to a robot's work envelope. Safety devices make use of physical safeguards available in a variety of forms. Some of them are:

- simple contact microswitches,
- restrained keys to the access doors,

- pressure mats,
- infrared light beams or light barriers, and
- vision systems.

The flashing red lights often seen within a work zone indicate that an apparently stationary robot is activated, but awaiting an input, or performing a time-delayed operation.

10.9 NON-INDUSTRIAL APPLICATIONS

The advances in robotics technology are not directed only for its use in industries. There has been almost a parallel growth of robotic technology in non-industrial environments. Robots are finding their way into research laboratories, energy plants, agriculture, hospitals, space, homes, textiles, services, education etc. It may be sometimes difficult to classify a particular application into whether it is industrial or non-industrial.

The applications of robots are only limited by the need and imagination of the developer and the end user. Robots have almost invaded in every walk of life. The technology is advancing rapidly and today if you think a robot cannot do a particular task or you read about it in fiction, it will become a reality tomorrow. Different applications of robots in some diverse segments are listed below. This list is in no way exhaustive.

Home Sector

Science fiction stories always considered robots as domestic slaves. This is going to be a reality in near future, though many of the current domestic applications are not much more than expensive toys. Some of the domestic applications possible are:

- Sweeping and cleaning.
- Cooking.
- Entertainment.
- Replacing pets.
- Garden maintenance.
- Security.

Health Care

The use of robots in human health care has a wide scope. Robotic technology is in use and is going to expand in health care for:

- Patient care and monitoring.
- Surgery.
- Rehabilitation—prosthetic limbs and robotic wheel chairs.

Microrobots can be injected into the human body to perform microsurgery.

Service Sector

Service robots can carry out diverse functions like:

- Traffic control.
- Fire fighting.
- Drive a vehicle.
- Sweep office rug, classrooms, streets etc.
- Manage shopping malls.
- Serve food in restaurant.
- Maintenance and repair.
- Disaster recovery.

Agriculture and Farms

Robot's use in agriculture and farming sector is very much limited today but is going to increase. They can be used to:

- Plough fields, sow seeds and transplant sapling etc.
- Pluck, sort and pack fruits.
- Breed livestock.
- Animal shearing.

Research and Exploration

With the help of robots, research in many inaccessible areas have been possible because of their greater capabilities to face hazardous environments and remote handling capabilities. These include:

- Space exploration.
- Under-sea exploration.
- Nuclear research.
- Geological exploration.

The list of tasks given above for different sectors can be expanded further to include following. A robot should

- Play football, cricket, etc.
- Knit a sweater.
- Change a tire, repair a puncture.
- Dispense gasoline.
- Dance.
- Play musical instruments.
- Polish diamonds.
- Make paintings.
- Talk and listen.

Robots have many potential applications, other than material handling, assembly, etc. in

- Pharmaceuticals industry.
- Textile industry.
- Chemical industry.
- Mining industry
- Construction industry.
- Energy sector.

Robots have potential to change our economy, our health, our living style, and the world we live in. New uses of robots bring hopes and possibilities, but also have associated dangers and risks. In section 1.2, three laws of robotics, postulated by Isaac Asimov in 1942 as a 3-point 'code of conduct' for the robots, were given. Looking at the developments of the past 60 years one needs to think – do these laws still apply, or do they need a change?

SOLVED EXAMPLES

Example 10.1 *Machine loading-unloading application*

The workcell of a processing application is illustrated in Fig. 10.14. The robot is deployed to pick the raw material from conveyor, load it into a machining center, unload the finished part after machining is complete and place it into a pallet. Define the sequence of steps required to carry out this application.

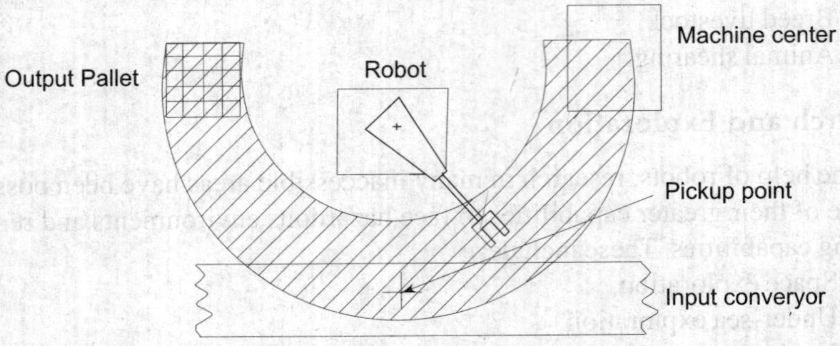

Fig. 10.14 *Workcell for machine loading-unloading application*

Solution The sequence of the activities (steps) performed by the robot to carry out the machine loading-unloading operation are described below.

1. Raw material is picked from pickup point on the input conveyor. It is assumed that the raw material is in proper position and orientation on the conveyor.
2. The robot end-effector approaches the raw material, grasps it and moves away from the conveyor. The raw material is moved to loading fixture and is placed in the fixture. The robot moves its arm away from machining center.
3. While machining is done on the part, the robot is idle.
4. On completion of the machining cycle at the machining center, the robot moves in and unloads the finished part from the machining center.
5. The finished part is placed in the pallet. This involves the palletizing operation.
6. The robot moves the end-effector back to pickup point.
7. During steps 4, 5 and 6 the machining center is idle.

All the operations of the machine loading-unloading are performed in sequential manner. A workcell controller has to perform the synchronization of operation of the conveyor, pallet, robot and machining center to carry out the tasks in the correct sequence during the workcycle. In the sequential operation, either the robot or the machining center is idle. One possible method to reduce machining center idle time is to use two robots one for loading and other for unloading. This may involve redesign or reorganization of the workcell.

Example 10.2 *Justification of Robots in Painting*

Rapid growth and the need for a precise and consistent painting method has prompted M/s ABC Ltd. to consider using robots in its paint operations. Shortcomings of the company's traditional method included a high rate of rejects (20% or more), inconsistent coating, paint wastage and high labour costs. In an interview conducted by a system analyst, the Director of the company told:

- "The decision to use robots is not an easy one."
- "It is nearly impossible to achieve consistency in applying paint manually on their high-volume line."
- "The paint presents a significant health hazard to employees – painters have to be completely shrouded by the environment, which is hot and uncomfortable for the manual painters. They also have to wear oxygen masks inside the paint booth, which are cumbersome."
- "We need to remove our employees from this painting environment."
- "It was time to find a better method."

Evaluate the feasibility of using robots by ABC Ltd. for its painting operations.

Solution Based on the information available, the team of system analyst, management, shop floor engineers and painting experts evaluated the propositions in Table 10.1 and awarded the points for each. The points awarded on a five-point scale are tabulated in Table 10.3.

Table 10.3 *Scoring for robot applications for painting operations at M/s ABC Ltd.*

Proposition	*Points awarded*
A.1 Monotonous/repetitive operation	5
A.2 Medium complex operation (based on required positioning and orientation moves which must be done to complete operation)	4
A.3 No complex judgment or decision making required	4
A.4 Cycle time is greater than 3 seconds	5
A.5 Need to position and orient part or tool	4
A.6 Shape, size or features change from piece to piece	4
A.7 Require high positioning repeatability/accuracy	4
A.8 Large payloads	3
A.9 Need to move to multiple positions or through points which are path dependent	3
B.1 Workstation is well organized (orderly and repeatable)	4
B.2 Production volume >20,000 and <1,000,000 pieces/year	4
B.3 Hazardous environment	5
B.4 Motion of robot need not be coordinated with other machines	4
B.5 Machines in workstation can receive parts automatically	3
B.6 Machines in workstation need not communicate with robot	3

From the scores in Table 10.3; the overall points for the application are computed as 59. Since this is greater than 50, the use of robot is justified for M/s ABC Ltd.

The company went in for the deployment of robots for the painting operations. After an economic analysis and justification, the turnkey solution implemented had following features:

- Two six-DOF robots, with a large work envelope and 5 kg pay load capacity, capable of operating in the harshest painting environment with a slender arm design, high repeatability and maximized reach.
- A programmable logic controller with software to control part style, colour and volume; synchronize the operations of movement of conveyors, parts and robots; control paint flow and spray guns;
- Paint mix metering system which regulates paint flow;
- High-volume, low-pressure spray guns;
- Motorized conveyor systems that shuttle parts in and out of the robotic painting area.

In operation, parts are shuttled into the enclosed paint booth via one of the two conveyors. One of the two robots is dedicated to applying a pre-coating to the part, which helps ensure a high-quality, consistent finish. The part then moves to the second robot, which applies the topcoat to the part. Coated parts are then conveyed to an oven for curing.

After deploying the above solution for the painting job at M/s ABC Ltd. following benefits were reported:

- The benefits of robotic painting became evident almost instantly.
- With manual spraying, there was about a 20 percent part reject rate. With robots, it became less than one percent.
- Through robotic painting, company is also able to effectively apply six mils of wet paint in one application, which speeded throughput and significantly reduced material costs.
- The system consistently processed 5,000 parts per day, robots being 24/7 worker, without requiring operator assistance.
- The company has been able to move its manual painters into safer, more productive jobs such as systems monitoring and maintenance.
- Use of robots for painting jobs allowed the company to meet the increasing demands for high-volume production without compromising quality.

EXERCISES

10.1 Which type of manipulator is best suited for (a) Machine loading and unloading application? (b) Assembly application?

10.2 What type of manipulator is generally used in welding applications?

10.3 What characteristics an arc-welding robotic system must have?

10.4 What are essential characteristics of a spot welding manipulator?

10.5 What is the minimum number of degrees of freedom for assembling

(a) Round peg in round hole
(b) Square peg in a square hole.
Justify your answer.

10.6 Why a robot is considered as a 24/7 worker?

10.7 Why robots are useful in industries?

10.8 As technology advances, will the human worker loose jobs to robotic workers? Comment.

10.9 By deploying robots in the industries there are multiple gains. Are we paying too great a human price for these short term gains?

10.10 What do you think of future of robots?

10.11 Write a short note describing the factory of future. What will be the role of humans in this factory?

10.12 Make a list of robot applications, other than those mentioned in this chapter, and classify them.

10.13 For the robot applications identified in Exercise 10.12, which ones cannot be performed by a robot if it has no compliance.

10.14 Consider the following case and answer the questions below it.

A local chemical factory has 550 employees. Because of strong competition and economical reasons, the owner decides to fire 90 workers and install 10 robots.

(a) Write your views on this proposal of the owner.
(b) Is it fair to fire the workers?
(c) Is owner obliged to retain the workers and use them for other tasks?
(d) If the chemicals in the factory are injurious to the workers and workers often have health problems, what will be your answer to (a) and (b) above?

10.15 Pick a small-scale manufacture in your city. Study his manufacturing operations and carry out justification analysis for using robots in this industry.

SELECTED BIBLIOGRAPHY

1. Suji Asami, "Robots in Japan — Present and Future", *IEEE Robotics and Auotmation Magazine*, ***1***(2), 22–26, June 1994.
2. Amitabh Bhattacharya, *Robotics and their Applications in India: A State of the Art Report*, Department of Science and Technology, New Delhi, 1987.
3. G. M. Bone and M. A. Elbestawi, "Dual-Sensor Based Robotic Deburring", *ASME Journal of Manufacturing Science and Engineering*, ***118***(3), 439–441, 1996.
4. R.A. Brooks, "New Approaches to Robotics," *Science*, ***1227***, September 13, 1991.
5. Jean-Paul Boillot, "Benefits of vision in Robotics Arc Welding," http://www.robotics.org/public/articles/.

6. R. C. Dorf, *Concise International Encyclopedia of Robotics: Applications and Automation*, Wiley, 1990.
7. Steven W. Holland, "The Next Big Thing in Automotive Robotics, "*Robotics World Magazine*, ***19***(1), Jan–Feb 2001.
8. K.B. Lim, "Introduction to the special issue of Design and Applications in Robots," *Robotica*, ***17***(1), 1–2, 1999.
9. G.C. Pettinaro, "Behaviour-based peg-in-hole," *Robotica*, ***17***(2),189–201, 1999.
10. R.D. Schraft, "Mechatronics and Robotics for Service Applications," *IEEE Robotics and Automation Magazine*, ***1***(4), 31–35, 1994.
11. Rob Spencer, "Ten years of Evolution in Robotics Industry," *Robotics World Magazine*, ***19***(5), June 2001.
12. Manuela M. Veloso, "Entertainment Robotics," *Communications of* ACM, ***45***(3), 59–63, March 2002.

Appendix A

Review of Linear Algebra and Trigonometry

The study of robot manipulation comprises of establishing the spatial relationships between the manipulator, manipulated object, and the workspace. The spatial geometry is required for developing the kinematics and dynamics of the manipulator. Vectors and matrices as well as vector and matrix operations are extensively required in spatial geometry for representing location and orientation of points in a 3-D space with respect to a reference frame.

In this Appendix a quick review and brush up of the required mathematics is presented.

A.1 NOTATIONS

In this text, usually upper case, bold face alphabets are used to denote vectors such as $\vec{\boldsymbol{P}}, \vec{\boldsymbol{Q}}$ and so on, and matrices such as $\boldsymbol{A}$, $\boldsymbol{B}$ and so on; lower case, bold face alphabets for unit vectors (e.g. $\hat{\boldsymbol{i}}$ $\hat{\boldsymbol{j}}$ $\hat{\boldsymbol{k}}$) and lower case alphabets for scalars. Coordinate frames are denoted as $\{1\}$ or $\{x\ y\ z\}$ for coordinate frame with X, Y, Z as principal axes. The association of frames to vectors and matrices is denoted by superscript, subscript, or both, for example ${}^1\boldsymbol{P}$ or ${}^2\boldsymbol{T}_3$.

A.2 VECTORS

A point in space is described with respect to a reference inertial coordinate frame as a vector from the origin of the frame to the point P. The components p_x, p_y, and p_z along the coordinate axes of the frame describe the vector as

$$ {}^1\boldsymbol{P} = {}^1p_x\hat{\boldsymbol{i}} + {}^1p_y\hat{\boldsymbol{j}} + {}^1p_z\hat{\boldsymbol{k}} \tag{A.1}$$

where $\hat{i}$, $\hat{j}$ and $\hat{k}$ are unit vectors on X, Y, and Z coordinate axes, respectively. Frame {1} or frame {x, y, z} is shown in Fig. A.1.

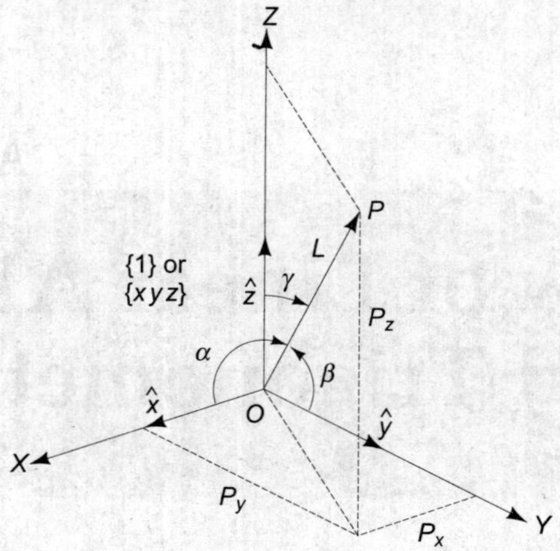

Fig. A.1 *A vector and orthogonal coordinate frame {1}*

The orientation of position vector ${}^1\boldsymbol{P}$ is expressed by the direction cosines as

$$\cos\alpha = \frac{{}^1p_x}{L} \qquad \cos\beta = \frac{{}^1p_y}{L}; \qquad \cos\gamma = \frac{{}^1p_z}{L} \tag{A.2}$$

where α, β and γ are respectively, the right-handed angles measured from the coordinate axes to the vector ${}^1\boldsymbol{P}$ and L is the magnitude of the vector defined as:

$$L = \left|{}^1\boldsymbol{P}\right| = \sqrt{({}^1p_x^2 + {}^1p_y^2 + {}^1p_z^2)} \tag{A.3}$$

In vector-matrix notation, the vector ${}^1\boldsymbol{P}$ can be represented by a 3×1 olumn matrix as

$${}^1\boldsymbol{P} = \begin{bmatrix} {}^1p_x \\ {}^1p_y \\ {}^1p_z \end{bmatrix} = \left[{}^1p_x \;\; {}^1p_y \;\; {}^1p_z\right]^T \tag{A.4}$$

where $[{}^1p_x \; {}^1p_y \; {}^1p_z]^T$ represents the transpose matrix.

Homogeneous Coordinates

In *homogeneous coordinates*, (n+1) dimensions are used to represent an $\boldsymbol{n}$-dimensional vector. The vector ${}^1\boldsymbol{P}$ in Fig. A.1 is represented in homogeneous coordinates as

$${}^1\boldsymbol{P} = \begin{bmatrix} a \\ b \\ c \\ \sigma \end{bmatrix} = [a \quad b \quad c \quad \sigma]^T \tag{A.5}$$

where $a = p_x\sigma$, $b = p_y\sigma$, and $c = p_z\sigma$. The parameter σ is called "*scale factor*". For example, a vector $\vec{P} = 3\hat{i} - 4\hat{j} + 7\hat{k}$ will be

$$\boldsymbol{P} = [3 \quad -4 \quad 7 \quad 1]^T \tag{A.6}$$

in homogeneous coordinate system with a scale factor $\sigma = 1$. Also,

$$\begin{aligned} \boldsymbol{P} &= [30 \quad -40 \quad 70 \quad 10]^T && \text{with } \sigma = 10 \\ \boldsymbol{P} &= [1.5 \quad -2 \quad 3.5 \quad 0.5]^T && \text{with } \sigma = 0.5 \end{aligned} \tag{A.7}$$

The scale factor relates the physical coordinates to the homogeneous coordinates. A scale factor of 1 ($\sigma = 1$) gives a homogeneous transformation, where transformed homogeneous coordinates are same as physical coordinates. In robotic applications scale factor of 1 ($\sigma = 1$) is used. $\sigma > 1$ is useful for scaling up and $0 < \sigma < 1$ is useful for scaling down numeric values. A homogeneous vector with $\sigma = 0$, for example $\boldsymbol{Q} = [a \;\; b \;\; c \;\; 0]^T$, represents an infinite vector, which gives direction or orientation of the vector. The null vector in homogeneous coordinates is $[0 \;\; 0 \;\; 0 \;\; \sigma]^T$, where σ is any positive nonzero scale factor.

A.3 VECTOR OPERATIONS

Vector operations are required in developing concise formulations of the statics, kinematics, and dynamics of manipulators.

A.3.1 The Dot Product

Let $\vec{P}$ and $\vec{Q}$ be two vectors, defined with respect to same coordinate frame, as

$$\vec{P} = a_1\hat{i} + b_1\hat{j} + c_1\hat{k} \tag{A.8}$$

$$\vec{Q} = a_2\hat{i} + b_2\hat{j} + c_2\hat{k} \tag{A.9}$$

In Eqs (A.8) and (A.9), the superscript for the coordinate frame is dropped for simplicity as both vectors refer to same frame. The *dot product* or *scalar product* of these vectors is defined as

$$\begin{aligned} \vec{P}\cdot\vec{Q} &= \left(a_1\hat{i} + b_1\hat{j} + c_1\hat{k}\right)\cdot\left(a_2\hat{i} + b_2\hat{j} + c_2\hat{k}\right) \\ &= a_1a_2 + b_1b_2 + c_1c_2 \qquad (\text{A.10}) \\ &= \boldsymbol{P}^T\boldsymbol{Q} = \boldsymbol{P}\boldsymbol{Q}^T \end{aligned}$$

The result of the dot product of vectors is a scalar.

In homogeneous coordinate representation, the vectors $\vec{P}$ and $\vec{Q}$ with σ_1 and σ_2 as scale factors for $\vec{P}$ and $\vec{Q}$, respectively, will be

$$\boldsymbol{P} = [a_1' \;\; b_1' \;\; c_1' \;\; \sigma_1]^T \tag{A.11}$$

$$Q = [a_2' \;\; b_2' \;\; c_2' \;\; \sigma_2]^T \tag{A.12}$$

where $$a_1' = a_1\sigma_1; \quad b_1' = b_1\sigma_1; \quad c_1' = c_1\sigma_1$$

and $$a_2' = a_2\sigma_2; \quad b_2' = b_2\sigma_2; \quad c_2' = c_2\sigma_2 \tag{A.13}$$

By using Eqs (A.11), (A.12), and (A.10), the dot product is obtained as

$$\vec{P}\cdot\vec{Q} = P^T Q = Q^T P = a_1'a_2' + b_1'b_2' + c_1'c_2' \tag{A.14}$$

Dividing this result by the product of the scale factors $\sigma_1\sigma_2$, the dot product in homogeneous coordinates, Eq. (A.14), is related to physical coordinates, Eq. (A.10), that is

$$\vec{P}\cdot\vec{Q} = \frac{a_1'a_2' + b_1'b_2' + c_1'c_2'}{\sigma_1\sigma_2} = a_1a_2 + b_1b_2 + c_1c_2 \tag{A.15}$$

The dot product of two vectors is a measure of the *orientation* between the two vectors. For an angle θ between the vectors,

$$\vec{P}\cdot\vec{Q} = |\vec{P}|\;|\vec{Q}|\cos\theta \tag{A.16}$$

Sor ne important and useful properties of dot product are:

1. $\vec{P}\cdot\vec{P} \geq 0$; the dot product of vector with itself is always nonnegative and is zero only for the zero vector, that is, $\vec{P}\cdot\vec{P} = 0$ implies $\vec{P} = 0$.
2. If $\vec{P}$ and $\vec{Q}$ have same direction $\theta = 0°$, then $\cos\theta = 1$ and $\vec{P}\cdot\vec{Q} = 0$ is equal to the product of the lengths of two vectors

$$\vec{P}\cdot\vec{Q} = |\vec{P}|\;|\vec{Q}| \tag{A.17}$$

   In particular if $\vec{P} = \vec{Q}$, then

$$\vec{P}\cdot\vec{Q} = |\vec{P}|\;|\vec{P}| = |\vec{P}|^2 \tag{A.18}$$

3. $\vec{P}\cdot\vec{Q} = \vec{Q}\cdot\vec{P}$; the dot product is commutative.
4. $(p\vec{P} + q\vec{Q})\cdot\vec{R} = p(\vec{P}\cdot\vec{R}) + q(\vec{Q}\cdot\vec{R})$; with p and q scalars, the dot product is a *linear function* of its first argument $\vec{P}$. In other words, the dot product of the sum is the sum of the dot products.
5. Two vectors $\vec{P}$ and $\vec{Q}$ are said to be *orthogonal* if and only if their dot product is null, that is,

$$\vec{P}\cdot\vec{Q} = P^T Q = 0 \tag{A.19}$$

   From Eq. (A.16), the angle between orthogonal vectors is $\pi/2$ radians. Hence, orthogonal vectors are perpendicular to each other. Such vectors are also known as *orthonormal* vectors.

6. The norm of a vector is denoted as $||\vec{P}||$ and is defined as

$$||\vec{P}|| = \sqrt{\vec{P}\cdot\vec{P}} = \sqrt{P^T P} \tag{A.20}$$

7. A *unit vector* $\hat{p}$ is a vector, whose ***norm*** is unity, that is, $\hat{p}^T\hat{p} = 1$. For a given vector $\vec{P}$ its unit vector is obtained by dividing each component by its norm:

$$\hat{p} = \frac{1}{||\vec{P}||}\vec{P} \tag{A.21}$$

The *Euclidean space* is the vector space whose dimension is three. This forms the basis of three orthogonal unit vectors of a coordinate frame.

A.3.2 Vector Cross Product

Another important operation with vectors, which is useful in the analysis of robotic manipulators, is cross product of vectors. The *vector cross product* or *vector product* of two vectors $\vec{P}$ and $\vec{Q}$ in the Euclidean space is the vector $\vec{R}$

$$\vec{R} = \vec{P}\times\vec{Q} = \begin{vmatrix} \hat{i} & \hat{j} & \hat{k} \\ a_1 & b_1 & c_1 \\ a_2 & b_2 & c_2 \end{vmatrix} = \begin{bmatrix} b_1c_2 - b_2c_1 \\ a_2c_1 - a_1c_2 \\ a_1b_2 - a_2b_1 \end{bmatrix} \tag{A.22}$$

or

$$\vec{R} = \vec{P}\times\vec{Q} = (b_1c_2 - b_2c_1)\hat{i} + (a_2c_1 - a_1c_2)\hat{j} + (a_1b_2 - a_2b_1)\hat{k} \tag{A.23}$$

$\vec{R}$ is a vector orthogonal to $\vec{P}$ and $\vec{Q}$.

For vectors $\vec{P}$ and $\vec{Q}$ in homogeneous coordinates, Eqs (A.11) and (A.12), the cross product in the vector-matrix notation will be

$$\vec{R} = \vec{P}\times\vec{Q} = \begin{bmatrix} b_1'c_2' - b_2'c_1' \\ a_2'c_1' - a_1'c_2' \\ a_1'b_2' - a_2'b_1' \\ \sigma_1\sigma_2 \end{bmatrix} \tag{A.24}$$

The length of the cross-product vector $\vec{R}$ depends on the angle between two vectors $\vec{P}$ and $\vec{Q}$, that is, for θ be the angle between $\vec{P}$ and $\vec{Q}$,

$$|\vec{R}| = |\vec{P}\times\vec{Q}| = |\vec{P}||\vec{Q}|\sin\theta \tag{A.25}$$

The cross product of two vectors $\vec{P}$ and $\vec{Q}$ can be considered as the result obtained by projecting $\vec{Q}$ on a plane perpendicular to the plane of $\vec{P}$ and $\vec{Q}$, rotating the projection by 90° in the positive direction about $\vec{P}$ and then multiplying the resulting vector by $|\vec{P}|$. Some characteristics of vector cross product are:

1. The cross product $\vec{Q} \times \vec{P}$ has the same magnitude as $\vec{P} \times \vec{Q}$ but in the opposite direction, thus,
$$\vec{Q} \times \vec{P} = -\left(\vec{P} \times \vec{Q}\right) \tag{A.26}$$
2. If the vectors $\vec{P}$ and $\vec{Q}$ are parallel, then θ is either 0° or $\pm 180^\circ$ and
$$\left|\vec{P} \times \vec{Q}\right| = \left|\vec{P}\right|\left|\vec{Q}\right| \sin\theta = 0 \tag{A.27}$$
Conversely, if the cross product is zero, then either at least one of the vectors is zero or vectors are parallel.
3. The cross product is distributed over addition, that is,
$$\vec{P} \times \left(\vec{Q} + \vec{R}\right) = \vec{P} \times \vec{Q} + \vec{P} \times \vec{R} \tag{A.28}$$
and
$$\left(\vec{Q} + \vec{R}\right) \times \vec{P} = \vec{Q} \times \vec{P} + \vec{R} \times \vec{P} \tag{A.29}$$
4. The cross product of a vector with itself is zero, that is,
$$\vec{P} \times \vec{P} = 0 \tag{A.30}$$

A.3.3 Differentiation of Vectors

Let point P in Fig. A.1 be moving with respect to frame {1}, that is, the vector ${}^1\boldsymbol{P}$ is time varying and its components are continuous function of time. Thus, the vector ${}^1\boldsymbol{P}(t)$ from Eq. (A.1) will be

$${}^1\boldsymbol{P}(t) = p_x(t)\hat{\boldsymbol{i}} + p_y(t)\hat{\boldsymbol{j}} + p_z(t)\hat{\boldsymbol{k}} \tag{A.31}$$

The derivative of vector ${}^1\boldsymbol{P}(t)$ with respect to time is defined by equation

$${}^1\dot{\boldsymbol{P}}(t) = \frac{d\,{}^1\boldsymbol{P}(t)}{dt} = \lim_{\Delta t \to 0} \frac{\Delta\,{}^1\boldsymbol{P}}{\Delta t} = \lim_{\Delta t \to 0} \frac{{}^1\boldsymbol{P}(t+\Delta t) - {}^1\boldsymbol{P}(t)}{\Delta t} \tag{A.32}$$

For the vector ${}^1\boldsymbol{P}(t)$ in Eq. (A.31), the first derivative is

$$\frac{d\,{}^1\boldsymbol{P}(t)}{dt} = \frac{dp_x(t)}{dt}\hat{\boldsymbol{i}} + \frac{dp_y(t)}{dt}\hat{\boldsymbol{j}} + \frac{dp_z(t)}{dt}\hat{\boldsymbol{k}} \tag{A.33}$$

since unit vectors $\hat{\boldsymbol{i}}$, $\hat{\boldsymbol{j}}$ and $\hat{\boldsymbol{k}}$ are constant vectors. The second derivatives of ${}^1\boldsymbol{P}(t)$ is

$$\frac{d^2\left({}^1\boldsymbol{P}(t)\right)}{dt^2} = \frac{d^2(p_x(t))}{dt^2}\hat{\boldsymbol{i}} + \frac{d^2\left(p_y(t)\right)}{dt^2}\hat{\boldsymbol{j}} + \frac{d^2(p_z(t))}{dt^2}\hat{\boldsymbol{k}} \tag{A.34}$$

The derivatives of dot and cross products of vectors can be obtained from the definition. From Eqs. (A.10), (A.22) and (A.32).

$$\frac{d}{dt}(\boldsymbol{P} \cdot \boldsymbol{Q}) = \frac{d\boldsymbol{P}}{dt} \cdot \boldsymbol{Q} + \boldsymbol{P} \cdot \frac{d\boldsymbol{Q}}{dt} \tag{A.35}$$

$$\frac{d}{dt}(\boldsymbol{P} \times \boldsymbol{Q}) = \frac{d\boldsymbol{P}}{dt} \times \boldsymbol{Q} + \boldsymbol{P} \times \frac{d\boldsymbol{Q}}{dt} \tag{A.36}$$

$$\frac{d}{dt}[\boldsymbol{P}\times(\boldsymbol{Q}\times\boldsymbol{R})] = \left[\frac{d\boldsymbol{P}}{dt}\times(\boldsymbol{Q}\times\boldsymbol{R})\right]+\left[\boldsymbol{P}\times\left(\frac{d\boldsymbol{Q}}{dt}\times\boldsymbol{R}\right)\right]+\left[\boldsymbol{P}\times\left(\boldsymbol{Q}\times\frac{d\boldsymbol{R}}{dt}\right)\right] \quad (A.37)$$

A.4 RIGHT-HANDED ORTHONORMAL COORDINATE FRAME

For a right-handed orthonormal Cartesian coordinate system with unit vectors $\hat{\boldsymbol{i}}, \hat{\boldsymbol{j}}$, and $\hat{\boldsymbol{k}}$ along the principal axes, see Fig. A.1, applying vector dot and cross products, following important and useful results are obtained easily:

$$\begin{aligned}
\hat{\boldsymbol{i}}\cdot\hat{\boldsymbol{i}} &= \hat{\boldsymbol{j}}\cdot\hat{\boldsymbol{j}} = \hat{\boldsymbol{k}}\cdot\hat{\boldsymbol{k}} = 1\\
\hat{\boldsymbol{i}}\cdot\hat{\boldsymbol{j}} &= \hat{\boldsymbol{j}}\cdot\hat{\boldsymbol{k}} = \hat{\boldsymbol{k}}\cdot\hat{\boldsymbol{i}} = 0\\
\hat{\boldsymbol{i}}\times\hat{\boldsymbol{i}} &= \hat{\boldsymbol{j}}\times\hat{\boldsymbol{j}} = \hat{\boldsymbol{k}}\times\hat{\boldsymbol{k}} = \boldsymbol{0}\\
\hat{\boldsymbol{i}}\times\hat{\boldsymbol{j}} &= -\hat{\boldsymbol{j}}\times\hat{\boldsymbol{i}} = \hat{\boldsymbol{k}}\\
\hat{\boldsymbol{j}}\times\hat{\boldsymbol{k}} &= -\hat{\boldsymbol{k}}\times\hat{\boldsymbol{j}} = \hat{\boldsymbol{i}}\\
\hat{\boldsymbol{k}}\times\hat{\boldsymbol{i}} &= -\hat{\boldsymbol{i}}\times\hat{\boldsymbol{k}} = \hat{\boldsymbol{j}}
\end{aligned} \quad (A.38)$$

A.5 MATRICES

Another essential mathematical tool for manipulator analysis is the matrices and the matrix algebra. These are discussed in this section.

A *matrix* is a rectangular array of real or complex numbers (called elements) arranged in m rows and n columns, that is

$$\boldsymbol{A} = \begin{bmatrix} a_{11} & a_{12} & \cdots & a_{1n} \\ a_{21} & a_{22} & \cdots & a_{2n} \\ \vdots & \vdots & \ddots & \vdots \\ a_{m1} & a_{m2} & \cdots & a_{mn} \end{bmatrix} \quad (A.39)$$

or
$$\boldsymbol{A} = [a_{ij}] \qquad \forall\ i = 1, 2, \cdots, m;\ \forall\ j = 1, 2, \cdots, n \quad (A.40)$$

where a_{ij} are the elements of matrix of *order m* by *n* or *dimension* of matrix is $m \times n$ and m and n are positive integers.

A matrix with single row (column) is called a *row* (*column*) matrix. Further, if $n = 1$, the matrix represents a (column) vector of dimension $(m \times 1)$ with elements a_i as the components of vector.

The *transpose* of an $(m \times n)$ matrix $\boldsymbol{A}$, denoted by $\boldsymbol{A}^T$ is a matrix with n rows and m columns, that is, for matrix $\boldsymbol{A}$ in Eq. (A.39), the transpose of $\boldsymbol{A}$ is $(n \times m)$, and is given by

$$\boldsymbol{A}^T = \begin{bmatrix} a_{11} & a_{21} & \cdots & a_{m1} \\ a_{12} & a_{22} & \cdots & a_{m2} \\ \vdots & \vdots & \ddots & \vdots \\ a_{1n} & a_{2n} & \cdots & a_{nm} \end{bmatrix} \quad (A.41)$$

In particular, transpose of a row matrix is a column matrix and vice-versa.

For example, a vector $\vec{P} = 3\hat{i} - 5\hat{j} + 11\hat{k}$ is represented by a (4×1) column matrix as $P = [3 \quad -5 \quad 11 \quad 1]^T$.

A *square* matrix is a matrix with equal number of rows and columns ($n = m$), that is, a $(n \times n)$ square matrix is

$$A = \begin{bmatrix} a_{11} & a_{12} & \cdots & a_{1n} \\ a_{21} & a_{22} & \cdots & a_{2n} \\ \vdots & \vdots & \ddots & \vdots \\ a_{n1} & a_{n2} & \cdots & a_{nn} \end{bmatrix} \tag{A.42}$$

or

$$A = [a_{ij}] \qquad \forall i, j = 1, 2, \cdots, n \tag{A.43}$$

A *diagonal* matrix is a $(n \times n)$ square matrix of order n whose all off-diagonal elements are zero, $(a_{ij} = 0)$ for $i \neq j$, that is,

$$A = \begin{bmatrix} a_{11} & 0 & \cdots & 0 \\ 0 & a_{22} & \cdots & 0 \\ \vdots & \vdots & \ddots & \vdots \\ 0 & 0 & \cdots & a_{nn} \end{bmatrix} = diag(a_{11}, a_{22}, \ldots, a_{nn}) \tag{A.44}$$

A *unit* or *identity* matrix, denoted by I is a diagonal matrix, whose diagonal elements are unity $(a_{ii} = 1)$ or

$$I = \begin{bmatrix} 1 & 0 & \cdots & 0 \\ 0 & 1 & \cdots & 0 \\ \vdots & \vdots & \ddots & \vdots \\ 0 & 0 & \cdots & 1 \end{bmatrix} = \left[a_{ij}\right] \qquad \begin{matrix} \text{with } a_{ij} = 1 & \text{for} & i = j \\ \text{and } a_{ij} = 0 & \text{for} & i \neq j \end{matrix} \tag{A.45}$$

A square matrix of order n is a *symmetric* matrix if its transpose is identical to itself, that is $A = A^T$. This implies for a symmetric matrix, $a_{ij} = a_{ji}$ for all i and j, that is,

$$A = \begin{bmatrix} a_{11} & a_{12} & \cdots & a_{1n} \\ a_{12} & a_{22} & \cdots & a_{2n} \\ \vdots & \vdots & \ddots & \vdots \\ a_{1n} & a_{2n} & \cdots & a_{nn} \end{bmatrix} \tag{A.46}$$

If the elements of a square matrix B have the property

$$b_{ji} = -b_{ij} \quad \text{for } i \neq j \text{ and} \quad b_{ii} = 0 \tag{A.47}$$

then the matrix is called a *skew-symmetric* matrix, that is

$$B = \begin{bmatrix} 0 & b_{12} & \cdots & b_{1n} \\ -b_{12} & 0 & \cdots & b_{2n} \\ \vdots & \vdots & \ddots & \vdots \\ -b_{1n} & -b_{2n} & \cdots & 0 \end{bmatrix} \tag{A.48}$$

The skew symmetric matrix is equal to negative of its transpose, that is,

$$B = -B^T \tag{A.49}$$

The vector product of two vectors $\vec{P}$ and $\vec{Q}$ can be expressed as a product of matrix operator $S(\vec{P})$ and $\vec{Q}$. Defining the skew-symmetric matrix operator $S(\vec{P})$ as

$$S(\vec{P}) = \begin{bmatrix} 0 & -p_z & -p_y \\ p_z & 0 & -p_x \\ p_y & p_x & 0 \end{bmatrix} \tag{A.50}$$

where p_x, p_y, and p_z are components of vector $\vec{P}$. The vector cross product of $\vec{P}$ and $\vec{Q}$ is given by

$$\vec{P} \times \vec{Q} = S(\vec{P})\vec{Q} = -S(\vec{Q}) = -S(\vec{Q})\vec{P} \tag{A.51}$$

The *rank* $\rho(A)$ of a $(m \times n)$ matrix A is the maximum integer r such that at least one nonnull minor of order r exists, that is

$$\rho(A) \le \min(m, n) \tag{A.52}$$

In other words, rank of a matrix is the order of largest submatrix of A with nonzero determinant.

A matrix with $\rho(A) = \min(m, n)$ is said to be *full-rank*. The rank of a matrix indicates the number of linearly independent rows (or columns) in the matrix.

A matrix with all elements equal to zero $(a_{ij} = 0 \text{ for all } i \text{ and } j)$ is called *null* matrix,

$$A = 0 \tag{A.53}$$

where 0 is $(m \times n)$ matrix.

A *partitioned* matrix is a matrix whose elements are matrices (blocks) of proper dimensions. A matrix with pq partitions A_{ij} for $i = 1, \ldots, q$ and $j = 1, \ldots, p$ is:

$$A = \begin{bmatrix} A_{11} & A_{12} & \cdots & A_{1p} \\ A_{21} & A_{22} & \cdots & A_{2p} \\ \vdots & \vdots & \ddots & \vdots \\ A_{q1} & A_{q2} & \cdots & A_{qp} \end{bmatrix} \tag{A.54}$$

where each A_{ij} is a matrix.

A.6 MATRIX OPERATIONS

Trace of Matrix

The *trace* of an $(n \times n)$ square matrix $\boldsymbol{A}$ is the sum of its principal diagonal elements,

$$Trace\ \boldsymbol{A} = \mathrm{Tr}(\boldsymbol{A}) = \sum_{i=1}^{n} a_{ii} \tag{A.55}$$

Some useful properties of trace are:

$$\mathrm{Tr}(\boldsymbol{A}+\boldsymbol{B}) = \mathrm{Tr}(\boldsymbol{A}) + \mathrm{Tr}(\boldsymbol{B})$$

$$\mathrm{Tr}(\boldsymbol{A}) = \mathrm{Tr}\left(\boldsymbol{A}^T\right) \tag{A.56}$$

$$\mathrm{Tr}(\boldsymbol{AB}) = \mathrm{Tr}(\boldsymbol{BA})$$

Equal matrices

Two matrices $\boldsymbol{A}$ and $\boldsymbol{B}$ are equal if they are of same dimension $(m \times n)$ and $a_{ij} = b_{ij}$ for all i and j.

In other words, if two matrices of same dimension are equal, their respective elements must be equal. That is, if $\boldsymbol{A} = \boldsymbol{B}$, then

$$a_{ij} = b_{ij} \qquad \forall i = 1,2,\cdots,n;\ \forall j = 1,2,\cdots,m$$

(A.57)

Matrix Addition

Sum (difference) of two matrices $\boldsymbol{A}$ and $\boldsymbol{B}$ of same dimensions is matrix $\boldsymbol{C}$ of the same dimension obtained by adding (subtracting) corresponding elements. Thus

$$\boldsymbol{C} = \boldsymbol{A} + \boldsymbol{B} \qquad \text{with} \quad c_{ij} = a_{ij} + b_{ij} \quad \forall i, j \tag{A.58}$$

and

$$\boldsymbol{C} = \boldsymbol{A} - \boldsymbol{B} \qquad \text{with} \quad c_{ij} = a_{ij} - b_{ij} \quad \forall i, j \tag{A.59}$$

For the matrix addition following properties hold

$$\boldsymbol{A} + \boldsymbol{B} = \boldsymbol{B} + \boldsymbol{A}$$

$$(\boldsymbol{A} + \boldsymbol{B}) + \boldsymbol{C} = \boldsymbol{A} + (\boldsymbol{B} + \boldsymbol{C})$$

$$\boldsymbol{A} + \boldsymbol{0} = \boldsymbol{A} \tag{A.60}$$

$$\boldsymbol{A} + (-\boldsymbol{A}) = \boldsymbol{0}$$

where $\boldsymbol{0}$ is the null matrix

Multiplication of Matrices

The product of a scalar c and a $(m \times n)$ matrix $\boldsymbol{A}$ is obtained by multiplying every element of $\boldsymbol{A}$ by the scalar c. Thus,

$$cA = Ac = \left[ca_{ij}\right] \qquad \forall i = 1,2,\cdots,m; \quad \forall j = 1,2,\cdots,n \tag{A.61}$$

Two matrices can be multiplied if and only if they are *conformable*, that is, number of columns of $\boldsymbol{A}$ is equal to number of rows of $\boldsymbol{B}$. Thus,

$$\boldsymbol{C}_{m\times n} = \left(\boldsymbol{A}_{m\times p}\right)\left(\boldsymbol{B}_{p\times n}\right) \tag{A.62}$$

The resultant matrix $\boldsymbol{C}$ has m rows and n columns. The elements of resultant matrix $\boldsymbol{C}$ of dimension $(m \times n)$ are given by

$$c_{ij} = \sum_{k=1}^{p} a_{ik}\, b_{kj} \qquad \forall i = 1,2,\cdots,m; \quad \forall j = 1,2,\cdots,n \tag{A.63}$$

Square matrices of order n are always conformable. The following properties hold for square matrix multiplications:

$$\begin{aligned} \boldsymbol{IA} &= \boldsymbol{AI} = \boldsymbol{A} \\ \boldsymbol{ABC} &= \boldsymbol{A}(\boldsymbol{BC}) = (\boldsymbol{AB})\boldsymbol{C} \\ \boldsymbol{A}(\boldsymbol{B}+\boldsymbol{C}) &= \boldsymbol{AB} + \boldsymbol{AC} \\ (\boldsymbol{A}+\boldsymbol{B})\boldsymbol{C} &= \boldsymbol{AC} + \boldsymbol{BC} \\ (\boldsymbol{AB})^T &= \boldsymbol{B}^T\boldsymbol{A}^T \end{aligned} \tag{A.64}$$

The following results are also important

$\boldsymbol{AB} \neq \boldsymbol{BA}$ in general.

If $\boldsymbol{AB} = \boldsymbol{BA}$, then matrices are said to be *commutative*.

$\boldsymbol{AB} = \boldsymbol{0}$ does not imply $\boldsymbol{A} = \boldsymbol{0}$ or $\boldsymbol{B} = \boldsymbol{0}$.

$\boldsymbol{AC} = \boldsymbol{BC}$ does not imply $\boldsymbol{A} = \boldsymbol{B}$.

The *derivative* of an $(m \times n)$ matrix $\boldsymbol{A}(t)$, whose elements $a_{ij}(t)$ are differentiable functions, is the matrix

$$\dot{\boldsymbol{A}}(t) = \frac{d\boldsymbol{A}(t)}{dt} = \left[\frac{da_{ij}(t)}{dt}\right] \qquad \forall i = 1,\ 2,\ \ldots,\ m \quad \forall j = 1,\ 2,\ \ldots,\ n \tag{A.65}$$

Determinant of Matrix

The *determinant* of a square matrix $\boldsymbol{A}$ of order n is the scalar denoted as $|\boldsymbol{A}|$ and given by

$$\begin{aligned} \det(\boldsymbol{A}) = |\boldsymbol{A}| &= \begin{vmatrix} a_{11} & a_{12} & \cdots & a_{1n} \\ a_{21} & a_{22} & \cdots & a_{2n} \\ \vdots & \vdots & \ddots & \vdots \\ a_{n1} & a_{n2} & \cdots & a_{nn} \end{vmatrix} \\ &= \sum a_{ij}(-1)^{i+j} \det(\boldsymbol{A}_{ij}) \qquad \forall i = 1,\ 2,\ \ldots,\ n \end{aligned} \tag{A.66}$$

where $\det(\boldsymbol{A}_{ij})$ is the determinant of the $(n-1)\times(n-1)$ matrix $\boldsymbol{A}_{ij}$ obtained by deleting i^{th} row and j^{th} column of matrix $\boldsymbol{A}$.

For example, for a third order matrix A, the determinant is

$$|A| = \begin{vmatrix} a_{11} & a_{12} & a_{13} \\ a_{21} & a_{22} & a_{23} \\ a_{31} & a_{32} & a_{33} \end{vmatrix}$$

$$= a_{11}\begin{vmatrix} a_{22} & a_{23} \\ a_{32} & a_{33} \end{vmatrix} - a_{12}\begin{vmatrix} a_{21} & a_{23} \\ a_{31} & a_{33} \end{vmatrix} + a_{13}\begin{vmatrix} a_{21} & a_{22} \\ a_{31} & a_{32} \end{vmatrix}$$

or $$|A| = a_{11}(a_{22}a_{33} - a_{23}a_{32}) + a_{12}(a_{23}a_{31} - a_{21}a_{33}) + a_{13}(a_{21}a_{32} - a_{22}a_{31}) \quad (A.67)$$

The following properties of determinant are useful:

(a) If a row (or column) of matrix A is zero, then $|A|$ is zero ($|A| = 0$).

(b) $|A| = |A^T|$ (A.68)

(c) For square matrices A and B, $|AB| = |A|\,|B|$.

(d) For identity matrix I, $\det(I) = 1$.

(e) If two adjacent columns (or rows) of matrix A are equal, $\det(A_{ij}) = 0$.

If the rows of a square matrix are linearly independent, then the determinant of that matrix is nonzero, and the matrix is said to be *nonsingular*. If the determinant of a square matrix is zero $|A|=0$, then matrix is said to be *singular* and the rows of the matrix are *not* linearly independent. Thus, the determinant can be used to test for matrix singularity.

Inverse of Matrix

The adjoint of a $(n \times n)$ square matrix A is the $(n \times n)$ matrix $Adj\,A$ defined as the transpose of A obtained by replacing each element of A by its cofactor in $|A|$, that is,

$$Adj\,A = \left[(-1)^{i+j} \det(A_{ij})\right]^T \qquad \forall\, i, j = 1, 2, \ldots, n \quad (A.69)$$

An $(n \times n)$ square matrix A is said to be invertible if and only if $\rho(A) = n$ and $\det(A) \neq 0$ (nonsingular matrix). The inverse matrix A^{-1}, termed as inverse of A is defined by the property

$$A^{-1}A = AA^{-1} = I \quad (A.70)$$

Since $\rho(I) = n$, A is said to be invertible if A^{-1} exits.
The inverse of A can be computed as

$$A^{-1} = \frac{1}{|A|} Adj\,A \quad (A.71)$$

The following properties hold for square matrices A and B of same order

$$
\begin{aligned}
(\boldsymbol{A}^{-1})^{-1} &= \boldsymbol{A} \\
(\boldsymbol{A}^{T})^{-1} &= (\boldsymbol{A}^{-1})^{T} \\
(\boldsymbol{AB})^{-1} &= \boldsymbol{B}^{-1}\boldsymbol{A}^{-1}
\end{aligned} \tag{A.72}
$$

If the inverse of a square matrix is equal to its transpose, $\boldsymbol{A}^T = \boldsymbol{A}^{-1}$, then the matrix is said to be orthogonal and Eq. (A.70) in this case is

$$\boldsymbol{A}^T\boldsymbol{A} = \boldsymbol{A}\boldsymbol{A}^T = \boldsymbol{I} \tag{A.73}$$

A.7 TRIGONOMETRIC IDENTITIES

Some trigonometric identities useful in robotic modeling are summarized below.

$$
\begin{aligned}
\sin(-\theta) &= -\sin\theta \\
\cos(-\theta) &= \cos\theta \\
\tan(-\theta) &= -\tan\theta
\end{aligned} \tag{A.74}
$$

$$
\begin{aligned}
\cos(\theta - 90^\circ) &= -\cos(\theta + 90^\circ) = \sin\theta \\
\sin(\theta - 90^\circ) &= -\sin(\theta + 90^\circ) = -\cos\theta
\end{aligned} \tag{A.75}
$$

$$\sin^2\theta + \cos^2\theta = 1 \tag{A.76}$$

$$
\begin{aligned}
\cos(\theta_1 + \theta_2) &= \cos\theta_1\cos\theta_2 - \sin\theta_1\sin\theta_2 = C_1C_2 - S_1S_2 = C_{12} \\
\cos(\theta_1 - \theta_2) &= \cos\theta_1\cos\theta_2 + \sin\theta_1\sin\theta_2 = C_1C_2 + S_1S_2
\end{aligned} \tag{A.77}
$$

$$
\begin{aligned}
\sin(\theta_1 + \theta_2) &= \sin\theta_1\cos\theta_2 + \cos\theta_1\sin\theta_2 = S_1C_2 + C_1S_2 = S_{12} \\
\sin(\theta_1 - \theta_2) &= \sin\theta_1\cos\theta_2 - \cos\theta_1\sin\theta_2 = S_1C_2 - C_1S_2
\end{aligned} \tag{A.78}
$$

$$
\begin{aligned}
\cos[(\theta_1 - \theta_2) - \theta_3] &= \cos(\theta_1 - \theta_2)\cos\theta_3 + \sin(\theta_1 - \theta_2)\sin\theta_3 \\
&= (C_1C_2 + S_1S_2)C_3 + (S_1C_2 - C_1S_2)S_3 \\
&= C_1C_2C_3 + S_1S_2C_3 + S_1C_2S_3 - C_1S_2S_3
\end{aligned} \tag{A.79}
$$

$$
\begin{aligned}
\sin[(\theta_1 - \theta_2) - \theta_3] &= \sin(\theta_1 - \theta_2)\cos\theta_3 - \cos(\theta_1 - \theta_2)\sin\theta_3 \\
&= (S_1C_2 - C_1S_2)C_3 - (C_1C_2 + S_1S_2)S_3 \\
&= S_1C_2C_3 - C_1S_2C_3 - C_1C_2S_3 - S_1S_2S_3
\end{aligned} \tag{A.80}
$$

$$
\begin{aligned}
\cos[(\theta_1 + \theta_2) - \theta_3] &= \cos(\theta_1 + \theta_2)\cos\theta_3 + \sin(\theta_1 + \theta_2)\sin\theta_3 \\
&= (C_1C_2 - S_1S_2)C_3 + (S_1C_2 + C_1S_2)S_3 \\
&= C_1C_2C_3 - S_1S_2C_3 + S_1C_2S_3 + C_1S_2S_3
\end{aligned} \tag{A.81}
$$

$$
\begin{aligned}
\sin[(\theta_1 + \theta_2) - \theta_3] &= \sin(\theta_1 + \theta_2)\cos\theta_3 - \cos(\theta_1 + \theta_2)\sin\theta_3 \\
&= (S_1C_2 + C_1S_2)C_3 - (C_1C_2 - S_1S_2)S_3 \\
&= S_1C_2C_3 + C_1S_2C_3 - C_1C_2S_3 + S_1S_2S_3
\end{aligned} \tag{A.82}
$$

$$
\begin{aligned}
\cos(\theta_1 + \theta_2 + \theta_3) &= C_1C_2C_3 - S_1S_2C_3 - S_1C_2S_3 - C_1S_2S_3 \\
\sin(\theta_1 + \theta_2 + \theta_3) &= S_1C_2C_3 + C_1S_2C_3 + C_1C_2S_3 - S_1S_2S_3
\end{aligned} \tag{A.83}
$$

$$\frac{\partial \sin\theta}{\partial\theta} = \cos\theta \text{ or } \frac{\partial S\theta}{\partial\theta} = C\theta \text{ or } \frac{\partial S}{\partial\theta} = C \tag{A.84}$$

$$\frac{\partial \cos\theta}{\partial\theta} = -\sin\theta \text{ or } \frac{\partial C\theta}{\partial\theta} = -S\theta \text{ or } \frac{\partial C}{\partial\theta} = -S \tag{A.85}$$

A.7.1 Some Useful Trigonometric Expressions

Using Eqs (A.77) and (A.78) following useful simplifications are obtained

$$C_1S_{12} - S_1C_{12} = S_2 \tag{A.86}$$

$$S_2C_{12} - C_2S_{12} = -S_1 \tag{A.87}$$

$$S_2S_{12} + C_2C_{12} = C_1 \tag{A.88}$$

$$C_1C_{12} + S_1S_{12} = C_2 \tag{A.89}$$

A.8 DETERMINATION OF ANGLES

Consider two equations in *sine* and *cosine* functions

$$\cos\theta_1 = r_{11} \tag{A.90}$$

$$\cos\theta_2 \sin\theta_1 = r_{12} \tag{A.91}$$

The solutions for θ_1 and θ_2 can be obtained as

$$\theta_1 = A\cos(r_{11}) \tag{A.92}$$

and

$$\theta_2 = A\cos\left(\frac{r_{12}}{\sin\theta_1}\right) \tag{A.93}$$

where *A*cos is *arc cosine* (or $\cos^{-1}$) function. The angles θ_1 and θ_2 obtained from Eqs (A.92) and (A.93) are not useful for the following drawbacks:

(i) The sign or quadrant of the angle is undefined as $\cos\theta = \cos(-\theta)$

(ii) The accuracy of the angle depends on the angle itself, that is,

$$\left.\frac{d\cos\theta}{d\theta}\right|_{(0^\circ,\ 180^\circ)} = 0 \tag{A.94}$$

(iii) In Eq. (A.93), the accuracy of θ_2 is low whenever θ_1 is close to 0.

(iv) The angle θ_2 is undefined when $\theta_1 = 0^\circ$ or $\pm 180^\circ$.

Similar situation will happen if *arc sine* ($\sin^{-1}$) function is used. Therefore, in determining the angles use of *arc sine* and *arc cosine* shall be avoided. The two input *arc_tangent* function (*A*tan2) eliminates the above problems and gives accurate angles. By supplying two arguments, the abscissa *x* and the ordinate *y* to the *A*tan2 function angles in the range $-\pi \le \theta \le \pi$ can be accurately determined. By examining the sign of both *x* and *y*, as shown in Fig. A.2 the quadrant of the angle is accurately known. The accuracy of the function is uniform over its full range and it can detect when either *x* or *y* is zero and returns correct result.

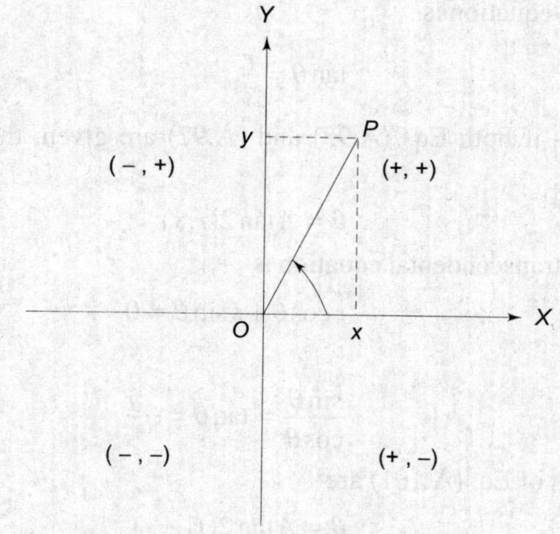

Fig. A.2 *The four quadrant Atan2 function*

A.8.1 The Atan2 Function

A two argument, four-quadrant version of *arctan* function is defined as "*Atan2*" and gives angles in the entire range $[-\pi, \pi]$. The common *arctan* (tan^{-1}) function can be used to implement four-quadrant *Atan*2 function as indicated in Table A.1. In Table A.1, depending on whether y is negative, zero, or positive, $sgn(y)$ function returns –1, 0, or 1 respectively.

Table A.1 *Implementation of two argument, four-quadrant* arctan *(Atan2) function*

x	*Atan2(y, x)*
Positive	$arctan(y/x)$
Zero	$[sgn(y)]\ \pi/2$
Negative	$arctan(y/x)+[sgn(y)]\ \pi$

The four-quadrant *arctan* function is very useful in inverse kinematic analysis. Some applications of this function are discussed below.
Let single equation be

$$\sin\theta = r \tag{A.95}$$

There are two solutions to this and are given by

$$\theta = \pm A\tan 2(\ r,\ \pm\sqrt{1-r^2}\ ) \tag{A.96}$$

Likewise, the single equation

$$\cos\theta = s \tag{A.97}$$

has two solutions given by

$$\theta = A\tan 2(\pm\sqrt{1-s^2},\ s) \tag{A.98}$$

Now, if the given equation is

$$\tan\theta = \frac{r}{s} \tag{A.99}$$

which is same as if both Eqs (A.95) and (A.97) are given, there is a unique solution

$$\theta = A\tan 2(r, s) \tag{A.100}$$

A more complex transcendental equation is

$$r\cos\theta + s\sin\theta = 0 \tag{A.101}$$

which gives

$$\frac{\sin\theta}{\cos\theta} = \tan\theta = -\frac{r}{s} \tag{A.102}$$

The two solutions of Eq. (A.101) are

$$\theta = A\tan 2(r, -s) \tag{A.103}$$

$$\theta = A\tan 2(-r, s) \tag{A.104}$$

Similarly, for the equation

$$r\cos\theta + s\sin\theta = t \tag{A.105}$$

the solutions are

$$\theta = A\tan 2(s, r) \pm A\tan 2(\sqrt{r^2 + s^2 - t^2}, t)$$

(A.106)

Finally, the solution for set of equations

$$r\cos\theta - s\sin\theta = t \tag{A.107}$$

$$r\sin\theta + s\cos\theta = u \tag{A.108}$$

is given by

$$\theta = A\tan 2(ru - st, rt + su) \tag{A.109}$$

BIBLIOGRAPHY

1. H Anton, *Elementary Linear Algebra*, 6th ed., Willy, N.Y. 1991.
2. A. Balestrino, G.De Maria and L. Sciavicco, B. Siciliano, "An Algorithmic Approach to Coordinate Transformation for Robotic Manipulators," *Advanced Robotics*, **2**, 315–404, 1988.
3. Daniel T. Finkbeiner, II, *Introduction of Matrices and Linear Transformation*, D.B. Taraporevala Sons & Co. Pvt. Ltd., Mumbai, 1968.
4. John B. Fraleigh, R.A Beauregard and V.J. Katz, *Linear Algebra*, 3rd ed., Addison Wesley, Reading, Mass, 1996.
5. G.H. Golub and C.F. Loan, *Matrix Computations*, 2nd ed., The Johns Hopkins University Press, 1989.

6. Francis B. Hildebrand, *Advanced Calculus for Applications,* Prentice Hall Inc., New Jersey, 1976.
7. R.M. Murray, Z. Li and S.S. Sastry, *A Mathematical Introduction to Robotic Manipulation*, CRC Press, Boca Raton, Fla, 1994.
8. R.P. Paul and H. Zhang, "Computationally Efficient Kinematics for Manipulators with Spherical Wrists Based on the Homogeneous Transformation Representation," *The International Journal of Robotics Research,* **5**(2), 32–44, 1986.

Appendix B

MATLAB AND SIMULINK*

This appendix describes the basics of MATLAB and its applications in robotics. Many software tools are available for engineering analysis of systems. These can be used for modeling, design, and simulation of systems. Some are for specific applications while others are general purpose.

A very powerful general-purpose tool that is often used for engineering systems is MATLAB. MATLAB is registered trademark of the MathWorks Inc., USA. This appendix is a brief introduction to MATLAB version 5.0 or later and SIMULINK as tools for analysis of robotic systems and illustrates their applications in robot kinematics, dynamics, control and so on.

MATLAB provides an interactive medium and many advanced mathematical tools as intrinsic functions. This greatly reduces the computational barrier that existed between the engineer and the computer—the higher level programming language like FORTRAN. The advance features of MATLAB such as interactive error tracing reduce the time needed to develop the problem solutions and most of the frustration of debugging a higher level programming language code is eliminated. The graphics is very simple but powerful in MATLAB and the graphics features can be used interactively.

MATLAB has a rich vocabulary, abundance of inline "help", and lucidly written and easily accessible "demo" programs. Users can also create their own functions and write online help for these. This Appendix is not a substitute for the manuals provided with the MATLAB. The manuals for 'The Student Edition' of MATLAB and 'The MATLAB User's Guide' should be read thoroughly and kept handy for reference. Other references at the end of the Appendix also provide valuable supplementary information. In this Appendix MATLAB basics are described and complementary material for MATLAB applications in robotics is included.

* MATLAB and SIMULINK are registered trademarks of The MathWorks, Inc.

B.1 GETTING STARTED

MATLAB consists of a large number of mathematical operators and commands. Many more commands and functions are provided in the auxiliary toolboxes, which work with MATLAB. The basic steps for using MATLAB are described in this section.

B.1.1 Loading MATLAB

There are several ways in which MATLAB can be started. Two common methods to load the MATLAB software are:

(i) *Double-click* on the shortcut icon corresponding to MATLAB on the desktop.

(ii) *Left-click*, in sequence, on the buttons/menus that popup: *Start→Programs→Matlab→MATLAB*.

B.1.2 MATLAB Command Window

When the MATLAB software starts, the screen changes to MATLAB screen with a MATLAB prompt ">>". This window is called *MATLAB Command Window* and is shown in Fig. B.1.

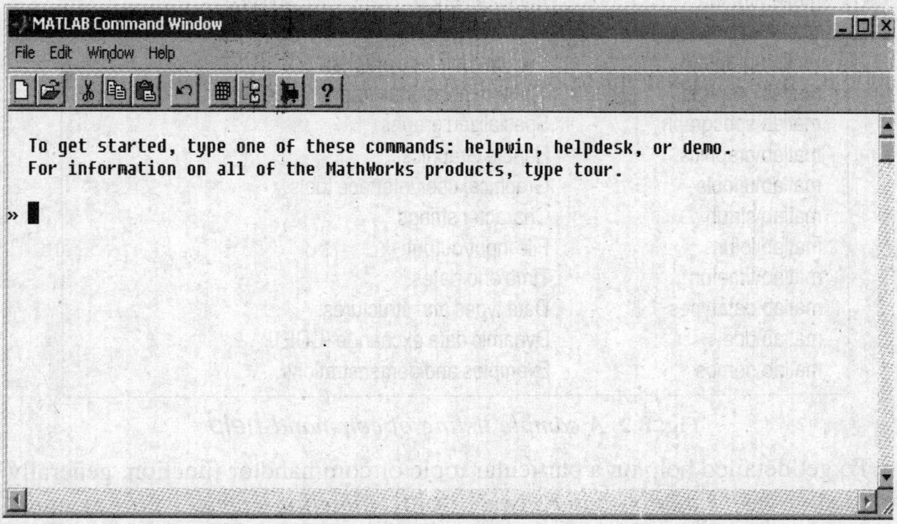

Fig. B.1 *The MATLAB Command Window*

All the commands are entered at this prompt and the output of the command entered is displayed either in the command window itself or in another window launched above the command window. To quit MATLAB type quit or exit after the prompt or press *Ctrl+Q*. Alternately, one can use Command Window's "close button" "☒" or *left-click* on *File → Exit MATLAB* to quit MATLAB.

MATLAB is case sensitive and all command verbs or function names are lower-case letters.

B.1.3 Getting Help

Learning the use of **help** and **demo** is an important first step in mastering MATLAB.

Typing help at the prompt, displays a summary list of available MATLAB commands (function-names) in alphabetical order, with a short description of each command. A sample of this list is shown in Fig. B.2.

```
>>help
HELP topics:
matlab\general      -   General purpose commands.
matlab\ops          -   Operators and special characters.
matlab\lang         -   Programming language constructs.
matlab\elmat        -   Elementary matrices and matrix manipulation.
matlab\elfun        -   Elementary math functions.
matlab\specfun      -   Specialized math functions.
matlab\matfun       -   Matrix functions - numerical linear algebra.
matlab\datafun      -   Data analysis and Fourier transforms.
matlab\polyfun      -   Interpolation and polynomials.
matlab\funfun       -   Function functions and ODE solvers.
matlab\sparfun      -   Sparse matrices.
matlab\graph2d      -   Two dimensional graphs.
matlab\graph3d      -   Three dimensional graphs.
matlab\specgraph    -   Specialized graphs.
matlab\graphics     -   Handle Graphics.
matlab\uitools      -   Graphical user interface tools.
matlab\strfun       -   Character strings.
matlab\iofun        -   File input/output.
matlab\timefun      -   Time and dates.
matlab\datatypes    -   Data types and structures.
matlab\dde          -   Dynamic data exchange (DDE).
matlab\demos        -   Examples and demonstrations.
```

Fig. B.2 *A sample listing of command* help

To get detailed help on a particular topic or command or function, generally referred to as *function-name*, type help <*function-name*>. For example >>help plot displays the available documentation on command plot.

Help in MATLAB can also be obtained by *left-click* on the '*help*' tab of the Command Window. This provides a *drop-down* menu for different types of help. Choosing *HelpWindow* launches a new window, the *MATLAB Help Window*, which gives the list of available MATLAB Help Topics. This window provides *scroll-bars* and buttons for *Back*, *Forward*, *Home*, *Tips*, *Close* and so on and many other aids. Highlighting any topic (by a *left-click*) and pressing *Enter* provides a classified detailed listing of commands in the topic with a brief one-line description of each command.

There are other related ways of getting online help in MATLAB. The following list gives the summary of some help related commands and their action.

>>help	list the available function-names or topics with a brief one-line description of function topic.
>>helpwin	launches the MATLAB Help Window discussed above.
>>help help	describes the use of help command.
>>help general	gives a classified list of general purpose MATLAB commands.
>>help *<function-name>*	gives detailed documentation on the function-name. For example, help exp gives help on exponentials.
>>help demo	lists names of various demo programs.
>>help diary	describes the use of diary command to save results appearing on the Command Window in a file for later printing or other use.
>>intro	illustrates various MATLAB functions by executing the source code contained in script file *intro.m*. It is a demo.
>>demo	initiates access to a lengthy set of demo programs illustrating features of MATLAB. To fully appreciate the functionally of MATLAB, users should run through all the demos.
>>type *<function-name>*	lists the entire source code for the function stored in the M-file for the *function-name*. For example, *type intro* lists the source code of the *intro* demo program stored in *intro.m* file. By studying this code, readers can quickly learn to use many MATLAB commands. For some built-in intrinsic functions the source code cannot be listed because it is stored in binary form.
>>lookfor *<keyword>*	instructs MATLAB to search for *keyword* in the first line of all M-files. For example, *lookfor integ* will display the first comment line of all files, which contain string "integ".

A very useful feature of MATLAB is that one can scroll through the commands entered at the prompt in Command Window by using *up* and *down arrow keys*. Any command on which cursor is present becomes the current command. It can be edited or changed as desired and then executed by pressing *return*.

B.2 COMPUTATIONS IN MATLAB

MATLAB is a very simple but powerful tool for all types of computations. In general, computations are carried out as simply as typing or writing on paper. Neither elaborate knowledge of a higher level programming language is required nor a complex program is to be written for data input, computation, and output.

B.2.1 Basic Mathematics with MATLAB

The mathematical operations are entered on the command window in the same way, as they would be written on paper. For example, typing $x = 5/2$ at the prompt as:

```
>>x = 5/2
```

yields the response

```
x =
   2.5000
```

and to solve the polynomial $y = \frac{1}{2}x + 3x^2 - 0.3(2x - 2)^3$, typing the polynomial at the prompt as:

```
>>y=1/2*x+3*x^2–0.3*(2*x–2)^3
```

yields the result

```
y =
   11.9000
```

The mathematical operations are carried out in the usual order of priority. A large number of mathematical functions are available in MATLAB. Figure B.3 is a list of some of the mathematical functions useful in robotics. The constant π is entered by typing pi.

B.2.2 Vectors in MATLAB

MATLAB deals with vectors and matrices with equal ease. A vector is defined by typing its elements in square brackets. For example, a vector $\boldsymbol{P}_1 = 2\hat{\boldsymbol{i}} - \hat{\boldsymbol{j}} + 4\hat{\boldsymbol{k}}$ is entered by typing:

```
>>P1 = [2  –1  4]
```

The display will be

```
P1 =
   2 –1 4
```

```
Trigonometric
    sin          -  Sine
    cos          -  Cosine
    tan          -  Tangent
    atan         -  Inverse tangent
    atan2        -  Four quadrant inverse tangent
Linear Algebra
    inv          -  matrix inverse
    det          -  determinant
    rank         -  matrix rank
    transpose    -  matrix transpose
    zero         -  zero array (m × n)
    eye          -  identity matrix
    size         -  size of matrix
    length       -  length of vector
Rounding and remainder
    fix          -  Round towards zero
    mod          -  Modulus(signed remainder after division)
    rem          -  Remainder after division
    sign         -  Signum
```

Fig. B.3 *Some of the elementary functions available in MATLAB*

To multiply vector $\boldsymbol{P}_1$ by –3, type

```
>>Q = -3*P1
```

The result is

```
Q =
  -6 3 -12
```

The sum of two vectors $\boldsymbol{P}_1$ and $\boldsymbol{P}_2 = -4\hat{\boldsymbol{i}} + 3\hat{\boldsymbol{j}} + \hat{\boldsymbol{k}}$ is obtained as

```
>>P2 = [-4  3  1]
P2 =
  - 4  3  1
>>Q = P1+P2
The result is
Q =
  -2  2  5
```

B.2.3 Matrices

Defining a matrix is as easy as defining vectors. The matrix elements are entered row wise with a semicolon (;) to separate the rows of the matrix. For example, the following 3 × 3 matrix

$$\boldsymbol{A} = \begin{bmatrix} 5 & -2 & 7 \\ -1 & 3 & 4 \\ 0 & 2 & 1 \end{bmatrix} \tag{B.1}$$

will be entered as

```
>>A=[5 -2 7; -1 3 4; 0 2 1]
```

the display will be

```
A =

   5  -2  7
  -1   3  4
   0   2  1
```

The transpose of matrix ***A*** can be easily found as

```
>>B = A'
B =

   5  -1  0
  -2   3  2
   7   4  1
```

The multiplication of these two matrices is obtained as

```
>>C = A*B
C =

  78  17   3
  17  26  10
   3  10   5
```

The determinant of matrix ***A*** is

```
>>D = det(A)
D =
  -41
```

The trace of the matrix ***C*** will be

```
>>trace(C)
ans =
  109
```

The matrix inverse is as easy, that is, $\boldsymbol{D} = [\boldsymbol{A}]^{-1}$ is

```
>>D = inv(A)
D =

   0.1220  -0.3902   0.7073
  -0.0244  -0.1220   0.6585
    0.488   0.2439  -0.3171
```

and

```
>>I = A*D
I =
  1  0  0
  0  1  0
  0  0  1
```

Illustrates the fact that a matrix times its inverse is the identity matrix. The applications of matrices are far beyond just matrix computations and MATLAB has functions for all common as well as complex operations on matrices, including symbolic operations.

B.2.4 Symbolic Operations

Symbolic matrix operations are often required in robotics and can be performed with MATLAB very easily. For example, a symbolic matrix ***R*** is manipulated as follows:

```
% define symbolic variable 'th' and matrix R
>> syms th
>> R=[cos(th) sin(th) 0;-sin(th) cos(th) 0;0 0 1]
R=
        [ cos(th),   sin(th),       0]
        [ -sin(th),  cos(th),       0]
        [       0,        0,        1]
>>
% get the transpose of R
>> RT=transpose(R)
RT=
        [ cos(th),    -sin(th),     0]
        [ sin(th),     cos(th),     0]
        [       0,          0,      1]
>>
% compute product of R and RT
>> Q=R*RT
Q=
        [ cos(th)^2+sin(th)^2,            0,      0]
        [ 0,          cos(th)^2+sin(th)^2,        0]
        [ 0,                              0,      1]
>>
>> simplify(Q)
ans =
        [ 1,             0,          0]
        [ 0,             1,          0]
        [ 0,             0,          1]
>>
% find inverse of R
>> simplify(inv(R))
ans =
        [ cos(th),       -sin(th),        0]
        [ sin(th),        cos(th),        0]
        [       0,              0,        1]
>>
```

Some of the commands and their description available for symbolic operations are given in Table B.1.

B.3 COMMAND FILES

For complex computations involving a series of commands, instead of typing the commands and inputs at the command prompt, MATLAB provides that the commands can be sequentially stored in a text file. This command file is referred to as M-file or MATLAB program and is stored with a suffix "*.m*" as *<filename.m>*. The commands in the M-file are executed by referring MATLAB to that file, that is, by typing the name of the M-file at the prompt.

To create a new M-file, *left-click* on the *open new* M-file icon. This launches the MATLAB Editor/Debugger, which is identical to Window's Notepad. Saving the file automatically gives it the "*.m*" extension. The M-file editor can also be invoked by typing>>edit for creating a new file, or by typing>> edit *<filename>* to open an existing M-file for editing. A M-file can also be created by using any text editor and saving it with a filename having "*.m*" extension.

A sample M-file for the matrix operations discussed above, in section B.2.3, would look like:

```
% Sample M-file for illustrating some matrix
% operations
% define a 3 x 3 matrix A
A=[5 -2 7; -1 3 4; 0 2 1]
% compute transpose of A
B = A'
% multiply A and B
C=A*B
% computing determinant of matrix A
D=det(A)
% Computing the trace (sum of diagonal elements) of
% matrix C
trace(C)
% computing matrix inverse
D=inv(A)
% verify A times its inverse(D)is identity matrix I
I=A*D
```

Let this file be saved as *MatOp.m*. Lines that start with % are comment lines; they are not interpreted by MATLAB as commands. To execute the *MatOp.m* (commands in it), type at the prompt

```
>>MatOp
```

After executing the commands in file *MatOp.m*, the command window will contain the following:

```
A =
   5  -2  7
  -1   3  4
   0   2  1
B =
   5  -1  0
  -2   3  2
   7   4  1
C =
  78  17   3
  17  26  10
   3  10   5
D =
 - 41
ans =
 109
D =
   0.1220  -0.3902   0.7073
  -0.0244  -0.1220   0.6585
    0.488   0.2439  -0.3171
I =
  1  0  0
  0  1  0
  0  0  1
```

To view all the above results you may have to scroll the command window up or down.

Users can create new commands or functions using M-file by defining a function-name and it's input-output in the first line of the M-file. Such a function can be used in subsequent computations as any other MATLAB function.

MATLAB also provides several programming constructs for using the MATLAB for more complex computations such as program flow control for repetitive computations, input-output control, special variables and constants, character string functions and so on. Some of these are tabulated in Table B.1 with a brief description of each.

Table B.1 *Some MATLAB programming and other commands and their description*

Command	*Description*
if	Conditionally execute statements.
else	Used with if.
for	Repeat statement a specific number of times.
while	Repeat statements an indefinite number of times.
	break Break out of for and while loops.
return	Return to invoking function
error	Display message and abort function.
input	Prompt for user input and read it.
pause	Wait for user response.
more	Controlled paged output in Command window.
print	Print graph or save graph to file.
fprintf	Formatted output to Command Window or file.
disp	Display a matrix or text.
pi	π, 3.1415926535897...
inf	∞, infinity.
flops	Count of floating point operations.
clock	Wall clock.
cputime	Elapsed CPU time.
strcmp	Compare string.
upper	Convert string to uppercase.
lower	Convert string to lowercase.
int2str	Convert number to string.
simplify	Simplifies each element of symbolic matrix
sym	Create symbolic object
syms	Shortcut for constructing symbolic objects
plot	Linear plot (2-D plot)

B.4 CONTROL SYSTEM APPLICATIONS

Control system analysis requires representation of transfer functions and block diagrams in MATLAB. Simple commands are used to define control system transfer functions and block diagrams and from these the system response in time domain or frequency domain can be easily obtained.

B.4.1 Transfer Function and System Response

Transfers functions are created either by defining the numerator and denominator polynomials of the transfer function or by defining poles, zeros, and gain of the transfer function.

The commands num and den are used define numerator and denominator of the transfer function, respectively. The coefficients of the numerator and denominator polynomials are entered in square brackets "[ ]" in descending power of s. To multiply two polynomials a command conv is used. After defining the numerator and denominator, printsys(num,den,'s') command is used to display the transfer function on the screen in the s-domain.

For example, to create and display the following transfer function on screen

$$G(s) = \frac{5(s+4)}{(s^2+3s+2)(s+2)} \tag{B.2}$$

following lines of MATLAB code can be used:

```
% commands to define the transfer function
% in Eq.(B.2)
num = 5*[1 4];
den = conv([1 3 2], [1 2]);
printsys(num,den,'s')
```

The display will be

```
num/den =
              5 s + 20
    ----------------------
    s^3 + 5 s^2 + 8s + 4
```

The response of system to different inputs can be obtained easily with MATLAB. For example, following set of commands give the response of a system whose transfer function is

$$G(s) = \frac{15}{s^2+4s+11} \tag{B.3}$$

to unit step input $u(t)$.

```
% Response of a system whose transfer function is
% given by Eq. (B.3)
num = 15;
den = [1 4 11];
printsys(num,den,'s');
step(num,den)
print -dtiff FigB_4
```

The step function produces the output as amplitude vs time graph of the system in another window. The output graph is shown in Fig. B.4. The last command print saves the output graph in *tiff* format in a file named *FigB_4.tiff*.

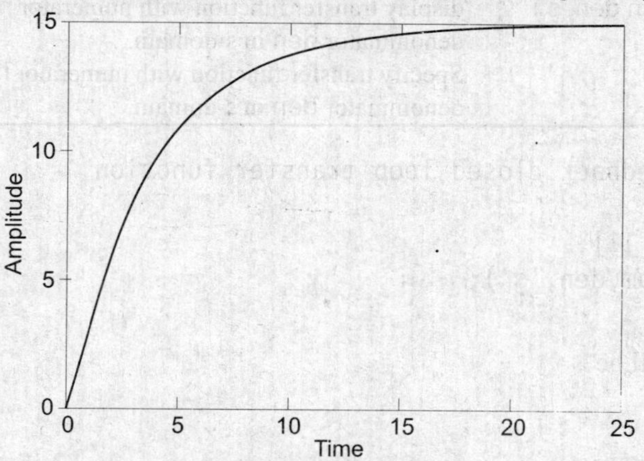

Fig. B.4 *The unit step response of system in Eq. (B.3)*

Some of the function-names of control system related commands are tabulated in Table B.2.

A control system block diagram consists of interconnected blocks, each with its own transfer function. MATLAB allows control systems to be built up from interconnected blocks.

B.4.2 Control System Block Diagram

For creating a unity feedback closed-loop system, define the open-loop transfer function of system, as described in previous section and use the command cloop. This will create the closed-loop transfer function and its response can be studied. For example, if the open-loop transfer function is given by Eq. (B.3), the unity feedback closed-loop transfer function will be obtained by the following MATLAB code.

Table B.2 *Some MATLAB functions for control system applications*

Function name	*Description*
[z,p,k] = tt2zp(num,den)	determine and display the zeros(z), poles(p) and gain(k) of the zero-pole-gain transfer function defined by num and den.
step	plot unit step response.
impulse	impulse input response.
lsim	response to arbitrary inputs.
gensig	generate input signal for lsim.
stepfun	generate unit-step input.
bode	Bode plot for frequency response.
nyquist	Nyquist plot.
rlocus	root locus for design.
cloop	define closed loop transfer function with unity feedback
feedback	closed loop transfer function with non-unity feedback.
series	connect two transfer functions in series
parallel	connect two transfer functions in parallel.
printsys(num,den,'s')	display transfer function with numerator num and denominator den in s-domain.
tf(num,den)	Specify transfer function with numerator num and denominator den in s-domain.

```
% Unity feedback closed loop transfer function
num = 15;
den = [1 4 11];
printsys(num,den,'s');
cloop;
```

The result will be

```
num/den =
       15
  ------------
  s^2 + 4 s + 12
```

If the feedback is not unity, the command **feedback** is used. For example, consider the control system block diagram shown in Fig. B.5

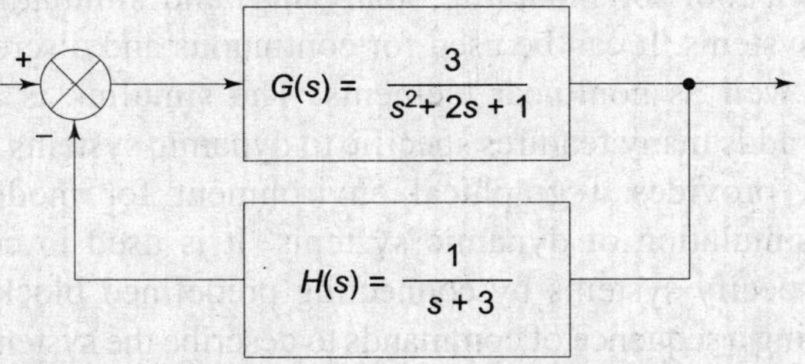

Fig. B.5 *Block diagram of closed-loop control system*

The MATLAB program to get the transfer function of the block diagram in Fig.B.5 will be:

```
% Transfer function and response of closed loop
% control system shown in Fig. B.5.
numg = [3];
deng = [1 2 1];
numh = [1];
denh = [1 1];
[nc,dc] = feedback(numg,deng,numh,denh);
printsys(nc,dc,'s');
step(nc,dc);
```

The result will be

```
num/den =
     3 s + 3
------------------------
s^3 + 3 s^2 + 3 s + 4
```

and the unit step response will be as shown in Fig. B.6. The numerator and denominator of the open-loop transfer function $G(s)$ are defined by **numg** and **deng**; and **numh** and **denh** define the feedback loop transfer function $H(s)$. The generated closed-loop transfer function is characterized by **nc** and **dc** as its numerator and denominator, respectively.

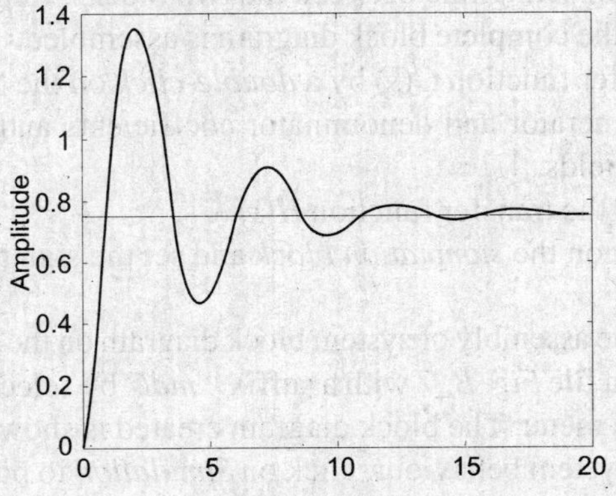

Fig. B.6 *Unit step response of the closed loop control system in Fig. B.5*

B.5 SIMULINK

SIMULINK is a tool for modeling, analyzing, and simulating physical and mathematical systems. It can be used for continuous and discrete time systems with linear as well as nonlinear elements. The simulink is an extension of MATLAB that adds many features specific to dynamic systems.

SIMULINK provides a graphical environment for model construction, analysis, and simulation of dynamic systems. It is used in conjunction with MATLAB to specify systems by connecting predefined blocks on the screen rather than writing a sequence of commands to describe the system block diagram.

SIMULINK is invoked by typing simulink at the MATLAB prompt or by *left-click* on the *New Simulink model* button in Command Window. This opens the *SIMULINK Control Window* in which the system can be assembled. Invoking SIMULINK also launches another window titled: *Library*: *simulink*, which contains a rich collection of building blocks for assembling systems grouped under categories like *Sources*, *Sinks*, *Linear* and so on.

The use of SIMULINK is illustrated by assembling of block diagram in Fig. B.5. To assemble the system follow the following steps:

- Click on the *Linear* in library window, this will open another window with different linear system blocks.
- Drag the transfer function icon (block) in to the SIMULINK Control Window (the working window) and drop it at appropriate place.
- Do the same for the feedback block.
- From the library window bring the summation block in to the working window.
- Open the *Sources* window from library window and select the step input block. Drag and drop it into working window.
- Similarly, from the *Sinks* library window drag the scope block into the working window.
- Connect the blocks placed in the working window by pressing the left mouse button, starting from output point of a block and moving the mouse pointer to input point of the next block to which it is to be connected. This draws the connecting line between the two blocks. Repeat this for all the blocks until the complete block diagram is assembled.
- Set the transfer function $G(s)$ by a *double-click* on the block for $G(s)$ and enter the numerator and denominator coefficients and block title in the appropriate fields.
- Similarly set the transfer function $H(s)$.
- *Double-click* on the *summation block* and set the sign to + – for negative feedback.

This completes the assembly of system block diagram on the screen. This block diagram is saved in a file *Fig. B_7* with a suffix "*.mdl*" by selecting *File* and *Save As* in the drop-down menu. The block diagram created is shown in Fig. B.7.

To simulate the system behaviour, click on *Simulation* to pull down its menu. Selecting *Start* will cause SIMULINK to create a graph window and display the output of the system in this window. If the output graph window does not appear, click on the output block of the block diagram. The graph produced is identical to the command mode response graph of the system, that is, output graph is same as Fig. B.6.

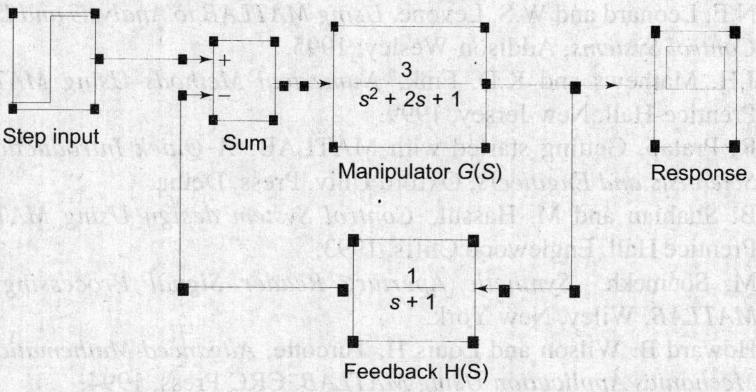

Fig. B.7 *Closed-loop system in Fig. B.5 assembled using SIMULINK*

B.6 TOOL BOXES

Toolboxes are specialized collections of M-files build specifically for solving particular class of problems. The toolboxes represent the efforts of some of the world's top researchers in fields such as controls, signal processing, communications, neuro-fuzzy systems, and others. A wide assortment of tool boxes are available for the MATLAB. One can choose the most appropriate toolbox for work to be done and also run demo for it.

Examples

Most of the examples in chapters have been worked out using MATLAB. The MATLAB code for these is developed for use under MATLAB version 5.0 or above. The intended operating system environment is either Microsoft Windows or UNIX X-windows. Many of the programs may run without changes with older versions of MATLAB.

All the MATLAB programs are available on the book's website. The website is organized corresponding to the chapters in the book. All common repeatedly used functions such as interactive data input, saving output in a file, etc. are put in a separate library. The code of examples used in this appendix is also included.

BIBLIOGRAPHY

1. J. Dongarra et.al., *LABACK: A Portable linear algebra library for high performance computers*, In Supercomputing 1990, IEEE Computer Society Press, 1990.
2. D.M. Etter, *Engineering Problem Solving with MATLAB*, Prentice-Hall Englewood Cliffs, 1993.

3. D. Hanselman, B. Littelfield *Mastering MATLAB — A Comprehensive Tutorial and Reference*, Prentice-Hall, New York, 1996.
4. Wisama Khalil and Denis Creusot, "SYMORO+: A System for the Symbolic Modeling of Robots," *Robotica*, ***15***, 153–161, 1997.
5. N.E. Leonard and W.S. Levone, *Using MATLAB to Analysis and Design Control Systems*, Addison-Wesley; 1995.
6. J.H. Mathews and K.D. Fink, *Numerical Methods Using MATLAB*, Prentice-Hall, New Jersey, 1999.
7. R. Pratap, Getting started with MATLAB: *A Quick Introduction for Scientists and Engineers*, Oxford Univ. Press, Delhi.
8. B. Shahian and M. Hassul, *Control System design Using MATLAB*, Prentice Hall, Englewood Cliffs, 1993.
9. M. Soumekh, *Synthetic Aperture Reader Signal Processing with MATLAB*, Wiley, New York.
10. Howard B. Wilson and Louis H. Turcotte, *Advanced Mathematics and Mechanics Application Using MATLAB*, CRC Press, 1994.
11. The *MathWorks Inc. MATLAB C Math Library User's Guide*, The MathWorks, Inc. USA, 1995.
12. *The MathWorks Inc. MATLAB User's Guide*, The MathWorks, Inc., USA; 1991.
13. *The MathWorks Inc.* The Students Edition of MATLAB for MSDOS Personal Computers, *The MATLAB Curriculum Series*, Prentice-Hall, Englewood Cliffs, NJ, 1992.
14. http://www.mathworks.com/ - The home page of MathWorks provides plethora of links to on-line information and demos of MATLAB toolboxes and SIMULINK.

Appendix C

Link Mass Distribution and the Moment of Inertia Tensor

This appendix reviews the description of the moment of inertia tensor and gives details to compute it for a link. To obtain the EOM using Lagrange-Euler or Newton-Euler formulation of the dynamic model, (see Chapter 6) the moment of inertia tensor is required. The computation of moment of inertia tensor for a link is illustrated as an example and inertia values are tabulated for common geometrical shapes of links.

C.1 MOMENT OF INERTIA TENSOR

For a rigid body (link) free to move in a 3-D space, there are an infinite number of possible rotation axes or coordinate frames. The mass of the body contributes inertia forces during motion of the body. To reflect all the inertial loads due to rotations about an axis, mass distribution of the body about that axis is required. A 4 × 4 *moment of inertia tensor* or *inertia tensor*, which characterizes the mass distribution properties of a rigid body at the origin of a frame of interest, is used to determine the inertial loads during rotations of the body.

Figure C.1 shows a rigid body with a frame attached to the body. The inertia tensor of the body with respect to a frame $\{x\ y\ z\}$ attached to it, is defined as

$$\boldsymbol{I} = \begin{bmatrix} \int x^2 dm & \int xy\, dm & \int xz\, dm & \int x\, dm \\ \int xy\, dm & \int y^2 dm & \int yz\, dm & \int y\, dm \\ \int xz\, dm & \int yz\, dm & \int z^2 dm & \int z\, dm \\ \int x\, dm & \int y\, dm & \int z\, dm & \int dm \end{bmatrix} \tag{C.1}$$

where dm is the mass of a small element of the body and the rigid body is considered to be composed of differential volume elements dV of material with mass density ρ such that $dm = \rho\, dV$. The differential element is located with a vector $\boldsymbol{r} = [x \quad y \quad z \quad 1]^T$ with respect to the frame $\{x\,y\,z\}$. The total mass of the body is m and each integral is taken over the entire volume V of the rigid body.

The moment of inertia tensor matrix in Eq. (C.1) is a symmetric matrix containing nine independent terms. These terms depend on the position and orientation of the body with respect to the frame and their computation is discussed next.

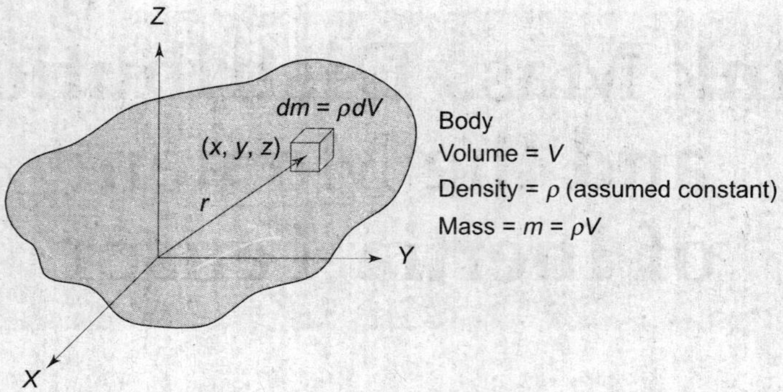

Fig. C.1 *A rigid body and a frame {x y z} fixed to it*

C.2 EVALUATION OF MOMENT OF INERTIA TENSOR

The *moment of inertia* of a body with respect to three coordinate axes of reference frame are defined as

$$
\begin{aligned}
I_{xx} &= \int_v (y^2 + z^2)dm \\
I_{yy} &= \int_v (z^2 + x^2)dm \\
I_{zz} &= \int_v (x^2 + y^2)dm
\end{aligned}
\tag{C.2}
$$

The moments of inertia are positive quantities. The *cross products of inertia*, which give the three off-diagonal terms of Eq. (C.1) are

$$
\begin{aligned}
I_{xy} &= -\int_v xy\,dm \\
I_{yz} &= -\int_v yz\,dm \\
I_{xz} &= -\int_v xz\,dm
\end{aligned}
\tag{C.3}
$$

The cross products of inertia can be either positive or negative. The *first moments of the body* are:

$$
\begin{aligned}
\int_v x\,dm &= m\,\bar{x} \\
\int_v y\,dm &= m\,\bar{y} \\
\int_v z\,dm &= m\,\bar{z}
\end{aligned}
\tag{C.4}
$$

where $\bar{r} = [\bar{x} \quad \bar{y} \quad \bar{z} \quad 1]^T$ represents center of mass or centroid of body in homogeneous coordinates and m is total mass of the body.

The diagonal integrals in Eq. (C.1) are simplified using Eq. (C.2) as

$$\int_v x^2\,dm = \int_v -\tfrac{1}{2}(y^2+z^2)dm + \int_v \tfrac{1}{2}(x^2+z^2)dm + \int_v \tfrac{1}{2}(x^2+y^2)dm$$

or
$$\int_v x^2\,dm = \tfrac{1}{2}(-I_{xx}+I_{yy}+I_{zz}) \tag{C.5}$$

Similarly,

$$\int_v y^2\,dm = \tfrac{1}{2}(I_{xx}-I_{yy}+I_{zz}) \tag{C.6}$$

$$\int_v z^2\,dm = \tfrac{1}{2}(I_{xx}+I_{yy}-I_{zz}) \tag{C.7}$$

Substituting Eqs. (C.3) – (C.7) in Eq. (C.1), the moment of inertia tensor $\boldsymbol{I}$ can be written as

$$\boldsymbol{I} = \begin{bmatrix} \tfrac{1}{2}(-I_{xx}+I_{yy}+I_{zz}) & I_{xy} & I_{xz} & m\bar{x} \\ I_{xy} & \tfrac{1}{2}(I_{xx}-I_{yy}+I_{zz}) & I_{yz} & m\bar{y} \\ I_{xz} & I_{yz} & \tfrac{1}{2}(I_{xx}+I_{yy}-I_{zz}) & m\bar{z} \\ m\bar{x} & m\bar{y} & m\bar{z} & m \end{bmatrix} \tag{C.8}$$

If the origin of frame {x y z} is made to coincide with the centre of mass or centroid of the link and rotations are considered about this frame, the moment of inertia tensor is called *centroidal moment of inertia tensor*. Obviously, the centroid is $\bar{r} = [0 \quad 0 \quad 0 \quad 1]^T$ and the centrodal inertia tensor is

$$\boldsymbol{I}_c = \begin{bmatrix} \tfrac{1}{2}(-I_{xx}+I_{yy}+I_{zz}) & I_{xy} & I_{xz} & 0 \\ I_{xy} & \tfrac{1}{2}(I_{xx}-I_{yy}+I_{zz}) & I_{yz} & 0 \\ I_{xz} & I_{yz} & \tfrac{1}{2}(I_{xx}+I_{yy}-I_{zz}) & 0 \\ 0 & 0 & 0 & m \end{bmatrix} \tag{C.9}$$

The moment of inertia tensor $\boldsymbol{I}_i$ for link $\boldsymbol{i}$, depends on the mass distribution of link $\boldsymbol{i}$ and not on its position or rate of motion, that is, moment of inertia tensor $\boldsymbol{I}_i$ is constant. The characteristics of moment of inertia tensor are summarized below:

- Moment of inertia is always positive.
- The products of inertia may have either positive or negative sign.
- If the axes of the reference frame are aligned with the *principal axes* of the body then the products of inertia are zero. For example, if *xy* is plane of symmetry, then $I_{xz} = 0$ and $I_{yz} = 0$ because for each plus *yz dm* or *xz dm* there will be a corresponding minus *yz dm* or minus *xz dm* and terms will cancel. In this case, the moment of inertia tensor is a diagonal matrix and the moments of inertia, which constitute the first three diagonal elements, are called *principal moments of inertia*. Expressions for the principal moments of inertia for common solid shapes are given in Table C.1. The

moment of inertia about another axis v parallel to **y**-axis is also included therein.

- With some care in frame assignments to the links, it is often possible to make moment of inertia tensor diagonal.
- The sum of the three moments of inertia is invariant under orientation changes of the coordinate frame.
- The *eigenvalues* of the moment of inertia tensor are the principal moments for the body. The principal axes are the *eigenvectors*.
- The moment of inertia tensor is symmetric positive definite matrix.

The geometry of actual links used in a manipulator is complex and it may be difficult to compute the moment of inertia tensor. In such a situation, an option is to measure rather than calculate the moment of inertia of each link.

C.3 MOMENT OF INERTIA TENSOR FOR A PRISM

A rectangular cross-section is often the common cross-section for links in a manipulator. The determination of moment of inertia tensor is illustrated for this geometry. Consider a prismatic link made from a material of uniform density ρ and having a rectangular cross-section, as shown in Fig. C.2. The link length, thickness and width are a, b, and c, respectively. The coordinate frame is attached at one end with the origin of the frame at the center of width c, as shown.

The total mass of link is m with $m = \rho\, abc$ and for the differential element mass is $dm = \rho\, dx\, dy\, dz$. The coefficients for the 4×4 moment of inertia tensor matrix are computed as follows:

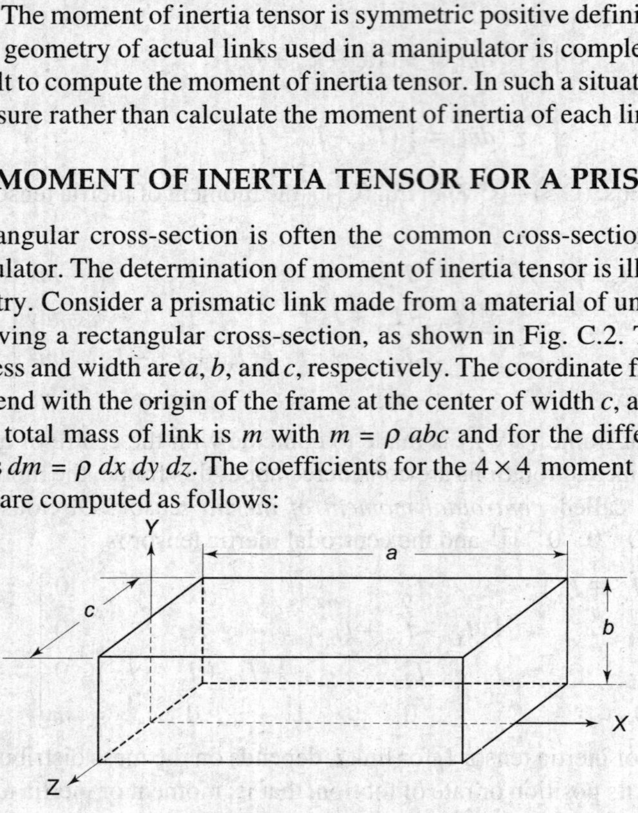

Fig. C.2 *Prismatic link with frame {x y z}*

First, moment of inertia with reference to coordinate frame axes are computed using Eq. (C.2). For **x**-axis, I_{xx} simplifies to

$$I_{xx} = \int_{-c/2}^{c/2} \int_0^b \int_0^a (y^2 + z^2)\, \rho\, dx\, dy\, dz$$

or
$$I_{xx} = \tfrac{1}{12} m(4b^2 + c^2) \tag{C.10}$$

Similarly, I_{yy} and I_{zz} for **y**-axis and **z**-axis are:

$$I_{yy} = \tfrac{1}{12} m(4a^2 + c^2) \tag{C.11}$$

and
$$I_{zz} = \tfrac{1}{3} m(a^2 + b^2) \tag{C.12}$$

Table C.1 *Moment of inertia for some common solid shapes about centroidal axes and about another axis*

Body shape	I_{zz}	I_{xx}	I_{yy}	I_{vv}
Right circular cylinder 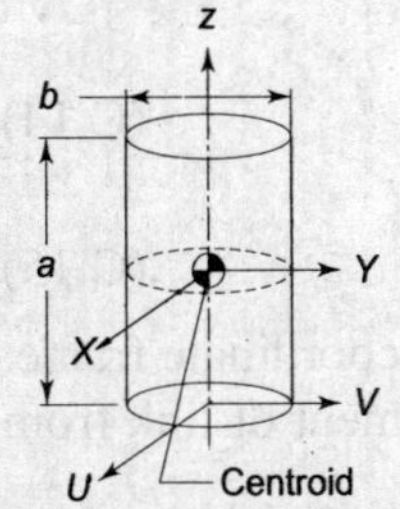	$\frac{1}{2} mb^2$	$\frac{1}{12} m(3b^2 + a^2)$	$\frac{1}{12} m(3b^2 + a^2)$	$\frac{1}{12} m(3b^2 + 4a^2)$
Slender bar ($b << a$) 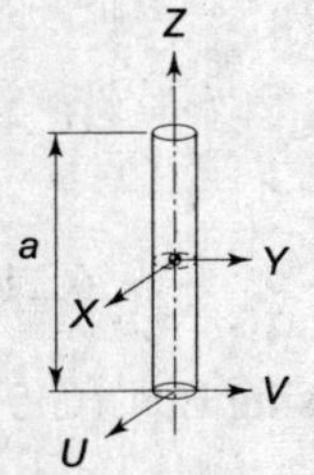	0	$\frac{1}{12} ma^2$	$\frac{1}{12} ma^2$	$\frac{1}{3} ma^2$
Prism or rectangular parallelepiped 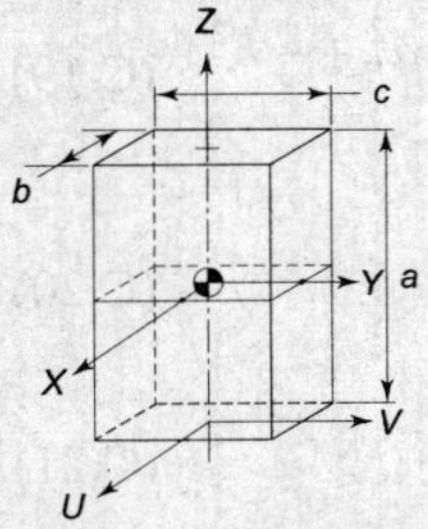	$\frac{1}{12} m(b^2 + c^2)$	$\frac{1}{12} m(a^2 + c^2)$	$\frac{1}{12} m(a^2 + b^2)$	$\frac{1}{3} m(a^2 + b^2)$
Thin cylindrical shell ($b << r$) 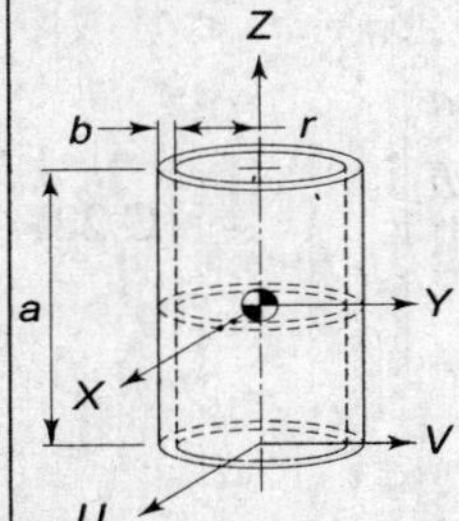	mb^2	$\frac{1}{12} m(6b^2 + a^2)$	$\frac{1}{12} m(6b^2 + a^2)$	$\frac{1}{6} m(3b^2 + 2a^2)$

Next, cross product of inertia are computed using Eq. (C.3). For I_{xy}

$$I_{xy} = -\int_{-c/2}^{c/2}\int_0^b\int_0^a xy\,\rho dx\,dy\,dz$$

or

$$I_{xy} = -\tfrac{1}{4}mab \tag{C.13}$$

Similarly,

$$I_{yz} = -\int_{-c/2}^{c/2}\int_0^b\int_0^a yz\,\rho dx\,dy\,dz = 0 \tag{C.14}$$

and

$$I_{zx} = 0 \tag{C.15}$$

Because of symmetry of link about xy-plane of the chosen coordinate frame, the above two cross products of inertia are zero. The first moment of link from Eq. (C.4) is

$$\int_v x\,dm = \int_{-c/2}^{c/2}\int_0^b\int_0^a x\,\rho dx\,dy\,dz = \tfrac{1}{2}ma \tag{C.16}$$

Similarly

$$\int_v y\,dm = \tfrac{1}{2}mb \tag{C.17}$$

and

$$\int_v z\,dm = 0 \tag{C.18}$$

The diagonal terms of moment of inertia tensor are obtained using Eqs. (C.5) – (C.7) as:

$$\begin{aligned}\int_v x^2\,dm &= \tfrac{1}{2}(-I_{xx} + I_{yy} + I_{zz})\\ &= \tfrac{1}{2}m(-\tfrac{1}{3}b^2 - \tfrac{1}{12}c^2 + \tfrac{1}{3}a^2 + \tfrac{1}{12}c^2 + \tfrac{1}{3}a^2 + \tfrac{1}{3}b^2)\\ &= \tfrac{1}{3}ma^2\end{aligned} \tag{C.19}$$

Similarly

$$\int_v y^2\,dm = \tfrac{1}{3}mb^2 \tag{C.20}$$

and

$$\int_v z^2\,dm = \tfrac{1}{12}mc^2 \tag{C.21}$$

The moment of inertia tensor is obtained by substituting the coefficients from Eqs. (C.10) – (C.21), in Eq. (C.1) to give

$$\boldsymbol{I} = \begin{bmatrix} \tfrac{1}{3}ma^2 & -\tfrac{1}{4}mab & 0 & \tfrac{1}{2}ma \\ -\tfrac{1}{4}mab & \tfrac{1}{3}mb^2 & 0 & \tfrac{1}{2}mb \\ 0 & 0 & \tfrac{1}{12}mc^2 & 0 \\ \tfrac{1}{2}ma & \tfrac{1}{2}mb & 0 & m \end{bmatrix} \tag{C.22}$$

C.4 MOMENT OF INERTIA TENSOR ABOUT ANY OTHER FRAME

Now consider that the moment of inertia is required for another coordinate frame attached to the body. To compute the new moment of inertia tensor, one method is to repeat the previous section integrals with different limits. Alternately, the *Parallel Axis Theorem* can be used.

The parallel axis theorem is useful in translating moment of inertia and the cross product of inertia from one coordinate frame to another. Consider two frames, one attached to the centroid of the body with its axes coinciding with the principal axes of the body. The second frame is attached at some other location on the body such that its axes are parallel to the principal axes of the body, that is, parallel to the axes of first frame.

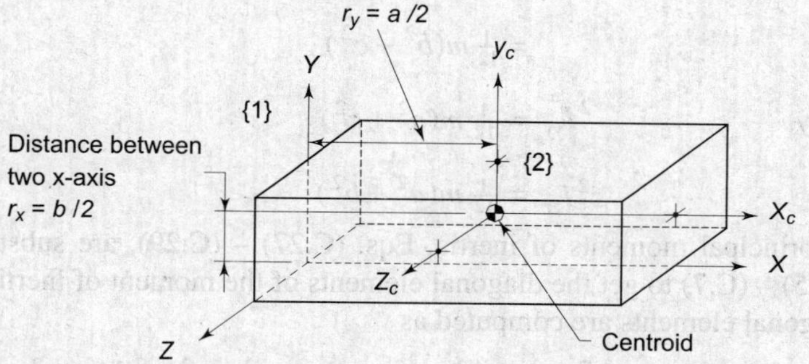

Fig. C.3 *A prism with a frame at centroid with axes aligned with principal axis of prism*

Let the frame attached to the centroid be frame $\{x_c\, y_c\, z_c\}$ or frame {2} and the other be frame $\{x\, y\, z\}$ or frame {1}, as shown in Fig. C.3. The inertia tensor in the centroidal frame is denoted by ${}^2\boldsymbol{I}$ and in other frame as ${}^1\boldsymbol{I}$. According to parallel axis theorem, the moment of inertia of the body about the two parallel axes $\boldsymbol{i}$, in two frames are related as

$$ {}^1I_{ii} = {}^2I_{ii} + mr_i^2 \tag{C.23} $$

where ${}^1I_{ii}$ and ${}^2I_{ii}$ are the moment of inertia about the principal axis $\boldsymbol{i}$ of the body, in frame {1} and frame {2}, respectively, and r_i is the distance between the two parallel axes $\boldsymbol{i}$ of the two frames.

Similarly, the cross products of inertia in frame {1}, and frame {2} are related as

$$ {}^1I_{ij} = {}^2I_{ij} - mr_ir_j \tag{C.24} $$

where ${}^1I_{ij}$ and ${}^2I_{ij}$ are the cross products of inertia in two frames and r_i is the distance between two parallel axes $\boldsymbol{i}$ in two frames, and r_j is the distance between the two parallel axes $\boldsymbol{j}$ in two frames, respectively.

C.5 APPLICATION OF PARALLEL AXIS THEOREM

Moment of inertia tensor for the prism, considered in Section C.3, about a frame fixed at the centroid of the prism and having its axes coinciding with the principal axes of the prism can be obtained using the parallel axis theorem. Consider the centroidal frame {2} and other frame {1} in Fig. C.3. The moment of inertia for frame {1} were computed in section C.3 and are given by Eqs. (C.10)–(C.21). The principal moment of inertia about **x**-axis of frame {2} is computed using the parallel axis theorem, Eq. (C.23), as

$$^2I_{xx} = {}^1I_{xx} - mr_x^2 \tag{C.25}$$

The distances between the axes of frame {1} and frame {2} are

$$r_x = \tfrac{1}{2}b;\quad r_y = \tfrac{1}{2}a \quad \text{and} \quad r_z = \tfrac{1}{2}\sqrt{a^2 + b^2} \tag{C.26}$$

Substituting the moment of inertia $^1I_{xx}$ from Eq. (C.10) and r_x from Eq. (C.26)

$$^2I_{xx} = \tfrac{1}{12}m(4b^2 + c^2) - m(\tfrac{1}{4}b^2)$$

$$= \tfrac{1}{12}m(b^2 + c^2) \tag{C.27}$$

Similarly, $$^2I_{yy} = \tfrac{1}{12}m(a^2 + c^2) \tag{C.28}$$

and $$^2I_{zz} = \tfrac{1}{12}m(a^2 + b^2) \tag{C.29}$$

The principal moments of inertia, Eqs. (C.27) – (C.29), are substituted in Eqs. (C.5) – (C.7) to get the diagonal elements of the moment of inertia tensor. The diagonal elements are computed as

$$\tfrac{1}{2}(-^2I_{xx} + {}^2I_{yy} + {}^2I_{zz}) = \tfrac{1}{2}\left[-\tfrac{1}{12}m(b^2 + c^2) + \tfrac{1}{12}m(a^2 + c^2) + \tfrac{1}{12}m(a^2 + b^2)\right]$$

$$= \tfrac{1}{12}ma^2 \tag{C.30}$$

Similarly, $$\tfrac{1}{2}(^2I_{xx} - {}^2I_{yy} + {}^2I_{zz}) = \tfrac{1}{12}mb^2 \tag{C.31}$$

and $$\tfrac{1}{2}(^2I_{xx} + {}^2I_{yy} - {}^2I_{zz}) = \tfrac{1}{12}mc^2 \tag{C.32}$$

The cross products of inertia for frame {2} are computed as follows:

$$^2I_{xy} = {}^1I_{xy} + m\,r_x r_y \tag{C.33}$$

Substituting $^1\boldsymbol{I}_{xy}$ from Eq. (C.13)

$$^2I_{xy} = -\tfrac{1}{4}mab + m\tfrac{1}{2}b\tfrac{1}{2}a = 0 \tag{C.34}$$

Similarly, $$^2I_{yz} = {}^2I_{zx} = 0 \tag{C.35}$$

The cross products of inertia are zero because frame {2} is attached at the centroid and the link is symmetric about the axes of the chosen frame. Similarly, the first moments about a frame attached to the centroid are also zero. Thus, the moment of inertia tensor for a frame with origin at the centroid and axes parallel to principal axes of prism is

$$
{}^2\boldsymbol{I} = \begin{bmatrix} \frac{1}{12}ma^2 & 0 & 0 & 0 \\ 0 & \frac{1}{12}mb^2 & 0 & 0 \\ 0 & 0 & \frac{1}{12}mc^2 & 0 \\ 0 & 0 & 0 & m \end{bmatrix} \tag{C.36}
$$

EXERCISES

C.1 Find the inertia tensor for a cuboid of uniform density r when the origin of the coordinate frame coincides with one corner of the cuboid.

C.2 A link is 0.5 m long with 10 mm by 20 mm cross-section. Find its inertia tensor assuming that the coordinate frame is attached at one end of the link with origin coinciding with center of cross section of the link.

C.3 An $a \times b$ rectangular cross-section link of length l and mass m is made of homogeneous material of constant density ρ. Determine the products of inertia, first moments, and the inertia tensor when the axes of coordinate frame are aligned with the principal axes of the link and origin of the frame is at the centroid. Is the inertia tensor diagonal? Explain why.

C.4 Find the inertia tensor for a link made with a tube of rectangular cross section and made from a material of uniform density. Assume that the frame for inertia tensor has its origin at the center of mass of the link and axes are perpendicular to the faces of the link.

C.5 Show that the inertia tensor of a slender member with length l and mass m acting at centre of mass with respect to a frame attached to the centroid is

$$
\boldsymbol{I} = \begin{bmatrix} \frac{1}{12}ml^2 & 0 & 0 & 0 \\ 0 & 0 & 0 & 0 \\ 0 & 0 & 0 & 0 \\ 0 & 0 & 0 & m \end{bmatrix} \tag{C.37}
$$

Assume that the principal axis of link coincides with $\boldsymbol{x}$-axis.

C.6 Consider a link of conical shape having mass m, as shown in Fig. EC.6. Find the moment of inertia tensor for this link about its center of mass with respect to a frame whose origin coincides with the centre of mass and axes are parallel to the principal axes of the cone.

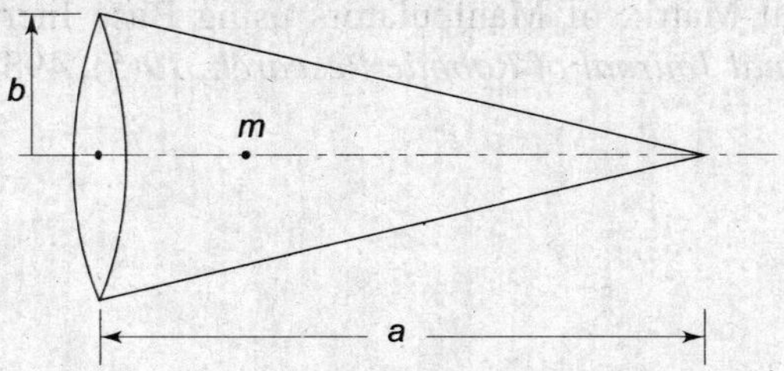

Fig. EC.6 *A conical solid*

C.7 Show that the inertia tensor of link in Ex. C.5 with respect to a frame attached at the distal end of the link is

$$I = \begin{bmatrix} \frac{1}{3}ml^2 & 0 & 0 & -\frac{1}{2}ml \\ 0 & 0 & 0 & 0 \\ 0 & 0 & 0 & 0 \\ -\frac{1}{2}ml & 0 & 0 & m \end{bmatrix} \tag{C.38}$$

The location of centroid with respect to frame attached at distal end is

$$r = \begin{bmatrix} -\frac{l}{2} & 0 & 0 & 1 \end{bmatrix}^T \tag{C.39}$$

C.8 The mass of a link used in a robot can be assumed as a point mass acting at the distal end of the link. If the link length is l and mass is m, show that inertia tensor with respect to a frame attached to the distal end is

$$I = \begin{bmatrix} 0 & 0 & 0 & 0 \\ 0 & 0 & 0 & 0 \\ 0 & 0 & 0 & 0 \\ 0 & 0 & 0 & m \end{bmatrix} \tag{C.40}$$

BIBLIOGRAPHY

1. S.H. Crandall and N.C. Dahl, *An Introduction to the Mechanics of Solids*, McGraw-Hill Kogakusha Ltd., Tokyo,
2. S. Doughty, *Mechanics of Machines*, John Wiley & Sons Inc., New York, 1988.
3. J. Grosjean, *Kinematics and Dynamics of Mechanisms*, McGraw-Hill, Singapore, 1991.
4. H. Mayeda and K. Ohashi, "Base parameters of dynamic models for general open loop kinematic chains," *Robotics Research*, **5**, 271–278, 1990.
5. K.R. Symon, *Mechanics*, Addison-Wesley Pub. Co. Inc., Reading, Massachusetts, 1964.
6. S. Timoshenko, and D.H. Young, *Engineering Mechanics*, McGraw-Hill, Book Co., Singapore, 1956.
7. Koji Yoshida and W. Khalil, "Verification of the Positive Definiteness of the Inertial Matrix of Manipulators using Base Inertial Parameters," *International Journal of Robotic Research*, **19**(5), 498–510, May 2000.

Appendix D

Derivation of Equations of Motion

This appendix gives the step by step computations of partial derivatives of Lagrangian to obtain the Equations of Motion (EOM), using Lagrange-Euler formulation developed in Chapter 6. The detailed evaluation of various coefficients of the dynamic equations is also explained.

D.1 DERIVATIVES OF LAGRANGE FUNCTION

The Lagrangian, Eq. (6.44), for a manipulator with ***n***-DOF is

$$\mathcal{L} = \frac{1}{2}\sum_{i=1}^{n}\sum_{j=1}^{i}\sum_{k=1}^{i} \mathrm{Tr}\left[\boldsymbol{d}_{ij} \boldsymbol{I}_i \boldsymbol{d}_{ik}^T \right] \dot{q}_j \dot{q}_k + \sum_{i=1}^{n} m_i \boldsymbol{g}\, {}^0\boldsymbol{T}_i\, {}^i\bar{\boldsymbol{r}} \tag{D.1}$$

where $\boldsymbol{d}_{ij} = {}^0\boldsymbol{T}_{j-1}\, \boldsymbol{Q}_j\, {}^{j-1}\boldsymbol{T}_i$ and $\boldsymbol{Q}_j$ is given by Eq. (6.27) or Eq. (6.30). The Lagrange-Euler formulation gives the generalized torque τ_i for actuator at joint ***i***, to drive link ***i*** of the manipulator as,

$$\tau_i = \frac{d}{dt}\left(\frac{\partial \mathcal{L}}{\partial \dot{q}_i} \right) - \frac{\partial \mathcal{L}}{\partial q_i} \tag{D.2}$$

To get the EOM, derivatives of $\mathcal{L}$ are required. The derivatives of $\mathcal{L}$ are calculated for link ***p*** as follows. From Eq. (D.1) the partial derivative of $\mathcal{L}$ with respect to $\dot{q}_p$ is

$$\frac{\partial \mathcal{L}}{\partial \dot{q}_p} = \frac{1}{2}\sum_{i=1}^{n}\sum_{k=1}^{i} \mathrm{Tr}\left[\boldsymbol{d}_{ip} \boldsymbol{I}_i \boldsymbol{d}_{ik}^T \right] \dot{q}_k + \frac{1}{2}\sum_{i=1}^{n}\sum_{j=1}^{i} \mathrm{Tr}\left[\boldsymbol{d}_{ij} \boldsymbol{I}_i \boldsymbol{d}_{ip}^T \right] \dot{q}_j \tag{D.3}$$

Equation (D.3) is simplified by using the fact that trace operator gives the sum of diagonal elements and, hence, $\mathrm{Tr}(\boldsymbol{A}) = \mathrm{Tr}(\boldsymbol{A}^T)$. Thus,

$$\mathrm{Tr}\left[\boldsymbol{d}_{ip}\boldsymbol{I}_i\,\boldsymbol{d}_{ik}^T\right] = \mathrm{Tr}\left[\boldsymbol{d}_{ip}\boldsymbol{I}_i\,\boldsymbol{d}_{ik}^T\right]^T = \mathrm{Tr}\left[\boldsymbol{d}_{ik}\boldsymbol{I}_i\,\boldsymbol{d}_{ip}^T\right] \qquad \text{(D.4)}$$

The second term in Eq. (D.3) is identical to right-hand side of Eq. (D.4) as dummy summation index j can be changed to k. Thus, Eq. (D.3) simplifies to

$$\frac{\partial \mathcal{L}}{\partial \dot{q}_p} = \sum_{i=1}^{n}\sum_{k=1}^{i}\mathrm{Tr}\left[\boldsymbol{d}_{ik}\boldsymbol{I}_i\,\boldsymbol{d}_{ip}^T\right]\dot{q}_k \qquad \text{(D.5)}$$

The $\boldsymbol{d}_{ij}$ terms are evaluated using Eqs. (6.29) and (6.46), that is

$$\boldsymbol{d}_{ip} = \frac{\partial\, {}^0\boldsymbol{T}_i}{\partial q_p} = \begin{cases} {}^0\boldsymbol{T}_{j-1}\,\boldsymbol{Q}_j\,{}^{j-1}\boldsymbol{T}_p & \text{for } p \le i \\ 0 & \text{for } p > i \end{cases} \qquad \text{(D.6)}$$

Since $\boldsymbol{d}_{ip}$ is 0 for $p > i$, this means that the summation from $i = 1$ to n in Eq. (D.5) is significant only from p to n. The final form of Eq. (D.5) is

$$\frac{\partial \mathcal{L}}{\partial \dot{q}_p} = \sum_{i=p}^{n}\sum_{k=1}^{i}\mathrm{Tr}\left[\boldsymbol{d}_{ik}\boldsymbol{I}_i\,\boldsymbol{d}_{ip}^T\right]\dot{q}_k \qquad \text{(D.7)}$$

The time derivative of Eq. (D.7) gives the first terms of Eq. (D.2). It evaluates as

$$\frac{d}{dt}\left(\frac{\partial \mathcal{L}}{\partial \dot{q}_p}\right) = \sum_{i=p}^{n}\sum_{k=1}^{i}\mathrm{Tr}\left[\boldsymbol{d}_{ik}\boldsymbol{I}_i\,\boldsymbol{d}_{ip}^T\right]\ddot{q}_k + \sum_{i=p}^{n}\sum_{k=1}^{i}\sum_{m=1}^{i}\mathrm{Tr}\left[\frac{\partial \boldsymbol{d}_{ip}}{\partial q_m}\boldsymbol{I}_i\,\boldsymbol{d}_{ip}^T\right]\dot{q}_k\dot{q}_m \qquad \text{(D.8)}$$

$$+\sum_{i=p}^{n}\sum_{k=1}^{i}\sum_{m=1}^{i}\mathrm{Tr}\left[\boldsymbol{d}_{ik}\,\boldsymbol{I}_i\left(\frac{\partial \boldsymbol{d}_{ip}}{\partial q_m}\right)^T\right]\dot{q}_k\dot{q}_m$$

The partial derivative of $\mathcal{L}$ with respect to q_p required in the second term of Eq. (D.2), is

$$\frac{\partial \mathcal{L}}{\partial q_p} = \frac{1}{2}\sum_{i=1}^{n}\sum_{j=1}^{i}\sum_{k=1}^{i}\mathrm{Tr}\left[\frac{\partial \boldsymbol{d}_{ij}}{\partial q_p}\boldsymbol{I}_i\,\boldsymbol{d}_{ik}^T\right]\dot{q}_j\dot{q}_k \qquad \text{(D.9)}$$

$$+\frac{1}{2}\sum_{i=1}^{n}\sum_{j=1}^{i}\sum_{k=1}^{i}\mathrm{Tr}\left[\boldsymbol{d}_{ij}\boldsymbol{I}_i\left(\frac{\partial \boldsymbol{d}_{ik}}{\partial q_m}\right)^T\right]\dot{q}_j\dot{q}_k + \sum_{i=1}^{n} m_i\, g^T \frac{\partial\, {}^0\boldsymbol{T}_i}{\partial q_p}\,{}^i\bar{\boldsymbol{r}}_i$$

The order of terms in second term of Eq. (D.9) is changed as explained above [Eq. (D.4)], which makes it identical to first term and the first summation in all the terms is significant from p to n [Eq. (D.6)]. Thus, Eq. (D.9) simplifies to

$$\frac{\partial \mathcal{L}}{\partial q_p} = \sum_{i=p}^{n}\sum_{j=1}^{i}\sum_{k=1}^{i}\mathrm{Tr}\left[\frac{\partial \boldsymbol{d}_{ij}}{\partial q_p}\boldsymbol{I}_i\,\boldsymbol{d}_{ik}^T\right]\dot{q}_j\dot{q}_k + \sum_{i=p}^{n} m_i\,\boldsymbol{g}^T\boldsymbol{d}_{ip}\,{}^i\bar{\boldsymbol{r}}_i \qquad \text{(D.10)}$$

D.2 THE JOINT EQUATION OF MOTION

Using Eqs. (D.2), (D.8), and (D.10), the torque for link $\boldsymbol{p}$ (or torque applied by actuator at joint $\boldsymbol{p}$) is obtained after simplifying as

$$\tau_p = \sum_{i=p}^{n}\sum_{k=1}^{i} \mathrm{Tr}\left[\boldsymbol{d}_{ik}\boldsymbol{I}_i\boldsymbol{d}_{ip}^T\right]\ddot{q}_k + \sum_{i=p}^{n}\sum_{k=1}^{i}\sum_{m=1}^{i} \mathrm{Tr}\left[\frac{\partial \boldsymbol{d}_{ip}}{\partial q_m}\boldsymbol{I}_i\boldsymbol{d}_{ip}^T\right]\dot{q}_k\dot{q}_m \quad \text{(D.11)}$$

$$+\sum_{i=p}^{n} m_i\,\boldsymbol{g}^T\boldsymbol{d}_{ip}\,{}^i\bar{\boldsymbol{r}}_i$$

This is the dynamic equation of motion for link p. The equation is independent of the order of summation and it can be written in a compact form for link $\boldsymbol{i}$ of an $\boldsymbol{n}$-DOF manipulator as

$$\tau_i = \sum_{j=1}^{n} M_{ij}\ddot{q}_j + \sum_{j=1}^{n}\sum_{k=1}^{n} h_{ijk}\,\dot{q}_j\dot{q}_k + G_i \quad \text{(D.12)}$$

which is obtained by exchanging dummy summation indices p and i for i and j, respectively. The coefficients in Eq. (D.12) are defined as:

$$M_{ij} = \sum_{p=\max(i,j)}^{i} \mathrm{Tr}\left[\boldsymbol{d}_{pj}\boldsymbol{I}_p\boldsymbol{d}_{pi}^T\right] \quad \text{(D.13)}$$

$$h_{ijk} = \sum_{p=\max(i,j,k)}^{n} \mathrm{Tr}\left[\frac{\partial \boldsymbol{d}_{pk}}{\partial q_j}\boldsymbol{I}_p\boldsymbol{d}_{pi}^T\right] \quad \text{(D.14)}$$

$$G_i = \sum_{p=i}^{n} m_p\,\boldsymbol{g}^T\boldsymbol{d}_{pi}\,{}^p\bar{\boldsymbol{r}}_p \quad \text{(D.15)}$$

and $\boldsymbol{d}_{ij}$ is given by Eq. (D.6).

D.3 THE DYNAMICS OF *n*-DOF MANIPULATOR

The equation of motion for all the links of a manipulator can be combined in to a single equation. From Eq. (D.12) manipulator dynamic equations in vector-matrix notation, are

$$\tau(t) = \boldsymbol{M}(\boldsymbol{q}(t))\,\ddot{\boldsymbol{q}}(t) + \boldsymbol{H}(\dot{\boldsymbol{q}}(t),\boldsymbol{q}(t)) + \boldsymbol{G}(\boldsymbol{q}(t)) \quad \text{(D.16)}$$

where

$\boldsymbol{q}(t)$ is $\boldsymbol{n} \times 1$ vector of generalized joint variables (displacements) for joints $i = 1, 2, ..., \boldsymbol{n}$; and is expressed as

$$\boldsymbol{q}(t) = [q_1(t)\ q_2(t)\ ...\ q_n(t)]^T \quad \text{(D.17)}$$

$\dot{\boldsymbol{q}}(t)$ is $\boldsymbol{n} \times 1$ vector of generalized joint velocities and is expressed as

$$\dot{\boldsymbol{q}}(t) = [\dot{q}_1(t)\ \dot{q}_2(t)\ ...\ \dot{q}_n(t)]^T \quad \text{(D.18)}$$

$\ddot{q}(t)$ is $n \times 1$ vector of generalized joint accelerations and is expressed as

$$\ddot{q}(t) = [\ddot{q}_1(t)\ \ddot{q}_2(t)\ ...\ \ddot{q}_n(t)]^T \tag{D.19}$$

$\tau(t)$ is $n \times 1$ vector of generalized torques applied at joints $i = 1, 2, ..., n$; and is expressed as

$$\tau(t) = [\tau_1(t)\ \tau_2(t)\ ...\ \tau_n(t)]^T \tag{D.20}$$

$M(q)$ is $n \times n$ inertial acceleration symmetric matrix whose elements are M_{ij} for $i, j = 1, 2, ..., n$; and are given by Eq. (D.13).

$H(\dot{q}, q)$ is $n \times 1$ nonlinear Coriolis and centrifugal force vector expressed as

$$\boldsymbol{H} = [H_1 \quad H_2 \quad \cdots \quad H_n]^T \tag{D.21}$$

where $$H_i = \sum_{j=1}^{n}\sum_{k=1}^{n} h_{ijk}\, \dot{q}_j \dot{q}_k \qquad \text{for } i = 1, 2, \ldots, n \tag{D.22}$$

The term h_{ijk} is given by Eq. (D.14) for $i, j, k = 1, 2, ..., n$.

$G(q)$ is $n \times 1$ gravity loading force vector with

$$\boldsymbol{G} = [G_1 \quad G_2 \quad \cdots \quad G_n]^T \tag{D.23}$$

where G_i is given by Eq. (D.15) for $i = 1, 2, ..., n$.

The dynamic model in Eq. (D.16) is the *state space equation* because the term $H(\dot{q}, q)$ has both position and velocity dependence.

The physical interpretation of each of these terms is given in Chapter 6. The evaluation of each term is explained with an example in the following pages.

D.4 EVALUATION OF COEFFICIENTS IN MANIPULATOR EOM

The evaluation of various terms defined in Eqs. (D.22)–(D.23) is illustrated with an example of a 3-DOF manipulator. The various terms of dynamic motion equations are obtained as follows.

D.4.1 The Inertia Matrix Coefficients

The inertia matrix for $n = 3$ is

$$\boldsymbol{M}(\boldsymbol{q}) = \begin{bmatrix} M_{11} & M_{12} & M_{13} \\ M_{21} & M_{22} & M_{23} \\ M_{31} & M_{32} & M_{33} \end{bmatrix} \tag{D.24}$$

The elements of inertia matrix, M_{ij} are given by Eq. (D.13).

For $i = j = 1$

$$\begin{aligned} M_{11} &= \sum_{p=\max(1,1)}^{3} \mathrm{Tr}\left[\boldsymbol{d}_{p1}\boldsymbol{I}_p\boldsymbol{d}_{p1}^T\right] \\ &= \mathrm{Tr}\left[\boldsymbol{d}_{11}\boldsymbol{I}_1\boldsymbol{d}_{11}^T\right] + \mathrm{Tr}\left[\boldsymbol{d}_{21}\boldsymbol{I}_2\boldsymbol{d}_{21}^T\right] + \mathrm{Tr}\left[\boldsymbol{d}_{31}\boldsymbol{I}_3\boldsymbol{d}_{31}^T\right] \end{aligned}$$

Similarly

$$
\begin{aligned}
M_{12} &= M_{21} = \mathrm{Tr}\left[\boldsymbol{d}_{22}\boldsymbol{I}_2\boldsymbol{d}_{21}^T\right] + \mathrm{Tr}\left[\boldsymbol{d}_{32}\boldsymbol{I}_3\boldsymbol{d}_{31}^T\right] \\
M_{13} &= M_{31} = \mathrm{Tr}\left[\boldsymbol{d}_{33}\boldsymbol{I}_3\boldsymbol{d}_{31}^T\right] \\
M_{22} &= \mathrm{Tr}\left[\boldsymbol{d}_{22}\boldsymbol{I}_2\boldsymbol{d}_{22}^T\right] + \mathrm{Tr}\left[\boldsymbol{d}_{32}\boldsymbol{I}_3\boldsymbol{d}_{32}^T\right] \\
M_{23} &= M_{32} = \mathrm{Tr}\left[\boldsymbol{d}_{33}\boldsymbol{I}_3\boldsymbol{d}_{32}^T\right] \\
M_{33} &= \mathrm{Tr}\left[\boldsymbol{d}_{33}\boldsymbol{I}_3\boldsymbol{d}_{33}^T\right]
\end{aligned}
\tag{D.25}
$$

D.4.2 The Coriolis and Centrifugal Coefficients

Equations (D.14), (D.21), and (D.22) are used for getting these terms. For $i = 1$ and $n = 3$,

$$
\begin{aligned}
H_1 &= \sum_{j=1}^{3}\sum_{k=1}^{3} h_{ijk}\dot{q}_j\dot{q}_k \\
&= h_{111}\dot{q}_1^2 + h_{112}\dot{q}_1\dot{q}_2 + h_{113}\dot{q}_1\dot{q}_3 + h_{121}\dot{q}_2\dot{q}_1 + h_{122}\dot{q}_2^2 \\
&\quad + h_{123}\dot{q}_2\dot{q}_3 + h_{131}\dot{q}_3\dot{q}_1 + h_{132}\dot{q}_3\dot{q}_2 + h_{133}\dot{q}_3^2 \\
&= h_{111}\dot{q}_1^2 + h_{122}\dot{q}_2^2 + h_{133}\dot{q}_3^2 + (h_{112} + h_{121})\dot{q}_1\dot{q}_2 + (h_{113} + h_{131})\dot{q}_1\dot{q}_3 \\
&\quad + (h_{123} + h_{132})\dot{q}_2\dot{q}_3 \\
H_2 &= h_{211}\dot{q}_1^2 + h_{222}\dot{q}_2^2 + h_{233}\dot{q}_3^2 + (h_{212} + h_{221})\dot{q}_1\dot{q}_2 + (h_{213} + h_{231})\dot{q}_1\dot{q}_3 \\
&\quad + (h_{223} + h_{232})\dot{q}_2\dot{q}_3 \\
H_3 &= h_{311}\dot{q}_1^2 + h_{322}\dot{q}_2^2 + h_{333}\dot{q}_3^2 + (h_{312} + h_{321})\dot{q}_1\dot{q}_2 + (h_{313} + h_{331})\dot{q}_1\dot{q}_3 \\
&\quad + (h_{323} + h_{332})\dot{q}_2\dot{q}_3
\end{aligned}
\tag{D.26}
$$

The coefficients h_{ijk} in Eq. (D.26) are obtained using Eq. (D.14).

D.4.3 The Gravity Loading Terms

From Eqs. (D.15) and (D.23), the gravity loading terms evaluate as

$$
\begin{aligned}
G_1 &= m_1\,\boldsymbol{g}\boldsymbol{d}_{11}\,{}^1\bar{\boldsymbol{r}}_1 + m_2\,\boldsymbol{g}\boldsymbol{d}_{21}\,{}^2\bar{\boldsymbol{r}}_2 + m_3\,\boldsymbol{g}\boldsymbol{d}_{31}\,{}^3\bar{\boldsymbol{r}}_3 \\
G_2 &= m_2\,\boldsymbol{g}\boldsymbol{d}_{22}\,{}^2\bar{\boldsymbol{r}}_2 + m_3\,\boldsymbol{g}\boldsymbol{d}_{32}\,{}^3\bar{\boldsymbol{r}}_3 \\
G_3 &= m_3\,\boldsymbol{g}\boldsymbol{d}_{33}\,{}^3\bar{\boldsymbol{r}}_3
\end{aligned}
\tag{D.27}
$$

D.4.4 Evaluation of d_{ij} Terms

In above equations, Eqs. (D.25) – (D.27), $\boldsymbol{d}_{ij}$ is involved, which is defined by Eq. (D.6), that is

$$\boldsymbol{d}_{ij} = \begin{cases} {}^{0}\boldsymbol{T}_{j-1}\,\boldsymbol{Q}_j\,{}^{j-1}\boldsymbol{T}_i & \text{for } j \le i \\ 0 & \text{for } j > i \end{cases}$$

where $\boldsymbol{Q}_j$ is given by Eq. (6.25) or Eq. (6.27).

The partial derivative of $\boldsymbol{d}_{ij}$ with respect to q_k, required for evaluating h_{ijk} in Eq. (D.14), is obtained as

$$\frac{\partial \boldsymbol{d}_{ij}}{\partial q_k} = \begin{cases} {}^{0}\boldsymbol{T}_{j-1}\,\boldsymbol{Q}_j\,{}^{j-1}\boldsymbol{T}_{k-1}\,\boldsymbol{Q}_k\,{}^{k-1}\boldsymbol{T}_i & \text{for } i \ge k \ge j \\ {}^{0}\boldsymbol{T}_{k-1}\,\boldsymbol{Q}_k\,{}^{k-1}\boldsymbol{T}_{j-1}\,\boldsymbol{Q}_j\,{}^{j-1}\boldsymbol{T}_i & \text{for } i \ge j \ge k \\ 0 & \text{for } i < j \text{ or } i < k \end{cases} \tag{D.28}$$

The proof of derivative in Eq. (D.28) is left to the reader. As an example, consider the evaluation of derivative of $\boldsymbol{d}_{11}$ with respect to q_1, that is, $i = j = k = 1$

$$\frac{\partial \boldsymbol{d}_{11}}{\partial q_1} = \frac{\partial({}^{0}\boldsymbol{T}_0\,\boldsymbol{Q}_1\,{}^{0}\boldsymbol{T}_1)}{\partial q_1} = {}^{0}\boldsymbol{T}_0\,\boldsymbol{Q}_1\,{}^{0}\boldsymbol{T}_0\,\boldsymbol{Q}_1\,{}^{0}\boldsymbol{T}_1 = \boldsymbol{Q}_1\,\boldsymbol{Q}_1\,{}^{0}\boldsymbol{T}_1 \tag{D.29}$$

Note that ${}^{0}\boldsymbol{T}_0$ is the identity matrix. The partial derivative $\dfrac{\partial \boldsymbol{d}_{ij}}{\partial q_k}$ interprets as the interaction effect of the motion of joint $\boldsymbol{j}$ and joint $\boldsymbol{k}$ on link $\boldsymbol{i}$.

D.5 CHARACTERISTICS OF DYNAMIC EQUATIONS

The dynamic equations of a manipulator, Eq. (D.16), include all inertial, centrifugal and Coriolis, and gravitational effects of link. These equations are coupled, nonlinear, second-order, ordinary differential equations. The time history of joint variables $\boldsymbol{q}(t)$ can be obtained by simultaneously integrating Eq. (D.16) for the known applied torques $\tau(t)$. This time history of joint variables can be used to compute the time history of the end-effector using the appropriate homogeneous transformation matrices of the kinematic model.

For a known trajectory planning, the time history of the joint displacements, joint velocities and joint accelerations is known. Using the known time history, Eqs. (D.12) to (D.23) can be utilized to compute the required joint torques $\tau(t)$ as a function of time.

The matrix structure of the equations of motion makes the LE formulation computationally intensive and highly inefficient. For $\boldsymbol{n}$-DOF manipulator, the number of required mathematical operations (multiplication and addition) to compute τ is estimated to be

$$\textit{Multiplications:}\ \left(\tfrac{128}{3}\right)n^4 + \left(\tfrac{512}{3}\right)n^3 + \left(\tfrac{844}{3}\right)n^2 + \left(\tfrac{76}{3}\right)n \tag{D.30}$$

$$\textit{Additions:}\ \left(\tfrac{98}{3}\right)n^4 + \left(\tfrac{786}{3}\right)n^3 + \left(\tfrac{673}{3}\right)n^2 + \left(\tfrac{107}{3}\right)n \tag{D.31}$$

There are several other approximate, analytical, iterative, and recursive formulations available today which are many times more efficient than LE formulation.

It is possible to design manipulators with minimized coupling effects, that is, choosing the parameters that make as many as possible M_{ij} and h_{ijk} coefficients to be zero. A manipulator, which has as many (or all) link parameters equal to zero and α_i equal to 0° or $\pm \pi/2$, is said to be *kinematically simple* manipulator. The dynamic equations of such a manipulator are simple and will require minimal computations.

BIBLIOGRAPHY

1. W. Book, "Recursive Lagrangian Dynamics of Flexible Manipulator Arms," *The International Journal of Robotics Research*, ***3***(3), 1984.
2. S.H. Crandall and N.C. Dahl, *An Introduction to the Mechanics of Solids*, McGraw-Hill Kogakusha Ltd., Tokyo.
3. S. Doughty, *Mechanics of Machines*, John Wiley & Sons Inc., New York, 1988.
4. J. Grosjean, *Kinematics and Dynamics of Mechanisms*, McGraw-Hill, Singapore, 1991.
5. K.R. Symon, *Mechanics*, Addison-Wesley Pub. Co. Inc., Reading, Massachusetts, 1964.
6. S. Timoshenko, and D.H. Young, *Engineering Mechanics*, McGraw-Hill, Book Co., Singapore, 1956.

Appendix E

List of Symbols

The scope and the interdisciplinary nature of the field of robotics comprising mechanical, electrical, control system any many other disciplines, requires a large number of notational symbols. The following tables summarize most of the symbols that appear in the text. The symbols are classified into six categories: parameters, functions, vectors and matrices, coordinate frames, operators and symbols and trigonometric symbols.

Table E.1 *Parameters*

Symbol	*Description*
$\boldsymbol{n}$	number of links/joints/degree of freedom
p_x, p_y, p_z	Components (or projections) of vector $\boldsymbol{P}$ on axes X, Y, Z
${}^1p_u, {}^1p_v, {}^1p_w$	Components of vector ${}^1\boldsymbol{P}$ along axes U, V, W
p_j	j^{th} component of vector $\boldsymbol{P}$; $j=x, y, z$
r_{ij}, a_{ij}	element (i, j) of a matrix
d_i	prismatic (linear) joint displacement
θ_i	revolute joint angle (angular joint displacement)
q_i	Generalized joint displacement variable (d_i or θ_i)
τ_i	Generalized joint torque or force
σ	homogeneous coordinates scale factor
L_i	link length
a_i	link length parameter for link $\boldsymbol{i}$
α_i	link twist angle for link $\boldsymbol{i}$

(Contd.)

Symbol	*Description*
C_i	centroid of link ***i***
$d\theta, dq$	infinitesimal joint displacement
δW	virtual work
dm_i	infinitesimal mass elements on link ***i***
m_i	mass of link ***i***
I_{kk}	cross product of inertia
I_{kl}	first moments of body $(k \neq l)$
$\mathcal{P}$	an open path
c	arc length of a curve
D	chord length, major diameter of object O.
p	degree of polynomial
ρ	density, radius of circle
P_s, P_g	end-points (start and goal) of a path
q^g	goal or target position
q^b	joint position at blend point
$q^j(t)$	joint position at j^{th} via or path point
$\dot{q}^j(t)$	joint velocity at j^{th} via or path point
$\dot{q}^{jl}$	joint velocity for linear segment between path points j and l
$P_j(t_j)$	j^{th} cubic, polynomial for segment j
r_1, r_2	roots of characteristics equation
q^s	starting or initial position
t	Time
t_j	duration of blend at path point j
t_0	start time of joint motion
t_b	blend duration (time)
T	trajectory traversal time, total travel time for a path
T_{jl}	travel time between path points j and l
T_m	travel time for segment m
t_g	goal or target time of joint motion
t_{jl}	travel time for linear segment between path points j and l
X	actual end-effector position trajectory
y_d	desired trajectory
i_a	armature current
R_a	armature resistance
K_b	motor back emf constant

(Contd.)

Symbol	Description
K_T	motor torque constant
e_b	back emf induced in armature
s	complex frequency
B	viscous friction coefficient
ζ	damping ratio
ω_n	undamped natural frequency
η	gear ratio
I	load inertia
K_d	derivative gain
K_I	integral gain
K_p	Proportional gain
e	error signal
e_{ss}	steady state error
A	Area of region R of image, image area
x_c, y_c	centroid, center of mass coordinates
$(\bar{x}_c, \bar{y}_c)$	coordinates of center of gravity
μ_{jk}	moment of image about (x_c, y_c), central moment
M_{jk}	moment of image of object O of order $j+k$
m	number of slots on optical encoder disk
β	orientation of image
Δx	pixel spacing in x direction
Δy	pixel spacing in y direction

Table E.2 *Functions*

Symbol	Description	Dimension
$\boldsymbol{P}(x, y)$	point in plane with coordinates (x, y)	2×1
$\boldsymbol{P}(x, y, z)$	point in space with coordinates (p_x, p_y, p_z)	3×1
${}^i\boldsymbol{P}$	vector $\boldsymbol{P}$ expressed in frame $\{\boldsymbol{i}\}$	3×1
${}^i\boldsymbol{D}_j$	$\boldsymbol{D}$ from frame $\{\boldsymbol{j}\}$ to frame $\{\boldsymbol{i}\}$	3×1
$\boldsymbol{R}_k(\theta), \boldsymbol{R}_{k,\theta}, \boldsymbol{R}(k,\theta)$	fundamental rotation matrix for rotation of θ about $\boldsymbol{k}$-axis	3×3
${}^i\boldsymbol{R}_j$	$\boldsymbol{R}$ from frame $\{\boldsymbol{j}\}$ to frame $\{\boldsymbol{i}\}$	3×3
$\boldsymbol{R}_{wvu}(\theta_1\theta_2\theta_3)$	$\boldsymbol{R}$ for Euler angle rotations	3×3
$\boldsymbol{R}_{xyz}(\theta_3\theta_2\theta_1)$	$\boldsymbol{R}$ for XYZ-fixed angle rotations	3×3
$\boldsymbol{T}(k,\theta)$	fundamental homogeneous transformation matrix	4×4
${}^i\boldsymbol{T}_j$	$\boldsymbol{T}$ from frame $\{\boldsymbol{j}\}$ to frame $\{\boldsymbol{i}\}$	4×4

(Contd.)

Symbol	Description	Dimension
${}^0\omega_i$, ω_i, $\dot{\theta}_i$	angular velocity of frame {*i*} relative to frame {0}	3×1
f_k, $f_{k-1,k}$	force on link ***k*** due to link (*k*–1)	3×1
$\dot{v}_i$	linear acceleration of frame {*i*} relative to frame {0}	3×1
${}^0v_i$, v_i	linear velocity of frame {*i*} relative to frame {0}	3×1
η_k, $\eta_{k-1,k}$	moment applied to link ***k*** by link (*k*–1)	3×1
$\dot{\omega}_i$	angular acceleration of link ***i***	3×1
$I_{\hat{k}}$	inertia tensor of link ***i***	4×4
$\dot{\bar{v}}_i$	linear acceleration of center of mass of link ***i***	3×1
$\bar{v}_i$	linear velocity of center of mass of link ***i***	3×1
$\mathcal{F}_i$	total external force acting at C_i	1×1
$\mathcal{N}_i$	total external moment acting at C_i	1×1
$\mathcal{K}(q,\dot{q})$	total kinetic energy	1×1
$\mathcal{P}(q)$	total potential energy	1×1
$\mathcal{L}(q,\dot{q})$	Lagrangian function: $\mathcal{H} - \mathcal{P}$	1×1
$X(s)$	Laplace transform of $x(t)$	1×1
$G(s)$	transfer function: $Y(s) / X(s)$	1×1
E	servo error vector	1×1
e	trajectory error vector: $e = r - q$	$n \times 1$
K_e	stiffness matrix of environment $: diag\{k_1, k_2, \ldots, k_n\}$	1×1
$I(x, y)$	image intensity at (x, y)	1×1
G^T	edge thresholding gradient function value	1×1
$G(x, y)$	local gradient function, local gradient of pixel (x, y)	1×1
2^d	number of gray levels with d bits per pixel	1×1
I^T	thresholding image intensity level	1×1

Table E.3 *Vectors and Matrices*

Symbol	Description	Dimension
I	identity matrix	$n \times n$
0	zero matrix	$n \times n$
A	a matrix with elements a_{ij}	$m \times n$
A^{-1}	inverse of A: $A^{-1}A = AA^{-1} = I$	$n \times n$
A^T	transpose of A: $A^T_{kj} = A_{jk}$	$n \times m$
det$A$, $\lvert A \rvert$	determinant of $n \times n$ matrix A	1×1
Tr(A)	trace of matrix $A = a_{11}+a_{22}+ \ldots +a_{nn}$	1×1
diag[A]	diagonal matrix with diagonal elements a_{ii}	$n \times n$
P, $\vec{P}$	a vector with components p_x, p_y and p_z	3×1
$x \times y$	cross product of vector x and y	3×1
$x \cdot y$	dot product of vectors x and y: $x^T y$	1×1
R	rotational transformation matrix, rotation matrix	3×3

(Contd.)

Symbol	Description	Dimension
$\boldsymbol{D}$	translation or displacement vector: $\left[d_x\ d_y\ d_z\right]^T$	3×1
$\boldsymbol{d}$	translation vector of end-effector	3×1
$\boldsymbol{T}$	Homogeneous transformation matrix	4×4
$\boldsymbol{T}_E$	end-effector orientation and position matrix	4×4
$\boldsymbol{n}$	normal vector with components n_x, n_y and n_z	3×1
$\boldsymbol{o}$	orientation or sliding vector with components o_x, o_y and o_z	3×1
$\boldsymbol{a}$	approach vector with components a_x, a_y and a_z	3×1
$\hat{\boldsymbol{u}},\hat{\boldsymbol{v}},\hat{\boldsymbol{w}}$	unit vectors along axes U, V, W	3×1
$\hat{\boldsymbol{x}},\hat{\boldsymbol{y}},\hat{\boldsymbol{z}}$	unit vectors along axes X, Y, Z	3×1
$\hat{\boldsymbol{n}},\hat{\boldsymbol{o}},\hat{\boldsymbol{a}}$	unit vectors in normal, orientation and approach $(\boldsymbol{n}, \boldsymbol{o}, \boldsymbol{a})$ directions	3×1
$\boldsymbol{q}$	generalized joint variable vector $=\left[q_1\ q_2\ \dots\ q_n\right]^T$	$n\times 1$
$\dot{\boldsymbol{q}}$	generalized joint velocity variable: $d\boldsymbol{q}/dt$	$n\times 1$
$\boldsymbol{v}$	Linear velocity: $d\boldsymbol{D}/dt$, $\left[v_x\ v_y\ v_z\right]^T$	3×1
ω	angular velocity. $d\theta/dt$, $\left[\omega_x\ \omega_y\ \omega_z\right]^T$	3×1
$\boldsymbol{V}$	Cartesian velocity vector $[v\ \omega]^T=\left[\dot{d}\ \dot{\theta}\right]^T$	6×1
$\mathcal{F}$	end-arm force vector $=\left[\boldsymbol{f}_{i,\,i+1}\ \ \eta_{i,\,i+1}\right]^T$	6×1
$\boldsymbol{J}_i(\boldsymbol{q})$	Jacobian matrix of link $\boldsymbol{i}$	6×1
$\boldsymbol{J}(\boldsymbol{q})$	manipulator Jacobian matrix $=[\boldsymbol{J}_1\ \boldsymbol{J}_2\ \dots\ \boldsymbol{J}_n]^T$	$6\times n$
${}^{i-1}\boldsymbol{r}_i$	position vector from origin O_{i-1} to origin O_i	3×1
$\boldsymbol{S}$	Skew-symmetric matrix	3×3
$\boldsymbol{g}$	gravitational acceleration vector, $\lvert\boldsymbol{g}\rvert$=9.8062 m/s^2	3×1
$\boldsymbol{r}$	reference position vector, distance from origin	1×1
${}^{i}\boldsymbol{r}_i$	vector to a point on link $\boldsymbol{i}$ relative to frame $\{\boldsymbol{i}\}$: $[x_i\ \ y_i\ \ z_i]^T$	3×1
${}^{i}\bar{\boldsymbol{r}}_i$	position vector of centroid of link $\boldsymbol{i}$ relative to frame $\{\boldsymbol{i}\}$: $[\bar{x}_i\ \ \bar{y}_i\ \ \bar{z}_i]^T$	3×1
$\boldsymbol{Q}_i$	partial derivative multiplier matrix for link $\boldsymbol{i}$	4×4
$\boldsymbol{I}_i$	inertia tensor of link $\boldsymbol{i}$ relative to frame at centroid $\{C_i\}$	4×4
$\boldsymbol{d}_{ij}$	coefficients for LE formulation (rate of change of ${}^{i}\boldsymbol{r}_j$)	4×4
τ	Joint torque vector : $[\tau_1\ \tau_2\ \dots.\ \tau_n]^T$	$n\times 1$
$\boldsymbol{M}(\ddot{\boldsymbol{q}})$	manipulator inertia matrix $\left[M_{ij}\right]$	$n\times n$
$\boldsymbol{G}(\boldsymbol{q})$	gravity loading vector (G_i)	$n\times 1$
$\boldsymbol{H}$, $\boldsymbol{h}(\dot{\boldsymbol{q}},\boldsymbol{q})$	velocity induced torques vector $[h_{ijk}]$	$n\times 1$

(Contd.)

Symbol	Description	Dimension
$\lVert \boldsymbol{x} \rVert$	norm of $\boldsymbol{x}$: $(\boldsymbol{x}.\ \boldsymbol{x})^{1/2}$	1×1
$\boldsymbol{K}_d$	derivative gain matrix	$n \times n$
$\boldsymbol{K}_i$	integral gain matrix	$n \times n$
$\boldsymbol{K}_p$	Proportional gain matrix	$\mathrm{n} \times \mathrm{n}$
$\boldsymbol{F}(c)$	a continuous vector function with path coordinate c	1×1

Table E.4 *Coordinate Frames*

Symbol	Description
$\{0\}$	base frame, inertial reference frame
$\{\boldsymbol{i}\}$	orthogonal coordinate frame $\boldsymbol{i}$ with axes X_i, Y_i, Z_i for link $\boldsymbol{i}$
$\{\boldsymbol{n}\}$	last coordinate frame for $\boldsymbol{n}$-DOF manipulator with axes X_n, Y_n, Z_n
U_i, V_i, W_i / X_i, Y_i, Z_i	principal axes of the orthogonal coordinate frame $\{\boldsymbol{i}\}$
$\{x\ y\ z\}$	orthogonal coordinate frame with axes X, Y, Z
$\{u\ v\ w\}$	mobile orthogonal coordinate frame with axes U, V, W
O_i	origin of frame $\{\boldsymbol{i}\}$
$\boldsymbol{k}_i$	$\boldsymbol{k}$-axis of frame $\{\boldsymbol{i}\}$, $k = x$, y or z; u, v or w etc.

Table E.5 *Operators and Symbols*

Symbol	Description
$\lvert\lambda\rvert$	absolute value of λ
$\lvert\overrightarrow{\boldsymbol{PQ}}\rvert$	magnitude of vector $\overrightarrow{\boldsymbol{PQ}}$
lim	Limit
$sign(\lambda)$	signum function: sign of λ
DOF	degree of freedom, $\boldsymbol{n}$
P	prismatic joint
R	revolute joint
3-D	three-dimensional
DWS	dexterous workspace
RWS	reachable workspace
EOM	equations of motion
LE	Lagrange-Euler
NE	Newton-Euler
MIMO	Multi Input Multi Output
PPD	partitioned proportional derivative
PD	proportional derivative control
PI	proportional integral
PID	proportional integral derivative
SISO	Single Input Single Output
ADC	analogue to digital conversion

Table E.6 *Trigonometric Symbols*

Symbol	Description
C_i	$C\theta_i = \cos\theta_i$
$C\theta_i$	$\cos\theta_i$
S_i	$S\theta_i = \sin\theta_i$
$S\theta_i$	$\sin\theta_i$
S_{ij}	$\sin(\theta_i + \theta_j)$
C_{ij}	$\cos(\theta_i + \theta_j)$
S_{ijk}	$\sin(\theta_1 + \theta_2 + \theta_3)$
C_{ijk}	$\cos(\theta_1 + \theta_2 + \theta_3)$
$V\theta_i$	$1 - \cos\theta_i = 1 - C\theta_i = 1 - C_i$
$A\cos(x)$	arc cosine or $\cos^{-1}(x)$
$A\tan(y/x)$, $\arctan(y/x)$	arc tangent or $\tan^{-1}(y/x)$
$A\tan 2(y, x)$	four-quadrant arctan(y/x)

Index

UNIVERSAL BOOK SHOW ROOM (EXTN.)
Is Now SURE BOOK CE...
50%/60% Return on Rs.
Amount Within ONE Year
26 MAY
If Syllabus and version is changed
the book will not be taken back.
Women's College Road, Koti, Hyd-95.
Phone: 23449924, 2...